Plant Chemosystematics

Plant Chemosystematics

J. B. HARBORNE

Department of Botany,
The University of Reading, England

and

B. L. TURNER

Department of Botany,
The University of Texas, Austin, Texas, U.S.A.

1984

ACADEMIC PRESS

(*Harcourt Brace Jovanovich, Publishers*)

London Orlando San Diego San Francisco New York
Toronto Montreal Sydney Tokyo São Paulo

ACADEMIC PRESS INC. (LONDON) LTD.
24–28 Oval Road
London NW1 7DX

U.S. Edition published by
ACADEMIC PRESS INC.
(Harcourt Brace Jovanovich, Inc.)
Orlando, Florida, 32887

British Library Cataloguing in Publication Data

Harborne, J. B.
Plant chemosystematics.
1. Botany—Classification
2. Biological chemistry
I. Title II. Turner, B. L.
581'.0 2 QK95

ISBN 0-12-324640-7
LCCCN 83-71882

Typeset by Bath Typesetting Ltd., Bath
and printed in Great Britain by
Thomson Litho Ltd., East Kilbride, Scotland

Dedicated to the memory of

Professor Ralph E. Alston

late Professor of Botany at the University of Texas at Austin, one of the major pioneers in the development of plant chemosystematics

Preface

Since we have dedicated this text to our late good friend and colleague, Professor Ralph E. Alston, it seems appropriate to relate briefly here our professional relationship with the man and his work and how this led to our mutual interest in the field of chemosystematics generally.

Professor Alston came to the University of Texas in 1957 with training as a physiological geneticist and only cursory interest in the field of systematics. His doctorate work dealt with flavonoids and their general control, a fairly narrow research specialty at the time. But the man was bigger than his field of specialization. His purview of things biological was quite exceptional and few, if any, scientific papers escaped his broad-ranging, imaginative but critical mind. This was important, for the "field" of chemosystematics in the mid-1950s consisted of a smatter of isolated papers in a range of botanical, chemical, and zoological journals by workers essentially unknown to each other. To recognize the *emergence* of this new field required wide interests and a sound training across a broad front of biological science.

An equally predisposing factor in Ralph's interest in chemosystematics was his great delight in field work. He loved the outdoors and preferred to be involved with problems in which he personally undertook field observations, population sampling, etc. This joy in field work developed relatively late in his short career, mostly after he became involved with the junior author on populational studies of natural hybridity in the legume genus *Baptisia*. Nevertheless, he was also a tireless laboratory worker, and was, in every way, a firm believer in getting at the molecular basis of biological problems.

The work on *Baptisia* seemed sufficiently impressive to Professor Alston so that he put almost full effort into this area of research. Beginning in 1959 until his premature death at the age of 42 in 1967, he lived and breathed this field (literally, for he spent much of his time working in an inadequately vented laboratory running hundreds upon thousands of two-dimensional paper chromatograms in his efforts to make certain that species, as populations, were understood, both as to the uniformity and variability of flavonoids). Because of his convictions and the satellitic, mostly verbal, stimulation of one of us (B.L.T.), these two workers contrived to write what might rightfully be called the first textbook on chemical systematics in the broad sense (Alston and Turner, 1963a). This was soon followed by a veritable deluge of texts on the subject, usually edited versions in which chemical experts were called upon to expand, chapter by chapter, upon their own particular chemical interests (Swain, 1963; Leone, 1964; Swain, 1966; Hawkes, 1968; Mabry *et al.*, 1968; Harborne and Swain, 1969; Harborne, 1970; Bendz and Santesson, 1974). A notable exception has been the six volume compendium of Hegnauer (1962–1973), but unfortunately this ambitious and remarkable compilation by a single individual was published over a ten year period in German, and further, it did not set out to cover the subject in an academic or conceptual sense. Gibbs (1974) has completed a less ambitious compendium with greater emphasis upon the taxonomic implications of microchemical data for flowering plants, especially at the ordinal level, and more recently, Smith (1976) has completed a text which covers a broad spectrum of conceptual concerns.

At the present writing, one cannot help but be impressed with the large number of biochemical articles having to do with *systematic problems* which appear almost daily in botanical (and other) journals. And *that* has been the most conspicuous change in the decade or so following the appearance of the text *Biochemical Systematics*: more and more the plant systematist is carrying out the chemical work necessary to help resolve his particular taxonomic problem.

And that is the purpose of this particular book, to review briefly, and hopefully succinctly, the state and potential of this approach to plant systematics up to the end of 1982. We make no pretence of covering the literature relating to zoological and prokaryotic systematics, for that would be a laborious undertaking requiring more sophisticated knowledge of relationships among these groups than we possess. However, the underlying principles and methodological approaches to such studies are virtually the same and, in this sense, the present text can be considered a revised version of *Biochemical Systematics*, in spite of the fact that only two chapters (2 and 3) are recognizable in the present version.

January, 1984

J. B. Harborne
B. L. Turner

Contents

Preface vii

Part I

Introduction

1 The Biochemical Background 3
2 Taxonomic Principles 7
3 Historical Perspective 34
4 Philosophical and Pragmatic Rationale 38

Part II

Secondary Metabolites

5 Plant Scents and Odours 49
6 Alkaloids and Other Plant Toxins 75
7 Plant Pigments 128
8 Hidden Metabolites 180
9 Environmental and Genetic Variability 216
10 Application of Chemistry at the Infraspecific Level 237
11 Application of Chemistry at the Specific and Generic Level 267
12 Application of Chemistry at the Familial Level 292
13 Chemical Documentation of Hybridization 313
14 Comparative Biosynthetic and Metabolic Pathways 343
15 Phytoalexin Induction 362
16 The Handling of Chemical Data 374

Part III

Macromolecular Approaches

17 Proteins 389
18 Nucleic Acids 439

19 Polysaccharides 461
20 Palaeobiochemistry 475
21 Epilogue and Future Perspectives 483

Bibliography 488
Index of Plant Genera and Species 545
Subject Index 555

Part I
Introduction

1

The Biochemical Background

The great advances in biochemistry which have come in the last few decades have impressed both the informed layman and the scientist. The scientist who has made an effort to acquire more than a passing acquaintance with the subject is appreciative of not only the elegance of method and the intellectual challenge of the field but, in addition, the implications, sometimes of even a philosophical nature, of these discoveries to other subdivisions of biology. For instance, the biochemical unity disclosed incidentally along with the elucidation of basic pathways of metabolism is as effective a support for Darwinian evolution as is comparative anatomy. Without a fossil record, and assuming that evidence from comparative anatomy were in some way unavailable, comparative biochemistry would have already established unequivocally the same concepts of evolution which now exist.

Four levels of biochemical unity may be recognized which, collectively, provide a framework for evolutionary theory. Starting with the most fundamental they are: (1) biochemical unity as expressed in the basic similarity of the hereditary material of all organisms; (2) biochemical unity as expressed in the group of co-enzymes which are essential to many of the basic biochemical processes; (3) biochemical unity as expressed in the similarity of metabolic pathways, particularly those involved in energy exchange, of different organisms; and (4) biochemical unity as expressed within major taxonomic groups in the common presence of certain structural components such as chitin, cellulose, and so on. At all of these levels there is also some degree of diversity. For example, while deoxyribonucleic acid is present in the chromosomes of diverse species, the same sequence of nucleotide subunits is unlikely to be expressed even in two individuals of a single species. All of this knowledge has a direct bearing upon phylogeny in its broadest meaning. At least, all of the facts have potential phylogenetic significance; those that emphasize unity, to relate species, and those that emphasize diversity, to separate species.

In recent years many books have been written about various aspects of the broad subject of biochemistry in relation to evolution. Typical among books by individual authors is that of Jukes (1966) entitled *Molecules and Evolution*. Molecular evolution has also been reviewed in such multi-author works as Ayala (1976). Comprehensive coverage is provided in a volume of a general treatise (Florkin and Stotz, 1974) in which biochemical evolution in plants is reviewed by T. Swain. Speculation about the origin of life and the role of symbiosis in plant evolution is also almost entirely centred on questions relating to changes at the molecular level (Margulis, 1981).

Dating back many years before the beginnings of enzyme chemistry and studies of metabolic pathways are numerous investigations of the distributions of various substances, initially in higher plants and now including fungi and bacteria as well. Such investigations often had pharmacological and other economic objectives, but some of the earliest workers were interested in correlations between the distributions of substances and the taxonomic treatments of the species investigated (for review, see e.g., Gibbs, 1963). Subsequent workers have continued to note such correlations or even to make a tentative taxonomic judgement based on their chemical results. Periodically, belief in the utility of biochemical data for systematic purposes has been reiterated. Although a wide range of biochemical characters are available (Table 1.1), biochemistry has not yet been responsible for any major advances in our knowledge of phylogenetic relationships. Yet, inexorable progress in the accumulation of biochemical data, many of which are already seen to be of phylogenetic importance, points to an obligatory integration of these data in systematics. The systematist does not have the prerogative of evaluating the purely chemical aspects of data, but he has a responsibility to be alert to progress in biochemistry, particularly when discoveries bear potentially upon phylogenetic considerations. Biochemistry relates to phylogeny at several levels, only one of which involves the taxonomic distribution of specific compounds. Certain approaches discussed in chapter 4 may seem to be remote or even irrelevant, but it is worth reiterating that no approach should be discouraged provided it is theoretically sound though its practical value may eventually prove to be slight.

It is not the purpose of this book to develop a case for the use of biochemical data in systematics since this has already been done. Rather, it is hoped to establish a better perspective concerning the place of biochemistry in systematics. There is a need for an exploration of some theoretical and intellectual aspects of the subject, the development of a basic rationale, an integration of certain chemical and biological aspects, an analysis of the advantages and limitations of the biochemical approach, a broad and essentially critical analysis of existing work. We have attempted to accomplish this series of objectives.

Table 1.1 Biochemical characters of application in plant systematics

Biochemical characters
Secondary Metabolites
Volatiles (plant scents and odours)
mono- and sesquiterpenes, aliphatic and aromatic volatiles, amines, sulphides, isothiocyanates
Defence agents (bitter principles and toxins)
alkaloids, cyanogens, non-protein amino acids, iridoids, sesquiterpene lactones, diterpenoids, steroids and triterpenoids.
Colouration (pigments)
anthocyanins, betalains, yellow and colourless flavonoids, quinonoids, carotenoids
Storage metabolites
fatty acids, sugars, polyols, cyclitols, alkanes, polyacetylenes
De novo *defence agents*
phytoalexins
Variations in Metabolism
Primary pathways
photosynthesis (C_3, C_4, CAM), lysine biosynthesis, tyrosine biosynthesis
Secondary pathways
shikimate isozymes
Degradative pathways
conjugation, detoxification, aromatic cleavage
Macromolecules
Proteins
variation in amino acid sequences, electrophoretic mobility, isoelectric focusing, serology, etc.
Nucleic acids
variation in base sequences, hybridization and base ratios
Polysaccharides
variation in sugar components, linkages, branching, etc.

We do not believe that biochemistry represents a panacea for all systematic problems. It is quite clear from many experiments reported in later chapters that the available biochemical characters (Table 1.1) can often do no more than provide supplementary data for phylogenetic considerations. However, profound and far-reaching new insight into phylogenetic relationships is potentially available in biochemistry especially at the macromolecular level and intensive study of DNA and protein sequences (see chapters 17 and 18) is already providing much new information in this area.

Nowadays, much is spoken and written about what is popularly known as molecular biology and its relationship to descriptive or classical biology. It is possible that some individuals regard these two categories as mutually exclusive. It is true that in this age one person rarely acquires eminence in both areas. However, there are many who can excel in performance in one area and be intellectually in contact with the other. It is the purpose of this book to contribute to an integration of these disciplines by providing the groundwork for a more effective utilization of biochemical data in systematics than has previously existed. One might go as far as to say that recent developments in molecular biology have re-emphasized once more the central role that systematics has in biology today.

2

Taxonomic Principles

I.	Introduction	7
II.	Taxonomic categories	9
III.	Phylogenetic concepts and taxonomic systems	17
IV.	Systems of classification	19
V.	Classification of vascular plants	25
VI.	Parallelism as a factor in classification	28
VII.	The fallacy of the fundamental character	30
VIII.	Conclusion	31

I. Introduction

Taxonomy is one of the oldest fields of biological science. Organisms, and their relationships to other organisms, have occupied man's thinking for hundreds, if not thousands, of years. In order to classify, even at the most elementary level, man had to recognize (or identify) organisms. To do this he was prone to observe, make comparisons and, to some extent, integrate data, and develop generalizations therefrom. It can be argued that taxonomy was almost synonymous with biology in its beginning as a science. The identity of organisms occupied the thinking of early biologists. To derive order out of the multitude of forms in existence, these biologists were primarily concerned with writing descriptions and giving names.

Many non-taxonomists, including biologists and other scientists, believe that the sole function of the taxonomist is to describe and name species. Although this is still an important function of taxonomy, it is not its beginning or end. Taxonomy, like other areas of biology, has kept pace with the mainstream of biological progress.

A well-trained worker in taxonomy today must have a broad background in the fundamental concepts and basic working techniques of a number of disciplines. He not only has to be familiar with the special disciplines of his own field, but also should have some familiarity with cytology, genetics, statistics, anatomy and, it is hoped, biochemistry. Without such breadth the worker is often confined to a rather narrow avenue with much diminished perspective. If he is to synthesize and integrate the data provided by classical methods and augment his knowledge with new kinds of evidence he must be, as he was in the beginning of the natural sciences, one of the better informed and widest-read of all biologists.

Taxonomic thought, as indicated in more detail below, changed radically with the advent of Darwinism. Taxonomists not only have incorporated various new morphological approaches (for example, embryology and palynology), but also have accepted enthusiastically the contributions from genetics and cytology. In the present text we are attempting to inform the interested taxonomic worker of some present trends and developments in biological thinking which are or may become relevant to taxonomy.

Certain biologists attempt to discredit taxonomy as a 'classical' or dead field. This is unfortunate since taxonomy offers a conceptual approach to biology at the organism level such as chemistry offers at the molecular level. Both taxonomy and chemistry are unifying fields. The former, based on evolutionary principles, provides a framework to account for morphological variation and its mechanisms at the organism and population level, while classical and theoretical chemistry provide a systematic framework to describe and in part comprehend variations in the organization of elementary particles.

Although the term taxonomy has long been used to cover systematic work in the inclusive sense, more recently a number of new approaches has occasioned the advent of new names, such as systematics, biosystematics, and so on. In the present text we have used the terms interchangeably and in the inclusive sense. Regardless of their appellation, all such workers are, in fact, taxonomists; perhaps a bit more modern by employing experimental procedures but otherwise attempting to solve the same problems, namely, to show relationships and to classify accordingly.

Placed in its proper perspective then, taxonomy becomes the framework or the ordered arrangement of innumerable observations and bits of information. This order is as useful for biochemical data as it is for morphological features. Indeed, it would seem almost indispensable for the former since the seemingly unlimited number of molecular configurations might lose much of their interpretative significance without such a foundation.

Taxonomists generally fall into one of two sorts: (1) those who are primarily interested in the biological units, particularly with respect to their identification, distribution and proper description; and (2) those who are less concerned with the names and descriptions of categories and more concerned with evolutionary histories, the relationships of categories one to the other, and the mechanisms of speciation. In taxonomy, as in most other fields, there are specialists, some who are involved with floristic work, some with identification, some with phylogeny, and some with evolutionary mechanisms. There is room for all, in spite of the fact that different approaches might seem to be more significant at different periods of time. Ultimately all of the information must be consolidated into any unified system of classification.

II. Taxonomic categories

A. Formal categories

There has been much misunderstanding about the nature of biological categories. Such terms as species, genus, tribe, family, order, and division have no specific meaning to most non-biologists and frequently disputed meaning among biologists. The categories may be regarded as highly arbitrary.

Fig. 2.1 The argument that species are arbitrary constructs of the human mind might lead to strange cage-fellows.

Any attempt by man to categorize natural variation must be arbitrary with respect to a terminological system. This does not mean that the natural entities which are being classified are, in themselves, arbitrary or subjective. If evolution is accepted as the general mechanism for the origin of extant taxa, it necessarily follows that the hierarchy of formal categories erected by man do stand in certain positions relative to each other.

It is often argued that the biological categories, in that they are classified by man, are completely subjective in nature. What is often overlooked here is that the subjectiveness is in applying the terminology; the objectiveness of the category under consideration, from a biological point of view, is real. If the biological entity were completely subjective, then, to use a far-fetched analogy, one might well expect individuals of the species *Homo sapiens* to be at least occasionally caged with *Gorilla gorilla* (cf. Fig. 2.1).

Nevertheless, at least a few professional plant systematists despair at attempts to define species. Thus, Levin (1979) states that 'Plant species lack reality, cohesion, independence and simple evolutionary or ecological roles', which is an overstatement, for most plant systematists have little hesitation in using the term 'species', describing their characters or pointing out to yet others their ecological place in this or that ecosystem. In fact, there is considerable concordance among plant systematists working in the field as to the *population* units that make up this or that genus. No doubt this has prompted Cronquist (1979) to codify taxonomic or phenetic species as '. . . the smallest groups that are consistently and persistently distinct, and distinguishable by ordinary means.' This is a reasonable working definition but one which tacitly assumes that the biological species concept (i.e. an interbreeding, self-sustaining, population unit that does not normally outcross or exchange genes with yet other population units at any one geographical site), is valid at least as to its theoretical foundation, or else no such unit might be found in nature. Grant (1981) has discussed the biological species concept in some detail and, suffice to say, we largely subscribe to his point of view.

Fortunately, however, most humans are not concerned with semantic problems and, though not trained in taxonomy, they find no difficulty in understanding the biological concept of a species, at least in practice.

The professional biological classifier has been said to arrive at his classification through a process popularly known as the taxonomic method. Several attempts have been made to define or otherwise explain the taxonomic method, but most definitions or descriptive attempts fall short of their mark. Although most taxonomists have a fairly good idea what is meant by this method, they find it difficult to express. Flake and Turner (1968) addressed the problem as follows:

> . . . organisms are assigned to a category according to the taxonomist's ability to discriminate between them. They are ranked subjectively, based on his best judgment as to what a species or some other category might be. In general, rank is assigned to groups according to the number of features which distinguish between the assemblages. When two or more assemblages are distinguished with difficulty one from the other, they are called varieties or subspecies of a single species; those which are relatively easily distinguished, but clearly related, are called species of the same genus; when two groups of assemblages can be readily recognized and each possesses a few peculiar kinds of characters, these are called genera, and so forth up the scale to where very unusual or wierd assemblages with a combination of many unique features might be called different families, orders, or even divisions.

This may sound non-scientific and rather arbitrary to those who have never attempted the exercise, particularly where the taxa in question are closely related and are sorted in the form of pressed plants or as preserved specimens in jars. But under natural conditions the problem is not so ill-posed. Features that mark the organisms are easily recognized, variation can be readily evaluated and this related to environmental features and, if the populations are large, literally thousands of individuals can be examined, albeit superficially, to test the validity of the classified assemblages.

Morphological characters, the basis for most taxonomic classifications, are known to be under genetic control. The taxonomist classifies a given population or group of populations as a specific or infraspecific taxon whenever he can detect an assemblage of individuals which seem to share a common gene pool (i.e. individuals that exchange genes among themselves preferentially). The number of genes that differ between two assemblages or gene pools determines the rank that will be assigned those particular taxa.

The determination of the number of genes associated with the character states distinguishing population assemblages is a laborious and time-consuming task, but this can be approached experimentally and such calculations have been attempted for at least a few taxa. Because of the problems inherent in the experimental approach, most taxonomists are content to work on the assumption that genetic similarities or dissimilarities are reflected by morphological similarities or dissimilarities: the more character states shared by two assemblages, the more closely related they are; the more distant their relationship, the fewer the character states shared.

If the character differences that distinguish between assemblages are indeed a reflection of genetic differences (and most experimental work indicates that this is so), then it should be possible to assign biologically meaningful numerical values to the attributes that characterize these assemblages. These data can then be used in computer programmes designed to cluster those individuals or populations (assemblages) which share the largest number of characters (Sneath and Sokal, 1973). Further, as will be indicated below, numerical treatments are useful in that some objective value can be assigned to the *distances* or *gaps* between assemblages and, as a result, classifications of a more nearly uniform nature can be achieved.

No new insights into the problem of assignment of rank to an assemblage (or taxon) will be offered in the discussion that follows; for whether two very distinct species are to be recognized as belonging to the same or different genera (or families!) is largely a matter of convenience or personal preference of the taxonomist. Numerical methods cannot resolve such a problem since the assignment of ranks is largely subjective. However, only rarely is the taxonomist faced with this kind of question. More often the problem resolves itself to *relating* one or more assemblages to other assemblages so that some spatial ordering among the assemblages (or taxa) is obtained. In short, using all the characters available to him, the taxonomist attempts, usually with rapid scanning by eye and mental assimilation of the visual pattern data, to form 'natural groups' and relate them one to the other by making multiple comparisons between and among all the individuals or assemblages under consideration. 'Natural groups' are biological aggregations of populations or individuals which, when compared on all their characters, are more closely related one to the other than they are to any other group or population of individuals.

Biological aggregations, once recognized as 'natural groups', are usually assembled into formal categories on purely phenetic grounds (i.e. without giving weight to the characters selected for study) or on phyletic grounds (where certain characters, usually selected *a posteriori*, are given more weight than others in arriving at the arrangement). The philosophy behind these two kinds of taxonomies has been tediously explored and it need only be noted here that most taxonomists accept phyletic treatments as preferable to those that are purportedly phenetic only.

Anderson (1957) attempted to evaluate the objectivity of the 'taxonomic method' (he used the term 'taxonomic intuition') by sending pressed plant material to several specialists in different parts of the world and asking these workers to classify the material as to the number of taxa involved, particularly as concerned their designation as genus, species and variety. The results of the study are significant in that most of the workers were in essential agreement as regards the *degree* of relationships expressed and, in particular, there was remarkable extent of agreement as to the generic status of the material considered. To most taxonomists the nature of this experiment would appear rather trivial. We think it can be fairly stated that most taxonomists working today who might be working with the same biological entities and using basically the same data will come to essentially the same conclusions with respect to the recognition and relative rank of the biological entities considered. The differences that one might expect are the actual hierarchies assigned to the categories recognized. For example, one worker might recognize 10 or 15 genera in a given family, while another might designate only a single genus for the same group, but recognize, instead, 10 or 15 species within this major taxon. They both agree as to the number of biological entities involved. The difference is one of rank which involves a subjective judgment. The biological status of these taxa would not be changed if they were called families or, for that matter, orders. However, one should understand that ideally, any changes in the nomenclature of the categories of a portion of a taxonomic system or arrangement should be followed consistently throughout that portion of the system under consideration.

It is evident that the taxon which lends itself most readily to experimental techniques, that is, the species, is also the taxon that is most likely to intergrade morphologically and genetically with some closely related taxon. Thus the species is the most difficult taxon for which to discern discontinuities and to establish parameters for recognition purposes. As one proceeds from the species to the genus, family, order, and so on, though the discontinuities between these various taxa becomes increasingly large, and consequently easier to circumscribe and identify, nonetheless the subjectiveness of these categories increases.

Or, stated another way, it is easier for the taxonomist to circumscribe and hence recognize the major taxonomic categories in spite of the fact that the lesser specific and infraspecific categories are better defined biologically and lend themselves to experimental genetic and population studies.

B. Experimental categories

The development of cytogenetics and its application to taxonomy made possible a quasi-experimental approach to plant classification. It was natural that early workers in this area of systematics felt that a panacea was in the making and that with detailed (cytogenetical) study much of the difficulty in defining or circumscribing formal categories would soon become a matter of the mere accumulation and application of such data. Unfortunately, this has not proven to be the case. It soon became apparent that sometimes obviously closely related taxa would not hybridize, whereas morphologically more distinct taxa hybridized with ease, often both in the experimental garden and in nature. Many studies conceived to establish genetic affinities between taxa of given groups more often succeeded in showing degrees of reproductive success or failure rather than demonstrating comparative genomic differences.

Such reproductive data are often difficult to obtain and, even where assembled, the data may contribute little to the solution of the species problem since, at least in the higher plant groups, taxa show all degrees of reproductive affinity, depending on the time and circumstances under which hybridization occurs (either artificial or natural).

Even such a promising criterion as chromosome number was often found to be a poor guide for the identification or circumscription of certain plant taxa. For example, populations, and even individuals within populations, tolerate a wide range of chromosome numbers. Although polyploids of a normally diploid entity are often ecologically, if not morphologically, distinct, they are sometimes interspersed *within* populations which appear to be fairly uniform from an ecological and morphological point of view. Examples of diploid and polyploid populations or individuals which can be distinguished in no other way than by their chromosome number are becoming increasingly common in the taxonomic literature, and this fact has understandably diminished the hopes of many workers who would wish to use cytogenetic data as the final criterion for categorical disposition (Lewis, 1980).

Fortunately, most workers, while recognizing the value of cytogenetic data for systematic purposes, have been aware of the taxonomic chaos that might ensue at the specific and infraspecific levels if any attempt were made to define rigidly the formal categories in terms of reproductive affinity. The formal categories, which are established by international agreement under an appropriate code, have been erected and modified subsequently by several generations of taxonomists. The taxa are usually circumscribed by discontinuities, and more often than not they are natural biological entities classsified according to their relative morphological similarities or differences (which presumably is a reflection of their genetic similarities or differences).

The 'experimental categories' (see below) are, in reality, no better defined than the formal categories and, as indicated above, they suffer an inherent classificatory deficiency in that they may or may not reflect relative genetic differences between and among taxa. Lewis (1957) has clearly set forward the value of experimental systematics from the standpoint of taxonomy by pointing out that while such approaches do not permit an objective definition of the species, they do provide an orientation for the concept. Hecht and Tandon (1953) have appropriately stated that:

> The delimitation of two species upon the basis of their failure to form a hybrid is untenable wherever single or few gene differences or simple structural heterozygosity leads to the formation of nonviable combinations. Incipient species may owe their origins to differences such as these, but the accumulation of further differences must follow before what was once a single species may be considered as two.

Lewis (1957), in a brief and excellent paper dealing with the relation of genetics and cytology to taxonomy, has stated:

> Highly interfertile geographical races of a species may be genetically far more different and phylogenetically much more distant than morphologically comparable but, intersterile populations. . . . Consequently, we should not attempt to reflect in our formal taxonomy evidence of discontinuity in the genetic system unaccompanied by corresponding genetic differentiation.

Unfortunately, too few of the early experimental workers recognized the limitations of their approaches, and, instead of accepting a modicum of rationale in the classical approaches, they were often overanxious to submerge or erect a species on the basis of rather limited or questionable cytogenetic data.

The most widely used series of experimental categories are the *ecotype*, *ecospecies* and *cenospecies* which are based on an ecological–genetic classification. Table 2.1 shows the characteristics and relationships of these informal groups. They are useful additions to the vocabulary in that they enable the experimental worker to describe more accurately the kinds of biological entities with which he is concerned. Information conveyed in this form avoids any cumbersome explanatory extrapolations to the formal categories. In addition to the experimental categories shown in Table 2.1, many additional informal descriptive terms have been proposed by numerous workers (Camp and Gilly, 1943; Grant, 1960; and others).

Table 2.1 Analytical key to the experimental categories. (After Clausen, 1951.)

<table>
<tr><th rowspan="2">Morphology</th><th rowspan="2">Ecology</th><th colspan="3">Genetic relationships</th></tr>
<tr><th>Hybrids fertile
Second generation vigorous</th><th>Hybrids partially sterile
Second generation weak</th><th>Hybrids sterile or none</th></tr>
<tr><td rowspan="2">Distinct</td><td>In distinct environments</td><td>Distinct subspecies (or ecotypes) of one species</td><td>Distinct species (ecospecies)</td><td rowspan="2">Distinct species complexes (cenospecies)</td></tr>
<tr><td>In the same environ-ment</td><td>Local variations of one species</td><td>Species overlapping in common territory</td></tr>
<tr><td rowspan="2">Similar</td><td>In distinct environments</td><td>Distinct ecotypes of one species</td><td colspan="2" rowspan="2">Genetic species only (autoploidy or chromosome repatterning)</td></tr>
<tr><td>In the same environment</td><td>Taxonomically the same entity</td></tr>
</table>

C. Biochemical categories

With the accumulation of chemical data from various plant groups it seems likely that some serious attempt will be made to erect a special nomenclature to deal with those categories so delimited. Tétényi (1958) has already proposed a series of infraspecific categories such as *chemovar*, *chemoforma* and *chemocultivar*, and so on to designate appropriate races or forms of chemically defined taxa. We are inclined to agree with Lanjouw (1958) 'that chemical strains or varieties formed in the wild should be treated as ordinary infraspecific units'; however, we doubt that these groups, unless accompanied by sufficient morphological divergence, should bear formal names according to the International Code of Botanical Nomenclature. It is already apparent that chemical components may show variation, just as do morphological features, and any effort to encourage a formalized nomenclature would only invite a deluge of names which would further extend the lists of synonymy and in other ways increase the nomenclatural burden. For the present, it appears wiser to develop informal descriptive categories, much as has been done by the cytogenetical workers. As an example, one could speak of the chemical races of a given taxon using the distinguishing constituents as adjectives—thus, cyanogenetic race or acyanogenetic race, and so forth. There seems to be little merit in a formal system along the line suggested above (see also Tétényi, 1968, 1970). If we are to believe in the biochemical individuality within *Homo sapiens* (Williams, 1956), there would be nearly as many formal 'varieties' or forms as there are people.

Nevertheless, as will be evident in reading chapter 10, *suites* of secondary compounds, which can be used to distinguish population units over a broad, or even local, area, undoubtedly occur and it will become increasingly tempting to apply formal recognition to such categories. In most cases, however, it seems likely that such patterns will also be accompanied by at least a few morphological features that might serve to delineate the populations concerned. If so, we see no reason not to accord such morphochemical units nomenclatural status, providing sufficient population sampling is undertaken to justify the action.

The field of biochemical systematics is only in its infancy and it is difficult to predict accurately its long-term effect on plant taxonomy. We are certain that it will add greatly to the data with which to develop further our system of classification. However, any changes in the nomenclatural system will surely be incidental to its more important contribution, that of providing a biochemical basis for showing relationships and ultimately the recognition and incorporation of *molecular evolution* into the over-all synthetic concepts of taxonomy.

III. Phylogenetic concepts and taxonomic systems

As early as 1926, Crow presented an excellent argument in defence of phylogenetic approaches to taxonomy, the following excerpt being typical:

> The relationships of organisms with one another are not theoretical interpretations at all, but descriptions of the actual facts of the relationship of parts of one organism to another. Phylogeny consists of theories and hypotheses formed from these facts. . . . Phylogeny can give little satisfaction to those who desire absolute truth, but those who hold a partial view to be better than none at all may find it an interesting study.

Theories and hypotheses, essential to analytical science but often rejected ultimately in the light of unfavourable subsequent evidence, are symbolic of progress, and failure of a new theory or a new hypothesis to emerge is perhaps indicative not of vitality but rather stagnation in that instance. No scientific discipline, unless it is purely descriptive, can afford to discourage or impugn the erection of rational hypotheses from available knowledge. Nevertheless, in systematic biology, which is an analytical science, those attempting to erect phylogenetic systems of classification, particularly those treating groups at higher taxonomic levels, often must defend not only their particular hypothesis, but even the utility of hypotheses *per se*. Doubtlessly many of those who object to phylogenetic classifications have, in part, acquired such an attitude as a result of the multiplicity of differing systems which have been proposed for particular groups. All of the systems are stoutly defended by their proponents and, among the comprehensive systems, all are constructed from more or less the same available data (Turner, 1977). Some taxonomists have even argued that systematists should not strive to arrange and classify plants on an evolutionary basis but rather should classify only on the basis of total similarities (such a system may be referred to as 'natural' even though not implicitly phylogenetic). However, such a position cannot possibly be defended on philosophical or even pragmatic grounds, and the writers consider it axiomatic that phylogeny is the intellectual forte of systematics.

Any hypothetical arrangement purporting to show phyletic relationships, whether based on cytogenetic, biochemical, morphological, or a combination of such data, although of limited value in itself, may be catalytic in the sense that it elicits further speculation and wider associations or suggests preferred additional investigation. In fact, it has succeeded if it has merely received sufficient attention to persuade its declaimers to crystallize their own position and reappraise the total evidence. Of course, a parade of tenuous and vacuous theories of trivial nature is to be discouraged, but most of this type are rather easily perceived by the competent systematist.

Prior to Darwin's publications there were few, if any, purportedly phylogenetic systems of classification proposed by the serious plant taxonomist for, so long as taxonomists accepted the idea of special creation, they were not likely to be concerned with phylogeny. Although several outstanding taxonomists during the 1800s classified plants by a 'natural' system, they often made no serious or conscious attempt to place the major taxa together according to their evolutionary relationships. For example, such outstanding workers as Bentham and Hooker, in their classic *Genera Plantarum*, placed the gymnosperms between the di- and monocotyledons instead of placing the latter two together as most phyletic workers have done since that time. Nonetheless, Bentham and Hooker's work remains to this day a useful system, mostly 'natural,' but not phylogenetic.

Much has been written about the speculation involved in numerous attempts by taxonomists to show phylogenetic relationships at various taxonomic levels. Most workers concede that it is possible to hypothesize with considerable assurance at the generic and specific level, mainly because these lower categories are suited to experimental, cytogenetic and population studies, but they also recognize that attempts to construct phylogenetic classifications at the higher taxonomic levels often involve highly subjective judgments. The fact that it becomes more difficult to position taxonomic groups with respect to each other at the higher taxonomic levels in no way invalidates the objectives sought, and the admission that this can be done at the lower levels, in principle at least, assures the worker that attempts to do this with the higher categories are fundamentally sound.

Some workers have despaired of ever achieving any stable or useful phylogenetic classification and have argued for a system that is both reasonably 'natural' and useful but without phyletic overtones. Whatever the argument against the incorporation of phylogeny in classificatory systems, it seems obvious that if plants are arranged in as close a phylogenetic order as possible, along somewhat practical lines, the taxonomist has performed a service, however small, to the biochemist interested in natural plant products, to the geneticist interested in making realistic crosses, or to the pharmaceutical worker in his efforts to locate new sources of drugs. In addition to these factors, as noted previously, phylogeny provides intellectual vitality to taxonomy. Finally, it should be noted that although other systems might be easier to erect, maintain and use for identification purposes, the utility often ends there.

Actually, most phylogenetic workers are cognizant of the speculative nature of their various systems, but many outside of the field are not fully aware of the tentative nature of differing and often contending systems. The fact that evidence is not available to prove or disprove one of two contending hypotheses concerning a particular relationship does not invalidate the system as a framework for future investigations. As new evidence accumulates, one of two competing systems may increase in favour. Indeed, the two may be replaced by a third which, while perhaps incorporating parts of both previous systems, may be substantiated with new evidence and information which were not available to previous workers.

IV. Systems of classification

Lawrence (1951) in an excellent treatment of the history of classification stated that:

> Many different classifications of plants have been proposed. They are recognizable as being or approaching one of three types: artificial, natural, and phylogenetic. An artificial system classifies organisms for convenience, primarily as an aid to identification, and usually by means of one or a few characters. A natural system reflects the situation as it is believed to exist in nature and utilizes all information available at the time. A phylogenetic system classifies organisms according to their evolutionary sequence, it reflects genetic relationships and it enables one to determine at a glance the ancestors or derivatives (when present) of any taxon. The present state of man's knowledge of nature is too scant to enable one to construct a phylogenetic classification, and the so-called phylogenetic systems represent approaches toward an objective and in reality are mixed and are formed by the combination of natural and phylogenetic evidence.

In the discussion below we will confine our attention primarily to those systems of a phylogenetic nature. Artificial systems are no longer used by the professional taxonomist and since, as indicated earlier, truly natural and truly phylogenetic systems are theoretically synonymous, there is little need to prolong a distinction between the two except in a historical–philosophical sense such as Lawrence has done.

After Darwin's work there began to appear numerous and varying systems of classification, nearly all of which were based on phylogenetic considerations. Turrill (1942) has perhaps justly criticized much of this speculation as has Lam (1959). The latter author, in particular, emphasized the necessity of fossil evidence before any substantial phylogenetic classification might be achieved, and he distinguishes between systems erected on the basis of 'static taxonomy' (proposed without palaeobotanical data) and systems based on 'dynamic taxonomy' (utilizing fossil data).

Since, in the case of most flowering plants, nothing resembling a progressive fossil sequence exists equivalent to the classic zoological examples (e.g., horse, ammonites, and so on), nearly all systems of classification for the group are based frequently on arbitrary principles as to what constitutes primitiveness or, in turn, advancement.

Numerous principles have been advanced, some of a contradictory nature depending on the point of view of the systematist (Just, 1948; Constance, 1955). For example, Engler and Diels (1936) considered that the majority of plants with simple unisexual flowers were primitive, while Bessey, Hutchinson, Cronquist and others have considered these same floral types indicative of advancement, the condition having developed by reduction processes from complete, bisexual flowers. The bases for some of the principles are well documented by extensive detailed correlative studies on both living and extinct groups, while other principles are based more or less on *a priori* judgment (for example, the assumption that free petals are more primitive than connate petals, and so on). It should also be emphasized that any evaluation of the various principles must be considered with respect to the group under examination. Thus Hutchinson (1959), in setting forth his views on the phylogeny of angiosperms, adopts the principle that 'the spiral arrangement of leaves on the stem and of the floral leaves precedes that of the opposite and whorled type.' However, Cronquist (1955, 1977a), in considering the phylogeny of the family Compositae, considered opposite leaves to be the primitive condition for the family, but this need not mean that he considers this to be a primitive character for the angiosperms generally. Similarly, Hutchinson's view that the herbaceous habit is primitive in the Ranunculaceae does not conflict with his supposition that woodiness is a primitive condition for the angiosperms generally.

Practically all of these principles concern morphological features, but it is not unlikely that as studies of molecular evolution or biochemical pathways develop there will be as many, if not more, principles formulated from purely chemical data.

Many of the early workers proposed classification systems which were accompanied by schematic diagrams showing the relative taxonomic positions of the taxa treated. Lam (1936) has written an excellent summary of such presentations, some of which are rather bizarre. Little advance in this type of symbolization has occurred since Lam's review of the subject. Most workers have presented their diagrams in a two-dimensional framework, mainly because fossil data are lacking to substantiate speculations in time. However, some workers, on the basis of several other kinds of evidence, have sought to reconstruct the chronological phyletic history of a given group and thus have added a third dimension, time, to their scheme. Diagrams of the sort mentioned have been constructed for taxonomic groups at all levels from the species to the division (Fig. 2.2 to Fig. 2.6). Most two-dimensional schemes are presented merely to show relative similarities and differences between taxa, although attempts are sometimes made to include the 'lines of evolution' for the taxa concerned, usually without time connotations.

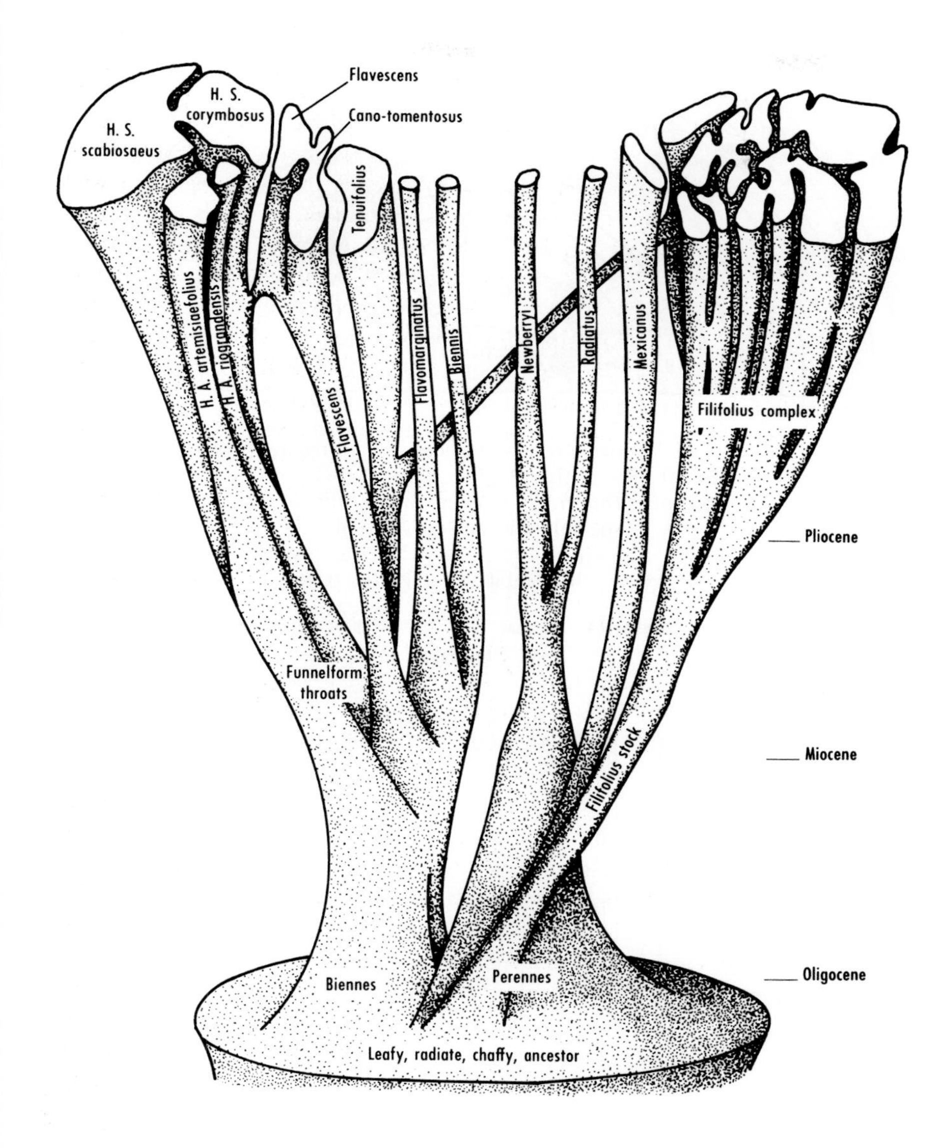

Fig. 2.2. Schematic representation of the suggested origin and evolution of present day *Hymenopappus* species (Turner, 1956).

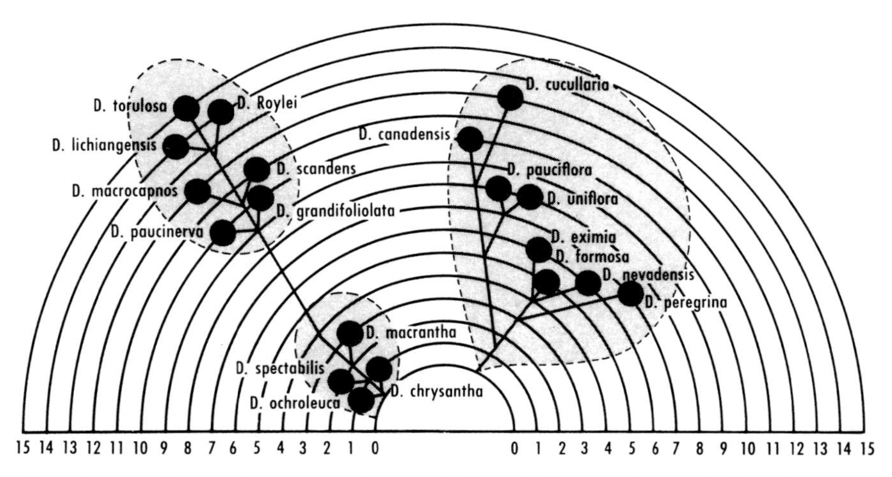

Fig. 2.3. Graph based on specialization indices indicating the probable phylogeny of *Dicentra*. Upper left: Subgenus *Dactylicapnos*. Lower left: Subgenus *Chrysocapnos*. Right: Subgenus *Dicentra*. Higher values indicate a greater degree of specialization (Stern, 1961).

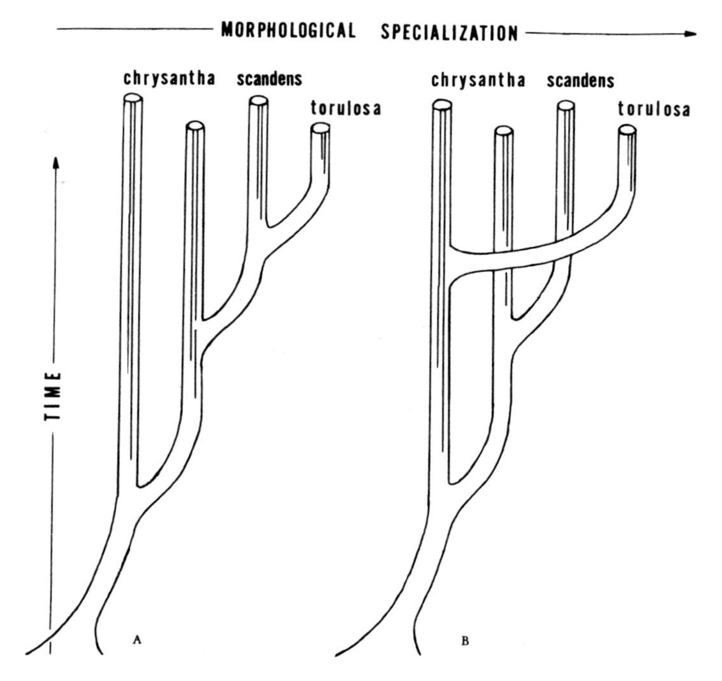

Fig. 2.4 Two possible phyletic interpretations of portions of the diagram shown in Fig. 2.3.

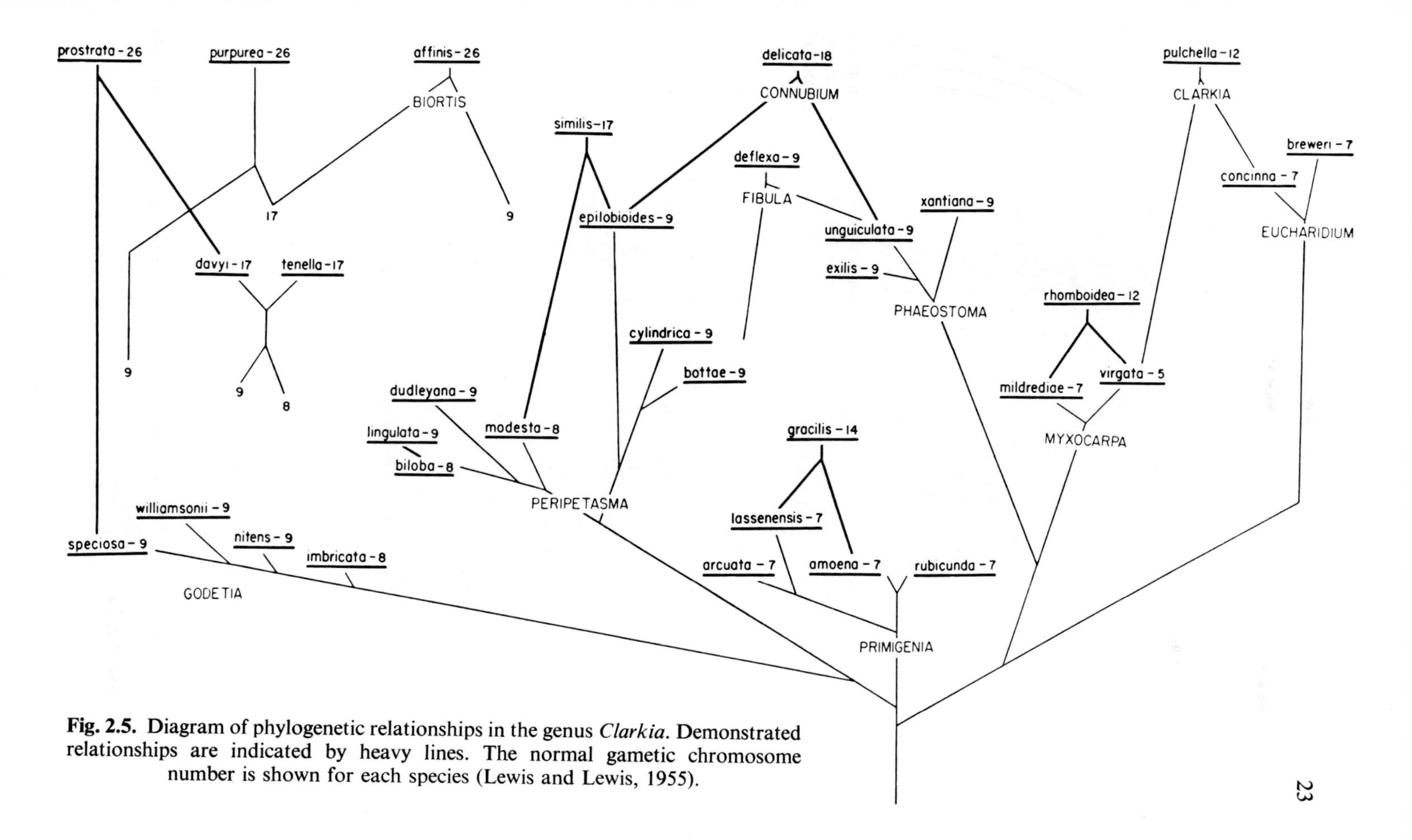

Fig. 2.5. Diagram of phylogenetic relationships in the genus *Clarkia*. Demonstrated relationships are indicated by heavy lines. The normal gametic chromosome number is shown for each species (Lewis and Lewis, 1955).

A. Two-dimensional phylogenetic diagrams

The two-dimensional presentation is popular because it is simple to construct and need not reflect phylogeny, though it would usually imply that the presentation was the best approximation from the data at hand. One popular form of the two-dimensional scheme is that shown for the genus *Dicentra* (Fig. 2.3). While phylogenetic lines are shown in this scheme and the relative positions of the various taxa, as determined from degrees of specialization, are indicated, the factor of time for the assumed branching is not indicated.

The diagram indicates that *Dicentra torulosa* is morphologically the most specialized, or advanced, and that *D. chrysantha* is the most primitive. In terms of position, as determined from the morphology of the characters selected, *D. torulosa* is closer to *D. scandens* than it is to *D. chrysantha*, but its actual phyletic relationship might be closer to *D. chrysantha*, its extreme specialization being a result of more rapid evolution from the phyletic line culminating in *D. chrysantha*. *Dicentra scandens* possibly diverged earlier from the *D. chrysantha* stock, but diverged at a much slower rate (Fig. 2.4a and 2.4b). As indicated by Stern (1961), 'the angles of divergencies, etc. are strictly diagrammatic and are not designed to denote constant rates of divergencies of evolution.'

The *Dicentra* diagram was constructed primarily from interpretations of exomorphic features. It is sometimes possible to construct two-dimensional phyletic diagrams with assurance, often with experimental support, when working with species groups where hybridization, autoploidy and amphiploidy have been major immediate factors in the speciation process. The diagram for the genus *Clarkia* by Lewis and Lewis (1955) is one of the better documented cases utilizing such information in conjunction with exomorphic features for phyletic evaluation. As is obvious from this diagram (Fig. 2.5), putative diploid species must have preceded the derived polyploids, but again the relative time of such divergencies is not shown in the diagram.

B. Three-dimensional phylogenetic diagrams

For higher plant groups, where fossil data are mostly lacking, three-dimensional schemes are usually purely speculative. Nonetheless, some monographers have ventured to reconstruct the phyletic past using geomorphological, phytogeographical, ecological and other lines of subtle evidence. If, for example, a North American genus with five species is critically studied and it develops that two of the species occupy mesic habitats which are believed to be floristically old (such as extant remnants of the Arctotertiary Geoflora; Chaney, 1938), while the remaining species occur in grassland and desert habitats (which, on palaeobotanical grounds, are believed to be more recently derived vegetational types; Axelrod, 1950, and others), then this information can be used to give relative time dimensions to any appropriate phylogenetic diagram. Phylogenetic schemes constructed from such data are often severely criticized, but, as indicated elsewhere, as a framework for future investigation they are often of definite value.

Time-dimensional phyletic diagrams have been proposed for the evolution of organic matter and organisms for the planet Earth, for the relationships between and within several families, for species within a genus (Fig. 2.2), and so on.

V. Classification of vascular plants

Because of the complex morphological variation of the vascular plants, this group has been the most extensively and successfully studied from a phylogenetic standpoint. This is particularly true of the flowering plants, and a number of systems of classifications, usually to the level of family, have been proposed for this group (Lawrence, 1951, for review; Cronquist, 1981; Hutchinson, 1959; Takhtajan, 1980; Thorne, 1976; and others). However, only a few phylogenetic systems have gained wide acceptance or attention, the more important being the systems of Engler, Bessey, and Hutchinson. Certain aspects of these three systems are discussed briefly below, mainly to acquaint the nontaxonomist with their nature and objectives.

A. Engler system

As indicated by Lawrence (1951, pp. 118–120), Engler 'attempted to devise a system that had the utility and practicality of a natural system based on form relationships and one that was compatible with evolutionary principles'. However, Engler considered the angiosperms to be polyphyletic, and his arrangements are more an attempt to show progressive complexity in structure rather than a phylogenetic sequence. This system has gained wide acceptance, primarily because of its broad and detailed coverage, and the plants in many of the world's major herbaria are arranged according to this system as are the treatments in numerous floras and texts. Engler's system is not ordinarily displayed in schematic form, mainly because its author did not claim his treatment to be phylogenetic (Turrill, 1942), and the system is recognized by most taxonomic workers as a useful but partly artificial arrangement.

B. Bessey system

Bessey was one of the most astute and prolific American taxonomists to put forward a system of classification for the higher plants. His system differed considerably from that of Engler in that, instead of emphasizing progressive specialization from the superficially simple flowers of both mono- and dicotyledons, such as Engler proposed, Bessey felt that progressive differentiation has proceeded along a number of lines, one of these being the loss of parts from a relatively simple but multicarpellate perfect flower such as is found in the families Ranunculaceae and Magnoliaceae. This system was not elaborated to the same degree as Engler's system and, in addition, it suffered certain shortcomings resulting from the fact that Bessey had only fragmentary knowledge of the families indigenous to other parts of the world. In any case, Bessey's system did not receive wide acceptance outside of the United States, although, as is apparent from the Hutchinson system (discussed below), the principles on which Bessey's system was erected have received wide approval elsewhere.

More recently Cronquist (1981) and Takhtajan (1980) have expanded many of the phylogenetic considerations of Bessey, creating systems that are becoming increasingly popular, at least as teaching models (cf. below). Ultimately, their views are bound to spill over into systematic arrangements, both in textbooks and floras, if not herbaria generally.

C. Other systems

Among the many published systems of angiosperm classification, one of the more interesting and controversial is that of John Hutchinson (1959), long-time researcher at Kew Gardens, London. Hutchinson's system of classification for the flowering plants was formulated on about the same principles as Bessey's system with one important exception: Hutchinson thought that there occurred early in the evolutionary history of the group a major phyletic dichotomy, resulting in an herbaceous offshoot which produced both the herbaceous dicotyledons and the predominantly herbaceous monocotyledons of today. The ancestral woody plexus was believed to have given rise to those dicotyledonous families with mainly woody species. When the herbaceous habit is found in otherwise essentially woody families such as the Leguminosae, it is assumed by Hutchinson to have an independent origin. The same is believed to be true for those semi-woody groups which occur in essentially herbaceous families (for example, *Clematis* sp. in the Ranunculaceae).

Hutchinson's scheme allows for the wide separation of what heretofore have been looked upon as fairly closely related taxa. He ascribes much of this similarity to convergent evolution (discussed below). Hutchinson contends that there is a considerable and fundamental phylogenetic gap between a buttercup and a magnolia tree and that, although the herbaceous habit has developed independently in several woody families, the preponderance of morphological evidence supports his arrangement.

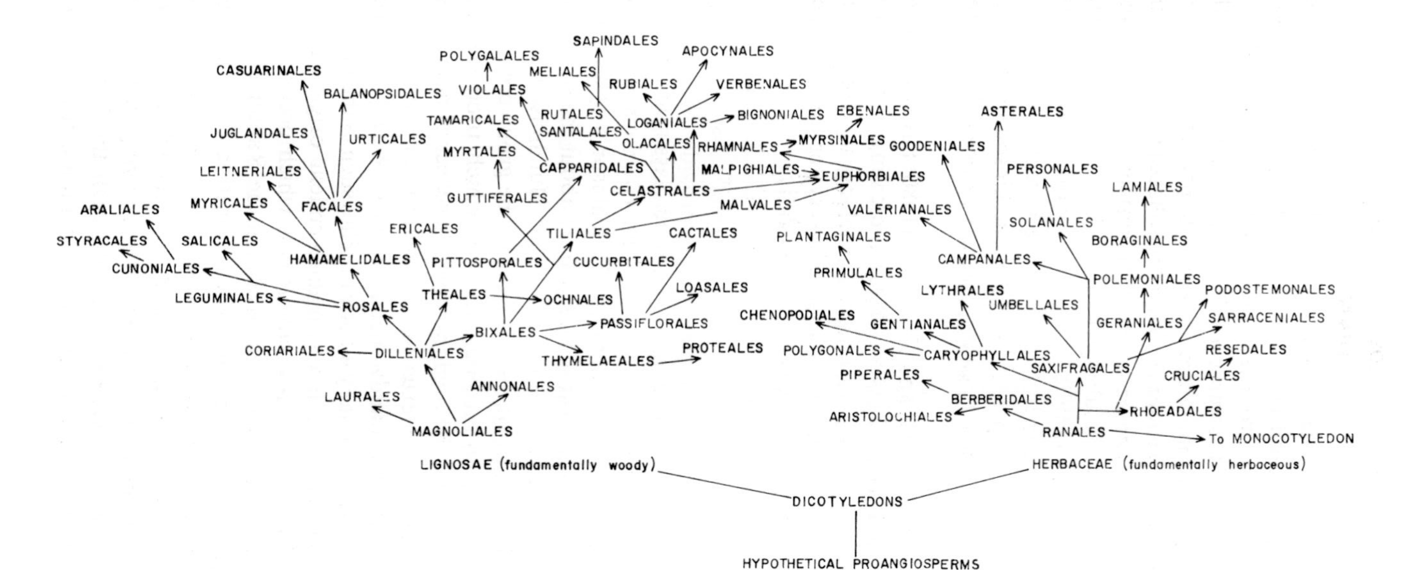

Fig. 2.6 Diagram showing the probable phylogeny and relationships of the Orders of Angiosperms as envisioned by Hutchinson (1959). Copyright 1959, Oxford University Press, 'The Families of Flowering Plants'

Figure 2.6 does not show the families within the orders recognized by Hutchinson, but this information is included in his original presentation. It is important to remember that his system, while agreeing in parts with yet others, is a novel hypothesis since it is predicated on an early phyletic dichotomy (herbaceous vs woody) that must necessarily change the arrangements of both orders and families. However, Hutchinson recognized deeply the hypothetical nature of his system for he stated in the preface of his later work:

> Botanical systems can never remain static for long, because new facts and methods of approach are liable at any time to modify them. Like other things in this changing world, that which seems to be a probability or even a certainty one day may quite well prove to be a fallacy the next.

Diagrams of the type mentioned above enable the interested worker to tell at a glance the presumed phyletic relationships within the groups concerned; however, it cannot be overemphasized that these are, at the most, hypothetical in nature and only in the rarest instances are they free of gross oversimplifications. For the experienced taxonomists such schemes may prove more irksome than instructive, but to the systematically inclined organic chemist (possibly even for specialists such as palynologists, embryologists, floristic cataloguers and so forth) they might provide some insight not apparent from the more formalized monographic treatments.

VI. Parallelism as a factor in classification

Grant (1959) attributes to two principal factors the main responsibility for the differing generic treatments accorded the phlox family (Polemoniaceae) by several workers on the basis of facts available. These are reticulate relationships following ancestral hybridization and parallelism in evolution. As indicated in Fig. 2.7, the two phenomena are often concomitant. Several workers have felt that convergence and parallelism *per se* make it difficult, if not impossible, to erect meaningful phylogenetic classification schemes, and some discussion of these phenomena will be included here.

Parallelism may occur as a result of hybridization and subsequent backcrossing (Fig. 2.7B). This type of parallelism, whether detected or not, would hardly affect classification systems since, in both the phylogenetic and typological approaches, the taxa concerned would be grouped in about the same relative systematic positions. The type of parallelism shown in Fig. 2.7A poses a more difficult problem, but, except where one or only several criteria are selected for emphasis over other kinds of data, such cases are apparently uncommon.

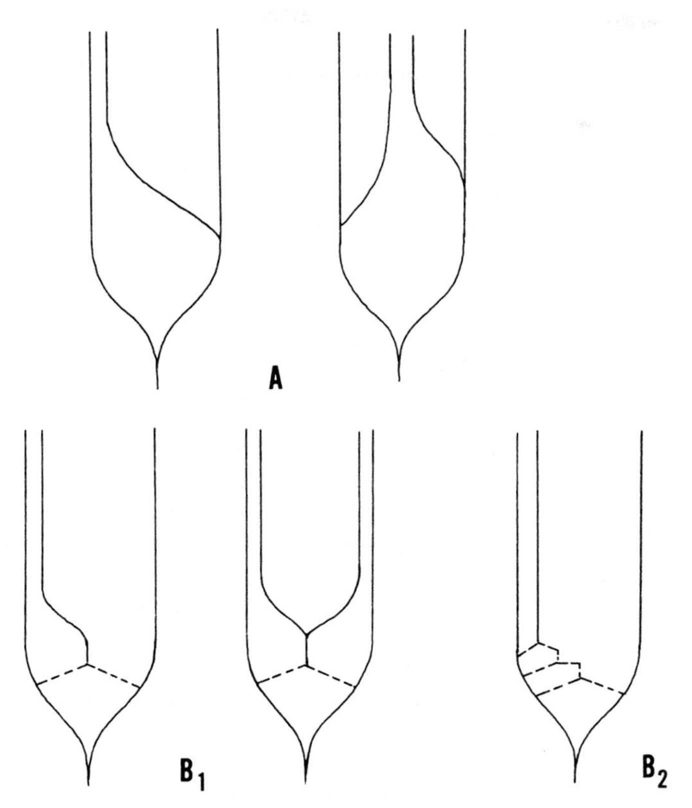

Fig. 2.7. Two ways by which parallelisms can develop in evolution. Parallel selection is assumed in each case. (A) Independent parallel mutations at homologous loci. (B) Hybridization followed by segregation in the direction of one or both parental species (B_1), or followed by backcrossing, viz. introgression (B_2). (From Grant, 1959: 'Natural History of the Phlox Family. Systematic Botany'. International Scholars, Sciences 1. Martinus Nijhoff, The Hague.)

The case for convergence in most closely related taxonomic groups usually rests on the quantitative features in one or at most only a few characters. If these characters are important 'key' characters (discussed below), then any systematic treatment based on such features is likely to be more artificial than natural, and to cite such examples as instances of erroneous phylogeny resulting from convergence and parallelism is to stretch the case. If two taxa have diverged sufficiently to be recognized by their phenotypic differences, reflecting multiple gene differences, then, on *a priori* grounds, the chance for absolute genetic convergence seems most unlikely in view of our present knowledge of mutation rates and the subtleties involved in the selective forces having to do with character fixation.

Several workers have mentioned what appear to be 'true' convergence and parallelism *for certain characters* of different taxa of higher plants (Bailey, 1944). An even more striking parallelism has been described for some chemical components of otherwise widely differentiated taxa. One rather striking example is the occurrence of the haemoglobin molecule in cells of fungi and in the root nodules of legumes (White *et al.*, 1959). Several additional examples of chemical convergence and parallelism will be discussed more fully elsewhere in the text.

The argument that convergence and parallelism make it impossible to achieve a meaningful phylogenetic system can be appropriately countered with the following remark from Crow (1926):

> The problem of the cause of convergence and parallel development is, of course, an extremely important one. But inasmuch as convergence itself was discovered by systematic and morphological investigations, and is itself a phylogenetic conclusion from the systematic and anatomical facts, the necessity of making more detailed study of phylogeny is all the more necessary. . . . To use the polyphyletic origin of a group formerly supposed to be a natural (monophyletic) one as an argument against the possibility of constructing a natural system is nothing more nor less than to use the conclusion of phylogeny to disprove phylogeny.

VII. The fallacy of the 'fundamental' character

Most workers today are aware that any ultimate system of classification must be based on the available data from all fields. To assemble these data is difficult enough, but to assess their phyletic significance often appears impossible. This is particularly true with respect to morphological features (as opposed to chromosomal or genetic data). For example, what genetic (or phylogenetic) significance does an inferior ovary versus a superior ovary have? How does one evaluate the genetic significance of separate carpels as opposed to fused carpels? Of course the answer is sometimes obvious when one is considering the mere presence or absence of a given character (other characters being similar), but when two taxa are separated by a combination of morphological features, all of which vary, both quantitatively and qualitatively, there is no simple solution.

Because of the complexities involved, many workers set arbitrarily certain 'fundamental' or technical characters to mark given groups. Consider the largest angiosperm family, the Compositae, with over 30 000 species. All of the species are more or less alike in that most contain an involucrate head, four or five united petals, a modified calyx-like structure (the pappus), an inferior ovary and two carpels, a single style with two branches, and so on. In spite of the extraordinary similarity of all of the species in this family, most workers have grouped the species into 13 or 14 tribes (Heywood *et al.*, 1977). The tribal groupings are mostly natural, but occasionally certain taxa are misplaced as to tribe, mainly because of the too rigid adherence to the so-called 'fundamental' features used to delimit the tribes initially. For example, the genus *Leucampyx* had long been placed in the tribe Anthemideae, because of the presence of chaff. However, more recent investigation has shown this genus to be unnaturally placed in the Anthemideae, since its most closely related taxon, *Hymenopappus*, an obvious prototype for the chaff-bearing *Leucampyx*, is apparently correctly placed in the tribe Heliantheae. The presence or absence of chaff in this case appeared to be sufficiently 'fundamental' to some workers to separate two very closely related taxa, not only into separate genera, but even into separate tribes. However, Turner and Powell (1977), on the basis of total data, united the groups in a single genus and suggested that their proper tribal disposition should be in the Heliantheae. Numerous similar cases could be cited.

Of course, the term 'fundamental' as applied to such characters is misleading. They are more appropriately called 'key' characters in that they usually furnish an easily observed, mostly constant, feature by which to recognize the affinities of a given taxon. It often takes the new student many years to appreciate this distinction, and even today some otherwise well-informed professional taxonomists still think of certain single characters as 'making' a given specimen and/or population a member of this or that species, tribe, family, and so forth.

Cronquist (1957) has appropriately emphasized this point in stating:

> Every taxonomic character is potentially important, and no character has an inherent, fixed importance; each character is only as important as it proves to be in any particular instance in defining a group which has been perceived on the basis of all of the available evidence.

Stated otherwise, there is no inherent value in a selected single character. As will be indicated in more detail elsewhere, this is as true for chemical characters as it is for megamorphic features.

VIII. Conclusion

With the development of rather rapid chromatographic techniques which allow rapid detection of numerous chemical constituents of organisms, it is now possible to make considerable *new* use of the many phyletic diagrams which have been prepared by various monographers. Most chemotaxonomic studies of a correlative nature have dealt with presumed phyletic relationships at the family level or higher, reflecting, no doubt, the textbook familiarity of such systems to many non-biologists. Interpretations of relationships at this level are perhaps no better than the data on which they are based, and at the present time these data are still quite limited.

With present knowledge and techniques, a more meaningful application of biochemical data towards classification schemes may be made at the family level and lower. Carefully constructed phylogenetic systems have been prepared for numerous generic groups, but only in a few instances (cf. chapters 11 and 12) has there been any concentrated effort to evaluate such systems with purely biochemical data. The hypothetical phyletic diagram for the genus *Hymenopappus* (Fig. 2.2) could be used profitably for the orientation of a purely chemotaxonomic study; for example, will biochemical data further support the basic dichotomy indicated by the Series *Biennis* and *Perennis*, or will new data come to light that might indicate a much more reticulate relationship between the species of these two series than is indicated? It might even be possible to test by chemical data the validity of some of the time speculation indicated in the *Hymenopappus* diagram. For example, it has been demonstrated in numerous instances that certain molecular configurations must occur before some more 'advanced' reaction is possible. If the latter molecular configuration was found only in the morphologically more advanced desert species, then this would correlate with the evidence from both morphology and palaeobotany as to the time of origin of desert habitats and the plant types which must have become adapted to such regions after or concomitant with their development. By the same reasoning, species that have retained certain hypothetical ancestral morphological features and ecological associations might be shown to have one or several of the metabolic precursors necessary for the molecular advancement indicated.

The approach to systematics of genera and lower categories using biochemical patterns has not been vigorously pursued, but as indicated in chapters 11, 12 and 13 it is capable of sufficient refinement that not only are species detectable but also *degrees* of hybridity for individuals from hybridized populations. Furthermore, it appears likely that with appropriate controls biochemical patterns can be constructed which permit rather objectively determined visual presentation of numerous chemical features for inter- and intrapopulational comparisons. Data obtained chromatographically can also be expressed mathematically with a minimum of interpretative effort so that considerable exactness in the presentation of relationship data can be achieved. Limitations involved in this type of comparison are obvious, of course, and further discussion will be devoted to evaluation of biochemical data in a later chapter.

The present categorization of vascular plants was developed by several generations of taxonomists, each generation adding observations and concepts to the preceding. Descriptive data were compiled for the lower taxa first and their significance and limitations determined before meaningful interpretations and circumscription of the higher categories could be made. Many errors were forthcoming in the extrapolations and interpretations incidental to its construction, but, over-all, the resulting taxonomic structure rests on a solid foundation of observational fact as opposed to mere conjecture.

Phylogenetic knowledge of both the major and minor categories of classification is certain to advance as our knowledge of biochemistry advances. To be sure, the ultimate proofs of the system must depend on the evidence from all fields, mainly palaeobotany, but we can no longer tacitly assume that '. . . a natural classification must in the main be based on external characters, simply on account of the much larger number of these and their much more restricted incidence' (Sprague, 1940). There is a wealth of biochemical data awaiting exploration and, while the gross examination of leaves and floral parts might be the most practical method for the classification of most plants today, the chemical approach is certain to add significantly to any ultimate phylogenetic system. Even at the level of identification there is a significant advantage to the biochemical approach, for if an exomorphic taxonomist were asked to identify a plant from a leaf or petal fragment he might despair, but given chemical data he might be able to identify the fragment to species.

3
Historical Perspective

I. A brief history of major developments in the field

The history of civilization, or indeed all time-dependent phenomena, can be divided into a number of major chronological periods according to the intellectual imagination or disposition of the recorder. Thus one might partition historic time into 100-year periods and graphically treat each unit with equal systematic coverage as if history were a straight line whose ascending time-event path was devoid of significant event fluctuations. Fortunately for students of history, most historians have found it more appropriate to divide recorded history into large or small time periods according to the importance or significance of the events surveyed.

Botanical historians have also recognized the special significance of certain contributions in making possible the development of new vistas in botany. Greene (1909) in his '*Landmarks of Botanical History*' emphasized the major early descriptive developments in taxonomic botany, particularly as related to specific individuals and their contributions to systematics. Beginning with prehistoric time, he recognized as foremost (1) the descriptive contributions of Aristotle and Theophrastus (followed by a long quiescence up to the fifteenth century), (2) the significance of the observations of the herbalists Tragus, Brunfels, Bauhin *et al.* of the sixteenth century, (3) the first distinction of the mono- and dicotyledons by John Ray in 1703, (4) recognition of sexual characters and their significance by Linnaeus and others in the mid-eighteenth century, and so on. Greene purposely selected the word 'Landmarks' in his published title since he recognized 'the impossibility of any such thing as a complete and faithful history of any period when once that period is past.'

Although such a treatment of botanical history might be sufficient to show the major descriptive phases, it seems that, from a dynamic-developmental point of view (in the historical sense), taxonomic history, beginning with Aristotle, can be logically divided into four or five major periods, each of which is terminated (or initiated as the case may be) by some major 'breakthrough' in scientific thought or through the development of techniques that have permitted the acquisition of new data (Table 3.1).

Table 3.1 The major historical or developmental periods of systematic biology

Period	Time	Characterization of the period
1. Megamorphic	ca. 400 B.C. to ca. 1700 A.D. (Beginning with Aristotle's time and continuing to Leeuwenhoek's invention of the microscope.)	A terminological-descriptive period characterized by the development of formal group concepts (e.g. families, genera, species, etc.) and the establishment of a descriptive language to define these groups better.
2. Micromorphic	ca. 1700 to ca. 1860 (Beginning with Leeuwenhoek and continuing to Darwin's published views on evolution.)	Leeuwenhoek's microscope and lens systems made possible the recognition of hitherto unknown microorganisms, the recognition of sexual features, and their significance and made possible the acquisition of new morphological data (viz., anatomical embryological, palynological, etc.).
3. Evolutionary	ca. 1860 to ca. 1900. (Beginning with Darwin's evolutionary theory and extending to the rediscovery of Mendel's laws of inheritance.)	Darwin's theory profoundly affected systematic thinking. Hereafter most classification systems were constructed on a phylogenetic basis.
4. Cytogenetic	ca. 1900 to ca. 1960(?). (Beginning with the rediscovery of Mendel's laws and extending to the present time.)	This period is characterized by the detailed application of cytogenetic data and population statistics to plant taxa, mostly at the generic, specific, and infraspecific levels. These techniques permitted the first truly experimental approach to systematics.

Period	Time	Characterization of the period
5. Biochemical	ca. 1950(?) to ——(?) (Beginning with the biochemical approach, made possible by the development of rapid and relatively simple techniques such as chromatography, and possibly extending to the determination of the sequences of subunits of polynucleotides such as DNA and RNA and of proteins. Techniques are already available whereby nucleotide and amino acid sequences can be analysed.)	Characterized in its early stages by the establishment of 'biochemical profiles' for various plant taxa and their comparative use in solving taxonomic problems; in later stages by a comparative biochemical approach that takes into consideration metabolic pathways, protein evolution, and comparative enzymology.

Different writers might recognize yet other 'breakthroughs' than those listed below, but we believe that few readers will argue about the impact of each on taxonomic practice and thought.

It should be obvious that the present treatment of taxonomic history in no way supposes that the valid techniques or methods of any prior period give way to those of another. Rather the methods and ideas of succeeding periods are usually superimposed on the pre-existing framework; and all are necessary (or at least have so far been found necessary) in our efforts to obtain an 'ultimate' phylogenetic system of classification.

These periods of botanical history have been treated extensively by a number of writers. Greene (1909) treated essentially the *Megamorphic Period*; Sachs (1890) treated, among others, the *Micromorphic Period*; a number of workers have recently reviewed the *Evolutionary Period* (Constance, 1955; Grant, 1977; among others); certain aspects of the *Cytogenetic Period* have been adequately reviewed by several workers (Stebbins, 1950; Clausen, 1951; Heslop-Harrison, 1953; Constance, 1955; Darlington, 1956; Lewis, 1957; Hedberg, 1958; and others), and it is probable that this period has not yet made its total contribution (i.e. in terms of broad principles and ultimate potential).

The following questions may be raised. Are we really at the beginning of a new period of taxonomic history? Will taxonomically oriented biochemical investigations yield data that make possible a better phylogenetic scheme? Will they give answers to taxonomic questions that previous methods did not permit? Will chemotaxonomy become as significant in the next half-century as cytotaxonomy has during the last? Is the time at hand for this molecular approach?

We believe that plant taxonomy has now entered this new phase of biochemical investigation. The purpose of the chapters that follow is to document (though selectively) the present state of our knowledge in this field, to give our interpretations of the significance of certain approaches already in use, to evaluate critically the limitations as well as potential of the field, and, finally, to develop philosophical concepts that might lead to increased activity and more important contributions in the future.

4

Philosophical and Pragmatic Rationale

Heywood (1966) in his usual, inimitable refreshing sort of subtle cynicism, makes the point that 'It is perhaps unfortunate, although inevitable, that descriptions such as chemotaxonomy or chemical taxonomy should be employed [in reference to the use of chemistry in systematic studies] since they give the impression that there is an approach to taxonomy in which chemical data are more important than are other classes of data.' This may be so in the mind of the ignoramus or taxonomic nóvice but it is no more reflective of logic than it is to assume that the term 'organic chemistry' suggests that the only approach to chemistry is through the study of organic molecules.

But there is an important point raised in Heywood's jostle, and that is the claim made by a few workers, both chemist and systematist, that there is something special about chemical characters that, when used in taxonomic studies, raise them a peg or two above the usually employed morphological characters. Davis and Heywood (1963) have attempted to assess critically this surmise and Heywood himself (1966) subsequently boils this more extended account down to the following, 'In general the value of chemical data in taxonomy stems from their stability, lack of ambiguity and resistance to change; that is when these characters can be established!' [Exclamation his; as if there was some intrinsic reason that this should be more difficult to accomplish for chemical characters than morphological ones.]

Consideration of this issue, the possible advantages of chemical characters over morphological ones for classification purposes, was raised as early as 1886 by Helen Abbott. She gives six reasons in favour of the chemical characters:

1. The disagreement among botanists themselves, depending on the insufficiency of the present methods of classification.

2. Chemical constituents, or the constructive elements of form, are intimately associated with the origin and progression of plant life, and are consequently better adapted for classification than organs and tissues because the component parts are less complex.
3. Because of the invariable composition and structure of given determinate chemical constituents.
4. The percentage of any given compound in a plant would gauge the progression or retrogression of a plant, species or genus, and would accentuate the characters of progression, adaption and filtration.
5. Variations in chemical constituents would be detected by analysis earlier than consequent variations of organs or tissues.
6. It is a law of internal influences controlling function and modifying forms rather than of external forces, hence a study of the elements of the innermost structure of plant life is a study of that law and of life itself.

However incomplete, this constitutes a rather remarkable evaluation, considering the state of the discipline during this period, and it appears that many subsequent workers have independently come to some of the same conclusions. Thus McNair (1965), in his numerous published papers from 1916 to 1945, fails to acknowledge her work by reference even once, although expressing several of these viewpoints himself.

Erdtman (1963) has perhaps presented the most lucid, relatively early, account of the value of chemical constituents for plant systematic studies. Being both a chemist with strong interdisciplinary interests in plant systematics and a gentleman (personal observation) he presents a balanced, seemingly dispassionate, discussion of the subject. Some of his more cogent views are expressed as follows (numbers mine):

1. '[Chemical constituents] are genetically controlled, and have the advantage over morphological ones in that they can be very exactly described in terms of definite structural and configurational chemical formulae.'
2. 'The elucidation of the structures and configurations of naturally occurring organic compounds paves the way to an understanding of their biosynthesis which is a matter of fundamental systematic importance.'
3. Since biochemical pathways are mediated by enzymes, Erdtman entertains the possibility that the latter '. . . will be found to be more important for the chemical classification of plants than the low molecular-weight 'secondary' products.'
4. 'The greatest virtue of the chemical method is that it is entirely independent of the classical biological methods'.

Erdtman does concede to some inherent limitations to the chemical approach noting that '. . . as a rule, only recent plants can be examined.

Moreover, the isolation and structural elucidation of plant constituents is often very difficult and time consuming.'

Most botanists coping with the problem of chemical versus morphological characters have displayed a vague sense of irritation that the question should be raised at all. At least most have expressed themselves unequivocally that chemical characters have no inherent advantage over morphological ones. Thus, Heslop-Harrison (1963) states:

> In assessing the relative value of chemical data as criteria for use in general-purpose classification, we need to know whether there is any quality in them which differentiates them as a class. In essence there is not. Biosynthetic pathways leading to particular compounds are expressions of the genome just as are morphological features; indeed so-called morphological features are all in some sense themselves expressions of biosynthetic pathways.

But he follows this with an intellectual hedge:

> It is conceivable, however, that chemical data may form particularly valuable taxonomic criteria because of qualities of consistency and readiness of assessment.

This perhaps reflects some sense of awareness that chemical characters might, after all, possess a little more potency than he might want to admit to.

In a subsequent reconsideration of these views, Heslop-Harrison (1968) was even less enthusiastic, stating that 'whatever their source, chemical data, as taxonomic criteria, are to be tested as indicators of similarity or difference in the particular context of each group, and evaluated on the same basis as any other characteristics of the organism.' There still smoulders within him, however, his earlier flicker of intellectualism in that he reckons 'they have many potential advantages, including for some classes of compounds, the particular merits of consistency, ease of assay and unambiguity'. Then he proceeds to cover even this flicker of optimism with a wet blanket:

> It is improbable in the extreme that comparative morphology should ever be superseded as the principal source of taxonomic criteria for the higher organisms, but if the proper aim of taxonomy is to produce classifications based upon the maximum correlation of attributes—Professor Cain's phenetic classifications—it is inevitable that chemical data will contribute increasingly. Whether this will lead to a massive re-shaping of existing classifications is an open question. My own opinion is that, should this happen, it will only be because the chemical data have been misused. The most satisfactory parts of our present classifications are based upon the correlation patterns shown by many attributes of organisms. The availability of new criteria is bound to alter the emphasis in some cases, particularly where existing taxonomy has been founded upon inadequate evidence; but for the better-known groups it is to me inconceivable that the addition of another random sample of genome-expressions should alter in any radical manner the patterns of similarity and difference manifest in the sample already taken into consideration in traditional taxonomic practice.

To be sure, this is only one opinion, but probably a prevailing one among those workers, presumably numerous, who might prefer a phenetic classificatory system to a phyletic one (cf. chapter 2) for surely if what you *think* you see is what you want then there is really no need to accept chemical data as anything but molecular debris cast on a crude taxonomic scale along with diverse assemblages of structural parts.

Heywood (1966) is more evasive. In summarizing a rather diversive account of 'character weighting' he states that '. . . there seem to be no grounds *at the moment* [italics added] for treating any class of data as specially privileged and therefore deserving of *a priori* weighting'. But he notes in the paragraph preceding this that 'At the level of what has recently been termed the 'informational molecules', that is, the proteins and nucleic acids, there is some justification for considering chemical information as more basic'.

As already indicated, Davis and Heywood (1963) were among the first purely plant taxonomists to evaluate the relative advantages and disadvantages of employing chemical characters in classification studies. After emphasizing that such characters might be used at all taxonomic levels, but with the admonition that they should not be regarded *a priori* as 'more important than other characters, either for the purposes of analysis or synthesis', they proceed to list approximately 12, implied or outright, disadvantages inherent in the use of chemical characters. Briefly stated, these are arranged in Table 4.1 in the order of their listing.

Table 4.1 Disadvantages of chemical characters, as listed by Davis and Heywood (1963).

1. Living plants are usually necessary for chemical analyses.
2. The chemical techniques employed [at that time] are often crude.
3. Chemical characters are affected by environmental variables.
4. Metabolic pathways leading to the production of chemical constituents are poorly known.
5. The adaptive significance of chemical characters are poorly known.
6. Those end-products that have been studied are not those that control morphogenesis.
7. Similar end-products may arise in groups which are unrelated.
8. The possibility exists that chemical and morphological features may have followed different evolutionary pathways.
9. Chemical variability occurs quite widely, such as chemical mutants within a species.
10. The protein content of a plant may change during its life and differ from organ to organ.
11. Many proteins are not detectable by methods so far employed.
12. Comparative chemistry is an aid to establishing consanguinity not (at least above the generic level) of phylogeny proper.

If one looks critically at this list of 'disadvantages' or limitations, it will be seen that nearly all are equally applicable to morphological features. For example, the limitations expressed in statements 3, 4, 5, 6, 7 [8 is presumably meant to read that chemical and morphological characters may have evolved such as to suggest different phyletic pathways; clearly, being part of the group under consideration, chemical features, however discordant, must necessarily be part of the evolutionary stream], 9 and 10 apply as well to morphological features. Thus, item 4 might read (perhaps with *better* justification) that 'metabolic pathways leading to the production of *morphological features* are poorly known', etc. Further, statement 2 of this list (Table 4.1), at least today, might be rephrased to appear as a chemical advantage for, as will be indicated elsewhere, the techniques for gathering chemical data (e.g. combined gas chromatography–mass spectroscopy with its computerized data bank; or amino acid sequence data) are surely more sophisticated both quantitatively and qualitatively, than that provided by eyeballs and ruler, the usual technique for assembling most morphological data (Turner, 1969, 1977). This leaves items 1 (the need for living plants) and 11 (the difficulty of detecting many proteins) as perhaps legitimate *current* disadvantages. However, many of the more stable compounds such as flavonoids and terpenoids are readily obtained from dry materials; in fact, most of the chemosystematic studies of flavonoids reported in this text were made from essentially dry herbarium specimens. But it is true that for most volatile compounds, and especially macromolecules, living material is needed.

This leaves item 12, which is a curious statement in that we must infer that comparative chemistry is perhaps useful as a phylogenetic approach *below* the generic level but not at the higher categorical levels. Why the same should not be true of morphological features is not clear unless, indeed, Davis and Heywood believe that phylogenetic insights much above the generic level might not be had at all, whatever the data.

The enterprise of plant systematics is sufficiently difficult so that *all* approaches to the subject are needed and should be encouraged (Constance, 1964) and rather than pit one approach against another it is perhaps better procedure, or at least psychology, merely to list their relative advantages and disadvantages. This has been done in Table 4.2, but it should be emphasized that the *degree* of a given advantage or disadvantage is an important consideration. Thus, one's capacity to rapidly view a multitude of morphological characters over a broad range of plant material far outweighs, *at the present time*, our capacity to provide, however eloquent, metabolic pathways leading to the structures of such chemical characters as flavonoids, alkaloids and terpenoids. Chemical constituents may provide more insight into a given taxonomic problem, but they nearly always utilize a disproportionate amount of the researcher's time, especially if mere phenetic classification of the higher categories is all that one is attempting. The advantages versus disadvantages tabulated in Table 4.2 could be expanded upon in an ever-expanding trivial sense, but the perceptive reader will have already grasped their relative merits without needless 'but-ifs' or extensions of additional examples.

Whatever the advantage or disadvantage of the chemical approach, as already noted in this chapter, most botanists who have thought about the subject, have tended to admit that macromolecules or 'information molecules' possess a quality not readily apparent in morphological characters. It is clear from comparative enzymology that such molecules have an evolutionary diversity of their own and that within their structural framework may be found real clues as to the cladistic (indeed chronistic!) assemblages which contain them. At least such assemblages (taxa) can be erected with greater statistical probability than might be possible through the use of purely exomorphic features. This topic will be discussed in more detail in chapter 17.

Table 4.2 Relative advantages and disadvantages of chemical versus exomorphic features in systematic research (assuming adequate training in the study of both).

Chemical	*Exomorphic*
Disadvantages	**Advantages**
1. Time-consuming to assemble in meaningful quantities	1. Relatively rapidly assembled
2. Living material usually needed for *in depth* studies (e.g. enzyme analyses)	2. Not needed
3. Expensive equipment often needed for structural elucidations	3. Not needed
4. Fossil molecular data from specific organs absent or nearly so	4. Fossil data rare; however fragmentary it's real
5. Chemical characters (e.g. DNA) not the focal point of selective forces	5. Selection acts upon phenotypes, not genotypes
Advantages	**Disadvantages**
1. Characters exactly described in terms of structure	1. Not so, less precise
2. Genetic basis for expression more readily determined	2. Less readily determined; if at all
3. Biosynthetic pathways, and hence probable position in an evolutionary sequence for the character concerned, more readily determined	3. Not readily determined; when proposed, evolutionary sequences highly subjective
4. Enzymes may be sought to establish relationships built on metabolic pathways, even when their end-products are absent	4. More difficult to establish relationship when characters are absent

Chemical	*Exomorphic*
5. In that they are more closely related to the gene assemblage leading to morphological expressions (DNA → RNA → protein → chemical products → morphological expression) they more readily reflect the genetic situation	5. More distant from the gene assemblage; more affected by terminal variables
6. In population work at the infraspecific and specific level, certain kinds of data (e.g. volatile compounds) can be assembled with precision, both quantitatively and qualitatively	6. Comparable population data, even if readily detectable at a morphological level, assembled with much difficulty both as to precision and time
7. Data is usually assembled and quantified without personal bias (i.e. selection of plants made without prior knowledge of compounds and quantitation machine produced)	7. Usually collected and quantified with much personal bias
8. Adaptive significance more readily determined, at least biosynthetically and experimentally	8. Determined with difficulty and only rarely experimentally
9. Discriminatory chemical characters more readily available at early developmental stages	9. Discriminatory characters (e.g. flowers) not readily available at early developmental stages
10. Hybridity and introgression, as determined from molecular complementation and unit-character distribution in back crosses, more readily determined	10. Not so readily determined because of genetic complexity of characters concerned
11. Macromolecular characters such as cytochrome *c* have built-in, evolutionary information, which can be used as an independent assessment in the recognition of cladistic groups	11. Morphological features *per se* do not possess such information or, if inferred from an extensive assemblage, it is fraught with much subjectivity
12. Certain macromolecules (e.g. cytochrome *c*, histones, etc.) appear to have value as chronistic markers	12. No such characters known
13. 'Genetic distance' among closely related populations, using isozymes, readily determined	13. Because of their complex genetic nature, morphological characters not comparable

Above all, as noted by both Erdtman (1963), a chemist, and Cronquist (1976), a taxonomist, the greatest virtue of the chemical method is that it is entirely independent of the morphological methods, which might or might not constitute an advantage on the chemical side, for were the chemists to have constructed our current classification schemes, the same 'advantage' (but surely with many howls of indignation) would then accrue to any newly developing school of morphosystematists.

In spite of the fine series of papers and review articles bearing on this subject that have appeared during the last decade, we still find the following statements, which constitute the closing remarks from Alston and Turner's text of 1963, to be the most intellectually satisfying, for it places all the pros and cons in some sort of time-perspective, while lending conceptual strength to the field of systematics as a whole.

> It hardly needs to be emphasized that knowledge of the genetic basis of a biochemical difference greatly increases the possibility that the systematic significance of the biochemical difference can be determined. Since phylogenetic relationship is based on evolutionary concepts which rest principally upon genetic mechanisms, then all differences, whether biochemical or morphological, ought to be expressed in genetic terms for maximal systematic utility. Up to now, only a minute proportion of either biochemistry or morphology is understandable in a genetic sense—biochemistry best in the more fundamental reactions (that is, amino acid synthesis), morphology in some of its more trivial expressions (that is, leaf shape, pubescence, aberrations of floral morphology, and so on). If we project the present situation into the future, we conclude that there is in the final analysis a much better chance of expressing specific biochemical differences in precise genetical terms (including characterization of the enzyme involved). Therefore, although the art of assessing the phylogenetic value of morphological data is farther advanced than the art of assessing the phylogenetic value of biochemical data, and we know far less at this time about variation in the chemistry of the plant, it is probably that in fifty years this situation will be reversed. Form is so subtly, delicately, and especially so indirectly regulated that its underlying genetics and biochemistry are likely to remain among the most intractable problems in biology for a long time. In fact, an understanding of morphogenesis requires first that its biochemical basis be understood.

Part II
Secondary Metabolites

5

Plant Scents and Odours

I. Introduction 49
II. The lower terpenoids 51
III. Aliphatic and aromatic volatiles 62
IV. Nitrogen-containing volatiles 66
V. Sulphur compounds 68
VI. Conclusion 73

I. Introduction

In the world of vegetation, smell is ever pervasive. Numerous plants give off or exude scents and odours from flower, leaf, fruit or root at some time during the life cycle and such smells may well be characteristic of individual species. The recognition of different plants by means of their volatile odours must date back many centuries. Scientists during the classical period of taxonomy used plant odours, consciously or unconsciously, to distinguish between closely related plants rich in smells, such as members of the mint family (Labiatae). Thus volatile leaf odours readily separate the pungent spearmint *Mentha spicata*, the apple-scented *M.* × *rotundifolia*, the musty-odoured horsemint *M. longifolia* and the peppery peppermint, *M.* × *piperita*.

The chemotaxonomic importance of plant odours, albeit crudely expressed, can also be dated back to the past. Thus, the characteristic smells present in dill, caraway, fennel, sweet cicely and coriander were used as a set of characters for grouping these plants together in the same family, even before the time of Linnaeus. As early as 1699, James Petiver was able to write in his herbal: 'the (family) Umbelliferae I generally observe to be endowed with a carminative taste *and smell*, are powerful expellers of wind—and of great use in treating the chollick'. Indeed, the curative value of these plants as herbs is based, in part, on the terpenoids present in the leaves or fruits of these umbellifers. Even today, the odours of members of the Umbelliferae can be useful, simple guides to identification in the field, especially in immature plants. Seedlings of wild carrot *Daucus carota* or wild parsnip *Pastinaca sativa* growing in the hedgerows are readily distinguished by the characteristically different odours released when fresh root tissues are gently crushed between the fingers.

The taxonomic use of plant odours *per se*, however, is limited by the difficulties of description and classification. How does one describe the odour of carrot or parsnip? Linneaus (1756) was one of the first to realise the need for a system of odour description and produced an odour classification, which is still used in modified form even today (Harper *et al.*, 1968). Linneaus divided odours into seven main types—aromatic, fragrant, musk-like, garlic-like, goat-like, foul and nauseating—and made further sub-divisions within these main classes. In spite of these and other attempts at classification, odour description still remained subjective, dependent on the ability of the individual to perceive different qualities of smell and, in the last resort, requiring a comparison with a few standard plants and the use of such terms as lemon-like or 'odour of new-mown hay'. It was not until the advent of chemical analyses of oils that it became possible to describe the infinite natural variation in plant odours in more objective terms.

One of the early attempts in modern times to apply knowledge of the chemistry of volatile constituents to plant taxonomy was that of Baker and Smith (1902) in Australia. These two chemists surveyed essential oils of many *Eucalyptus* species and reported the fact that some, but not all, species could be readily distinguished by a particular terpene make-up. The major contribution of these authors, in historical terms, however, was not the finding that odoriferous constituents could be used to separate species, but rather the recognition that certain species could be further sub-divided into 'chemical races' due to both qualitative and quantitative differences in their essential oil constituents (see chapter 10).

A more important pioneering study at the species level was that of Mirov, who examined the volatile oils present in the oleoresin of the majority of the 100 or so known pine species. In summarizing a lifetime of study on the genus *Pinus*, Mirov (1965) was able to report that, as taxonomic markers, pine terpenoid patterns could be invaluable for separating many species from each other. The method was not foolproof, since not all species in the genus possessed a characteristic oil pattern. An added bonus of the work, however, was the realization that knowledge of essential oils added immeasurably to the understanding of the geography and evolutionary history of the genus.

In recent years, Mirov's work on *Pinus* has been extended to many other conifer genera, particularly by von Rudloff and Zavarin and their co-workers. We now have an immense fund of information on the essential oil patterns in *Abies*, *Juniperus*, *Picea*, *Pseudotsuga* and other genera. Fewer studies have been devoted to the angiosperms, but volatile patterns have been analysed in relation to taxonomy in such families as the Compositae, Hypericaceae, Labiatae, Orchidaceae and Umbelliferae. Not all plant smells are pleasing and distasteful odour principles, based on different chemicals, have been recognized in families such as the Cruciferae and Alliaceae where surveys have indicated their value as taxonomic markers.

In this chapter, therefore, it is intended to consider the major classes of organic compounds, which contribute to odour and scent in the plant kingdom. A brief account will be given of their natural distributions and of methods of analysis and their taxonomic potential will also be discussed.

II. The lower terpenoids

A. Chemistry and distribution

The major components of the volatile steam-distillable 'essential oil' fractions responsible in plants for characteristic odours are the terpenoids. These particular chemicals are commercially important as the basis of natural perfumes, of spices and flavourings in the food industry and of certain medicinal preparations. Minor constituents, co-occurring with the terpenoids in essential oils, include aliphatic and aromatic volatile compounds, nitrogen-containing volatiles and sulphur compounds; the chemistry and distribution of these other substances will be mentioned in later sections.

Chemically, the terpene volatiles can be divided into two classes, the monoterpenoids and the sesquiterpenoids, C_{10} and C_{15} isoprenoids, which differ in volatility and boiling-point range. Thus, monoterpenoids boil between 140 and 180°C, whereas sesquiterpenoids tend to have boiling points of 200°C or more. Monoterpenes are derived biosynthetically from the condensation of two C_5 mevalonate-derived units and by definition have 10 carbon atoms, joined together in a structure indicating their origin from the head-to-tail union of two branched isoprenoid residues. Monoterpenoids can be subdivided into three groups, depending on whether they are acyclic (e.g. geraniol, the odour principle of geranium), monocyclic (e.g. limonene, a principle of lemon and other *Citrus*) or bicyclic (e.g. α- and β-pinene, commonly present in pine oil). Within each group, the monoterpenes may be simple unsaturated hydrocarbons (e.g. limonene) or may have other functional groups and be alcohols (e.g. menthol, a major principle of spearmint), aldehydes or ketones (e.g. menthone, a major principle of peppermint). More complex lactone derivatives are known, such as nepetalactone, one of the major odour constituents of catmint, *Nepeta cataria* (Labiatae), a plant to which the domestic cat and other felines are attracted because of the odour. Fig. 5.1 gives the structures of some representative monoterpenoids.

The study of natural terpenoids is complicated by the factor of isomerism and mixtures of isomers are often isolated from plants. Geometric isomerism is possible with unsaturated acyclic monoterpenoids, and *cis* and *trans* isomers may exist. For example, geraniol (Fig. 5.1) is a *trans* isomer; the *cis* isomer is known as nerol and both isomers may be present in different proportions in the same oil. Most terpenoids have asymmetric carbon atoms, are optically active and again can exist in isomeric forms. Such

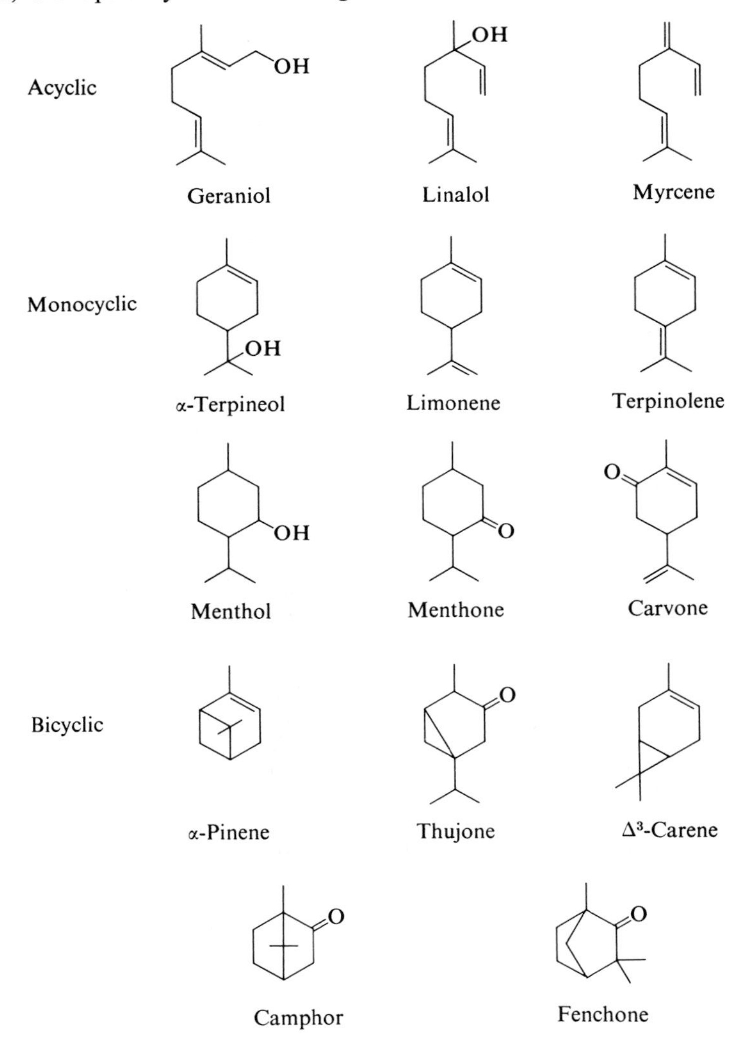

Fig. 5.1 Structures of some typical plant monoterpenoids.

optical isomers can be distinguished by measuring the optical rotation of the pure oil. Limonene, as usually isolated, is the dextrorotatory form (+)-limonene (optical rotation $[\alpha]_D^{20} + 124°$). However, (—)-limonene ($[\alpha]_D^{20} - 101°$) is also known and may occur as such or mixed with the (+)-form. Loss of optical activity (or racemization) can occur during isolation and purification, so that it is important to handle essential oils under mild conditions as far as possible.

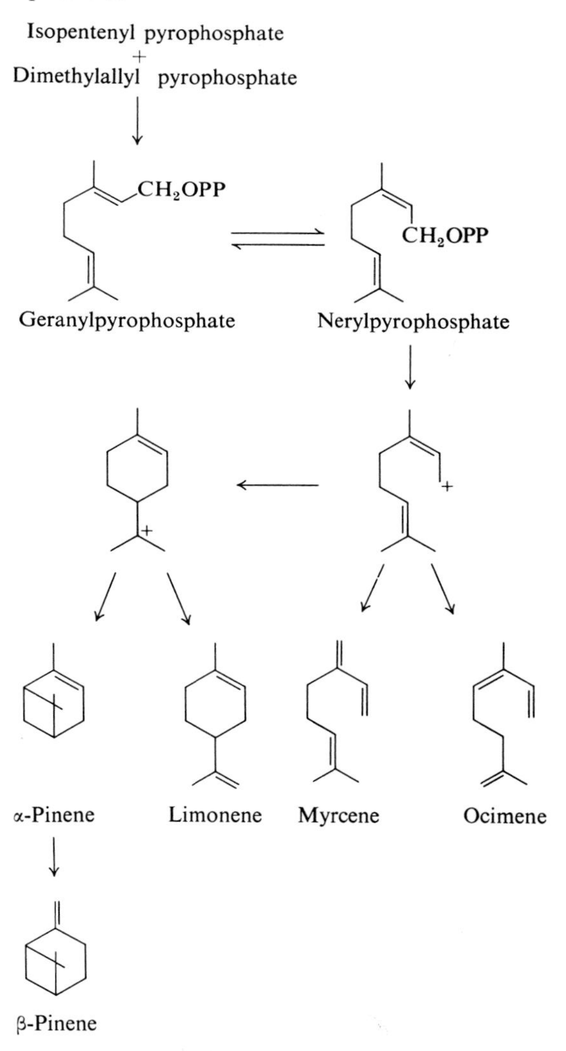

Fig. 5.2 Biosynthetic origin of common monoterpenes.

Simple monoterpenes are very widespread in nature and occur not only in higher plants, but also in bryophytes, algae and fungi. In gymnosperms and angiosperms, they are common components of the essential oils of leaves and regularly occur as mixtures of varying complexity. Compounds such as α- and β-pinene, limonene, linalol, Δ^3-carene, β-phellandrene and myrcene are often found together in leaf oils and it is their quantitative variation which may separate one species from another. These common terpenes are closely related biosynthetically and can be formed by simple modification of the two main C_{10} terpenoid precursors, geranyl and neryl pyrophosphates (Fig. 5.2). Flower and fruit oils tend to have more specialized monoterpenoids, such as keto derivatives. Whatever the tissue, complex mixtures are the rule. Although there may be 10 or 15 easily detectable major components in a given essential oil, there may be a hundred or more other terpenoids in trace amounts.

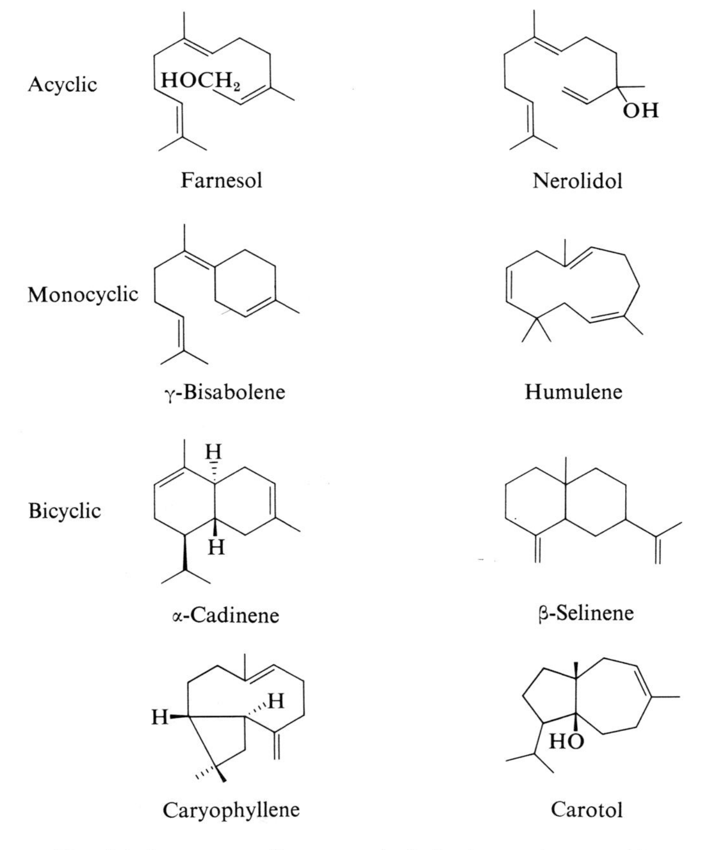

Fig. 5.3 Structures of some typical plant sesquiterpenoids.

Like the monoterpenoids, the sesquiterpenoids are mevalonate derived, but their structures are based on three, not two, C_5 units so that they have C_{15} skeletons. They can be subdivided according to whether they are acyclic (e.g. farnesol, nerolidol), monocyclic (γ-bisabolene, humulene) or bicyclic (β-selinene, carotol) (Fig. 5.3). Indeed, there are over 1000 sesquiterpenoids with well defined structures, classified into some 100 skeletal types. Only a very few can be mentioned here. Some occur fairly widely (e.g. caryophyllene, γ-bisabolene) but most are quite restricted (e.g. carotol, present in *Daucus carota* and a few related umbellifers). An important, but less volatile, group of sesquiterpenoids are their lactone derivatives, which are discussed later in chapter 6 of this book.

The major functions of volatile terpenoids in plants are ecological. Thus, they make a major contribution to floral fragrances and are important in attracting animal pollinators to plants by odour. They may also be involved, more subtly, in other plant–animal interactions and can even be used by animals in defence. For example, high concentrations of α- and β-pinene are very pungent and are sequestered by an insect such as the sawfly during feeding on pine needles. They are then utilized by the sawfly as a defensive discharge against its predators. Terpenoids may also be important in the plant itself and may, by allelopathy, protect one plant species from competition by another (cf. Harborne 1982).

B. Detection of terpenoids in plants

Fresh leaf material is preferable for oil extraction since terpenoids are not only volatile, and thus may be lost during drying, but they are also labile and may undergo irreversible changes if leaves are dried in an oven. It is possible to carry out oil analyses on dried leaf material, though if comparable results are desired, it is important that all samples are treated the same way. Volatile compounds have been examined in leaves from herbarium sheets with successful results, at least as far as qualitative patterns are concerned. Remarkably enough, Harley and Bell (1967) were able to obtain a weak essential oil profile from a piece of *Mentha* leaf taken from a herbarium sheet prepared by Linneaus as long ago as 1810.

For floral analyses, fresh flowers are essential and, in cases where the fragrance is very weak, some concentration of oils may be necessary by collecting the floral essence into a cold trap in liquid nitrogen (see Bergström, 1978). With fruits and seeds, freshly harvested material is best. Some changes occur with ageing of the seed; for example, alterations in oil patterns have been noted in nutmeg *Myristica fragrans* seed following storage (Sanford and Heinz, 1971).

The classic procedure for separating essential oils is by steam distillation, the volatiles being recovered from the distillate by extraction into ether. Such a procedure is rarely used in survey work, since volatiles are readily released from leaf tissue simply by crushing in the presence of diethyl ether, petroleum or acetone. The process of crushing may produce different amounts of leaf aldehyde, hex-2-en-1-al, as an artifact of the wound process. Mature seed may be powdered in a mill and then extracted with minimal amounts of ether. With some seed, the presence of 'fixed oil' in the extracts may interfere with thin-layer chromatography (TLC) analyses of the terpenoids.

Before applying chromatographic separation, the nose should be used to determine whether or not volatiles are present in a particular extract. Much can also be done, with experience, in distinguishing the presence of particular components in an oil by smell. It is clear, for example, that Penfold and Morrison (1927), in confirming the earlier work of Baker and Smith (1902) on chemical races in *Eucalyptus dives*, were able to do this in the field simply by smell. Thus, they write: 'The observation was made of two trees growing together, indistinguishable from one another by both botanist and bushman, but each containing a different essential oil. On crushing the leaves between the fingers, one yielded the typical phellandrene–piperitone odour, whilst in the other the odour of cineole–phellandrene–terpineol was most pronounced'.

A variety of analytical techniques have been applied to essential oils but the apparatus of choice is indubitably gas–liquid chromatography (GLC). Indeed, this technique might have been tailor-made for their separation. Its routine use in almost all chemotaxonomic studies of plant volatiles (von Rudloff, 1969) has, however, obscured the fat that TLC can also be applied to terpenoid analyses (Stahl, 1969) with very satisfactory results. TLC only requires simple, inexpensive apparatus and heating, which is necessary in GLC, can be avoided. An advantage of TLC is that specific components in essential oil mixtures (e.g. the ketones carvone and pulegone) can be readily located by use of special spray reagents on the developed plate. Disadvantages of TLC include the low sensitivity compared to GLC, the loss of the very volatile terpene hydrocarbons during chromatography and the difficulty of accurate quantification.

The great attraction of GLC is that a single application of an oil extract to a GLC column immediately produces both qualitative and quantitative data, in the form of the now familiar GLC trace (Fig. 5.4), which records the different components of the oil as they emerge in turn from the silicone-coated column after separation, and pass through a detector. The column must be heated in an oven to facilitate the process and temperature programming (increasing the temperature with time) is needed to give the best separations. On non-polar columns, hydrocarbons such as limonene and α- and β-pinene usually emerge first, followed by the sesquiterpenes and then the oxygenated derivatives. In the case of very complex mixtures, use of more than one column may be necessary in order to resolve closely related pairs of oil constituents.

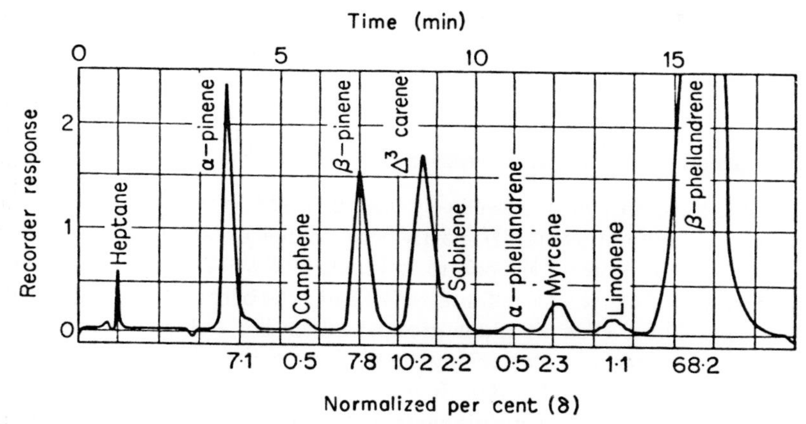

Fig. 5.4 Gas–liquid chromatography separation of terpenes in pine resin. Separation on 10% oxydipropionitrile column on 60–80 mesh acid-washed chromosorb W (from Smith, 1964).

The most important extension of the normal GLC procedure to chemosystematic studies is the procedure known as 'direct injection' of plant material. In this method, developed independently by Karlsen and Svendsen (1966), von Rudloff (1969) and Harley and Bell (1967), samples of dry leaf material are placed directly into the inlet port of the GLC apparatus and the heat of the inlet oven volatilizes the oil, which then passes directly onto the column into the flow of carrier gas. This procedure not only avoids the labour of preparing suitably purified oil extracts for column injection, but also minimizes artifact production during extraction. Finally, the whole procedure can be automated so that successive plant samples can be passed through a GLC apparatus in turn at the rate of one an hour.

In the direct injection procedure, only small bits of plant material are needed and it is possible to obtain an analysis on a simple conifer needle (von Rudloff, 1969) or a single mint leaflet weighing 2 mg (Richardson, 1978). Ironically, the fact that such a tiny amount of tissue is needed is responsible for one disadvantage of this method. Thus there may be quantitative variation in oils from one leaf to another in the same plant and a single sample may not be representative. This disadvantage is avoided when a much larger sample of leaves is pooled, extracted and analysed in the more usual way. This disadvantage seems to limit the application of direct injection to analyses of conifers (von Rudloff, 1969) but is less important when analysing herbaceous plants (Richardson, 1978).

Identification of the terpenes separated by GLC is achieved, in the case of known compounds, by co-injection with standards and GLC comparisons on different columns. Further confirmation is possible by gas chromatography–mass spectroscopy (GC–MS). The scope and limitations of GLC in chemosystematic studies of terpenoids are excellently summarized by von Rudloff (1969).

C. Taxonomic utility of plant terpenes

Terpenoids have only been extensively used as taxonomic markers in studies of gymnosperms and this must be partly because they are so widely distributed in the leaves of conifers. By contrast, much less work has been done on terpene patterns in angiosperm families. This may be because of their irregular distribution. Thus, even in plant groups which are rich in essential oils, individual members may be lacking them. For example, in a survey of 34 species of *Plectranthus* (Labiatae), 18 were found to have excellent oil profiles, with up to 32 components, but the remainder of the species failed to give any response in terms of leaf volatiles (Aye, 1974).

Compared with other secondary constituents, terpenoids do tend to show more intraspecific variation. Chemical races have been detected in individual species of almost every plant genus that has been studied extensively. Some variations occur at the population level; such differences in pattern are often correlated with geographical or ecological factors, so that the results of such surveys have been important in extending our knowledge of plant population structure (cf. chapters 9 and 10).

From the taxonomic viewpoint, essential oil studies have been mainly useful as an aid in defining the species, for detecting hybridization in natural populations, in confirming the presence of geographical races and in confirming generic and tribal limits. A few selected examples will be given here; these subjects are also considered in more detail in later chapters. The utilization of volatile leaf oils in gymnosperm taxonomy has recently been reviewed at length by von Rudloff (1975b).

1. *Definition of the species*

One simple case in *Pinus* may be taken to exemplify the general possibilities of an approach through the terpenoids. There have been difficulties in *Pinus* in determining the circumscription of certain species solely by morphology, and Mirov's investigations of oleoresin turpentines in the group have occasionally provided a key. A case in point is the taxonomy of the Eastern Mediterranean pines: plants referred to as *P. brutia*, *P. elderica*, *P. pityusa* and *P. stankewiczii* have either been lumped together or sunk into *P. halepensis*. Mirov *et al.* (1966a,b), however, found that while *P. halepensis* (sensu strictu) contains 95% α-pinene in the oleoresin, turpentines of the other four taxa never have more than 80% α-pinene and contain, in addition, 5–25% β-pinene, 10–30% Δ^3-carene and traces of nine other terpenes. The respective turpentines also differ in optical rotation, that of *P. halepensis* being dextrorotatory and those of the other four laevorotatory. These chemical differences clearly establish the need for maintaining at least two species names for these plants and for separating *P. halepensis* as a distinctive taxon.

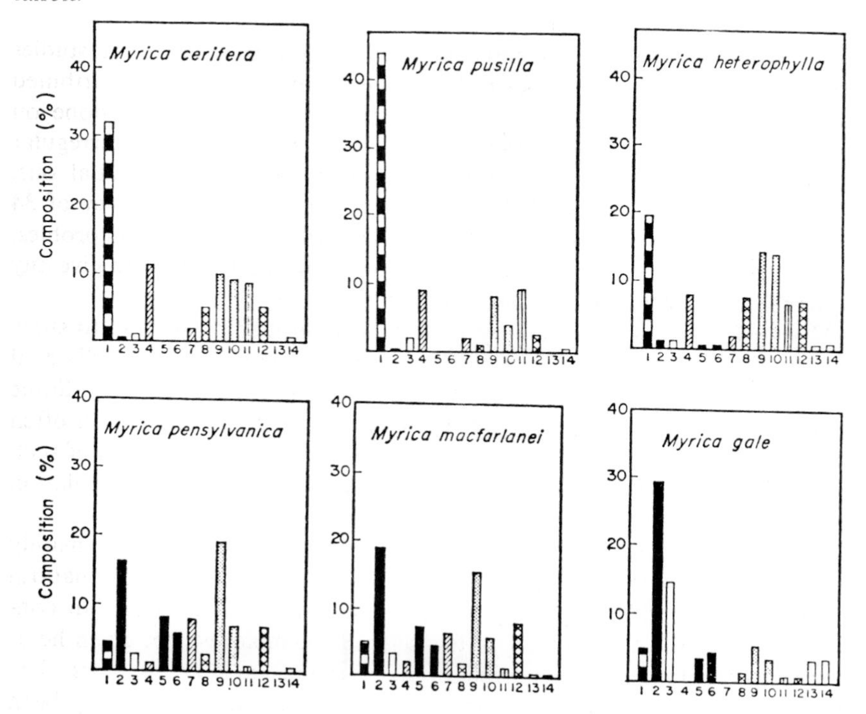

Fig. 5.5 Leaf essential oil patterns in *Myrica* species. Key: 1, α-pinene; 2, myrcene; 3, limonene; 4, cineole; 5, *cis*-ocimene; 6, *trans*-ocimene; 7, γ-terpinene; 8, linalol: 9, caryophyllene; 10, α-humulene; 11, α- and β-selinene; 12, nerolidol; 13, γ-eudesmol; 14, α- and β-eudesmol (from Halim and Collins, 1973).

It may also be possible to use oil patterns for defining species in angiosperms. Thus, preliminary surveys in *Bothriochloa* (de Wet and Scott, 1965) *Monarda* (Scora, 1967) and *Salvia* (Emboden and Lewis, 1967) show that species-specific patterns in leaf volatiles are the rule rather than the exception. In *Myrica* also, leaf-oil patterns are complex and highly species-specific. Here, analyses have been employed to settle taxonomic confusion about two taxa of uncertain affinities. Oil patterns (Fig. 5.5) showed clearly that one '*M.*

pusilla', was in fact synonymous with *M. cerifera*, and that the other '*M. macfarlanei*', thought to be a hybrid *M. pensylvanica* × *M. cerifera*, was pure *M. pensylvanica* (Halim and Collins, 1973).

2. *Identification of hybrids*

Hybridization may occur in natural populations of conifers and detection may be difficult by morphology alone. Oil studies have helped to confirm hybrids in natural stands where two related species grow sympatrically. Thus Mirov (1965) was able to recognize hybrids between the jackpine, *Pinus banksiana* and the lodgepole pine *P. contorta* since the oil contains terpenes characteristic of both parents: α- and β-pinene from jackpine and β-phellandrene from lodgepole pine. Similarly, von Rudloff and Holst (1968) found that the Rosendahl spruce (*Picea glauca* × *mariana*), a natural hybrid between white and black spruce, has a terpene composition intermediate between the parents (Fig. 5.6). Again, Lawrence *et al.* (1975) confirmed natural hybridization between *Cupressus sargentii* and *C. macnabiana*, two sympatric species of the Californian coastal ranges, by analysing leaf oils and finding monoterpene patterns intermediate between the two parents in some plants.

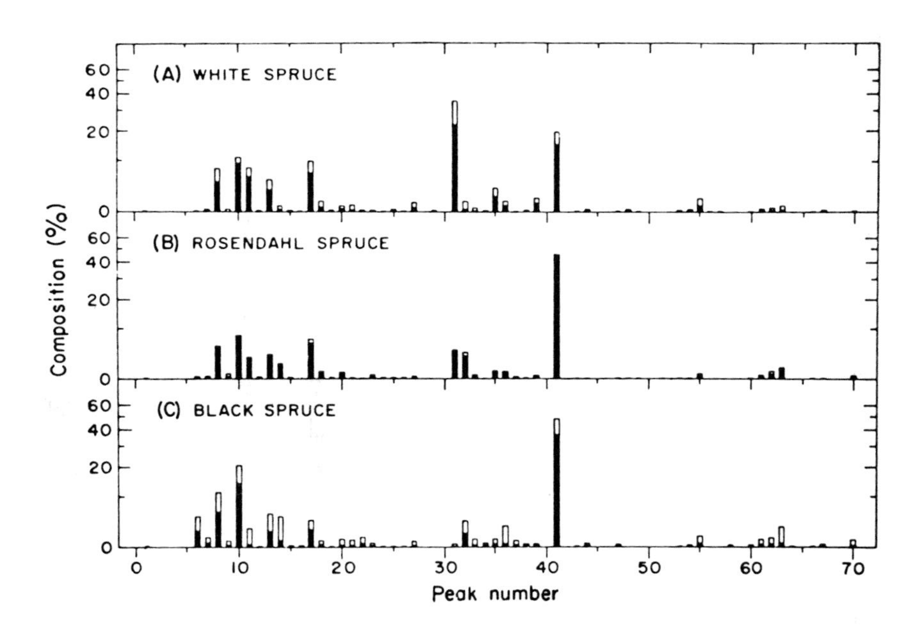

Fig. 5.6 Terpene leaf patterns in white (A) and black (C) spruce and in the hybrid, Rosendahl spruce (B) (von Rudloff and Holst, 1968).

Perhaps the most dramatic example of the power of terpene analysis to solve problems of hybrid identification in conifers is the negative case of *Juniperus*. Here, GLC analyses of leaf oils in plants thought on other grounds to be hybrids between the two allopatric species *J. ashei* and *J. virginiana* showed only oil patterns characteristic of one or other of the putative parents. This work is described in more detail in chapter 13.

3. *Generic and tribal limits*
Essential oils have rarely been used as taxonomic markers at the higher levels of classification. In the case of the conifers, leaf terpenoids are occasionally useful for distinguishing genera, but in general show few clear cut correlations with generic or familial limits (von Rudloff 1975b). The situation in the angiosperms may be more promising. Thus in the Dipterocarpaceae, Bisset *et al.* (1966) have shown that *Dipterocarpus* (42 spp. analysed) differs in sesquiterpene pattern in wood resins from *Doona* (5 spp. analysed), α-gurjunene being characteristic in the first and β-elemene of the second of these two genera. Again, in the Umbelliferae, different genera in the tribe Caucalideae can be distinguished by differences in oil content of the fruits (Table 5.1). The oil results (Williams and Harborne, 1972) indicate a close relationship between the genera *Daucus* and *Pseudorlaya* and other chemical results (see below) support this. Significantly, these two genera were at one time subtribally separated from the other four genera.

Table 5.1 Distribution of monoterpenes and sesquiterpenes in the fruits of the Caucalideae

	Genus[a]					
Terpenoid	*Torilis*	*Orlaya*	*Pseudorlaya*	*Daucus*	*Caucalis*	*Turgenia*
α-Pinene	−	−	+	+	−	−
β-Pinene	−	−	+	+	+	−
Limonene	−	−	+	+	−	−
Biphenyl	+	+	−	−	−	−
Geranyl acetate	−	−	−	+	−	−
Caryophyllene	−	+	−	+	−	−
Carotol	+	−	−	+	−	+
Unknown *S*1	+	−	−	+	−	−
Unknown *S*2	−	+	−	+	−	−
Unknown *S*3	+	+	−	+	+	+
Unknown *S*4	−	−	−	+	−	−
Unknown *S*5	−	−	−	+	−	−

[a]Based on an analysis of fruit essential oils from five *Torilis* spp., eleven *Daucus* spp., and three *Orlaya* spp.; remaining genera analysed from a single accession. Only major components are included. Unknowns *S*1–*S*5 are different sesquiterpenoids.

Some of the sesquiterpenes found in the Caucalideae, e.g. carotol, probably characterize the tribe. At least, carotol has not been found in the related tribe, the Laserpitieae. Twenty-two species from seven genera of the latter tribe have been analysed for oil constituents in fruits (Adcock and Betts, 1974). Although there are some similarities in major oil constituents (α-pinene, β-pinene and limonene are common), there are also clear-cut differences; myrcene, *p*-cymene and perillal are only recorded in Laserpitieae. Such differences at tribal level might be exploited in future taxonomic revisions of this oil-rich family.

III. Aliphatic and aromatic volatiles

A. Aliphatic essential oil constituents

It is not generally realized that many other organic molecules besides terpenoids may contribute to the volatile odours and smells of plant tissues. Aliphatic constituents may be particularly prominent in flower and fruit volatile fractions and they can often be the major odour principles. Such aliphatic compounds are either simple long-chain hydrocarbons or their oxygen derivatives (Table 5.2).

Table 5.2 Some aliphatic odour principles in plants

Structure	Name[a]	Typical source
Hydrocarbons		
$CH_3(CH_2)_{13}CH_3$	Pentadecane	*Magnolia acuminata* flower
$CH_3(CH_2)_{15}CH_3$	Heptadecane	*Ophrys* sp. flower
$CH_3CH_2CH{=}CH(CH_2)_2CH{=}CHCH_2OH$	Nonadien-1-ol	*Cucumis sativus* (cucumber) fruit
Aldehyde		
$CH_3(CH_2)_8CHO$	Decanal	*Coriandrum sativum* (coriander) fruit
Fatty acid esters		
$CH_3(CH_2)_{10}CO_2CH_3$	Methyl dodecanoate	*Magnolia grandiflora* flower
$CH_3(CH_2)_7CH{=}CH(CH_2)_7CO_2CH_3$	Methyl oleate	*Ophrys* sp. flower
$CH_3(CH_2)_4(CH{=}CH)_2CO_2C_2H_5$	Ethyl decadienoate	*Pyrus domestica* (pear) fruit

[a]The first part of the name indicates the number of carbon atoms (pentadec = 15, etc.), the second part the type of compound (ane = saturated hydrocarbon, ol = alcohol, al = aldehyde, oic = acid).

A typical range of aliphatic volatiles are those reported in orchid, *Ophrys*, flowers by Bergström (1978). Hydrocarbons present include compounds from C_{12} to C_{19}, namely dodecane, tridecane, tetradecane, pentadecane, heptadecane and nonadecane. Some mono-unsaturated hydrocarbons are also present, together with related alcohols, aldehydes and carboxylic acids. The acids are normally esterified, as in methyl oleate (Table 5.2). In *Ophrys* sp., these substances are important attractants to male *Andrena* and *Eucena* spp. bees, which pollinate the bee-shaped flowers during pseudocopulation; many of the same volatiles have also been detected in the bees (Bergström, 1978).

Another group of plants where hydrocarbons are known to be important contributors to floral fragrance are *Magnolia* species. These plants are largely pollinated by beetles, who are attracted to the flowers by odours distasteful to human noses. Thus in *Magnolia acuminata*, the major floral odour constituent is pentadecane, a substance with an unpleasant smell. Thien *et al.* (1975) in exploring the chemotaxonomic possibilities of floral odours in *Magnolia*, have found that all of eight species studied have distinct odour profiles and all can be clearly distinguished by GLC analyses of their volatiles (Table 5.3). It will be seen that a variety of different chemicals, besides pentadecane, contribute to odour, even in such a small group of related species. Each species has a different profile so that there are considerable possibilities here of using floral fragrances in taxonomic comparisons.

Table 5.3 Odour profiles of some magnolia flowers

Magnolia species	Odour class	Major volatiles identified[a]
M. acuminata	Weak, distasteful	Pentadecane (84%) and other hydrocarbons
M. tripetala	Weak, distasteful	Acetophenone (59%) and cinnamyl alcohol (13%)
M. grandiflora	Fruity, pleasant	Methyl dodecanoate (24%)
M. pyramidata	Fruity, pleasant	Methyl decanoate (45%)
M. fraseri	Fruity, pleasant	Methyl decanoate (35%) and terpenes
M. ashei	Faint, sweet	Terpenoids: 50% β-elemene, 14% γ-gurjunene
M. virginiana	Sweet, fruity	Methyl decanoate (5%) and terpenes
M. macrophylla	Sweet, fruity	Methyl tetradecenoate (16%) and terpenes

[a]Data from Thien *et al.* (1975). Only a few of the major constituents are shown, to indicate some of the differences between species. % figures refer to percentage of total volatiles.

Aliphatic volatiles are widely present in plant fruits as well as in flowers. Fruits are characterized by organic ester odours, such as those of amyl acetate (banana), ethyl 2-methylbutyrate (apple) and ethyl decadienoate (pear). Fruit volatiles can be very complex mixtures of many constituents; no less than 144 components have been detected in the odour of strawberries (Nursten, 1970). Variations in aliphatic constituents undoubtedly are responsible for many of the differences in fruit odours generally.

B. Aromatic odour constituents

A considerable number of miscellaneous aromatic constituents have been recorded in plant odours. Most of these have some degree of taxonomic restriction in their natural distribution. The structures of a few representatives are given in Fig. 5.7; those shown are either benzoic acid derivatives or phenylpropanoids. Among the former are vanillin, the odour principle of vanilla pod and the methyl ester of salicylic acid, the 'oil of wintergreen' first recorded in leaves of *Gaultheria procumbens* (Ericaceae). Among phenylpropanoids are coumarin, the 'smell of new-mown hay', methyl cinnamate, the odour of cinnamon, and myristicin, in the odour of nutmeg.

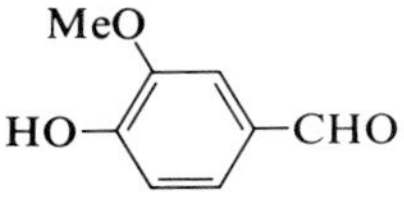

Vanillin

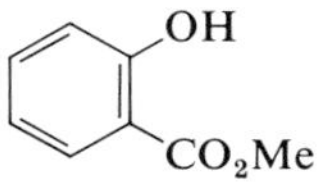

Methyl salicylate

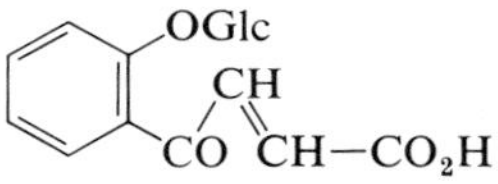

O-Glucosyloxycinnamic acid (coumarin precursor)

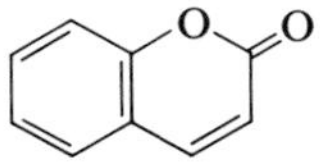

Coumarin

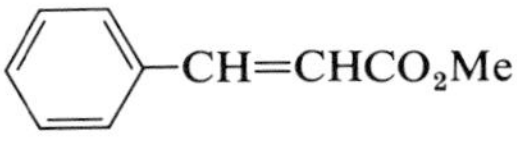

Methyl cinnamate

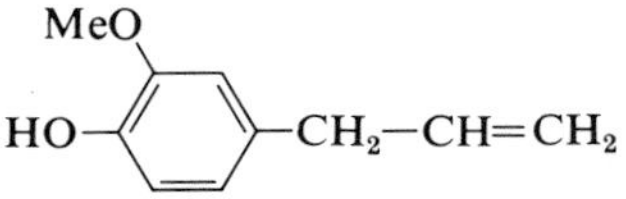

Eugenol

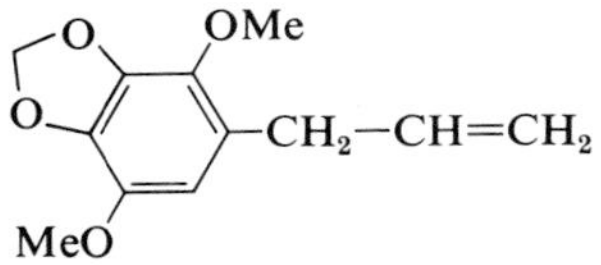

Parsleyapiole

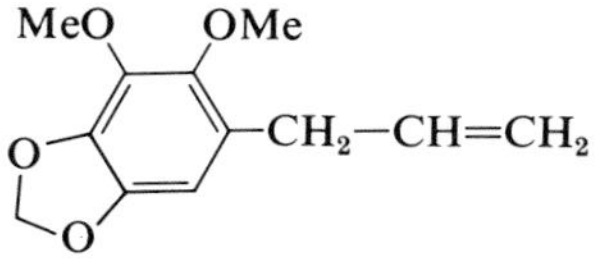

Dillapiole

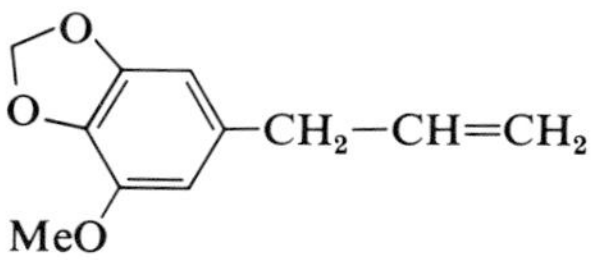

Myristicin

Fig. 5.7 Some aromatic volatiles of plants.

One group of plants where aromatic compounds make a major contribution to floral fragrance are the orchids of Central and South America. Here, chemistry of the scents has been studied in relation to pollination ecology and to the pheromone biology of the euglossine bees which are specific pollinators of these plants. The considerable speciation observed among these bees and the orchids they pollinate can be related through the characteristic odours of the different orchids, which are then used in turn by the various bee species as specific sex pheromones, by which the male attracts the female (Dodson, 1975).

Some 150 orchid species from 25 genera have been analysed by GLC for their scents (Dodson *et al.*, 1969) and at least 50 components have been variously detected. Most species produce at least 7–10 major chemicals, some more and some less. Among aromatics identified are benzyl acetate, eugenol, methyl salicylate and methyl cinnamate. With these and other scents, it is important to remember that the percentage composition of volatile constituents, as revealed by GLC, is not necessarily a good guide to the actual smell, since a very minor constituent may be the most important contributor to the total fragrance. This is true in the flowers of *Catasetum roseum*, which smell of cinnamon (methyl cinnamate) but in which the odour principle only occurs as 1% of the total floral oil.

Taxonomically, GLC analyses of orchid flower odours show considerable promise. Although the orchid species fall into odour groups according to their particular pollinator, there are still detectable quantitative differences in odour patterns between species of the same group (Dodson and Hills, 1966). The relationship between flower odour and pollinator is so close that it is possible to predict from the GLC profile of the volatiles which bee species will pollinate a particular orchid species. Thus, a field naturalist could conceivably find it easier to identify an orchid simply by recognizing its specific pollinator, leaving the taxonomic work to be done by the insect.

A family rich in aromatic volatiles based on the phenylpropanoid C_6–C_3 carbon skeleton is that of the Umbelliferae. Some of the characteristic aroma constituents of root, leaf or fruit in this family are of this type. Two species-specific constituents are parsleyapiole, from *Petroselinum crispum* and dillapiole from *Anethum graveolens* (see Fig. 5.7). One more widespread constituent is myristicin, first isolated from *Myristica fragrans* (Myristicaceae) but more typical in its occurrence in umbellifers. This is a biologically interesting molecule, since it reputedly has hallucinogenic activity in man (Shulgin, 1966). Its distribution in fruits of the Umbelliferae has been determined in some 100 species representing 50 genera (Harborne *et al.*, 1969). It is simply detected in fruit extracts by TLC in benzene or toluene on silica-gel plates; it gives a brown colour after the plate is sprayed with vanillin–H_2SO_4 and heated (see Harborne, 1973b). The survey showed it to be present in 14 genera, all of the subfamily Apioideae. As a character that contributes to flavour and odour, its concentration in fruits may be affected by breeding and cultivation, so that taxonomic significance should be limited to surveys on wild species. Thus it was readily detected in fruits of cultivated parsnip *Pastinaca sativa* and cultivated fennel *Foeniculum vulgare*, but not in fruits of wild accessions of these two taxa. It acts as an interesting taxonomic marker in the tribe Caucalideae, where its occurrence in fruits is restricted to two closely related genera, *Daucus* and *Pseudorlaya*. In *Pseudorlaya*, it occurs in both known species, whereas in *Daucus*, it is restricted, remarkably enough, to *D. glochidiatus* and *D. montanus*, the only two polyploid members of the genus.

One other noteworthy aromatic volatile of Umbelliferae is coumarin, which actually occurs in plant tissue in bound form as the corresponding *o*-hydroxycinnamic acid glucoside (see Fig. 5.7) and is enzymically released, when tissue is crushed. Its odour, that of new-mown hay, is familiar because it is released during hay-making or whenever grass is cut. Particularly high concentrations of precursor occur in sweet vernal grass, *Anthoxanthum odoratum*, but it is also well represented in legume pasture plants, e.g. *Melilotus alba*. In spite of its fragrant 'heady' odour, coumarin is actually bitter in taste and it can be a repellent to grazing cattle.

As a taxonomic marker, coumarin is so widely distributed that it is a character of little general significance. It has been recorded not only in many angiosperms but also in ferns and in fungi. At lower levels of classification, however, it may be of more diagnostic significance. Thus, in legumes, it occurs regularly in *Melilotus* species and also occasionally in the related genus, *Trigonella*. Interestingly, its presence in the latter genus is restricted to certain sections, the species of which are similar to *Melilotus* both in phytoalexin response (see chapter 16) and in morphology (Ingham and Harborne, 1976). Its pattern of occurrence thus confirms a considerable heterogeneity in what is widely recognized as a taxonomically difficult plant group.

IV. Nitrogen-containing volatiles

The presence in plants of unpleasant, offensive aminoid odours is such a striking feature that the fact is often reflected in the species name: hogweed, *Heracleum sphondylium*; stinking hellebore, *Helleborus foetidus*; stink iris, *Iris foetidissima*; and so on. Because of their highly repugnant odours, their chemistry has not been intensively studied but there is increasing evidence that most unpleasant smells in plants are due to nitrogen-containing constituents. Indeed, distasteful odours in flowers represent a chemical mimicry by which the plant produces a smell of decaying faeces to deceive carrion and dung flies to transfer their attention to the flower heads for pollination purposes. The chemistry of these volatiles, not surprisingly, is similar to that of the breakdown products of carrion and dung.

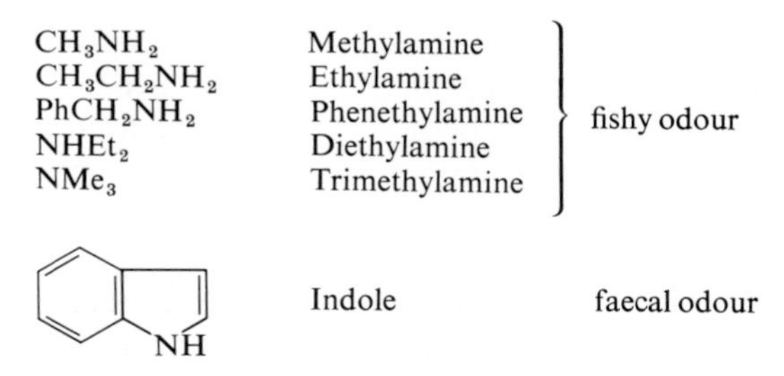

Fig. 5.8 Nitrogen-based volatiles detected in plant aminoid odours.

Major constituents of aminoid odours are the fishy-smelling aliphatic amines. These are usually primary amines, such as ethylamine, propylamine and phenethylamine, or secondary amines such as diethylamine, or occasionally tertiary amines such as trimethylamine (Fig. 5.8). Difficulties in surveying plant tissues for these volatile amines have recently been resolved by Ilert and Hartmann (1972) who have developed an identification technique based on steam distillation of fresh flowers, derivatization of the amines with dinitrofluorobenzene, followed by separation by TLC. By these means, these authors were able to detect amines in about half of 52 species of flowering plants belonging to 30 families. Amine-rich odours were detected, for example, in *Malus* and *Crataegus* species (Rosaceae), *Staphylea colchica* (Staphyleaceae) and *Lychnis coronaria* (Caryophyllaceae). Most amine-positive species contained from three to ten different amines, so that there were a variety of patterns of possible taxonomic utility (Hartmann *et al.*, 1972).

Trimethylamine is particularly unpleasant and is used as a standard for a fishy odour. It is responsible, in part, for the evil smell of hogweed inflorescences. It also occurs in leaves, and has been detected in those of the stinking goosefoot, *Chenopodium vulvaria*, at the remarkably high concentration of 400 p.p.m. (Cromwell and Richardson, 1966). Trimethylamine, interestingly enough, has quite a wide distribution in plants, occurring in lower concentration in seaweeds, lichens and fungi.

Finally, mention must be made of the aromatic nitrogen-based compound, indole, which has a faecal odour and has also been found in repulsive plant odours. It has been characterized in *Sauromatum guttatum* and is also probably present in *Arum dioscorides*, *Dracunculus vulgaris* and *Lysichitum americanum* (Chen and Meeuse, 1971). It could not, however, be detected in related plants, such as cuckoo pint, *Arum maculatum* or *A. orientale*, which are just as evil smelling at the time of flowering. Interestingly monoamines such as ethylamine, isobutylamine and phenethylamine have been found in *A. maculatum*. There is thus evidence here for chemical variation in odour production in members of the Araceae, which might be exploited for chemotaxonomic purposes.

V. Sulphur compounds

In addition to the ubiquitous non-volatile sulphur amino acids cysteine and methionine, there are a number of more volatile sulphur-containing organic metabolites in plants of restricted distribution. Their presence in plant tissues is generally advertised by an unpleasant, sometimes lachrymatory, smell. In fact, there are two distinct classes which are of taxonomic interest: the mustard oils (or isothiocyanates) of the Cruciferae with pungent odours and bitter tastes; and the organic disulphides of the Alliaceae with obnoxious 'rotten egg' smells. Both groups are important to man, since the flavours of mustard, radish, onion, chive and garlic are due in part to the presence of these substances.

Both groups of sulphur volatiles occur *in vivo* in bound form and while the odoriferous elements may be released slowly from the living plant, the full force of their smell is only apparent after tissue has been crushed and enzymes present have freed the volatile from the parent compound. Their presence or absence is thus particularly easy to monitor during preliminary surveys. Since their chemistry and natural distributions are so distinct, the two classes of volatile compounds require separate consideration here.

A. The isothiocyanates

The isothiocyanates or mustard oils were first studied in relation to their occurrence in the Cruciferae and their presence in these plants is as characteristic of the family as the distinctive petal pattern, with the cruciform-like flower. Every species of crucifer that has been tested contains mustard oils. The *in vivo* precursors are the mustard oil glycosides or glucosinolates, substances with important taste properties but which are essentially non-volatile. The acrid flavour of mustard oil is released when the fresh crucifer tissue is crushed, since the glucosinolates are accompanied in the plant by the appropriate hydrolytic enzyme, a thioglucosidase known as myrosinase. The major products of release are glucose, sulphate and the isothiocyanate, formed by hydrolysis and rearrangement of the thioglucoside according to the scheme shown in Fig. 5.9. Under some circumstances, fragmentation of the glucosinolate can give rise to elemental sulphur, together with the corresponding nitrile RCN.

$$\underset{\text{Glucosinolate}}{R-C\begin{matrix}\nearrow NOSO_3^- \\ \searrow SGlc\end{matrix}} \xrightarrow[\text{myrosinase}]{\text{enzyme}} \underset{\text{Isothiocyanate}}{R-N{=}C{=}S} + \text{glucose} + SO_4^{2-}$$

Typical glucosinolates:

Glucocapparin:	$R = CH_3$
Sinigrin:	$R = CH_2{=}CH-CH_2$
Glucoibervin:	$R = MeS(CH_2)_3$
Glucotropaeolin:	$R = PhCH_2$
Sinapine:	$R = p\text{-}OH-C_6H_4$

Fig. 5.9 Structure of plant glucosinolates.

When screening plants for isothiocyanates, detection, in theory, can be accomplished via the parent glucosinolates, the volatile mustard oils or the presence of the specific enzyme myrosinase. In practice, most attention is given to the isothiocyanates, which can be separated by GLC, after steam distillation of crushed tissue. These compounds can also be converted to stable thiourea derivatives, which give blue colours with Grote's reagent and which can be separated by paper chromatography or by TLC (see Harborne, 1973b).

Isothiocyanates also give a 'false' positive reaction with picrate paper, a test usually thought to be specific for prussic acid (HCN) (see chapter 6). It is thus possible to monitor the distribution of isothiocyanates in plants, by means of this simple colour test (the yellow paper changing to red or brown, depending on concentration), as long as cyanogenic glycosides are absent. The two classes of glycoside do not co-occur in Cruciferae and this procedure has been employed successfully to determine quantitative variations in mustard oil content in natural populations of *Brassica oleracea* (Mitchell and Richards, 1978).

At least 70 glucosinolates are known, all with the same basic structure. The majority are aliphatic derivatives, with the remainder having aromatic substituents of one sort or another (Fig. 5.9). Glucosinolates occur in most plant parts (roots, stems and leaves) but may be concentrated in the seeds, which are often a convenient source for purposes of isolation (Kjaer, 1976). The best known glucosinolate is sinigrin, a major constituent of cabbage, which yields allyl isothiocyanate on enzymic hydrolysis.

Sinigrin and its related isothiocyanate have been implicated as feeding attractants to various cabbage pests, including the cabbage white butterfly

Pieris brassicae and the cabbage aphid *Brevicoryne brassicae* (van Emden, 1972). The major role of these substances in the Cruciferae, however, must be as feeding deterrents to herbivores and there is evidence that they are toxic to most Lepidoptera (see Erickson and Feeney, 1974). Glucosinolates have also been implicated as antifungal agents; they are responsible, at least, for resistance to downy mildew in certain cabbage varieties and wild cabbage plants (Greenhalgh and Mitchell, 1976).

Glucosinolates are formed in plants from one or other of the protein amino acids, along a pathway related to that utilized for cyanogenic glycoside synthesis (Kjaer and Larsen, 1973). A switch mechanism probably operates at the point of bifurcation in the pathways (Fig. 5.10), since the two classes represent alternative forms of toxin, so that they never actually co-occur in the same plant.

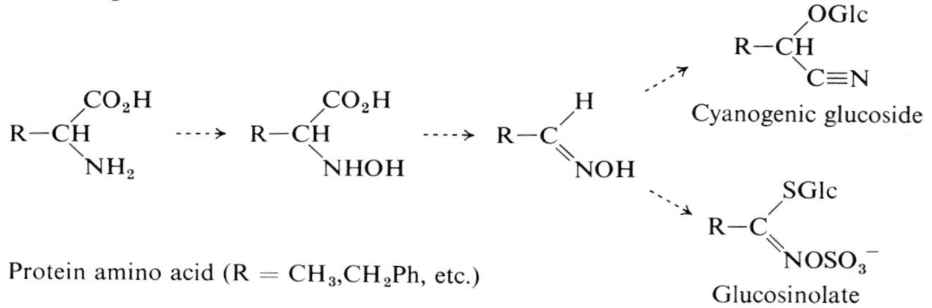

$R{-}CH(CO_2H)(NH_2)$ ----> $R{-}CH(CO_2H)(NHOH)$ ----> $R{-}C(H){=}NOH$ ----> $R{-}CH(OGlc)(C{\equiv}N)$ Cyanogenic glucoside

$R{-}C(H){=}NOH$ ----> $R{-}C(SGlc){=}NOSO_3^-$ Glucosinolate

Protein amino acid ($R = CH_3, CH_2Ph$, etc.)

Fig. 5.10 Pathways of biosynthesis to glucosinolates and cyanogenic glycosides.

Taxonomically, glucosinolates are limited to dicotyledonous angiosperms, where they have a discontinuous distribution, with major occurrence in one group of related families—the Cruciferae, Moringaceae, Capparaceae and Resedaceae. At one time these four families were included in the same order, the Rhoeadales, with the Papaveraceae and Fumariaceae, but the absence of glucosinolates from the latter two families was a significant factor in showing that the order was really heterogeneous. Another important chemical feature—the presence of isoquinoline alkaloids throughout the Papaveraceae and Fumariaceae—separates the two groups. Indeed, this strong chemical evidence has been used, together with data from morphology and serology, to replace the order Rhoeadales by a new system, in which two new orders—the Capparales and Papaverales—are recognized (Table 5.4: see Merxmuller and Lenz, 1967 and Cronquist, 1968).

Although uniform in possessing glucosinolates, the families of the newly created Capparales do differ in the chemical nature of their thioglucosides. In particular, the Capparaceae is distinct in having mainly the alanine-derived methyl glucosinolate. This and other evidence points to the Capparaceae as the more primitive family in the group, with the other families having more complex glucosinolates being derived from it.

Table 5.4 Revision of the families of the Rhoeadales on chemical evidence

Old classification	Families	Presence (+) or absence (−) of: Isothiocyanate	Isoquinoline alkaloids	Revised classification
order Rhoeadales	Cruciferae	+	−	order Capparales (plants with centrifugal stamens)
	Capparaceae[a]	+	−	
	Resedaceae	+	−	
	Moringaceae	+	−	
	Tovariaceae	+	−	
	Papaveraceae	−	+	order Papaverales (plants with centripetal stamens)
	Fumariaceae	−	+	

[a]Earlier spelt as Capparidaceae

The glucosinolate character has already proved to be of value for taxonomic purposes in the case of a family of uncertain affinities, the Stegnospermaceae. This small family was originally included in the order Centrospermae, but its position here has been questioned since it lacks the betalain pigments characteristic of this order and also the typical Centrospermae P-type plastids (Goldblatt *et al.*, 1976). A phytochemical search of the Stegnospermaceae showed the presence, for the first time, of glucosinolates and on this basis and that of chromosomal data, the family is now aligned in the Capparales.

That the glucosinolate character has been independently achieved in the angiosperms generally is clear from the fact that the compounds are known to occur in at least six families which are quite unrelated to the Capparales. These families are the Caricaceae, Euphorbiaceae, Gyrostemonaceae, Limnanthaceae, Salvadoraceae and Tropaeolaceae (Ettlinger and Kjaer, 1968). Detailed studies of glucosinolates have been carried out in at least two of these other families.

In Caricaceae, Tang *et al.* (1972) have shown that benzyl isothiocyanate is the only acrid oil present. GLC analyses of macerated seeds indicate that species of *Carica* (6 spp.) and *Jarilla* (1 sp.) are relatively rich in this isothiocyanate, with amounts in the range 1.37–1.96% dry wt. By contrast, seeds of all three species of *Jacaratia* tested contain only 2–4 p.p.m., a concentration difference of 1 : 6000. Since *Carica* and *Jacaratia* are otherwise difficult genera to distinguish, this quantitative chemical character promises to be of value for purposes of identification.

Benzyl isothiocyanate also occurs in *Tropaeolum* (Tropaeolaceae) but is accompanied by at least three aliphatic isothiocyanates (Kjaer *et al.*, 1978). In a study of seed volatiles of nine species, these authors were able to divide them into three groups: three with benzyl isothiocyanate alone; five with two additional branched chain isothiocyanates; and *T. peregrinum* with the extra presence of ethyl isothiocyanate. A number of other minor volatiles were also detected and there seems to be a range of chemical characters available in these mustard oils for assessing relationships in *Tropaeolum*, a large and complex genus.

B. Organic sulphides

Organic sulphides are readily apparent when present in plant tissues because of their highly obnoxious odours. Some sulphides are formed exclusively in marine algae and others are restricted to certain groups of higher plants. The only group where their taxonomic implications have been considered is the genus *Allium*. The compounds identified here include both mono- and disulphides with methyl, *n*-propyl and allyl substituents (Fig. 5.11). They appear to be present in bound form and, like the isothiocyanates, are released as a result of enzymic hydrolysis when the tissue is damaged. Sulphides may be detected in the volatiles of both the bulbs and the aerial parts of the plant.

Most attention has been given to the cultivated *Allium*: *A. cepa* (onion), *A. chinense* (rakkyo), *A. porrum* (leek), *A. sativum* (garlic) and *A. schoenoprasum* (chive). Over 50 constituents have been identified variously in these plants and the characteristically different odours are probably related to the different proportions of the main mono- and disulphides (Fig. 5.11). Trisulphides are also present in onion and garlic, while the obnoxious-smelling methanthiol (MeSH) has been detected in onion, leek and garlic. The lachrymatory factor of onion is also a sulphur compound, either thiopropanal S-oxide or the related sulphenic acid (Johnson *et al.*, 1971).

In the first study of wild *Allium*, Saghir *et al.* (1966) determined the amounts of methyl, *n*-propyl and allyl sulphides in 25 North American species and related their results to sectional groupings based on cytology and morphology.

Me—S—S—Me	Dimethyldisulphide
*n*Pr—S—S—*n*Pr	Di-*n*-propyldisulphide
CH_2=CH—CH_2—SS—CH_2—CH=CH_2	Diallyldisulphide
CH_2=CH—CH_2—SS—Me	Methylallyldisulphide
CH_2=CH—CH_2—S—CH_2—CH=CH_2	Diallylmonosulphide
CH_3—CH_2—CH=S=O	Thiopropanal S-oxide

Fig. 5.11 Some organic sulphides of *Allium*.

In some sections, species patterns were uniform, but unfortunately, in others, considerable divergencies were noted in the sulphides of closely related taxa. In a more successful survey of Old World species of *Allium*, Bernhard (1970) found reasonably good correlations between sulphide profiles and sectional classification. However, sampling was limited and only some 25 species were examined. From the work done to date, it is possible to suggest that analyses of sulphides in *Allium* could be useful in future taxonomic revisions of the genus. In general, however, wider surveys are still needed to establish fully the validity of this approach to *Allium* systematics.

VI. Conclusion

There are at least seven classes of organic substances which have been implicated in odour production in plants (Table 5.5). Plant volatiles are readily separated and identified by GLC, a chromatographic procedure that produces

Table 5.5 The major classes of volatiles in higher plants

Class	Odour type	Organ distribution	Taxonomic occurrence
Monoterpenoids and sesquiterpenoids	Usually fragrant, but may be pungent	Leaves especially; but also fruits, flowers and roots	Most gymnosperms; many angiosperm families
Aliphatic	Sweet and fruity; can be distasteful	Flowers and fruits	Widespread, notably Orchidaceae and Magnoliaceae.
Aromatic	Usually pleasant	Flowers and fruits; also leaves	Widespread, notably Orchidaceae and Umbelliferae
Nitrogen-containing	Foetid, fishy, faecal	Flowers; also leaves	Restricted, notably Araceae
Isothiocyanates	Acrid	All parts	Restricted, notably Cruciferae
Sulphides	Obnoxious	Bulbs, leaves	Restricted, notably Alliaceae

both qualitative and quantitative data for taxonomic purposes. Although volatile fractions have been analysed in a large number of plants, the substance or substances present in the fraction which are responsible for the actual odour of a given plant are not always known and much further work is needed to relate organic chemistry to odour constitution. Nevertheless, the volatile profile of a particular plant is usually complex and is nearly always characteristic of that plant. Thus it can be employed for chemotaxonomic purposes if similar profiles are available for related taxa.

The most widespread group of odour substances are the volatile terpenoids and they have already proved to be of considerable importance as taxonomic markers in gymnosperms. The isothiocyanates of the Cruciferae have also been shown to have an interesting and taxonomically significant distribution pattern among angiosperm families. The utility of the other classes of volatile have yet to be established in chemotaxonomic practice, but most have been shown to have real potential.

6

Alkaloids and Other Plant Toxins

I.	Introduction	75
II.	Alkaloids	77
III.	Cyanogens	90
IV.	Non-protein amino acids	97
V.	Iridoids	104
VI.	Sesquiterpene lactones	109
VII.	Diterpenoids	114
VIII.	Triterpenoids and steroids	119
XI.	Conclusion	125

I. Introduction

Most, if not all, higher plants contain organic constituents in their tissue which are potentially toxic to animal browsers. Such toxins may be poisonous to plant life as well but are present in a safe bound form as glycoside or are sequestered in the vacuole and only released when tissue is damaged. The presence of mammalian toxins in plants is extremely well documented but relatively few wild species (such as ragwort, deadly nightshade, Jimson weed, etc.) are specifically regarded as dangerous to man or his livestock. However, many more species are potentially toxic, in the broadest sense, and if not dangerous to man, may poison birds, fish or insects. McIndoo (1945), for example, lists over 1000 species in the North American flora which contain insecticidal constituents. There are also plants (nettle, poison ivy) whose defence consists of chemicals which cause external injury or skin allergy in herbivores. It is fortunate for man that many poisonous plants advertise their harmful nature by a disagreable taste signal. Plant toxins are often bitter and can be detected, and avoided, by tasting the leaves. Indeed, the correlation with bitterness is such that in the case of plant alkaloids it is possible to taste leaf extracts as a preliminary screen for alkaloid-containing plants. In view of the potency of some alkaloids, however, it is important not to ingest the poisons present in the leaf sap while carrying out such surveys!

The ability to recognize poisonous plants was clearly vitally important to our ancestors, who exploited a much wider variety of plants for dietary purposes than we do today. Primitive man of necessity became a chemotaxonomist of no mean ability. Besides avoiding such plants as food, early man also made use of these plant extracts for other purposes, notably in arrow poisons for slaughtering game and as a murder weapon. The death of the Greek philosopher Socrates by the administration of an extract of hemlock, *Conium maculatum*, leaves is often quoted as an early use of alkaloids for judicial poisoning. More constructive uses were also made of many of the same poisonous plants, due to the intense physiological effects of the crude extracts on the central nervous system. Many, if not all, of the plants used in early herbals were those with toxic agents present and were plants which we now know to contain alkaloids.

During the nineteenth century in the early days of organic chemistry, many of the active constituents of herbal plants were isolated and studied. The organic structures were eventually fully determined and the substances synthesized in the laboratory. In spite of the subsequent development of synthetic medicinals, many plant drugs are still widely used in medical treatment today (see Farnsworth and Bingel, 1977). This is true, for example, of hyoscyamine from *Atropa belladonna*, morphine and codeine from *Papaver somniferum* and of emetine from *Uragoga ipecacuanha*.

The extensive pharmaceutical use of alkaloid-containing plants in medicine has provided the incentive during the present century of many plant surveys for alkaloids. Today, it is probably true that more plant species have been screened for the presence or absence of alkaloids than for any other type of secondary constituent. Thus, the chemotaxonomy of alkaloids is well developed (see Hegnauer 1963, 1966) and it is widely recognized that certain angiosperm families are very rich in alkaloids, while others are very poor. The predictive value of the alkaloid character is considerable and studies of species related to classical sources of alkaloid have nearly always led to the discovery of new substances. For example, following on from the key discovery of the pain-killing morphine in *Papaver somniferum*, phytochemists have screened every other available *Papaver* species and have invariably found a variety of alkaloid bases to be present (Santavy, 1970).

Although alkaloids are the best known group of plant toxin, other related nitrogen-containing molecules of plants are also poisonous. There are, for example, the cyanogenic glycosides, which give prussic acid on hydrolysis. These substances also have a bitter taste as in bitter almond fruits and their presence in plants is revealed by the 'odour of bitter almonds' as the HCN is released when tissues are broken. In spite of the toxicity, it is interesting that one particular cyanogenic glycoside, laetrile (amygdalin), from peach seeds, is being employed currently in fringe medicine as an anti-cancer agent.

Other classes of nitrogenous compounds in plants which are toxic include nitro derivatives, present in *Astragalus*, non-protein amino acids, commonly present in legume seeds, and several proteins.

Yet other groups of plant toxin lack nitrogen and are simply derivatives of carbon, hydrogen and oxygen; it is not always realized, for example, that many plant terpenoids are highly toxic substances. The terpenoid-based cardiac glycosides (or cardenolides) are amongst the most poisonous of all organic plant constituents. Among other terpenoid classes are also many well known toxins. This is particularly true of sesquiterpene lactones, of diterpenoids and of saponins. All these toxins have a limited distribution in plants and hence are of taxonomic interest.

The function in the plant of toxic alkaloids and terpenoids has often been questioned in the past with little conclusive result. However, there has been a growing realization during the last 20 years that these substances are important to the plant in providing protection from overgrazing by herbivores of all kinds (see Harborne 1982 and references therein). The co-evolutionary importance of toxins has been particularly emphasized by Ehrlich and Raven (1965) in their analyses of plant–butterfly interactions. These authors have suggested that the diverse chemical patterns of plant toxins are related to constant modifications during evolutionary time of plant defence following variations in insect feeding pressures.

This functional role of plant toxins must be taken into account when considering their chemotaxonomic importance. The anti-feedant role explains, in part, the often sporadic distribution of a particular class of toxin. Irregular patterns may be simply the result of a diversity of toxins being synthesized in a particular group of plants. The fact that one toxin may replace another, in evolutionary terms, means that no one class of toxin is likely to be universal in its occurrence. This limits the utility of toxins as taxonomic markers; it is not possible to make broad chemical comparisons if a particular character is absent from many of the taxa under consideration. All the compounds considered in this chapter are of restricted occurrence. The alkaloids, for example, are only present in some 15–20% of higher plant species. Their utility as taxonomic markers is, thus, generally limited to the lower levels of classification.

In the present chapter, nitrogenous toxins will be considered first and then attention will be given to the main groups of non-nitrogenous toxic constituents.

II. Alkaloids

A. Chemistry and distribution

The alkaloids comprise the largest single class of secondary plant substances and some 6000 different structures have been characterized to date. New alkaloids are being continually discovered and it is likely that on average at least one new compound is reported daily. There is no one definition of the term 'alkaloid' which is completely satisfactory, but alkaloids generally include 'those basic substances which contain one or more nitrogen atoms, usually in combination as part of a cyclic system'. Alkaloids are often poisonous to vertebrates but this is not a universal attribute. Perhaps their most characteristic property is that they are physiologically active in Man, usually at very low concentration. On ingestion, they may be anaesthetic, analgesic, hallucinogenic or hypnotic. Some of the most active alkaloids—cocaine, morphine, atropine, colchicine—are widely used in medicine.

As already mentioned, many, but by no means all, alkaloids are bitter-tasting and they can be detected in fresh plant tissues by this means. The alkaloid quinine from cinchona bark is one of the bitterest substances known, being significantly bitter at a concentration in water of 10 μM. Most alkaloids are colourless crystalline solids, although a few (e.g. nicotine) are liquids at room temperature. They are usually optically active and though normally present as either the (+)- or (—)- form, mixtures of both isomers occasionally occur naturally in plants.

Chemically and biosynthetically, alkaloids are a very heterogeneous group, ranging from simple monocyclic compounds like coniine, a major alkaloid of hemlock, *Conium maculatum*, to the pentacyclic structure of strychnine, the toxin of *Strychnos nux-vomica.* Many alkaloids are terpenoid in structure, e.g. solanine, the steroidal alkaloid of the potato, *Solanum tuberosum*, and are best considered from the biosynthetic standpoint as modified terpenoids. Many other alkaloids are aromatic and have the basic nitrogen in an aromatic ring system, as in the isoquinoline alkaloids, e.g. berberine, the basic principle of *Berberis* root. Less commonly, the basic nitrogen may be in a side chain, as in colchicine, the tropolone alkaloid of the autumn crocus, *Colchicum autumnale.* A range of typical alkaloid structures is shown in Fig. 6.1.

Some amines and purines, which strictly speaking should be classified under other headings, are sometimes included with alkaloids because they have a similar physiological activity. This applies to most aromatic plant amines, such as mescaline, which is hallucinogenic in the same way that the true alkaloid lysergic acid diethylamide (LSD) is. It also applies to certain purines, such as the methylpurine derivative caffeine, the diuretic principle of coffee, tea and cocoa. In order to avoid confusion when considering alkaloids as taxonomic markers, Hegnauer (1966) suggested that distinction should be made between three categories of alkaloid:

(1) *true alkaloids*, amino acid derived and with a heterocyclic nitrogen atom, e.g. strychnine;
(2) *proto-alkaloids*, amino acid derived but which lack a heterocyclic nitrogen atom, e.g. mescaline;

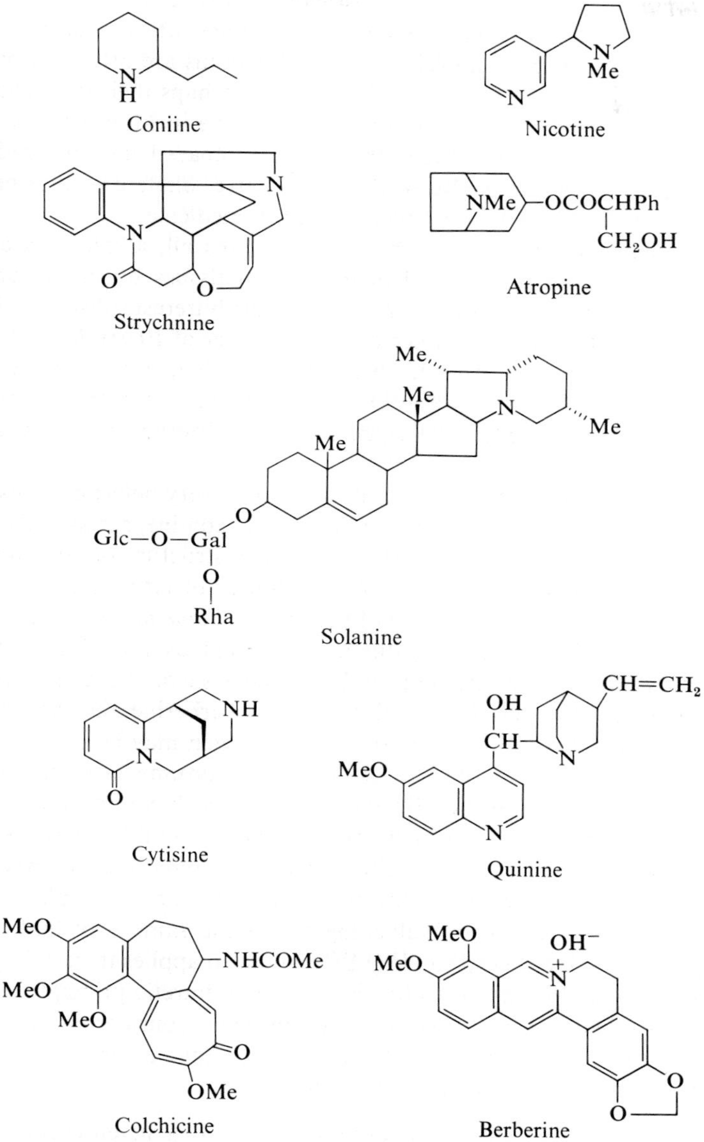

Coniine

Nicotine

Strychnine

Atropine

Solanine

Cytisine

Quinine

Colchicine

Berberine

Fig. 6.1 Structures of some common plant alkaloids.

(3) *pseudo-alkaloids*, nitrogen-containing but in which the carbon skeleton is acetate-derived, e.g. coniine, or terpenoid-derived, e.g. solanine.

Such a system of nomenclature is useful, but, in the light of more recent biosynthetic data, is probably too simple and could do with some revision.

The most common precursors of alkaloids are protein amino acids, although the actual biosynthesis is more complex than this statement suggests. Morphine, the alkaloid of opium poppy, for example, is formed from two molecules of tyrosine, one of which is actually incorporated into its synthesis in modified form as dopamine (see Fig. 6.2). A major intermediate is (—)-reticuline, but this undergoes a variety of reactions including oxidative carbon–carbon coupling and finally demethylation before morphine is produced. Even a simple alkaloid like coniine requires several steps for its synthesis (Fig. 6.3). Although theoretically derivable from the amino acid lysine, coniine is actually formed from the acetate-derived 5-oxo-octanoic acid. The key reaction in coniine synthesis is the transamination of the related aldehyde in the presence of the amino acid alanine. The transaminase catalysing this reaction has been characterized and differs from the more common transaminases of primary amino acid metabolism. It should be noted, moreover, that piperidine alkaloids closely similar to coniine are indeed lysine-derived. This is true of isopelletierine, which occurs in the pomegranate (*Punica granatum*) of the Punicaceae, a family distantly related to the Umbelliferae. These biosynthetic discoveries exemplify the fact that similar alkaloids may be formed by different pathways in different families and so cannot be considered to be chemotaxonomically analogous. Further information on the biosynthetic pathways to the alkaloids can be obtained from the books of Geissman and Crout (1969), Mann (1978) and Cordell (1981).

Alkaloids are characteristic constituents of only one major plant group: the angiosperms. In general, they are infrequent in or absent from gymnosperms, ferns, mosses and lower plants. Exceptionally, alkaloids occur among simple vascular plants in *Equisetum* and *Lycopodium* and among gymnosperms in the families Taxaceae and the Cephalotaxaceae. Several fungi (e.g. *Amanita*, *Claviceps*) are also remarkable in synthesizing various hallucinogenic alkaloids. Very recently alkaloids have been reported in animals (Tursch *et al.*, 1976) and the defence toxin of the ladybird *Coccinella septempunctata* has been identified as the alkaloid *N*-oxide, coccinelline. Although this substance is probably synthesized *de novo* by the ladybird, other reports of alkaloids in animal tissues are due to ingestion and storage of dietarily derived plant molecules (see Rothschild, 1973).

Within the angiosperms, alkaloids are very unevenly distributed, occurring in some 15–20% of species surveyed. Although some families are exceedingly rich—Papaveraceae, Rutaceae, Ranunculaceae, Magnoliaceae—others are

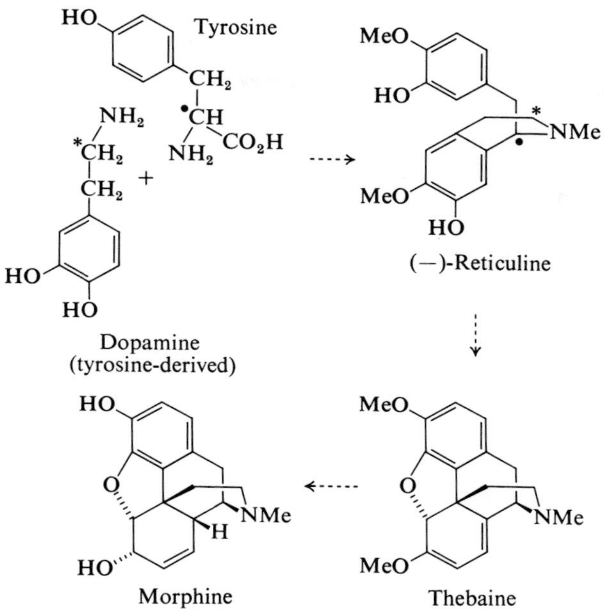

Tyrosine

Dopamine (tyrosine-derived)

(−)-Reticuline

Thebaine

Morphine

Fig. 6.2 Outline pathway of the biosynthesis of morphine from two molecules of tyrosine.

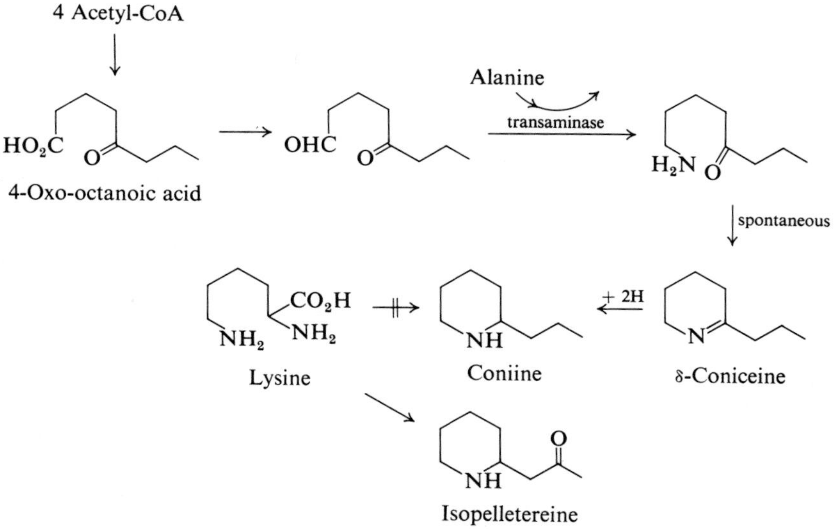

4 Acetyl-CoA

4-Oxo-octanoic acid

Alanine

transaminase

spontaneous

$+ 2H$

Lysine

Coniine

δ-Coniceine

Isopelletereine

Fig. 6.3 Outline of biosynthetic pathway to coniine and isopelletereine.

especially poor—Cucurbitaceae, Ericaceae, Umbelliferae, Primulaceae. If one arbitrarily defines a family that is rich in alkaloids as one in which 50 or more different structures have been reported, then there are some ten dicotyledonous and two monocotyledonous families which are in this category (Table 6.1). In addition, however, it should be noted (Hegnauer, 1966) that certain orders of families are alkaloid-rich, these being the Centrospermae, Gentianales, Papaverales, Ranunculales, Rosales, Rutales and Tubiflorae. Most of these groups are predominantly tropical in origin and there is some evidence, reviewed by Levin (1976), that tropical plants tend to be richer in alkaloids than temperate ones. Also, the actual toxicity (on a weight basis) of the alkaloids in tropical families tends to be greater than that of the alkaloids found in temperate plants (Levin and York, 1978).

Table 6.1 Plant families with over fifty alkaloids present

Family	Typical source	Typical alkaloid(s)	Chemical type(s)
Dicotyledons			
Apocynaceae	*Catharanthus roseus*	Ajmalicine	Benzyliso-quinoline
Compositae	*Senecio jacobaea*, ragwort	Senecionine	Pyrrolizidine
Leguminosae	*Cytisus laburnum*, broom	Cytisine	Quinolizidine
Loganiaceae	*Strychnos nux-vomica*	Strychnine	
Menispermaceae	*Archangelisia flava*	Berberine	Protoberberine
Papaveraceae	*Papaver somniferum*, opium poppy	Morphine, codeine, thebaine	Morphine
Ranunculaceae	*Aconitum napellus*	Aconite	Diterpene
Rubiaceae	*Cinchona officinalis*, cinchona bark	Quinine	Quinoline
	Uragoga ipecacuanha roots	Emetine	Emetine-type
Rutaceae	*Skimmia japonica*	Skimmianine	Quinoline
Solanaceae	*Solanum tuberosum* potato	Solanine, chaconine	Steroidal
	Atropa belladonna, deadly nightshade	Atropine	Tropane
	Nicotiana tabacum, tobacco	Nicotine	Pyridine
Monocotyledons			
Amaryllidaceae	*Narcissus pseudonarcissus*, daffodil	Galanthamine, tazettine	
Lilliaceae	*Colchicum autumnale*, autumn crocus	Colchicine	Tropolone

Most surveys for alkaloids are carried out on leaf tissue, but it must be remembered that these substances may be present in any or all parts of the plant. In many herbs (e.g. tobacco), there is evidence that the alkaloids are manufactured in the root and transported to the leaf, so that at certain stages in the life cycle, the root may be richer in alkaloid than the leaf. Fruits and seeds may have the highest alkaloid content; this is true of some Solanaceae (e.g. *Atropa belladonna*) and Leguminosae (e.g. *Cytisus scoparius*). By contrast, alkaloids may be present in the leaves of such plants in relatively low yields. Because of such variation in location and yield, there is a problem in defining an alkaloid-containing plant. Hegnauer (1963) has suggested that such a plant might be one in which the alkaloid level in some tissue (usually leaf) is at least 0.01 % dry weight.

Detailed listings of the known alkaloids and their plant sources are given in Pelletier (1970), Raffauf (1970) and Glasby (1975–1977). Valuable summaries of the systematic distribution by families have been compiled by Hegnauer (1963, 1966) and by Seigler (1977a). Their biology and metabolism is reviewed by Walker and Nowacki (1978).

B. Detection

Many procedures are available for detecting alkaloids in plant tissues and there is a continued need for improvements in such techniques, since there must still be many plant species, particularly in tropical rain forests, which have yet to be tested for these basic constituents. Although varying in detail, the methods used are essentially similar and depend on extracting the alkaloid fraction from the tissue (usually dried whole plants) with weak acid, removing the debris and basifying with concentrated ammonia. The alkaloid precipitate is recovered into chloroform and the crude extract is finally dissolved in aqueous acid and tested against a number of reagents, a precipitate providing a positive response. The most popular reagents are that of Dragendorff, a mixture of bismuth subnitrate and potassium iodide, and one based on potassium iodoplatinate.

Preliminary testing can be carried out in the field on fresh material (see e.g. Lawler and Slaytor 1969; Hartley *et al.*, 1973) but such tests are usually followed up by more extensive laboratory experiments on dried tissue. In some cases, it appears that positive tests in the field are not supported by more extensive tests on dried tissue; the latter failure may be due to alkaloid losses during the drying process. Such losses are particularly likely in the case of plants which are difficult to dry (e.g. orchids) and Lüning (1967) regards it as essential to use fresh tissue for alkaloid testing in the Orchidaceae. On the other hand, in the Rubiaceae and Menispermaceae, excellent alkaloid responses have been obtained from herbarium samples (e.g. Phillipson and Hemingway, 1975; Phillipson, 1982).

A special problem with screening plants for alkaloids is the low concentrations usually encountered and it is normally advisable to use large leaf samples (equivalent to 100 g dry wt) for successful results. Also, because of the chemical heterogeneity of alkaloids, no one test is infallible. Furthermore, the most commonly used reagent, Dragendorff's, is not entirely specific for alkaloids. As a consequence, there are undoubtedly some spurious positive results in the literature. Hultin and Torssell (1965), in devising a screening procedure for the Swedish flora, regarded six different reagents as being necessary (cf. Cromwell, 1955) and were only satisfied that alkaloid was present if the plant extract responded positively to all six tests.

Further chemical identification requires the isolation and separation of the alkaloids present for spectral and melting-point comparisons with standards. Classically, alkaloids were separated as their hydrochlorides or picrates. Nowadays, they are usually purified as the free bases by TLC or GLC (for details, see Stahl, 1969; Harborne, 1973b).

C. Chemotaxonomic utility of alkaloids

1. *General*

Alkaloids have been proposed as possible taxonomic markers during the course of a variety of angiosperm studies but it is clear that there are a number of problems associated with their utility in solving specific taxonomic problems. Because of their limited distribution overall, they cannot be employed at the higher levels of classification, but only within those orders and families where they are well represented. Even in such cases it is important to restrict discussion to particular classes of alkaloid, since the alkaloid character as such has little meaning. Problems of divergence and parallelism occur with alkaloids, as with other chemical characters (see Hegnauer, 1963, 1966). There are also difficulties of definition in terms of biosynthetic origin, but these are slowly being solved as more information comes to hand from labelling experiments.

Finally, there is the problem of defining alkaloid occurrence in quantitative terms. Hegnauer (1963) has pointed out the importance of distinguishing between alkaloid accumulation and alkaloid synthesized in trace amounts, possibly as a by-product of other metabolic pathways. This point can be illustrated by reference to nicotine, which accumulates in quantity in the leaf of tobacco *Nicotiana tabacum* and also in leaves of all other *Nicotiana* species tested. As a chemical marker, it is rather easy to detect, since it is steam-volatile (and thus easily separated from most other nitrogenous constituents) and also it produces a sensitive response to reaction with cyanogen bromide and aniline. By using this procedure, nicotine has actually been detected as *a trace constituent* in a variety of other plant sources, including species from such different genera as *Equisetum*, *Lycopodium*, *Asclepias*, *Acacia* and *Mucuna*. Clearly, these are all unrelated occurrences of no obvious taxonomic significance. However, such reports do not necessarily diminish the value of nicotine accumulation as a characteristic chemical feature of the genus *Nicotiana*, which separates it from most other genera in the Solanaceae.

2. *Alkaloids at the lower levels of classification*

Although relatively little use has been made of alkaloids in the taxonomic revision of species, these substances have considerable potential in terms of the chemical complexity that is often present. Many different alkaloids may occur characteristically within a single taxon. Thus the periwinkle, *Catharanthus roseus* (Apocynaceae), has yielded more than 72 different structures (Taylor and Farnsworth, 1975). This plant has been extensively screened because its alkaloids have valuable curative properties. Two of the bases present, leukoblastine and leukocristine, are in current clinical use for the treatment of human cancer.

The possibilities of using Apocynaceae alkaloids for taxonomic purposes have not been considered in the case of the periwinkle, but have been investigated in species of the closely related *Vinca*. Aynilian *et al.* (1974) studied the alkaloid pattern of *Vinca libanotica*, a species of doubtful validity because of its close morphological similarity to *V. herbacea*. They found that, on chemical grounds, it was readily distinguished from *V. herbacea* and other *Vinca* spp. and suggested that it merited recognition as a separate species. As in many alkaloid investigations, only one sample of each taxon was studied, so that this suggestion needs confirming with a wider sampling before it can be taken seriously by taxonomists.

The above example indicates the necessity of studying samples of different geographical origin before defining the alkaloid chemistry of a given species, since regional races may occur, especially among widespread variable taxa. It is interesting that in one extensive study of *Cinchona* where alkaloid content was examined in relation to geography, rather clear taxonomic conclusions could be drawn from the results. Turner (1970) has highlighted a simple but nevertheless significant survey by Camp (1949) of the concentrations of four *Cinchona* alkaloids (quinidine, cinchonidine, quinine and cinchonine) in the bark of 120 trees representing 11 populations. Sampling was conducted in trees growing along the north–south transect in South America along the Andes from Colombia to Peru—a distance of 500 miles. Camp was mainly interested in *Cinchona* bark as a source of quinine, an important drug in the treatment of malaria, and he confined his analyses to total crystallizable alkaloid (TCA). As shown in Fig. 6.4, most of his tree samples fell into one of three main groupings. It will be seen, however, that intermediate values to the main groupings were observed in *Cinchona* populations of intermediate latitude. There was thus no sharp cut-off between the three main population samples. Trees of these three main populations might originally have been classified as different species, but it is clear from this survey that what is actually present in South America is a single clinal complex of three intergrading population units. In other words, what originally might have been divided into three can now be simplified to a single taxon.

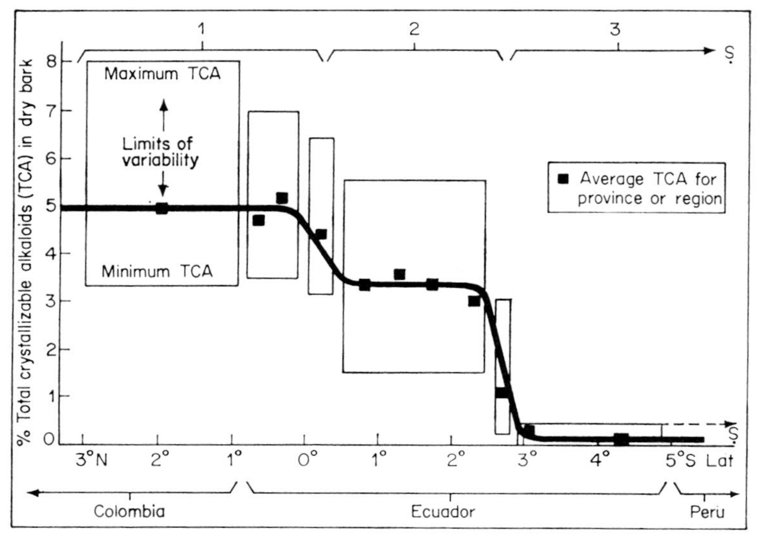

Fig. 6.4 The cline in total crystallizable alkaloids (TCA) in *Cinchona* tree barks collected in South America along the crest of the Andes.

Another drug plant in which much geographical sampling for alkaloids has been accomplished is the opium poppy, *Papaver somniferum* (Mothes, 1976), the latex of which contains *inter alia* the three related structures: morphine, codeine and thebaine. The three compounds are simply related chemically in that codeine is the monomethyl ether of morphine and thebaine the dehydrodimethyl ether (see Fig. 6.1). Biosynthetically, these alkaloids are formed in the reverse order. Thebaine is the first to be synthesized and this then undergoes selective *O*-demethylation to codeine and eventually to morphine.

Although there is variation in alkaloid content and type in *P. somniferum*, all plants studied before the above pathway was elucidated contained the three alkaloids in varying amounts. The discovery of the biosynthetic pathway during the 1950s suggested that mutant forms of *Papaver* should exist which only produce thebaine, due to a genetic block in the final demethylations to morphine. Since such plants would be of considerable medicinal interest, a search of *P. somniferum* and related species was conducted during the 1960s for a thebaine-rich taxon. Such a species was found by East

German scientists (see Mothes, 1976) and, although it was first thought to be *P. orientale*, it was eventually recognized as *P. bracteatum*. A strain of the latter, which produces 100% thebaine, is now being grown on an agricultural scale. Taxonomically, it is interesting that alkaloid chemistry clearly distinguishes between the two species, *P. somniferum* and *P. bracteatum*, alkaloid synthesis always proceeding in the one as far as the addictive alkaloid morphine, whereas the other stops synthesis at the stage of thebaine, a non-addictive and safe alkaloid. The possibility, of course, still exists that a mutant *P. somniferum* with only thebaine might one day be found.

3. *Alkaloids at the higher levels of classification*

The benzyltetrahydroisoquinoline alkaloids are particularly widely distributed in more primitive angiosperms, in no less than 18 dicotyledonous orders, and they represent one of the few groups of alkaloid which appear to have a taxonomically significant natural distribution. Their overall occurrence is shown in Fig. 6.5, in the phyletic framework of Takhtajan's (1969) system of angiosperm classification. These alkaloids in general appear to represent a primitive chemical feature, concentrated in woody families but retained during evolution as a relict in a number of more advanced taxa. They are present in concentration in five orders—the Magnoliales, Laurales, Ranunculales, Papaverales and Rutales—which are certainly not phenetically related but which on other evidence are believed to be phyletically linked. These are homologous occurrences in the sense that the pathway of biosynthesis is known to be the same in Magnoliaceae, Ranunculaceae and Papaveraceae.

In particular, Cronquist (1968) has suggested the possible development of the Papaverales from the Ranunculales. A detailed comparison of the isoquinoline alkaloids of these two orders on a biosynthetic basis adds rather convincing chemical support (Cagnin *et al.*, 1977) for this suggestion. Cronquist (1968) is also on record as suggesting the development of the Rosales and Guttiferales from the same Ranalean line. Surveys for benzylisoquinolines among these two orders rather negates this view, since neither of them contain this class of alkaloid. Other chemical evidence (see Fig. 6.6) also weakens this part of Cronquist's phyletic scheme (see Kubitzki, 1969).

As mentioned above, isoquinoline alkaloids of the *Magnolia* type do appear in Rutaceae (Rutales), which is undoubtedly the most advanced family to retain these compounds in quantity. Within the family, benzylisoquinolines are confined essentially to three genera, *Xanthoxylum* (including *Fagara*), *Phellodendron* and *Toddalia* and are nowhere else (Price, 1963; Waterman, 1975). These genera are each placed in the Englerian system in a different subfamily or tribe of the Rutaceae. The alkaloid chemistry, which is supported by distribution patterns of furanocoumarins and of limonoids, clearly links these three genera, which are otherwise well separated. As

Waterman (1975) writes 'since it is most improbable that these alkaloids have arisen independently of each other within the family, some phyletic line not expressed in Engler's classification must exist between these three plant groups'.

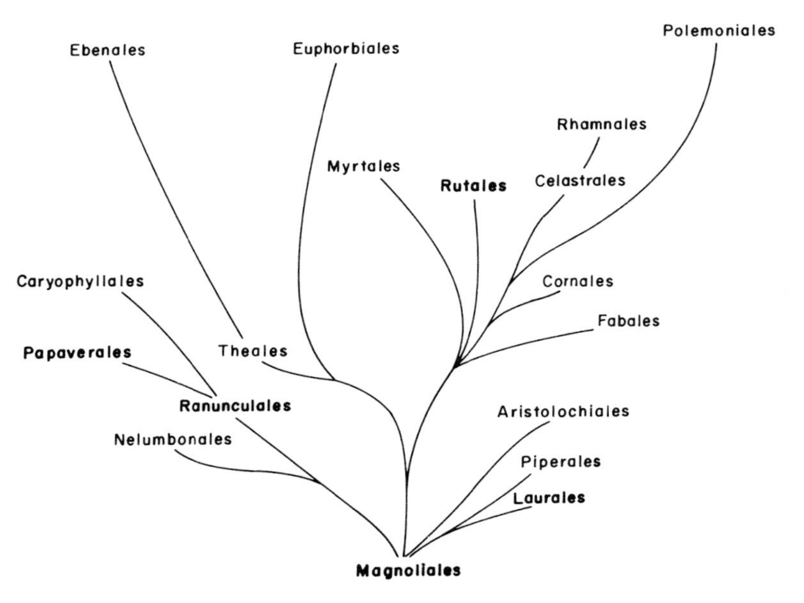

Fig. 6.5 Probable phylogenetic relationships between isoquinoline-synthesizing plant orders. Bold names indicate major occurrences.

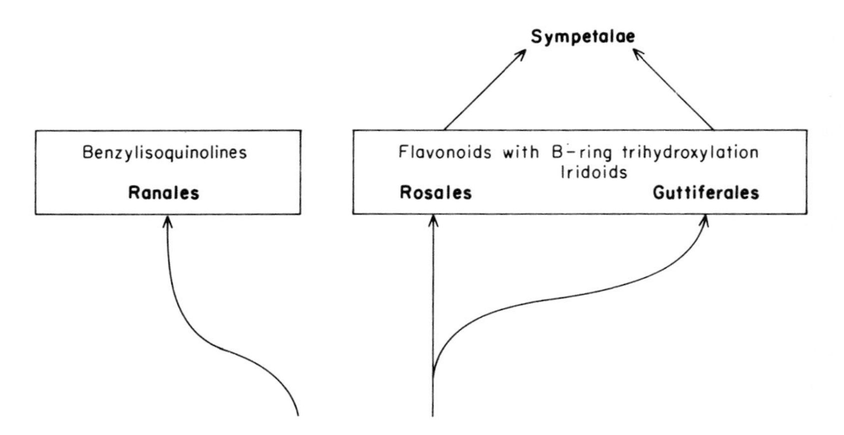

Fig. 6.6 Kubitzki's (1969) modification of Cronquist's (1968) Ranales → Rosales plus Guttiferales scheme.

Several other classes of alkaloid also have distribution patterns which reflect taxonomic relationships between families. The betalain alkaloids occur very consistently in families of the order Centrospermae but are otherwise absent from angiosperms. Because these substances are pigments and their distribution is related to that of the anthocyanins, their chemotaxonomy will be considered later in chapter 7. Again, complex indole alkaloids have a significant distribution, occurring mainly in three families—Loganiaceae, Apocynaceae, Rubiaceae—which are all sympetalous and which are often placed fairly near each other in most systems of classification. However, it must be pointed out that the Asclepiadaceae, which is especially closely related morphologically to Apocynaceae, completely lacks such alkaloids. Indeed, asclepiads rarely have alkaloids and the few that are present are derived biosynthetically from phenylalanine, rather than from indole. Since most morphological workers would agree that the Asclepiadaceae is a more advanced family than the Apocynaceae, it is reasonable to conclude that the indole alkaloids are more primitive in the order Gentianales, to which Takhtajan assigns these several families, their absence from Asclepiadaceae being due to phyletic loss over time.

Most other classes of alkaloid have more varied distribution patterns in angiosperms and there seems to be less phyletic or other significance in their occurrences. The tropane alkaloids, for example, occur in considerable abundance in just one family, the Solanaceae, and have also been detected (although the compounds are of slightly different structural type) in the related Convolvulaceae (Romeike, 1978). Unfortunately, there are also occurrences in eight other families, which by no stretch of the imagination can be considered to be related to the Solanaceae. These are the Cruciferae, Dioscoreaceae, Elaeocarpaceae, Erythroxylaceae, Euphorbiaceae, Orchidaceae, Proteaceae and Rhizophoraceae. The present apparently disparate records of tropane alkaloids in the angiosperms may only reflect our considerable ignorance of their proper distribution, since they may eventually be found to occur more widely and some link between these different sources may eventually become apparent.

Again, the pyrrolizidine alkaloids, like the tropanes, have a scattered distribution among angiosperms (Culvenor, 1978). They occur widely in only one family, the Boraginaceae, but are also highly characteristic in their occurrence in the tribe Senecioneae of the Compositae and in the genus *Crotalaria* of the Leguminosae. There are also scattered reports in Apocynaceae, Celastraceae, Orchidaceae, Ranunculaceae, Rhizophoraceae, Santalaceae, Sapotaceae and Scrophulariaceae. Their taxonomic value is therefore strictly limited to the lower levels of classification. Here, they have some value. For example, the presence or absence of pyrrolizidine alkaloids in taxa which are borderline members of the Senecioneae has been used in determining whether to include or exclude them from the tribe. Thus, the genus

Adenostyles once placed in the tribe Eupatorieae, clearly belongs in the Senecioneae since it possesses both this alkaloid type and also a class of sesquiterpene lactone (furanoeremophilanes) characteristic of the tribe (see Culvenor, 1978).

To summarize, the various alkaloid classes have a rather variable distribution, family by family, within the flowering plants and their occurrences, as yet, offer only limited insight into familial and ordinal relationships. Nonetheless, the situation is one of considerable potential and undoubtedly alkaloids will become of greater systematic interest as more information accrues.

III. Cyanogens

A. Chemistry and distribution

That bitter almonds from the tree *Prunus amygdalus* are capable of releasing a toxic gas when crushed has been known from time immemorial. It has also been known for many years, at least since 1800, that release of this gas, hydrogen cyanide (HCN), is related to the presence in intact plants of substances called cyanogenic glycosides, which yield HCN on either enzymic or non-enzymic hydrolysis. Cyanogenesis is present in a number of food and fodder plants (e.g. cassava, clover, *Sorghum*) and cyanide poisoning in man or livestock may follow consumption of such plant tissues. Besides the danger of acute poisoning, there is increasing evidence that dietary ingestion of sub-lethal doses of cyanogen can cause neurological and other damage in vertebrates (Conn, 1978).

The ease of detection of HCN, by its smell of bitter almonds or more reliably (and safely) by picrate paper, has meant that many surveys for cyanogenesis have been carried out. The ability of tissues to release HCN was used early on as a taxonomic marker in the Rosaceae for distinguishing the tribe Amygdaleae (all positive) from the tribe Chrysobalaneae (now family Chrysobalanaceae) (all negative) (Lindley, 1830). By contrast with these botanical advances, chemical studies of the cyanogenic glycosides have been relatively restricted and, even today, only some 30 such compounds have been fully characterized (Seigler, 1977b).

HCN is released from cyanogenic glycosides according to the scheme shown in Fig. 6.7. The first step is normally enzymic and cyanophoric plants contain a specific hydrolase for carrying this out; in short, the enzyme will act on endogenous substrate when leaf or other tissue is crushed. The second step, the liberation of HCN and a substituted aldehyde or ketone from the unstable cyanohydrin intermediate, occurs spontaneously.

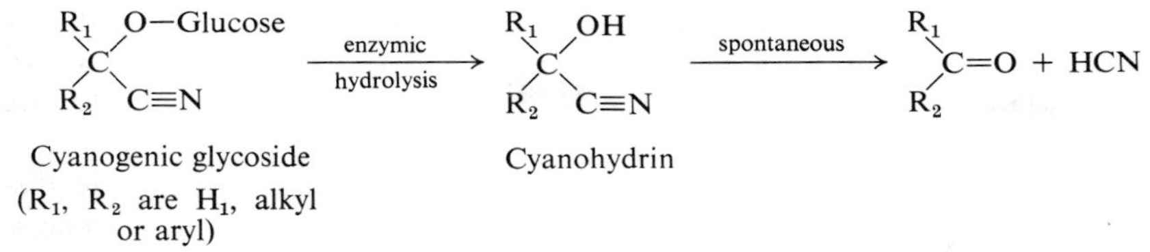

$$R_1R_2C(O\text{-Glucose})C\equiv N \xrightarrow[\text{hydrolysis}]{\text{enzymic}} R_1R_2C(OH)C\equiv N \xrightarrow{\text{spontaneous}} R_1R_2C{=}O + HCN$$

Cyanogenic glycoside (R_1, R_2 are H_1, alkyl or aryl) — Cyanohydrin

Fig. 6.7 Enzymic release of hydrogen cyanide from cyanogenic glycosides.

Cyanogenic glycosides are formed biosynthetically from protein amino acids through a series of intermediates. The pathway from the amino acid valine to one of the commonest cyanogenic glycosides, linamarin, is shown in Fig. 6.8. In its natural occurrence, linamarin is practically always accompanied by lotaustralin, which is derived biosynthetically by the same pathway from the amino acid isoleucine. These two cyanogens occur together in many plants, including flax *Linum usitatissimum*, clover *Trifolium repens* and bird's foot trefoil *Lotus corniculatus*. A third aliphatic amino acid, leucine, can also be a precursor of cyanogen, giving rise, for example, to dihydroacacipetalin, a glycoside of *Acacia* species. In this group of cyanogens may also be included gynocardin, a cyanogen with a cyclopentene ring from *Gynocardia odorata* (Flacourtiaceae), which may be derived biosynthetically from a similarly substituted glycine derivative.

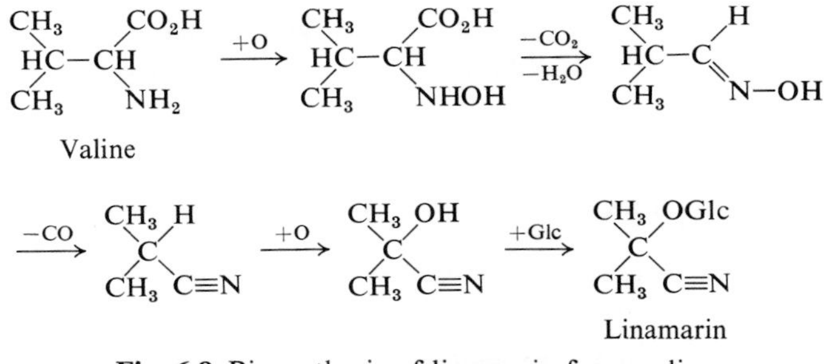

$$(CH_3)_2CH{-}CH(CO_2H)NH_2 \xrightarrow{+O} (CH_3)_2CH{-}CH(CO_2H)NHOH \xrightarrow[-H_2O]{-CO_2} (CH_3)_2CH{-}CH{=}N{-}OH$$

Valine

$$\xrightarrow{-CO} (CH_3)_2C(H)C\equiv N \xrightarrow{+O} (CH_3)_2C(OH)C\equiv N \xrightarrow{+Glc} (CH_3)_2C(OGlc)C\equiv N$$

Linamarin

Fig. 6.8 Biosynthesis of linamarin from valine.

Besides the aliphatic cyanogens, there are a larger number of aromatic cyanogenic glycosides, formed biosynthetically from either phenylalanine or tyrosine. A phenylalanine-derived glycoside is amygdalin, the toxin of bitter almonds, which is unusual in containing the disaccharide sugar gentiobiose, instead of the much more common glucose. An example of a tyrosine-derived cyanogen is dhurrin, which occurs widely in *Sorghum* tissue.

To the cyanogenic glycosides proper must be added a relatively new group of cyanogenic compounds called cyanolipids. These also give HCN on hydrolysis, but differ in that a fatty acid moiety is released on hydrolysis instead of sugar. All known cyanolipids are derived biosynthetically from the amino acid, leucine. Not all cyanolipids, however, are true cyanogens; some are clearly related in structure but lack an oxygen function adjacent to the cyanide group and cannot give HCN on hydrolysis. Structures of representative cyanogenic glycosides and cyanolipids are shown in Fig. 6.9.

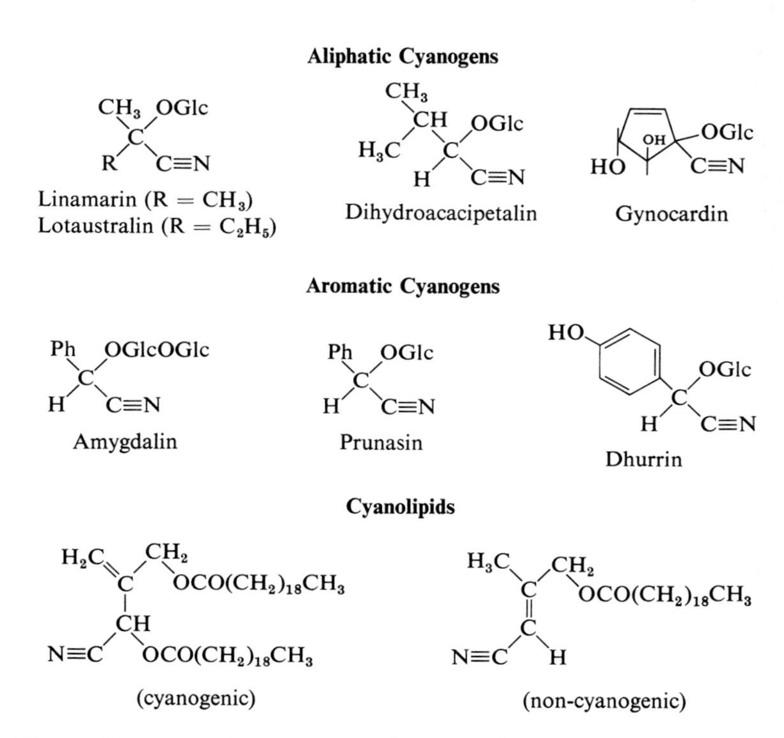

Fig. 6.9 Representative structures of cyanogenic compounds of plants.

Cyanogenic compounds are widely distributed in nature. According to a recent review, over 2000 species representing some 100 families of vascular plants are known to be cyanophoric (Hegnauer, 1977). They occur in fungi and ferns as well as in gymnosperms and angiosperms. They have also been recorded as defensive agents in several insects. Cyanogenesis is well known to be polymorphic in a number of characteristic sources. This is so in the tree *Prunus amygdalus* of which bitter and sweet (cyanogenic negative) varieties are known and in herbs such as *Lotus corniculatus* and *Trifolium repens*. Population variation in the latter two species has been extensively documented and the genetics of cyanogenesis has also been established. It appears, at least in *L. corniculatus*, that the variations in frequency of cyanogenesis in different populations are closely correlated with the pressures of herbivore grazing by slugs and snails; other ecological factors, however, may also be involved (Jones, 1972; Jones *et al.*, 1978).

There seems little doubt that cyanogenesis is generally important in plants as a defence against browsing animals. In bracken fern, where the character again is polymorphic, cyanogenesis protects against feeding by deer (Cooper-Driver and Swain, 1976). Experiments in *Sorghum* sp. have shown that insects are also repelled from feeding by cyanogens. The locust, for example, is very sensitive to HCN released in *Sorghum* and is specifically deterred from feeding by HCN, rather than by the benzaldehyde simultaneously released during hydrolysis of the main glycoside, dhurrin (Woodhead and Bernays, 1977).

From the taxonomists' viewpoint, the fact that the character is occasionally polymorphic means that it is essential to sample a species as widely as possible before recording it as cyanogenic or acyanogenic. Since many of the literature reports of cyanogenesis in plants are based on limited sampling, it should be remembered that some apparently negative results could be in error.

A general account of cyanogenesis and cyanide formation in plants and animals has recently been published by Vennesland *et al.* (1981).

B. Detection of cyanogenesis

Cyanogenesis is readily and simply detected in plant tissues by means of the well known picrate colour test. The test consists of crushing a few pieces of fresh leaf in a test tube and suspending over them a filter paper soaked in picric acid, held in place by a cork stopper. As HCN is released, the yellow picrate colour changes to red or brown. This reaction depends on the activity of endogenous enzyme liberating HCN from the glycoside also present in the tissue. In fact, the enzyme may occasionally be absent, since it is genetically controlled by a gene which is independent of those controlling glycoside synthesis. For this reason, it is important to leave the test set up for at least 24 hours, since a delayed positive result will be observed even in the absence of enzyme due to slow non-enzymic release of HCN from the cyanogen.

The picrate test is reasonably specific for HCN but, as mentioned elsewhere (see p. 69), 'fake' positive results may be produced in crucifer plants by isothiocyanate released from glucosinolate present. In practice, the difference between HCN and isothiocyanate should be very apparent, due to the difference in smell. The presence of HCN can also be confirmed by means of a number of other more specific colour reactions (see Fikenscher and Hegnauer, 1977), among which the Feigl-Anger test with a sensitivity of 1 μg HCN is the most popular.

The picrate test depends on cyanogenic glycoside being present in significant quantity, in amounts of more than 20 mg per kg fresh wt.; it is thus a test for the *accumulation* of cyanogen in plants. Traces of HCN (about 0.2 mg per kg fresh wt.) can be shown to be present in acyanogenic species (plants responding negatively to picrate paper) and it is possible that all plants contain minimal amounts of prussic acid. Indeed, a role for trace quantities of HCN has been proposed as part of the control mechanism in nitrate reduction (Gewitz *et al.*, 1974) or in nitrate assimilation (Solomonson and Spehar, 1977). If HCN has a physiological function in plants, it is certainly not stored in any quantity and this represents quite a different situation from that in which cyanogen accumulates in tissues. In fact, HCN in any quantity is highly toxic to plants and, as pointed out by Hegnauer (1977), accumulation of cyanogen or other toxic material always involves the co-evolution of mechanisms of tolerance to the toxic effects that might be exerted by the compounds concerned. This co-evolution involves *inter alia* production of enzymes capable of rapidly turning over the said toxin.

Although it is a simple matter to detect cyanogenesis, it is much more difficult to identify the glycoside responsible for the release of HCN from a given plant. Both the isolation and characterization of these substances present problems. However, nuclear magnetic resonance (NMR) spectral techniques have recently proved of especial value for purposes of identification. The chemical characterization of cyanogens is fully covered in a recent review (Seigler, 1977b).

C. Taxonomic utility of cyanogenic glycosides

At one time, it appeared that cyanogenesis was a rather sporadic chemical character in plants and of barely any taxonomic utility. Indeed, this was Hegnauer's conclusion in a review published in 1960. However, only 17 years later, the same author (Hegnauer, 1977) completely reversed this earlier appraisal and concluded that 'if adequately and carefully used, cyanogenesis may prove to be of considerable value to a natural classification of plants, even at the higher systematic levels', This reassessment was largely due to biosynthetic developments and the realization that cyanogens fall into five main biogenetic groupings. Their distribution then becomes much more meaningful. Data on the natural occurrence of these five classes of cyanogen have recently been collected together in the important review of Hegnauer (1977). The same author has also discussed the subject in various contexts in earlier reviews and the present summary is based essentially on his findings.

The importance of cyanogens at the higher levels of classification was considered in a paper of 1973, in which Hegnauer compared cyanogenic distribution in the monocotyledons with that in dicotyledons. In this analysis, the cyanogenic compounds were divided into the five groupings represented by the following substances: prunasin (phenylalanine-derived), taxiphyllin (tyrosine-derived), linamarin (valine-derived), gynocardin (cyclopentene substituted) and cyanolipid (fatty acid substituted). The results of the analysis showed that the evolution of cyanogenic types followed exactly the same pathway in the superorder Magnoliidae (Polycarpicae plus Papaveraceae) as it did in the Liliatae of the Monocotyledoneae. This striking metabolic similarity favours the assumption that the Liliatae evolved from ancestors resembling present day Magnoliidae. Alternatively, this important new evidence could be used to support the views of Huber (1977), based on anatomical considerations, that the monocotyledons as a group and the Magnoliidae of the dicotyledons are much closer phyletically than is usually allowed.

On the basis of more recent findings, cyanogens appear to be characteristic markers even at the family level. Thus, cyclopentene-substituted glycosides, e.g. gynocardin, are found exclusively in two closely related families, Flacourtiaceae and Passifloraceae; this is useful new evidence since the ordinal classification of these and related families is still highly confused (cf. Hegnauer, 1977). Again, cyanolipids occur exclusively (in seed oils) in just one family, the Sapindaceae (Seigler and Kawahara, 1976); significantly, cyanolipids could not be detected in seeds of three related families, Melianthaceae, Meliosmaceae and Hippocastanaceae.

Below the family level, one of the most important uses of the cyanogenic character has been in the Rosaceae. As already mentioned, HCN was employed as a diagnostic character by Lindley (1830) and Endlicher (1836) for separating the tribe Amygdaleae (HCN + ve) from the tribe Chrysobalaneae (HCN — ve). Although the Chrysobalaneae have subsequently been raised to family rank as Chrysobalanaceae and the Amygdaleae are now referred to the Prunoideae, this early use of HCN emphasizes the consistency in occurrence of cyanogenesis in the family. Subsequently many further surveys have been carried out and the character has been found to vary significantly according to the plant part studied. The present situation is summarized by Hegnauer (1976).

The distribution of cyanogens in the Rosaceae, as at present ascertained, is complex but a striking feature is the close correlation between cyanogenesis as a leaf character and the basic chromosome numbers of the different subfamilies. Cyanogenesis is present in five groups: Spiraeoideae ($x = 9$), Prunoideae ($x = 8$), Kerrieae ($x = 9$), Maloideae ($x = 17$) and Dryadoideae (members with $x = 9$). It is, by contrast, absent from those members of the Dryadoideae with $x = 7$ and from the Rosoideae ($x = 7$). This and other evidence supports the view that Rosoideae evolved by aneuploidy from Spiraeoideae, losing the ancient mechanism of cyanogenesis in the process.

As already mentioned, the distribution of cyanogenesis in leaf of the Rosaceae differs from that in the seed, and the nature of the cyanogen also differs, the leaf (and bark) usually having prunasin, the seed having amygdalin. As a seed character, cyanogenesis is more restricted than in leaf and only occurs in seeds of Maloideae and Prunoideae; significantly, here it is restricted to taxa that bear fleshy fruits.

Finally, at the species level, cyanogenesis can be useful taxonomically in the right circumstances. As already mentioned, one significant occurrence of cyanogenesis is in *Lotus corniculatus*, where it is polymorphic in both leaf and flower. The character is present in leaves of other related *Lotus* and appears to divide members of the *L. corniculatus* aggregate clearly into two equal groups. Positive species are *L. japonicus*, *L. corniculatus*, *L. krylorii* and *L. tenuis*, whereas *L. alpinus L. borbasii*, *L. uliginosus* and *L. caucasicus* are negative (Hegnauer, 1976). Its presence or absence is of interest here in relationship to the origin of the tetraploid *L. corniculatus*, which could have originated by auto- or allopolyploidy from any of the six diploid species (*L. caucasicus* is tetraploid). In fact, both autopolyploidy from *L. tenuis* and allopolyploidy involving *L. tenuis* and *L. uliginosus* have been proposed as possible mechanisms. The chemical evidence (see Fig. 6.10) from cyanogenic studies and from the determination of the presence or absence of proanthocyanidin rather strongly supports the latter view. The cyanogenic character is probably a primitive feature in angiosperms and would be expected to be present in at least one of the putative parents of *L. corniculatus* as it is in *L. tenuis*. Furthermore, the absence of cyanogenesis from three of the other diploid species, including *L. uliginosus*, make it very unlikely that any of these three could have given rise to *L. corniculatus* by autotetraploidy. The proposal that *L. corniculatus* is derived as indicated (Fig. 6.10) from *L. uliginosus* and *L. tenuis* by allopolyploidy is also supported by cytogenetic considerations.

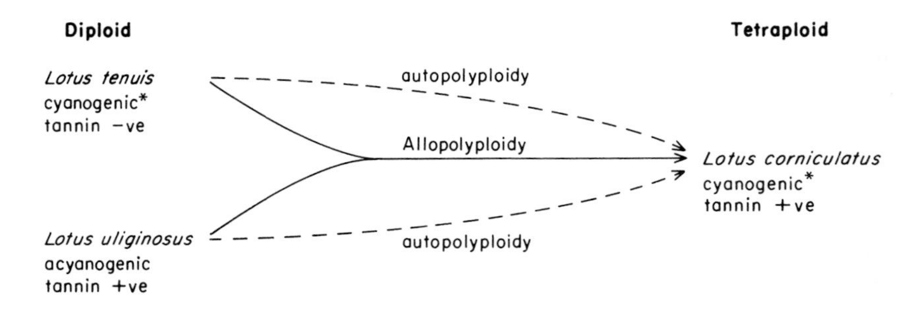

Fig. 6.10 Possible origins of the tetraploid *Lotus corniculatus*. * = character is polymorphic; tannin = proanthocyanidin.

IV. Non-protein amino acids

A. Chemistry and distribution

The non-protein or 'unusual' amino acids are a group of plant toxins especially associated with seeds and which are particularly common in members of one large plant family, the Leguminosae. There is increasing experimental evidence (see Bell, 1978) that they are functionally important in such seeds as a protection from insect and other herbivores. They are probably multi-functional and because of their richness in nitrogen, they also serve as a storage form of ammonia nitrogen in seeds, the nitrogen being liberated during the early metabolism of the germinated seedling (Rosenthal, 1982).

As toxins, non-protein amino acids may affect a wide spectrum of living organisms. Seeds of certain *Lathyrus* species are eaten by farm animals, and occasionally by man during periods of famine in Asian countries, and the toxicity of the amino acids present is expressed in a neurological disease called lathyrism. The symptoms include loss of muscular co-ordination and paralysis of the legs, the condition in extreme cases causing death (Bell, 1972). One of the compounds implicated in the classical lathyrism of *Lathyrus* spp. is the unusual α-amino-β-oxalylaminopropionic acid. Another example of a toxic amino acid is hypoglycin A, which occurs in unripe fruits of akee, *Blighia sapida*, and is the agent responsible for a condition called hypoglycaemia produced by eating these fruits. Blood sugar levels are seriously lowered and occasionally the disease is fatal in man.

As the name implies, each of the non-protein amino acids is structurally related to one or another of the 20 protein amino acids (see Fig. 6.11) and usually exists as a zwitterion, with both basic and acidic properties. Since they are structural analogues, they are anti-metabolites. Their harmful effects on various life forms are in part due to the fact that they are so closely similar in overall shape to protein amino acids that they can be absorbed from the free amino acid pool into protein synthesis. Here, they may replace the protein amino acid from which they are derived and thus enzymes may be formed with reduced catalytic activity and impaired function. Non-protein amino acids may also interfere seriously with the synthesis and metabolism of those protein amino acids which they resemble.

There must be at least 300 non-protein amino acids known at the present time. The structures of over 200 were illustrated in a comprehensive review of Fowden (1970) and a considerable number of new compounds have been described since (see Bell, 1980). In general, they are biosynthesized by pathways similar to those leading to the protein amino acids. In simple cases, one modification in synthesis may produce the non-protein analogue.

Azetidine 2-carboxylic acid, for example, is almost certainly formed by a pathway similar to that used for proline synthesis (Fig. 6.11).

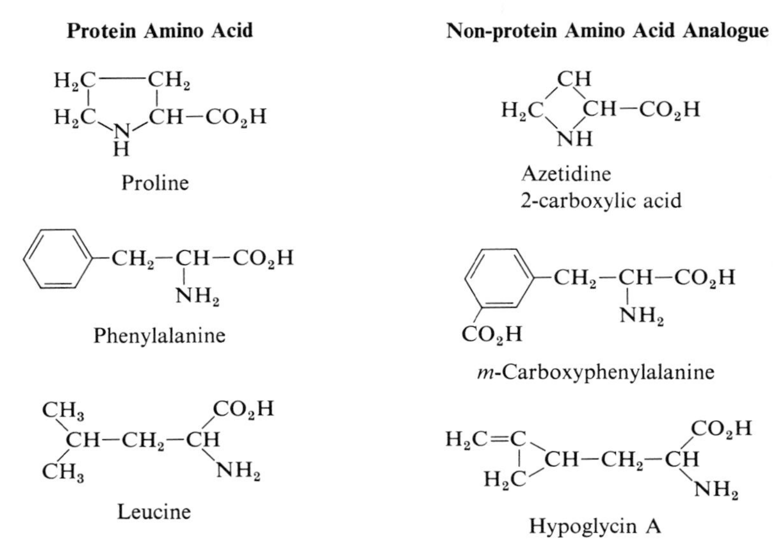

Fig. 6.11 Three examples of non-protein amino acids, showing the protein amino acids to which they are related.

Unusual amino acids are widely, albeit sporadically, distributed in the plant kingdom, having been recorded in all groups of higher plants. Besides the Leguminosae, they are regularly found in a number of other angiosperm families, including the Cucurbitaceae, Euphorbiaceae, Iridaceae, Liliaceae, Rosedaceae and Sapindaceae. They are recorded in the gymnosperms, in Cycadaceae. It is not known whether they occur more generally in the flowering plants, since few surveys have been attempted at the higher levels of classification. These non-protein amino acids also occur in algae, fungi and bacteria but usually the structures are slightly different. One such compound unique to the fungi is ibotenic acid, an insecticidal principle of the fly agaric, *Amanita muscaria*.

B. Detection

Non-protein amino acids are present in the water-soluble fraction of plant extracts and are isolated and analysed by procedures similar to those used for the common protein amino acids. In seeds, they are often present in high concentration and they may occasionally crystallize out from a crude aqueous extract; this happens with lathyrine from *Lathyrus tingitanus* seed. A major problem in isolation is separation from the co-occurring protein amino acids; this is usually achieved by ion-exchange chromatography or electrophoresis.

For screening purposes, two-dimensional paper chromatography or TLC is valuable, since non-protein amino acids have distinctive R_F values and appear as 'new' spots, against the standard pattern given by the 20 protein amino acids (see Fig. 6.12). Often, they display different colours after reaction with ninhydrin, the standard detection agent for amino acids. Instead of the usual purple colour, they may appear as green, brown or red spots. For routine screening of seed extracts, high-voltage electrophoresis is also widely employed since, by this procedure, the unusual acids are normally well separated from the common ones. Canavanine is particularly readily detected by electrophoresis at pH 7 and can also be specifically detected by the magenta colour it gives with Fearon's reagent (Bell, 1958; see also Harborne, 1973b).

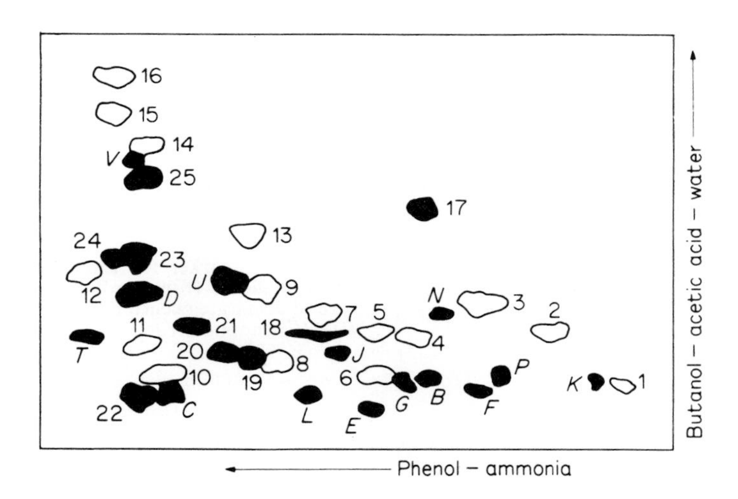

Fig. 6.12 Two-dimensional paper chromatographic profile of amino acids of Cucurbitaceae. Key: protein amino acids are open spots; non-protein amino acids are solid spots. Identified compounds are numbered, unidentified spots are lettered.

The routine screening of legume seeds has recently been automated by Bell and his co-workers, using a specially modified amino acid analyser. This procedure gives both quantitative and qualitative results and the results are stored in a computer, so that profiles of different species can be compared automatically (Charlwood and Bell, 1980).

C. Taxonomic utility

The comparative phytochemistry of non-protein amino acids has been reviewed by Fowden (1970, 1973a,b), who has indicated that they can be useful taxonomic markers at subgeneric, generic or tribal levels in the various families where they are consistently present. Eight of the more useful acids are listed in Table 6.2, together with a brief indication of their taxonomic value. Most attention has been given to the significance of their occurrences in the Leguminosae and it would seem most appropriate here to illustrate the taxonomic importance of these particular substances in respect of their various records in this one family. Patterns at the generic level are of particular interest in the cases of *Acacia*, *Lathyrus*, *Vicia* and *Phaseolus*; the widespread occurrence of canavanine throughout the subfamily Papilionoideae is also of considerable systematic significance.

Table 6.2 Some non-protein amino acids of taxonomic significance in plants

Amino acid[b] and source	Taxonomic significance
$NH_2CONH\text{-}CH_2\text{-}CH(NH_2)\text{-}CO_2H$ ALBIZZINE in seeds *Albizia julibrissin* (glutamine)[a]	Characteristic of *Acacia* spp; but notably absent from the series *Gummiferae* (Seneviratne and Fowden, 1968); also in *Albizia*, *Mimosa*.
(azetidine ring, NH)–CO_2H Azetidine 2-carboxylic acid in rhizomes Solomon's seal *Polygonatum officinalis* (proline)[a]	Taxonomic marker for Liliaceae and also in Amaryllidaceae (Fowden and Steward, 1957); several disparate occurrences, e. g. in sugarbeet *Beta* (Chenopodiaceae)
$H_2N(HN{=})C\text{–}NH\text{–}O(CH_2)_2\text{–}CH(NH_2)\text{–}CO_2H$ Canavanine in seeds of Jackbean, *Canavalia ensiformis* (arginine)[a]	Only found in one subfamily, the Papilionoideae, of the Leguminosae (Bell *et al.*, 1978); tribal occurrences closely follow phyletic relationships
$(HO)_2C_6H_3\text{–}CH_2\text{–}CH(NH_2)\text{–}CO_2H$ Dopa in seeds of *Mucuna pruriens* (tyrosine)*	Taxonomic marker as *major* seed constituent (6–9% dry wt) of genus *Mucuna* (Bell and Janzen, 1971): also in lesser amount in other legumes, e.g. *Vicia faba*.

Compound	Comment
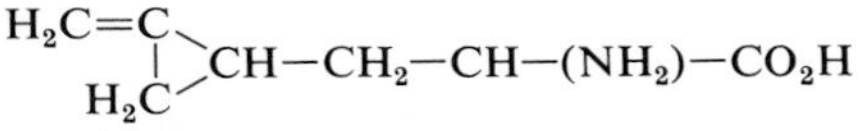 $H_2C=C$, H_2C (ring) $CH-CH_2-CH-(NH_2)-CO_2H$ Hypoglycin A in unripe fruits of akee, *Blighia sapida* (isoleucine)*	Characteristic in both Sapindaceae and Hippocastanaceae (Fowden, 1973a, b); presence in both families supports the view that these taxa could be merged into one family.
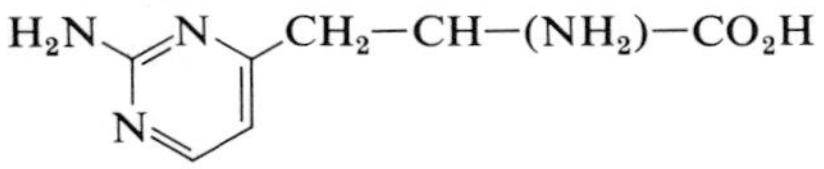 H_2N, N, N (pyrimidine ring) $CH_2-CH-(NH_2)-CO_2H$ Lathyrine in seeds of *Lathyrus tingitanus* (arginine)	Characteristic of *Lathyrus*, absent from closely related *Vicia* (Bell, 1966).
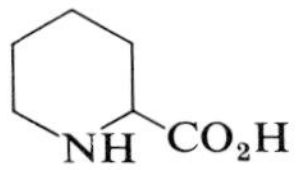 NH CO_2H (piperidine ring) Pipecolic Acid in seeds of *Phaseolus vulgaris* (proline)[a]	Characteristic of *Phaseolus*, absent from closely related *Vigna* (Bell, 1971): also fairly widespread in plants generally.
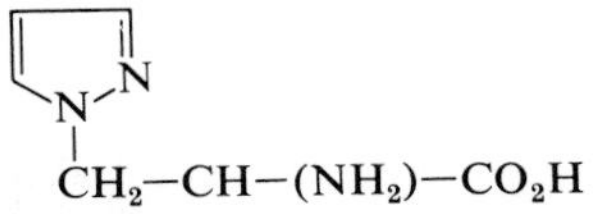 N, N (pyrazole ring) $CH_2-CH-(NH_2)-CO_2H$ β-(Pyrazol-1-yl)alanine in watermelon seed, *Citrullus lanatus* (histidine)*	Characteristic of Cucurbitaceae, in three of the seven tribes of the subfamily Çucurbitoideae (Dunnill and Fowden, 1965).

[a]Indicates the most closely related protein amino acid
[b]Unusual amino acids belong to the same series of optical isomers, the L series, as the protein amino acids, and all compounds should strictly speaking be designated as L-albizzine, L-canavanine, etc.

The genus *Acacia* comprises some 850 to 900 species, approximately 700 of which are native to Australia. The remainder occur in tropical and subtropical regions of Africa, Asia and America. Taxonomically, there are difficulties because of the variable morphology and the sheer size of the genus. Bentham's (1864) major classification of the genus has been revised more than once, most recently by Vassal (1972). The amino acids in the seeds of at least 104 species have now been determined (Seneviratne and Fowden, 1968; Evans *et al.*, 1977) and very clear-cut correlations emerge between series and sectional classification and amino acid content (data summarized in Table 6.3). The most distinctive marker is *N*-acetyldjenkolic acid, which occurs in species of the series *Gummiferae* and nowhere else. There is a whole series of different acids which separate the species into four main biochemical groupings (Table 6.3). As Fowden (1970) writes 'these differences are so marked in practice that one would feel justified in assigning a species to the *Gummiferae* on the basis of amino acid composition data alone'.

Table 6.3 Non-protein amino acids as taxonomic markers in the genus *Acacia*

Series and sections[a]		Amino acid pattern[b]
Gummiferae	(13)	*N*-Acetyldjenkolic acid plus 6 other markers
Phyllodineae	(41)	Albizzine plus 3 sulphur amino acids
Botyrocephalae	(7)	
Pulchellae	(12)	
Vulgares		
sect. aculeiferum	(14)	Albizzine, 3 sulphur amino acids plus α-amino-β-oxalylaminopropionic acid and other related structures
sect. monacantha	(14)	Two unidentified acids: also low concentrations of djenkolic acid

[a]Series classification after Bentham, sectional classification after Vassal (numbers of species examined are given in parentheses).
[b]Data from Seneviratne and Fowden (1968) and Evans *et al.* (1977).

Apart from the purely taxonomic correlations, there are also different patterns according to the geography of the plants. One of the most interesting findings is that the Australian pattern occurs exceptionally in just one non-Australian plant, *Acacia heterophylla*, which is native to the Mascarene Islands, off the east coast of Africa. This discovery provides a striking biochemical link between Australia and the Mascarenes and it adds support to the now widely accepted view that these two habitats once formed part of the same land mass, known as Gondwanaland, at a much earlier stage in the life of this planet (Bell and Evans, 1978).

Taxonomically interesting patterns of amino acids have also been observed in the genera *Lathyrus* and *Vicia* (Bell, 1966). *Lathyrus* spp. contain a series of seven distinctive acids, including lathyrine, whereas seeds of *Vicia* spp. contain a set of six additional acids, including canavanine (see Table 6.2). Taxonomically, the two genera are very closely related and there is no one morphological character which decisively separates them. This is a situation where chemistry clearly has potential for identification purposes. There are also lesser variations in amino acid pattern within the genera, which may prove to be of importance in sectional classification.

Similar surveys have been carried out on seeds of the beans *Vigna* and *Phaseolus* (Bell, 1966). Here, pipecolic acid (see Table 6.2), which is absent from *Vigna* spp. proper, occurs in eight of ten *Phaseolus* species. The two anomalous taxa *P. mungo* and *P. aureus* are on morphological grounds currently thought to belong more properly to the genus *Vigna* than to *Phaseolus*. Indeed, they have subsequently been transferred to *Vigna* (Verdcourt, 1970) and the re-assignment of these two bean taxa to *Vigna* from *Phaseolus* has recently been supported by serological evidence (Chrispeels and Baumgartner, 1978).

Finally, a word must be said about canavanine, systematically the most intriguing of all the many non-protein amino acids found in legume seeds. It has a unique occurrence, since it is entirely confined in the whole of the plant kingdom to this one major group, the subfamily Papilionoideae. Its restriction to this subfamily within the Leguminosae supports chromosomal and other phytochemical evidence for a marked separation of the Papilionoideae from the subfamilies Mimosoideae and Caesalpinioideae. Its distribution within the Papilionoideae has been mapped out in a number of surveys, the most extensive and recent being that of Bell *et al.* (1978). From the phenetic viewpoint, the distribution pattern is completely enigmatic, since it may be present throughout certain genera (e.g. *Canavalia*, *Centrosema*, *Ononis*) but may be very variable in others (e.g. *Vicia*). It is restricted to about half the tribes of the Papilionoideae, but is not consistently present in any of the tribes for which it is positive.

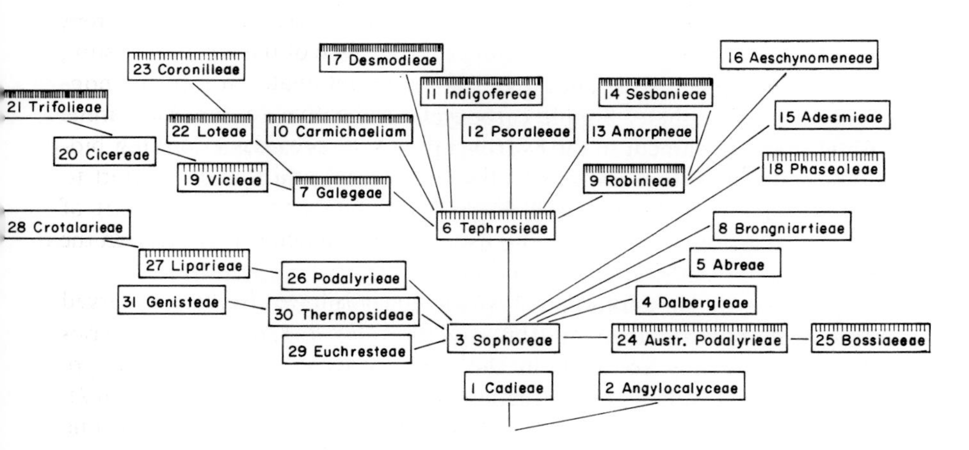

Fig. 6.13 Supposed relationships of tribes of Papilionoideae, simplified and showing occurrence of canavanine. Close hatching indicates canavanine present in all or most genera and open hatching indicates presence in some genera (from Bell *et al.*, 1978).

The systematic significance of canavanine becomes much clearer, however, if its distribution is considered in a phyletic framework. When this is done (Fig. 6.13), the tribal distribution then fits fairly closely to other evolutionary trends in the subfamily derived from morphological and anatomical characters. The canavanine data, indeed, highlight one possible anomaly in this scheme, the positioning of the Australian Podalyrieae and the Bossiaeeae. This and other evidence actually support a realignment in which the Bossiaeeae is shown to originate from the Tephrosieae and the Australian Podalyrieae hence from the Bossiaeeae. This is a reversal of the situation shown in Fig. 6.13 but is likely to become the preferred system. Following this rearrangement, canavanine can then be seen as an advanced feature in the Papilionoideae, which could have arisen only once at some median stage in tribal evolution and then spread throughout most of the more highly advanced taxa.

V. Iridoids

A. Chemistry and distribution

The iridoids are a group of bitter tasting monoterpenoid lactones, which are widely present in plants, being found in about 70 families grouped in some 13 orders. They have only recently come into prominence as taxonomic markers in plants; their chemical instability delayed structural identification and it was not until the early 1960s that their chemistry was fully established (Bate-Smith and Swain, 1966). Iridoids are derived biosynthetically from monoterpene precursors and, like the monoterpenes, are formed from geranyl pyrophosphate. Simple carbocyclic iridoids are lactone derivatives, with glucose attachment to the hydroxyl group of the lactone ring (Fig. 6.14). A typical iridoid is loganin, which occurs for example in *Strychnos nuxvomica* fruits to the extent of 4–5% dry weight.

A second group of iridoids can be distinguished in which the five-membered ring of carbocyclic iridoids is opened, giving rise to seco-iridoids, which have, as a result, an additional aldehydo function. The seco-iridoid derived from loganin is secologanin (for structure, see Fig. 6.14), a widespread substance especially common in the Caprifoliaceae. Seco-iridoids have a special role in plants as biosynthetic precursors of terpenoid alkaloids. For example, condensation of the aldehydo group of secologanin with the amino acid tryptophan produces an alkaloid such as corynantheine, a major base of *Corynanthe johimbe* (Rubiaceae). In general, seco-iridoids tend to co-occur with their related alkaloids.

Within each of the two groups of iridoids, represented by loganin and secologanin respectively, there are many derivatives known. Aucubin is a dihydroxy compound, first isolated from *Aucuba japonica* (Cornaceae) but also well represented in species of the Scrophulariaceae. Asperuloside is a double lactone, named after the source *Asperula odorata* but actually first obtained from madder root, *Rubia tinctorum*. Acylated derivatives are also known such as catalposide, the *p*-hydroxybenzoyl ester of catalpol, from *Catalpa* species (Fig. 6.14).

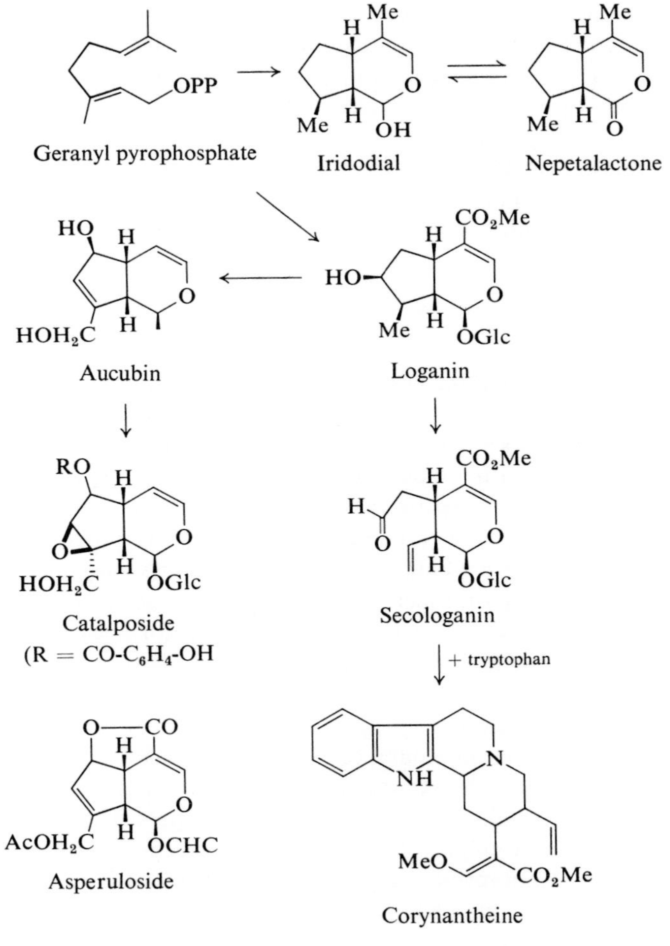

Geranyl pyrophosphate

Iridodial

Nepetalactone

Aucubin

Loganin

Catalposide
(R = CO-C_6H_4-OH

Secologanin

+ tryptophan

Asperuloside

Corynantheine

Fig. 6.14 Structures of iridoids and their biosynthetic origins.

A few iridoids without glucose attachment are present in plants, a notable example being nepetalactone, the active principle of catmint *Nepeta cataria* and the substance that causes members of the cat family to have a peculiar fascination for this plant. The actual role of nepetalactone in *Nepeta* spp. appears to be that of an insect feeding deterrent (Eisner, 1964). Structures related to nepetalactone occur in defence secretions of ants, stick insects and beetles. Such iridoids are usually volatile, occurring in plants in the essential oil fraction, and are generally atypical. All surveys of plants for iridoids are based on the detection of the non-volatile derivatives.

Iridoid glucosides are restricted in their occurrence to the dicotyledons and are mainly found in more advanced families. They are not recorded in the monocotyledons, in gymnosperms or in any lower plant group. Their main purpose seems to be one of providing feeding deterrence, although they are also antimicrobial substances. Some iridoid-containing plants have been used medicinally (e.g. eyebright, *Euphrasia*), but in large concentrations these substances may be toxic. Bread prepared from wheat contaminated with weed seeds containing iridoids (e.g. *Rhinanthus*) has caused human illness and death.

B. Detection

There is no single infallible test for detecting the presence of iridoids in plants, so that it is necessary to apply several diagnostic procedures in order to be sure of detecting them. A preliminary indication is the bitter taste of the leaf sap, coupled with the blackening of the tissue during drying. Blackening can be hastened by heating the leaf in acid solution and is due to the release of the free aglycone, which undergoes oxidative polymerization to a black precipitate; intermediate blue or green colours may also be noted. Plants containing aucubin, asperuloside and catalposide respond in this way. There are other colour tests which can be applied for specific compounds. There is the sensitive Trim and Hill test where a plant extract made from fresh or dried tissue is boiled with acid copper sulphate, when a blue, green or red colour may develop.

A more general chromatographic procedure has been developed by Wieffering (1966) in which extracts are chromatographed on paper in several solvents and the iridoids detected as coloured spots after spraying with antimony chloride in chloroform or anisaldehyde in sulphuric acid. Even this procedure may not reveal every type of iridoid. When isolated in quantity, some iridoids may be obtained in a crystalline form but many are amorphous and have to be converted into a crystalline acetate before purity can be assured. Structural identification using modern spectral procedures is reviewed by Plouvier and Favre-Bonvin (1971).

C. Taxonomic utility

Up to about 1960, most of the data on iridoid glucosides in plants were derived from chance discoveries of these compounds in medicinal species or resulted because chemists were investigating the cause of a bitter taste in one plant or a characteristic darkening reaction in another. In spite of the incomplete nature of these data, Hegnauer (1964) in a review was able to discern that all the plant families in which iridoids appeared were to some extent related. Many belonged to the Tubiflorae, as defined by Engler, and all could be related back phyletically to the Rosiflorae. Bate-Smith and Swain (1966) on analysing the same data also suggested that all the families with the ability to synthesize iridoids might eventually be recognized as constituting a natural assemblage.

Since then, many more occurrences of iridoids have been noted. In a recent comprehensive review of the iridoid character in plants, Jensen *et al.* (1975b) were able to confirm these earlier findings, supporting the view that all 13 orders in which iridoids had been identified are monophyletic. The fact that the presence or absence of iridoids cuts across the traditional Englerian order, the Tubiflorae, has led these authors to reject this order as a natural group. The close relationship between iridoid distribution and taxonomy can be appreciated by reference to the documented map of these substances, when laid out according to the recent system of Dahlgren (1975). Figure 6.15 shows that all these orders are clustered together in the same area of the system. More detailed consideration of the occurrence of particular iridoid structures also reveals many correlations between chemistry and the taxonomy of individual orders (Jensen *et al.*, 1975a,b).

From the purely systematic viewpoint, the fact that iridoids characterize a cluster of morphologically related orders and families means that the character might be employed to place families of otherwise uncertain affinities within the cluster. One such family is the monogeneric Fouquieriaceae, which is conventionally placed near the Tamaricaceae in the Parietales but which has also been linked by various taxonomists with no less than five other orders. Following an earlier report of Bate-Smith (1964), Dahlgren *et al.* (1976) established that iridoids such as loganin occurred in these plants. On this basis, and after full consideration of many biological and other chemical features, these authors concluded that Fouquieriaceae is more closely allied to taxa of the Ericales, one of the more uniform iridoid-producing orders (Fig. 6.15), than to those of any other order.

The iridoid character has also proved to be of interest at the family level, especially in the Cornaceae (Bate-Smith *et al.*, 1975), Labiatae, Rubiaceae and Scrophulariaceae (Kooiman, 1969, 1970, 1972; see also Hegnauer and Kooiman, 1978). Perhaps the most striking data have been obtained in the

Labiatae where iridoid occurrences correlate well with differences in pollen type within the family. A survey of 194 species representing 40% of the known genera showed that subfamilies and tribes with binucleate tricolpate pollen consistently contained iridoids in the leaves, those with trinucleate hexacolpate pollen consistently lacking these lactones. Not surprisingly, in view of their common biosynthetic origin with the monoterpenes, the presence of iridoids in this family is inversely related to the accumulation of essential oils in the leaves. More interestingly, from the functional standpoint, the iridoid character is linked to resistance to the rust fungus *Puccinia menthae*, since taxa lacking iridoids are generally susceptible (Kooiman, 1972).

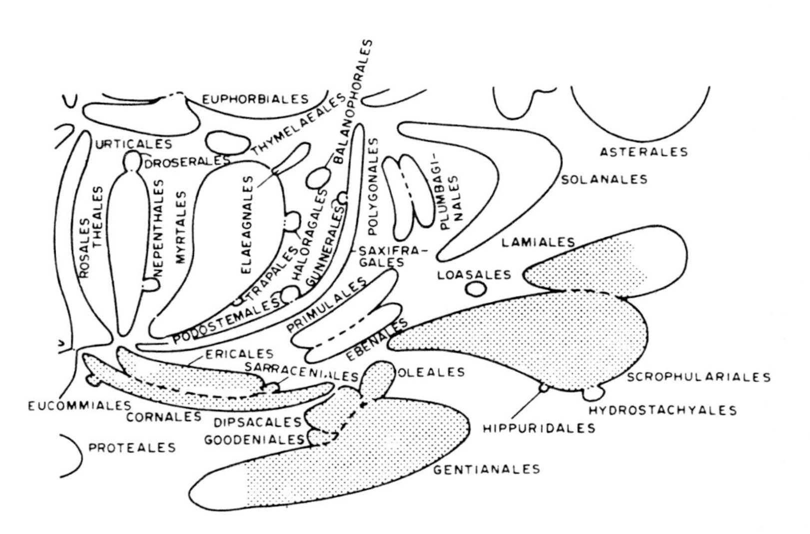

Fig. 6.15 Plant orders containing iridoids (dotted areas), illustrated from Dahlgren's scheme for the angiosperms.

Finally, iridoid characters have also been explored at lower levels of classification. Complex mixtures of acylated catalpol derivatives have been isolated, for example, from the genus *Veronica* (Scrophulariaceae), the patterns in 43 species being related in part to sectional classification in the genus (Grayer-Barkmeijer, 1973). Chemical variation with geography occurred in one species *V. teucrium*, but in general iridoids seemed to show little infraspecific variation. Very consistent iridoid patterns at the species and subspecies level have also been noted in *Lamiastrum* (Labiatae) (Wieffering and Fikenscher, 1974). However, more studies are needed to establish the degree of polymorphism that can be expected with these terpenoid glucosides.

VI. Sesquiterpene lactones

A. Chemistry and distribution

Of the more than 1400 known sesquiterpene lactones, some 1340 (about 90%) are from the Compositae (Herz, 1977; Fischer *et al.*, 1979; Seaman, 1982). The latter is, however, one of the largest of angiosperm families, with over 1400 genera and 25 000 species (Heywood *et al.*, 1977) and one in which relationships are highly complex and where chemical characters are clearly important in the process of systematic revision.

These lactones are C_{15} terpenoids, derived via the mevalonate pathway from three C_5 isopentenyl pyrophosphate units. They are commonly referred to as bitter principles, which indicates they may generally serve as anti-feedants in plants, at least to mammalian grazers. Two notable bitter-tasting lactones are lactupicrin, the principle of wild lettuce leaf, and absinthin, the bitter substance of *Artemisia absinthum* which also appears in the liqueur absinthe derived from the same plant. Other important biological activities of sesquiterpene lactones are anti-tumour properties (e.g. bakkenolide-A from *Petasites*) and allergenic effects (e.g. parthenin from *Parthenium hysterophorus*), causing contact dermatitis in man (Rodriguez *et al.*, 1976). Some are quite toxic (e.g. vermeerin from *Geigeria*) and are responsible for cattle and sheep poisoning. The anti-herbivore activity of such lactones as glaucolide-A from *Vernonia* species has been demonstrated by Burnett *et al.* (1978).

The vast majority of known sesquiterpene lactones are based variously on the 15 skeletal types illustrated in Fig. 6.16 and, although it has not yet been proved, are probably biogenetically related as indicated in this figure. Thus, they are formed by cyclization and oxidation from the same C_{15} precursor, *trans,trans*-farnesyl pyrophosphate to give the 10-membered carbocyclic germacranolides. These germacranolides are then further transformed to give, most commonly, the 5/7 and 6/6 carbocyclic ring systems of the guaianolides and eudesmanolides. Other rearrangements and cleavages lead to the biogenetically advanced types such as the helenanolides and ambrosanolides. There is some evidence that the formation of the key lactone ring is one of the last stages in synthesis, but this has yet to be proved by labelling experiments. Introduction of other functional groups commonly present (e.g. alcohol, keto, acyl groups) presumably also takes place at a late stage in synthesis. Typical structures of composite sesquiterpene lactones are shown in Fig. 6.17.

In addition to their major occurrence in composites, sesquiterpene lactones have been recorded in the Magnoliaceae (*Liriodendron*, *Michelia*), Lauraceae (*Neolitsea*, *Lindera*, *Laurus*) and Umbelliferae (*Ferula*, *Laser*, *Laserpitium*, *Smyrnium* and *Melanoselinum*). However, except for a few of the compounds

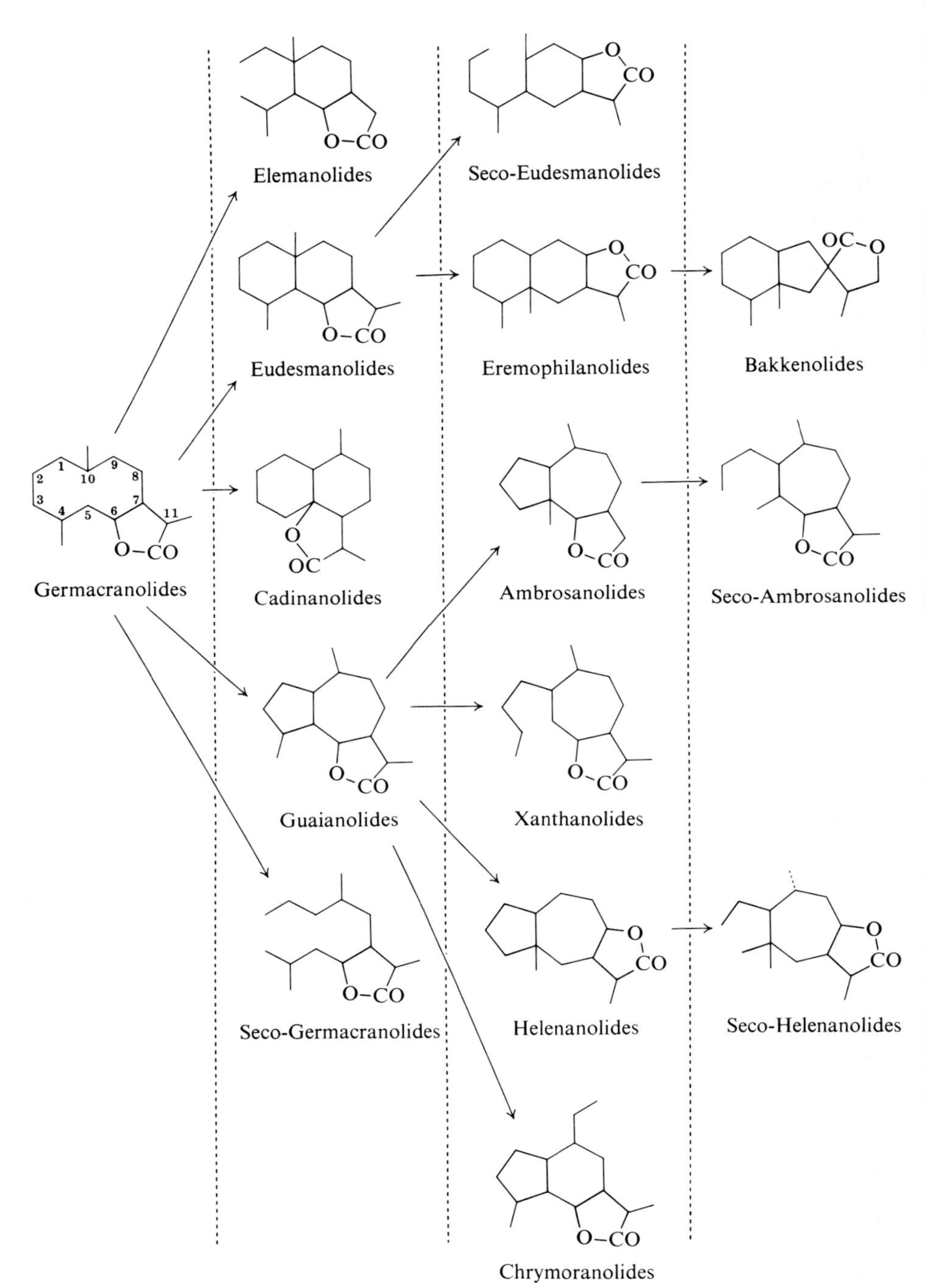

Fig. 6.16 Biogenetic scheme for sesquiterpene lactones which occur in the Compositae (from Herz, 1977).

isolated from the Umbelliferae, the lactones reported in these other families are only of the biogenetically more primitive types (see Fig. 6.16). Representatives of one of the most primitive lactone types, the eudesmanolides, have also been found in three liverworts and in a basidiomycete.

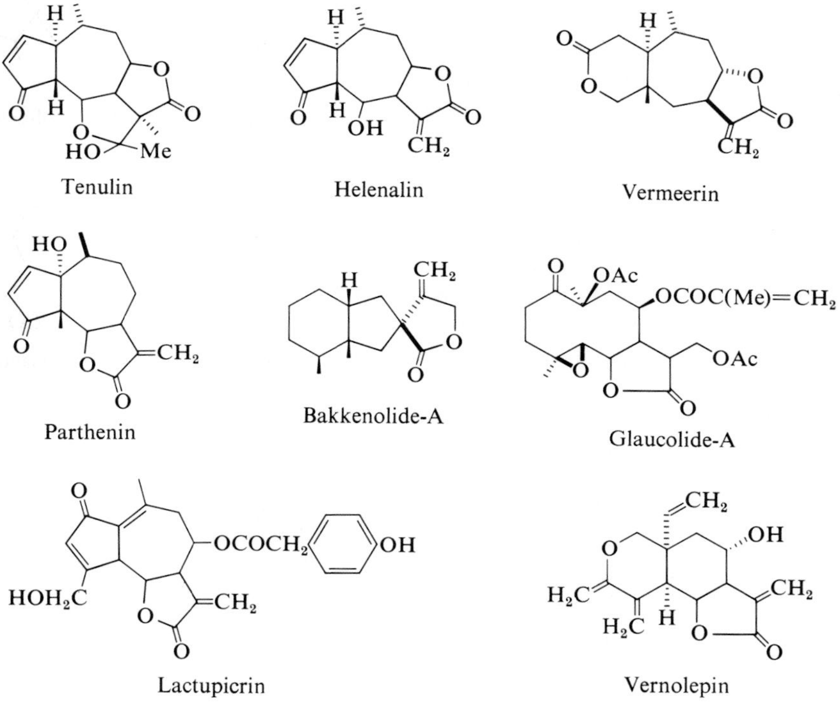

Tenulin Helenalin Vermeerin

Parthenin Bakkenolide-A Glaucolide-A

Lactupicrin Vernolepin

Fig. 6.17 Structures of typical sesquiterpene lactones

Sesquiterpene lactones are usually located in glands on the leaves and inflorescences and may be associated with glandular trichomes, as in *Parthenium hysterophorus*. They have also been reported from most other plant parts, including roots, wood, flowers and root barks.

B. Detection

A general procedure for the detection of sesquiterpene lactones in plants has been outlined by Mabry (1970). It consists, in outline, of extracting dried tissue with chloroform. This extract, after purification on a silica-gel column, is analysed by TLC and NMR spectroscopy. Many lactones give bright colours when TLC plates, after development, are sprayed with concentrated

H_2SO_4 and heated. However, there is no one single colour test that unambiguously indicates the presence of a lactone in a crude plant extract. Methods are being developed, however, for more certain TLC detection of sesquiterpene lactones in plant tissues (see Drozdz and Bloszyk, 1978).

In large-scale isolation, chromatography on silica gel is an essential prerequisite and the compounds thus separated are often readily crystallized. They are then identified on the basis of melting point, rotation and NMR and mass spectroscopy. A valuable compilation of the physical properties of most of the known sesquiterpene lactones is that of Yoshioka *et al.* (1973).

C. Taxonomic utility

Systematically, the sesquiterpene lactones are markers of interest only in respect of their presence in many species of the Compositae. The distribution patterns here have been discussed in several recent reviews (e.g. Herz, 1977; Seaman, 1982) and detailed lists of occurrences on a tribal basis are available in a recent Symposium volume (Heywood *et al.*, 1977). In spite of much effort by many phytochemists, numerous genera and species still remain uninvestigated. The known distribution of sesquiterpene lactones at the tribal level is summarized in Table 6.4 and from this it will be seen that the presence or absence of the major biogenetic types (Fig. 6.16) provides a key for separating most of the tribes. It appears that the more primitive tribes may have both simple patterns (e.g. Vernonieae) or complex mixtures (e.g. Heliantheae). Sesquiterpene lactone synthesis is thus a character well established in the early development of the family and one which has gradually disappeared with evolutionary advancement. Thus, lactones are absent from, or poorly represented in, the most highly advanced tribes (e.g. Mutisieae, Astereae, Tageteae, Arctoteae–Calenduleae).

Chemical races in sesquiterpene lactones have been observed in a number of composite species, but especially in members of the ragweed genus *Ambrosia*. Chemical variation in some *Ambrosia* spp. can be related to morphological or ecological factors (see Mabry, 1970) but in others, e.g. *Ambrosia chamissonis*, no correlation could be established between the various chemical races present and leaf morphology (Payne *et al.*, 1973). Also in *Artemisia tridentata*, another species with chemical races, there was no apparent relationship between chemistry and changes in chromosome number (Kelsey *et al.*, 1975).

The presence of chemical races, albeit in a very small number of species, points to the importance in taxonomic work of screening several populations of a given taxon. Bierner (1973a,b), for example, used lactone characters in *Helenium* to suggest that *H. quadridentatum*, which had been merged with *H. elegans* by Rock (1956), should be retained as a species in its own right, since it differed in having helenalin instead of tenulin, the lactone present in *H. elegans* proper. His conclusions have to be taken with some caution, since sampling was limited to five populations of *H. elegans* and two of *H. quadridentatum*. One may also note that the chemical difference between the two lactones is a relatively minor one (see Fig. 6.16).

Table 6.4 Distribution of sesquiterpene lactone biogenetic types in tribes of the Compositae

Tribe	Generic frequency	Lactone types present[a]
Eupatorieae	4/50	I, II and III
Vernonieae	4/50	I and II
Astereae	1/100	II
Inuleae	5/300	I – IV
Heliantheae	24/250	I – IV
Helenieae	11/60	I – IV
Tageteae	0/15	None
Senecioneae	4/50	I, III and IV
Anthemideae	10/50	I, II and III
Arctot.-Calend.	1/50	II
Cynareae	8/50	I and II
Mutiseae	1/55	II
Cichorieae	6/75	I and II

[a]Type I = germacranolides, type II = eudesmanolides, guaianolides, etc.
Type III = eremophilanolides, helenanolides, etc.
Type IV = bakkenolides, seco-ambrosanolides, seco-helenanolides (for structures, see Fig. 6.13).

The taxonomic importance of lactone patterns has been explored in several genera, in some depth, with interesting conclusions. In *Artemisia*, the patterns in different species can be correlated with evolutionary changes in floral biology (see Greger, 1977). In *Ambrosia*, patterns are useful in determining the geographical origins in certain species (Mabry, 1970). Finally, in *Vernonia*, the available data provide insight into phylogenetic relationships. While American species have germacranolides such as glaucolide-A, African species have a different lactone pattern based, for example, on the eudesmanolide vernolepin (see Fig. 6.17 for structures). These results suggest that most of the New World species belong to a different line from that of Old World taxa. Supporting evidence for this view is provided by flavonoid studies (Harborne and Williams, 1977) and by chromosomal data (Jones, 1977). However, Turner (1981) has recently reported on a few North American species of *Vernonia* which can be directly related to the African species and which possess a sesquiterpene chemistry of the African groups (Gershenzon *et al.*, 1984).

VII. Diterpenoids

A. Chemistry and distribution

The diterpenoids comprise a chemically heterogeneous group of about 800 structures, all with a C_{20} skeleton based on four isoprene units and derived biosynthetically from geranylgeraniol. They are notable for the enormous variation in skeletal type and for their occurrence in both normal and antipodal stereochemical series. Most diterpenoids are of very limited distribution; their most obvious occurrence is in plant resins and latexes, where they make a major contribution to the sticky nature of such materials. These resin compounds have a protective function in nature in that they are exuded from wood of trees or the stems of herbs and are hostile to animal and microbial predators. Leaf-cutting ants, for example, notably avoid attacking plants protected by latex. Gymnosperm resins are especially rich in di- and tricyclic diterpenoids such as abietic and agathic acid. The various 'copal' resins of legume trees are also a rich source of diterpenoids (Ponsinet *et al.*, 1968), the bicyclic hardwickic acid being one example.

Toxicity to animals is especially associated with certain diterpene types. Thus, there are a group of grayanotoxins, typified by grayanotoxin I, which occur in leaves of Ericaceae, for example, in species of *Kalmia*, *Leucothoe* and *Rhododendron*. The toxins occasionally occur also in the flowers of *Rhododendron* spp. and can be carried by bees to contaminate their honey, making it poisonous. Many Euphorbiaceae (*Croton*) and Thymelaceae (*Daphne*) contain toxic diterpenes (e.g. phorbol and mezerein) which have the additional property of being highly irritant and hence co-carcinogenic. Another property of a number of diterpenoids is intense bitterness; this is true of columbin (of *Jateorhiza*, Menispermaceae) and of marrubiin (of *Marrubium*, Labiatae), for example. The structures of the diterpenoids mentioned above are collected in Fig. 6.18.

The ability to synthesize diterpene structures is universal to plants, since phytol, the acyclic parent compound of the series, is present in ester attachment in the molecule of chlorophyll, and hence occurs in all green plants. The capacity to modify considerably the parent diterpene precursor into the tetracyclic structure of the gibberellins is also widespread (Fig. 6.19). The growth hormone, gibberellic acid (or GA_3) is universal, as far as is known, in the plant kingdom. In fact, there is not just a simple gibberellin in plants, but rather a series of some 60 compounds, all with the same basic structure. Of these, only gibberellic acid is recognized as being commonly present and it is assumed that it probably always occurs. The other members of the series have only been found in very few sources so far. Although there may well be a chemotaxonomic element in the distribution of these other gibberellins, the problems involved in detecting them, because of their extremely low concentrations, has so far largely precluded the use of such structural variation. However, differences in gibberellin substitution patterns have been observed in the seeds of legumes and of cucurbits (Sponsel *et al.*, 1979).

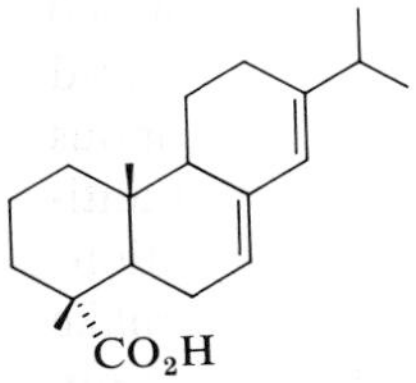

Abietic acid
(Pine resins)

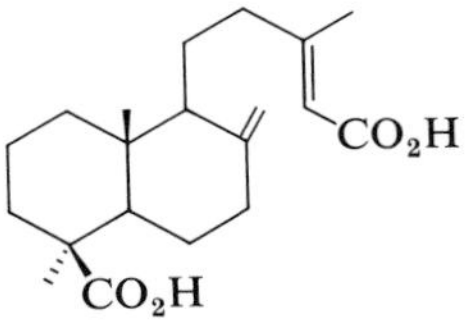

Agathic acid
(*Agathis*, Araucariaceae)

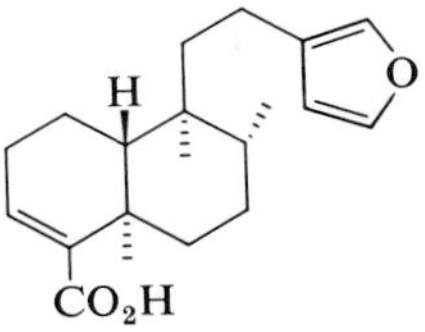

(−)—Hardwickic acid
(*Hardwickia pinnata*, Leguminosae)

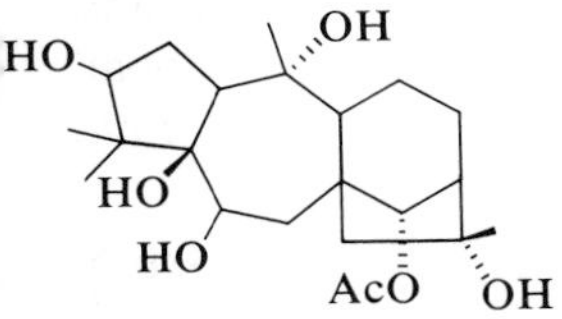

Grayanotoxin I
(*Rhododendron*, Ericaceae)

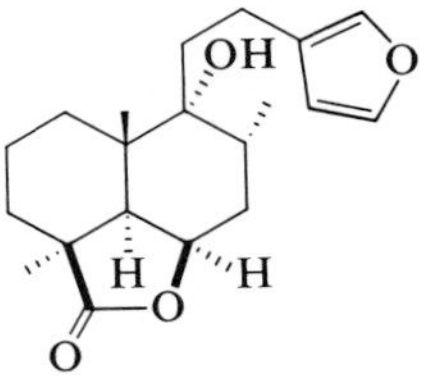

Marrubiin
(*Marrubium vulgare*, Labiatae)

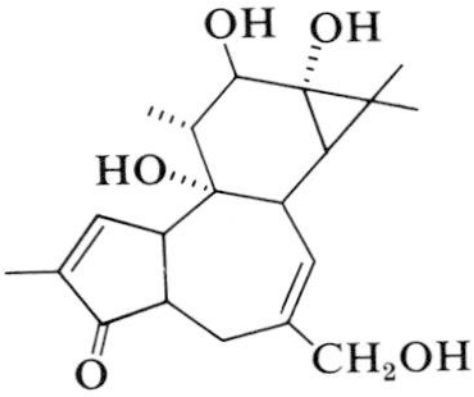

Phorbol
(*Croton* spp., Euphorbiaceae)

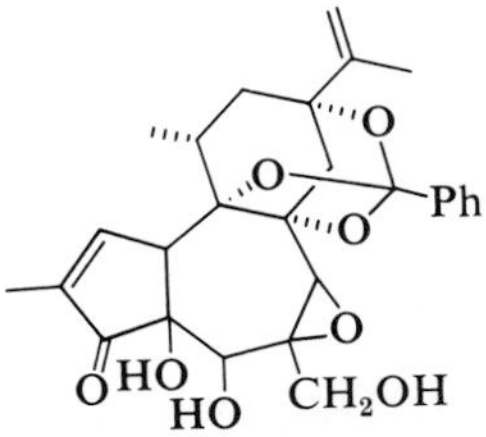

Mezerein
(*Daphne mezereum*, Thymelaceae)

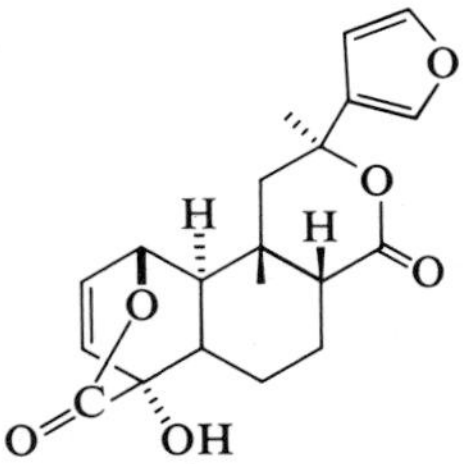

Columbin
(*Jateorhiza palmata*, Menispermaceae)

Fig. 6.18 Structures of typical diterpenoids of plants.

Apart, therefore, from phytol and gibberellic acid, the remaining diterpenoids are very restricted in occurrence and usually occur within one or a few plant families. Families in which diterpenoids are regular constituents include among the gymnosperms, Pinaceae, Cupressaceae and Podocarpaceae and among the angiosperms, Ericaceae, Euphorbiaceae, Labiatae, Leguminosae (especially subfamily Caesalpinioideae) and Thymelaceae. There are also records of diterpenoids in individual species in many other angiosperm families. The literature on diterpenoids is scattered, but a useful general review is that in Nakanishi *et al.* (1974). Several more detailed reviews are also available (e.g. Hanson, 1968; Ourisson, 1974).

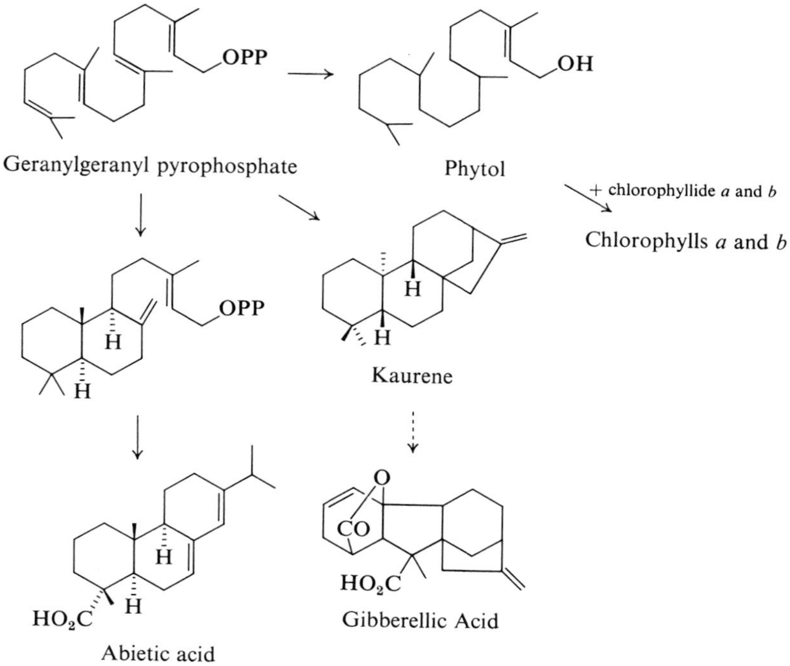

Geranylgeranyl pyrophosphate

Phytol

+ chlorophyllide *a* and *b*

Chlorophylls *a* and *b*

Kaurene

Abietic acid

Gibberellic Acid

Fig. 6.19 Outline pathway of the biosynthesis of different diterpenoids.

B. Detection

In general, diterpenoids are separated by GLC and TLC using the same procedures as for the lower terpenoids (see chapter 5). Diterpenes are less volatile, so that slightly different conditions may be needed. Further identification is based largely on infrared (IR), NMR and mass spectroscopy. No general screening procedure is available, since there is no specific colour test for diterpenoids. They are detected, after TLC, by techniques that are general for all terpenoids.

Typical GLC conditions for separating diterpene hydrocarbons are described by Aplin *et al.* (1963), who surveyed eleven such compounds in leaf of various Podocarpaceae. Diterpenoid phenols have been surveyed in gymnosperm resins by Gough (1966), who converted the acid groups to methyl ester form, before GLC separation.

C. Taxonomic utility

Taxonomically diterpenoids have principally been of use at the lower levels of classification, and mostly among gymnosperms. One typical study of resins in Cupressaceae suggested the need for a taxonomic reassessment of *Tetraclinis articulata* (Gough, 1966). This species is usually considered a Southern Hemisphere floristic element, although it occurs in North Africa remote from the other Southern genera. Chemical examination of its resins (Table 6.5) shows that it differs markedly in the resin acid content from *Widdringtonia*, its nearest neighbour from South Africa, and also from *Callitris*, a genus native to Australia, with which it was once classified. The chemical data thus strongly indicate that it should be considered as belonging to the Northern Hemispheric elements of the Cupressaceae. In particular, it has a closely similar diterpenoid pattern to that of Eurasian *Cupressus* species (see Table 6.5).

Table 6.5 Distribution of diterpenoid alcohols and acids in some members of the Cupressaceae

Sub-family	Genus	Distribution of diterpenoids							
		1	2	3	4	5	6	7	8
Northern	*Juniperus*	−	(+)	−	+	(+)	−	−	−
	Arceuthos	−	+	−	+	+	−	−	−
	Eurasian *Cupressus*[a]	+	+	−	+	+	−	−	−
	Tetraclinis	+	+	+	+	+	+	−	−
Southern	*Widdringtonia*	−	−	−	+	−	−	−	+
	Callitris	−	−	−	−	−	−	+	−
	Austrocedrus	+	−	−	+	−	−	−	−

Key: 1, torulosic acid; 2, agathic acid; 3, 12β-acetoxysandaropimaric acid; 4, ferruginol; 5, totarol; 6, torulosyl acetate; 7, callitrisic acid; 8, unidentified.

[a]American *Cupressus* species are similar except that they lack torulosic acid.

Another example indicating the value of diterpenoid data in taxonomic work concerns the long-standing horticultural problem of the true identity of a juvenile conifer known as *Chamaecyparis obtusa* cv. Sanderi. This problem was recently solved simply by GLC analyses of foliage extracts of the taxon and putative relatives. Indeed, comparison with 47 species of Cupressaceae from the Northern Hemisphere showed that the material concerned was identical to the juvenile form of *Thuja orientalis* and was quite different from *Chamaecyparis* (Fig. 6.20). The juvenile forms have the same four major diterpene acids, present in the same decreasing order of concentration: communic acid (C), sandaraco–pimaric acid (B), isopimaric acid (D) and Δ^8-isopimaric acid (A). Traces of three diterpenoid phenols, sempervirol, totarol and ferruginol, were also present. It is notable that in this study all other taxa examined had different profiles. Furthermore, the identification of the juvenile plant as *T. orientalis* was consistent with all other available data (Gough and Welch, 1978).

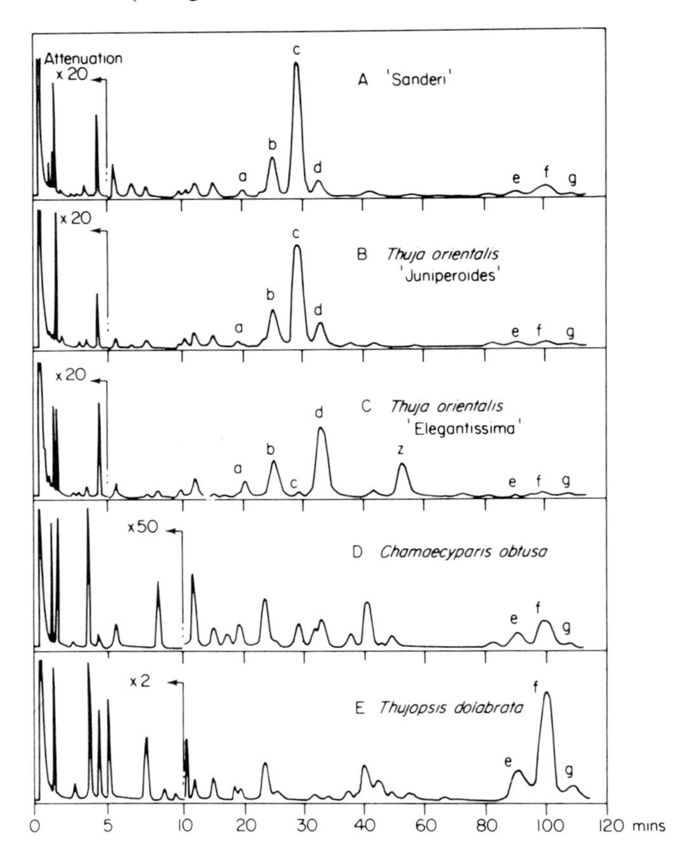

Fig. 6.20 GLC profiles of leaf diterpenoids of 'Sanderi', a problem taxon, and of *Thuja orientalis* and *Chamaecyparis obtusa*.

Use of diterpenoids as taxonomic markers in angiosperms has not been developed very far yet. Some 40 diterpenoids have been isolated variously from resins in members of the tribe Amherstieae of the Leguminosae, but little systematic significance has yet been established for them (Ponsinet *et al.*, 1968). Ourisson (1974) has pointed out the close chemical similarity in diterpenoids found in Euphorbiaceae and in Thymelaceae. Two examples shown in Fig. 6.18 are phorbol and mezerein respectively. A morphological relationship does exist between *Daphne* (Thymelaeceae) and members of the Euphorbiaceae and these chemical data would provide an argument for placing the two families nearer to each other than is usually allowed.

VIII. Triterpenoids and steroids

A. Chemistry and distribution

Triterpenoids are compounds with a carbon skeleton based on six isoprene units and which are derived biosynthetically from the acyclic C_{30} hydrocarbon, squalene. They have relatively complex cyclic structures, most being either alcohols, aldehydes or carboxylic acids. They are colourless, crystalline, often high melting, optically active substances, which are generally difficult to characterize because of their lack of chemical reactivity. Biologically, however, they are extremely active and some are very poisonous substances.

Triterpenoids can be divided into at least four groups of compounds: true triterpenes, steroids, saponins and cardiac glycosides. The latter two groups are essentially triterpenes or steroids which occur mainly as glycosides. There are also the steroidal alkaloids, but these were covered in an earlier section along with the other alkaloids.

Many triterpenes are known in plants and new ones are regularly being discovered and characterized. So far only a few are known to be of widespread distribution. This is true of the pentacyclic triterpenes α- and β-amyrin and the derived acids, ursolic and oleanolic acids (see Fig. 6.21). These and related compounds occur especially in the waxy coatings of leaves and on fruits such as apple and pear and they may serve a protective function in repelling insect and microbial attack. Triterpenes are also found in resins and barks of trees and in latex (*Euphorbia*, *Hevea*, etc.).

Certain triterpenes are notable for their taste properties, particularly their bitterness. Limonin, the lipid-soluble bitter principle of *Citrus* fruits, is a case in point. It belongs to a series of pentacyclic triterpenes which are bitter, known as limonoids and quassinoids. They occur principally in the Rutaceae, Meliaceae and Simaroubaceae (Connolly *et al.*, 1970) and are of chemotaxonomic interest (Dreyer, 1966).

Another group of bitter triterpenes are the cucurbitacins (e.g. cucurbitacin D, Fig. 6.21), which at one time appeared to be characteristic components of seeds of the Cucurbitaceae. Recently, cucurbitacins have been detected in four other families, i.e. in *Iberis* (Cruciferae) (Curtis and Meade, 1971), in *Gratiola* (Scrophulariaceae) (cf. Lavie and Glotter, 1971) in *Anagallis arvensis*, Primulaceae (Yamada *et al.*, 1978) and in *Purshia tridentata* (Rosaceae)

(Dreyer and Trousdale, 1978). Thus, they may prove more widely distributed than originally thought.

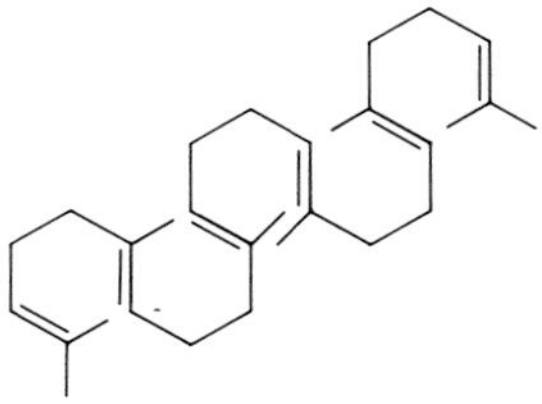

Squalene

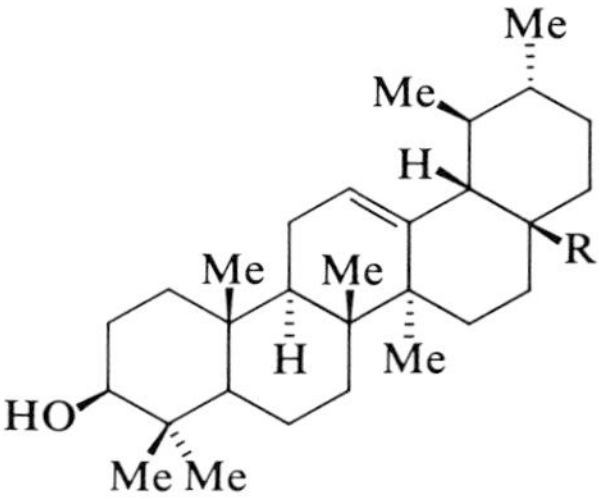

α-Amyrin (R = Me)
Ursolic acid (R = CO_2H)

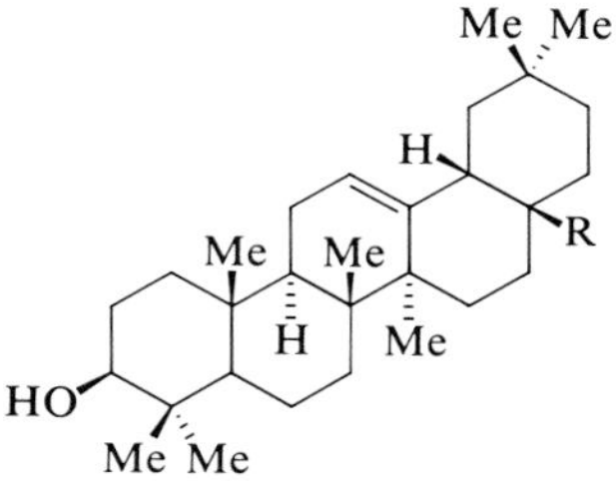

β-Amyrin (R = Me)
Oleanolic acid (R = CO_2H)

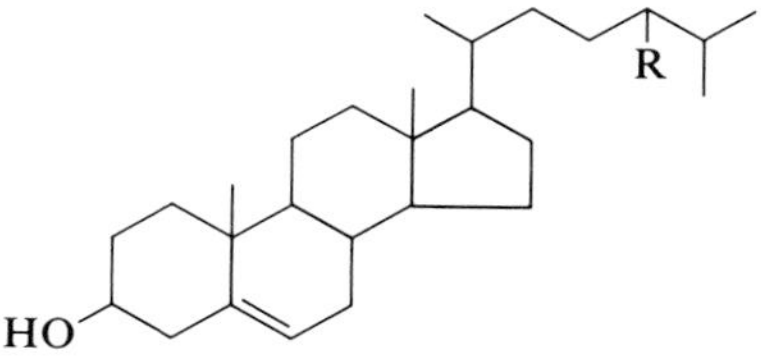

Sitosterol (R = Et)
Campesterol (R = Me)

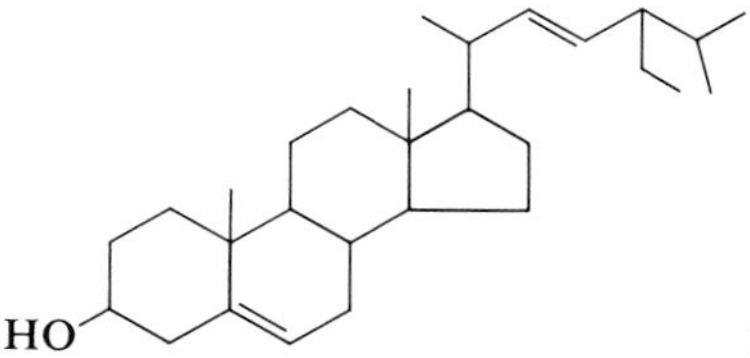

Stigmasterol

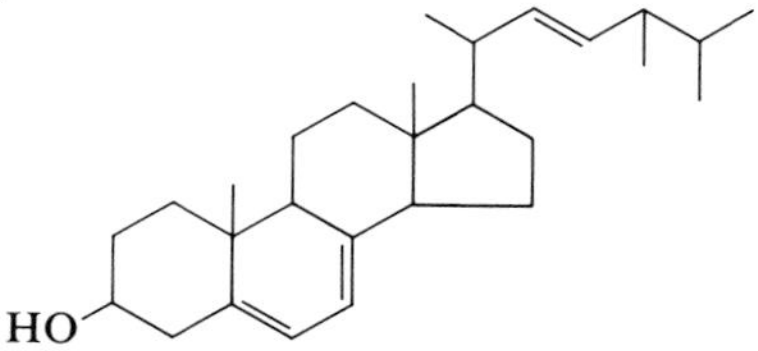

Ergosterol

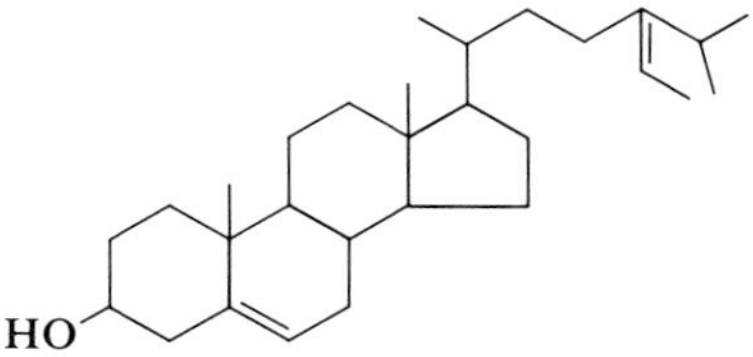

Fucosterol

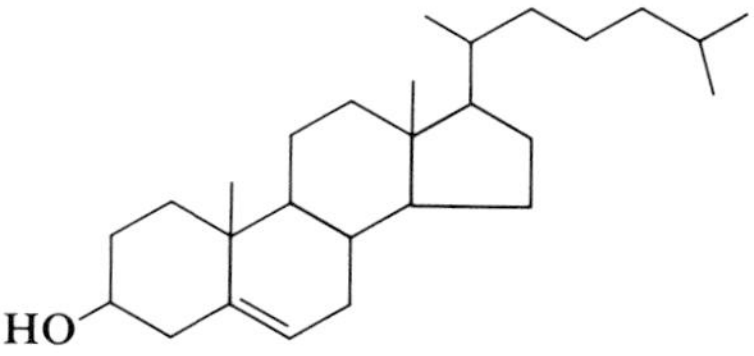

Cholesterol

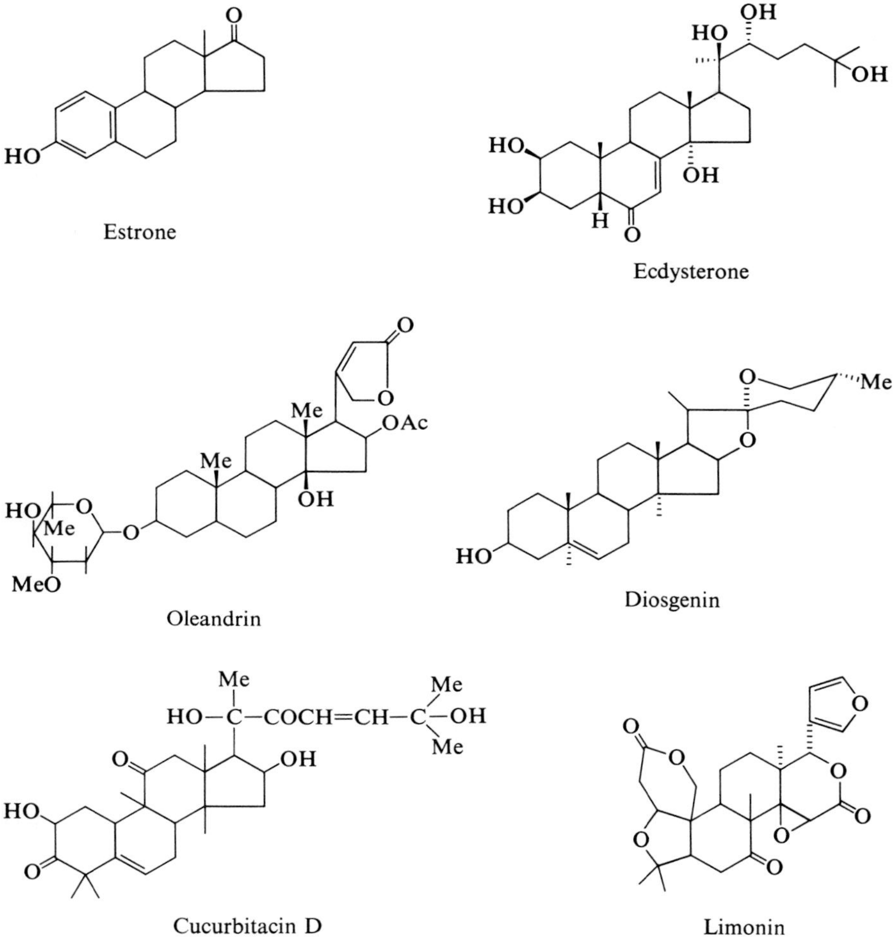

Estrone

Ecdysterone

Oleandrin

Diosgenin

Cucurbitacin D

Limonin

Fig. 6.21 Structures of some typical plant triterpenoids.

Sterols are triterpenes that are based on the cyclopentane perhydrophenanthrene ring system. At one time, sterols were mainly considered to be animal substances (as sex hormones, bile acids, etc.) but in recent years, an increasing number of such compounds have been detected in plant tissues. Indeed, three so-called 'phytosterols' are ubiquitous in occurrence in higher plants: sitosterol (formerly known as β-sitosterol), stigmasterol and campesterol. These common sterols occur both free and as simple glucosides. Sitosterol in particular is thought to be an essential component of plant cell membranes. A less common plant sterol is α-spinasterol, an isomer of stigmasterol found in spinach, alfalfa and senega root (*Polygala senega*). Certain sterols are confined to lower plants; one example is ergosterol, found in yeast and many fungi. Others occur mainly in lower plants but also appear occasionally in higher plants, e.g. fucosterol, the main steroid of many brown algae and also detected in the coconut.

Phytosterols are structurally distinct from animals sterols, so that recent discoveries of the latter in plant tissues are most intriguing. One of the most remarkable is of the animal oestrogen, oestrone, in date palm seed and pollen (Bennett *et al.*, 1966). Pomegranate seed is another source, but the amounts present are very low; according to Dean *et al.* (1971), there is only 4 μg of oestrone per kg of tissue. Less remarkable perhaps, is the detection of cholesterol as a trace constituent in several higher plants, including date palm, and in a number of red algae (Gibbons *et al.*, 1967). Finally, the occurrence of insect moulting hormones, the ecdysones, in plants must be mentioned, since they provide a fascinating insight into the way plants may have evolved in order to protect themselves from insect predation. Ecdysones were discovered in plants by Nakanishi (1968) and have subsequently been found in a range of plant tissues, with high concentrations being present in a number of ferns (e.g. bracken, *Pteridium aquilinum*) and gymnosperms (e.g. *Podocarpus nakaii*). Ecdysones also occur occasionally in angiosperms; they have been detected, for example, in *Helleborus* species (Ranunculaceae), where they co-occur with saponins and bufadienolides (Hardman and Benjamin, 1976). In all, at least 40 phytoecdysones have been recorded in plants (Nakanishi *et al.*, 1974).

Saponins are glycosides of both triterpenes and sterols and have been detected in over 70 families of plants (Tschesche and Wulff, 1973; Mahato *et al.*, 1982). They are surface-active agents with soap-like properties and can be detected by their ability to cause foaming and to haemolyse erythrocytes. The search in plants for saponins has been stimulated by the need for readily accessible sources of sapogenins which can be converted in the laboratory into animal sterols of therapeutic importance (e.g. cortisone, contraceptive oestrogens, etc.). Compounds that have been used to date include hecogenin from *Agave* and yamogenin from *Dioscorea* species. Saponins are also of economic interest because of their occasional toxicity to cattle (e.g. saponins of alfalfa) or their sweet taste (e.g. glycorrhizin of liquorice root). The glycosidic patterns of the saponins are often complex; many have as many as five sugar units attached and glucuronic acid is a common component. Steroidal saponins are especially characteristic among dicotyledons in *Digitalis* (Scrophulariaceae) and otherwise occur mainly in four monocotyledonous families, Amaryllidaceae, Bromeliaceae, Dioscoreaceae and Liliaceae. By contrast, triterpenoid saponins are rare in the monocotyledons but are widely present in dicotyledons, having been recorded from at least 60 families (Paris, 1963).

The last group of triterpenoids to be considered are the cardiac glycosides or cardenolides; here again there are many known substances, with complex mixtures occurring together in the same plant. A typical cardiac glycoside is oleandrin, the toxin from the leaves of the oleander, *Nerium oleander* (Apocynaceae). An unusual structural feature of oleandrin (see Fig. 6.21) and many other cardenolides is the presence of special sugar substituents, sugars indeed which are not found elsewhere among plants. Most cardiac glycosides are toxic and many have pharmacological activity, especially on the heart. Rich sources are members of the Scrophulariaceae (*e.g. Digitalis*), Apocynaceae, Moraceae and Asclepiadaceae. Special interest has been taken in the cardiac glycosides of *Asclepias*, because they are absorbed by larvae of the Monarch butterflies feeding on these plants and are then used by the adult butterflies as a protection from predation by blue Jays (Rothschild, 1972). The butterfly is unharmed by these toxins, which on the other hand act as a violent emetic to the bird.

B. Detection

A widely used colour test that can be applied to crude plant triterpenoid fractions is the Liebermann–Burchard reaction (acetic anhydride–concentrated H_2SO_4), which produces a blue-green colour with most triterpenoids. Simple tests are also available for detecting saponins in plants. The formation of persistent foaming after shaking an aqueous plant extract in a test tube is a good indication that saponin is present. They are also frequently detected by their ability in crude extracts to haemolyse blood. Such simple tests should, however, always be confirmed by one or more appropriate TLC procedures.

All types of triterpenoid are separated and purified by similar methods, based mainly on TLC or GLC (after derivatisation). Identification is based on melting point, optical rotation, GC–MS, IR and NMR spectroscopy. TLC is nearly always carried out on silica gel and detection can be achieved by means of the Carr–Price reagent, 20% antimony chloride in chloroform. A range of colours are produced, after heating the sprayed plates at 100°C. The Liebermann–Burchard reaction can also be used, and the sensitivity is good (2–5 μg level).

C. Taxonomic utility

The comparative biochemistry of triterpenoids has been extensively studied not only in higher plants but also in bryophytes, algae and fungi. In many cases, taxonomic implications have been drawn from their distribution patterns. In this treatment, it is only possible to choose a few examples from the many studies that have been carried out.

At the family level, one promising group for future chemotaxonomic exploration is the grasses, or Gramineae, since they contain a series of rare triterpene methyl ethers. Ohmoto *et al.* (1970), in a general survey of 56 species, using both seed and leaf, isolated 28 different triterpenes, many as methyl ethers belonging to the fernane or arborane skeletal types. These authors found some correlations between triterpene methyl ethers and tribal and generic limits, but it is now fairly clear that these markers are most useful at the species level. Such characters have already been exploited for taxonomic purposes within the genera *Chionochloa* (Russell *et al.*, 1976), *Cortaderia* (Purdie and Connor, 1973) and *Saccharum* (Smith and Smith, 1978). In *Saccharum* leaf waxes, for example, there are six triterpene methyl ethers and they have a distribution in different clones which clearly reflect the chromosome number and geographic origin. *Saccharum officinarum* or sugar-cane is distinctive both qualitatively, in containing arundoin, and quantitatively, in having higher concentrations than other species; thus cultivation seems to have increased the capacity for triterpene synthesis in the genus.

Triterpenoid markers have also been studied at the generic level in the family Dipterocarpaceae (Bisset *et al.*, 1966; Bandaranayake *et al.*, 1977). The presence or absence of twelve triterpenoids have been recorded in bark, timber or resin of 92 species drawn from four genera (Table 6.6). Ashton (1972), on morphological grounds, submerged the genus *Doona* into *Shorea*. The chemical differences in triterpenoids, however, are quite striking and would argue for continued recognition of *Doona* as a separate genus of the family.

Triterpenoids of the limonoid type are widespread in the Rutaceae and have an interesting distribution here. Thus, Dreyer (1966) detected limonin or related structures in 26 *Citrus* species, in three related genera, *Poncirus*, *Microcitrus* and *Fortunella* (all of the subfamily Aurantioideae), and in six genera of two other subfamilies, Toddalioideae and Rutoideae. By contrast, they have not been reported in the four other subfamilies, members of which are morphologically distinct from those mentioned above.

Triterpenoids with a significant distribution pattern at the genus level are the steroidal sapogenins present in *Dioscorea* root. Akahori (1965), when analysing 12 species grown in Japan, discovered that species with opposite leaves, stems twining to the right, and edible roots all lacked saponins. By contrast, species with alternative leaves, stems twining to the left, and no bulbils all contained diosgenin. The only anomalous species, *D. bulbifera*, belonging to the second group but lacking disogenin, was distinct morphologically in having globose tubers.

Table 6.6 Distribution of triterpenoids in Dipterocarpaceae

	Genus			
Triterpenoid	*Dipterocarpus* (41)[a]	*Doona* (6)	*Shorea* (37)	*Stemonoporus* (8)
Dammarenediol 20S	+	+	+	—
Dipterocarpol	+	—	+	—
Dammaradienone 20S	+	—	—	—
Hydroxydammarenone	—	+	—	—
ψ-Taraxasterol	—	+	—	—
β-Amyrin	—	+	+	—
Shoreic acid	—	—	+	—
Dammarenolic acid	—	—	+	—
Ursolic aldehyde	—	—	+	—
δ-Amyrenone	—	—	—	+
α-Amyrin	—	—	—	+
Ursolic acid	—	—	—	+

[a]Values in parentheses indicate the number of species examined
Data from Bisset *et al.* (1966) and Bandaranayake *et al.* (1977)

Reichstein (1965) found triterpenoids of interest at the species level when examining plants of *Acokanthera* (Apocynaceae) for cardiac glycosides. He noted that three subspecies of *A. schimperi* contain acovenoside A, whereas a fourth related taxon, *A. oubaio*, contained oubain instead. The chemical differences between these two cardenolides are that the former lacks the CH_2OH and OH groups present at $C_{(10)}$ and $C_{(11)}$ in oubain and has as sugar 3-*O*-methylrhamnose instead of rhamnose. These differences mean that oubain is much more potent as an arrow poison. Reichstein mentions that the natives of East Africa are able to select without difficulty the oubain-containing taxon, in spite of the fact that morphological differences between the taxa, which occur together, are relatively few.

IX. Conclusion

There are seven major classes of toxin in plants (Table 6.7) which are of interest in plant taxonomy. The largest group are the alkaloids, but other important nitrogen-based toxins are the non-protein amino acids and the cyanogenic glycosides. The largest class of non-nitrogenous toxin are those based on the triterpenoid skeleton and of these the most toxic members are the cardiac glycosides. Although the cardiac glycosides have long been known to be poisonous, at least in significant doses, the harmful effects in man of contact with sesquiterpene lactones have only been realized quite recently.

Table 6.7 The major classes of plant toxins

Class (numbers of known structures)	Toxic activities	Organ distribution	Taxonomic occurrence
Nitrogenous toxins			
Alkaloids (6000)	Many poisonous, hypnotic, paralytic, etc.	Leaves; also roots and fruits	Almost exclusive to angiosperms, in 20% of families
Cyanogens (30)	Arrests respiration by blocking cytochrome system	Mainly leaves; and also other tissues	Widespread but sporadic
Non-protein amino acids (300)	Neurotoxic, hypoglycaemic, etc.	Mainly seeds	Especially Leguminosae, also several other families
Non-nitrogenous toxins			
Iridoids (100)	Insecticidal, antimicrobial	Mainly leaves	In some 70 dicotyledonous families.
Sesquiterpene lactones (600)	Allergenic, cytotoxic	Leaf trichomes	Almost exclusive to the Compositae
Diterpenoids (600)	Co-carcinogenic, irritant	Mainly resins	Common in gymnosperms, restricted in angiosperms
Triterpenoids and steroids[a] (3000)	Haemolytic, cardioactive	Leaves, roots, bulbs	Widespread

[a]Includes many subclasses, such as saponins, cardiac glycosides, phytosterols and tetracyclic triterpenes

Many of the compounds mentioned in this chapter, although hidden away within the leaf or other tissue, are apparent to the observer through taste, touch or smell. Thus, they may impart a bitter or disagreeable taste to leaf or seed, when present in these tissues in any quantity. Their presence may also be apparent, especially in the case of alkaloids, in that ingestion of the plant may produce intense physiological effects. These effects may be pleasant (e.g. hallucinogenic, analgesic, etc.) or unpleasant (e.g. hypnotic, paralytic, etc.). Colour and other simple tests have been devised for screening plants for toxins, but such tests are usually fallible to some degree and they need to be combined with more rigorous procedures, based on TLC and spectroscopic analysis. Electrophoresis is a useful procedure for screening plants for certain classes of alkaloid and also for non-protein amino acids.

From the taxonomic viewpoint, complete chemical characterization of the toxins of a given plant is now essential. It is not enough to know that alkaloid is present; it must be established that a particular type of alkaloid, formed along a particular biosynthetic pathway, occurs in the plant in question. Only then can toxins be used meaningfully for systematic comparison. Chemical identification is also vital in the case of cyanogens; it is not enough simply to record a plant species as responding positively to the picrate paper test. Knowledge that the known cyanogenic glucosides belong to one or other of five biogenetic classes has increased their systematic potential considerably and new reports have to be fitted into this framework of present advance.

Almost all the compounds discussed here are of limited distribution. One group of terpenoids, the sesquiterpene lactones, only have a significant occurrence in one plant family, the Compositae. The taxonomic value of these and the other toxins is limited to the lower levels of classification, where they may provide tribal, generic, sectional or species characters. In spite of extensive attention paid to the chemotaxonomy of plant alkaloids, it is difficult to point to any major advance in understanding systematic relationships that has emerged from alkaloid characters. Nevertheless, they have considerable potential. In the case of the cyanogenic glucosides, it is possible to mention at least one family, the Rosaceae, where they represent important markers at tribal and generic limits. Non-protein amino acids have proved to be interesting taxonomically in almost every family in which they accumulate. This is particularly so in the case of the Leguminosae, where very extensive and detailed surveys of seed tissue have been developed and are being continued.

Of the various non-volatile terpenoids mentioned in this chapter, the triterpenoids are probably the most widely used as taxonomic markers, largely because of their regular presence in most angiosperm families. Diterpenoids have so far only been exploited systematically within the gymnosperms, and here largely at the generic or specific level. By contrast, the iridoids, which are confined to the dicotyledons, have proved to have been of taxonomic value at ordinal and familial, as well as generic, levels. Finally, the sesquiterpene lactones, because of their limited natural occurrence, clearly have not yet proved to be outstanding taxonomic markers. Although their distribution within the Compositae is still largely enigmatic (Seaman, 1982), yet it would be difficult to ignore these fascinating molecules in any modern approach to the systematics of this diverse family.

7
Plant Pigments

I.	Introduction	128
II.	Anthocyanin pigments	130
III.	Yellow flavonoids	140
IV.	Colourless flavonoids	146
V.	Betalain pigments	159
VI.	Quinone pigments	163
VII.	Carotenoid pigments	172
VIII.	Conclusion	177

I. Introduction

In spite of the infinite variety of plant colours in nature, pigmentation of flowers or other tissues has only had limited application in classical taxonomy. Even today in formal descriptions of new taxa, there may be no record of flower colour or of any other special colouring that may be present. By contrast, popular floras with colour plates (e.g. McClintock and Fitter, 1956) may use flower colour both for general classification and also for keying out species within a genus or family. Probably the main reason why flower colour has not been much employed by classical taxonomists for purposes of identification, or as a hierarchical feature, is simply because of its instability on the herbarium sheet. It is an unreliable character in dried plants since colours are frequently lost during drying or else a significant shift in colour may take place. Even when extracted from fresh plant tissue, pigments can be unstable in solution and present a problem to phytochemists studying them.

Another reason why flower colour, in particular, has not often been adopted in the past for classificatory purposes is the genetic variation that may be apparent. Flower colour variation is a characteristic feature of certain plant species in natural habitats. Mutation from a coloured to a white form may occur (e.g. in campion or bluebell) or various shades of cyanic colour may be present (e.g. in comfrey, *Symphytum officinale*). The problem of infraspecific flower colour variation, however, has often been exaggerated. Such colour variation is only really marked in plant species after they have been brought into cultivation and where colour forms have been deliberately selected and preserved. There is little evidence, in fact, to indicate that colour, or the pigments responsible for colour, is any worse than any other biological character from the view point of genetic instability.

Probably the most valid objection to the uncritical use of colour description as a taxonomic marker lies in the fact that the same colour may be produced in different plants by different chemical structures. The most dramatic example is the case of members of the order Centrospermae, where purple colours, which are usually based on anthocyanin pigments, are formed in most of these plants by structurally unrelated betacyanin pigments. Spectrally and visually, the two classes of purple pigment are difficult to tell apart; it is only after detailed chemical analysis (e.g. the use of electrophoresis) that it is possible to distinguish them. The taxonomic significance of the replacement of anthocyanin by betacyanin in most families of the Centrospermae will be discussed in section V of this chapter.

Other cases of 'chemical mimicry' in flower pigmentation are also known. Yellow flower colour is a characteristic feature of the family Compositae and one might suppose it to be uniform chemically throughout the family. In fact, the pigments responsible for yellow do vary considerably. Yellow flowers may contain carotenoids, mixtures of carotenoids and yellow flavonoids, or pure yellow flavonoids. The yellow flavonoid may be a chalcone, an aurone, a flavone or a flavonol. Furthermore, there may be considerable intricacy in the distribution of these yellow pigments variously in different parts of the flower head.

Although colour description can thus be an unreliable guide in classification, the chemical analysis of the pigments responsible for colour does provide a useful approach to systematic problems. Mixtures of pigments are the rule rather than the exception so that a range of chemical characters may be available for taxonomic comparison. In fact, anthocyanins and related flavonoids have been widely surveyed in flower and fruit and much information is now available on their distribution in plants. Included with the anthocyanins are the structurally related flavones and flavonols. Although only weakly coloured in the visible region, these substances absorb strongly in the ultraviolet region and some insects respond to their presence in flowers.

Besides pigments obviously present to the naked eye, colour may be present hidden in tissues within the plant, under the bark, or in underground organs, in roots and tubers. For example, quinone pigments are common in rhisomatous rootstock of sedges (Cyperaceae), whereas anthocyanins are frequent in roots of *Anthemis* and other Anthemideae. Another type of 'hidden' colour is the presence of colourless substances in plants which give rise to colour during extraction or when extracts are heated or treated with acid. One class mentioned here are leucoanthocyanidins (or proanthocyanidins), which yield cyanidin on acid treatment. It is important to distinguish between cyanidin so formed and its production on hydrolysis of visually coloured anthocyanin pigments.

Historically, the chemotaxonomy of higher plant pigments has lagged behind related studies in lower plants. In algae, the various classes have long been distinguished on the basis of their colours: red for the Rhodophyta, green for the Chlorophyta, blue–green for the Cyanophyta and so on. These differences in outward appearance are reflected in the complex chemistry of their pigments. Even the green chlorophylls, which are constantly the same throughout the higher plants, show some chemical variation in different algal groups (Stewart, 1974). Lichenologists have also long been dependent on pigment analyses for purposes of identification and classification. A range of special depsidone pigments are present in lichens, together with anthraquinones (see C. F. Culberson, 1969) and their utility in classification has been frequently reviewed (see, e.g., Hale, 1974).

Many different pigments have been identified in higher plants; an authoritative account of their chemistry and biochemistry is provided in the text of Goodwin (1976a). Other more specialized reviews will be mentioned later in this chapter. For convenience, plant pigments will be considered under the following main headings: flavonoids, divided into anthocyanins, yellow flavonoids and colourless flavonoids; betalains; quinonoids; and carotenoids. These divisions represent biosynthetic differences and also functional variations. The role of different pigments in flowers in relation to pollination ecology must be borne in mind when considering taxonomic aspects. Anthocyanin flower pigmentation has clearly been subject to processes of natural selection, so that evolutionary trends can be discerned among angiosperms towards blue colour in temperate floras and scarlet colour in tropical floras. The relationship between the chemistry of flower pigments and pollination vectors has been summarized elsewhere (Harborne, 1982). In the present chapter, the chemistry and distribution of the different classes of pigment will be reviewed and methods of detection will be briefly considered, their chemotaxonomic utility will be mentioned and various examples given to illustrate their chemotaxonomic potential.

II. Anthocyanin pigments

A. Chemistry and distribution

The anthocyanins are the most important and widespread group of colouring matters in plants. These intensely coloured water-soluble pigments are responsible for nearly all the pink, scarlet, red, mauve, violet and blue colours in the petals, leaves and fruits of higher plants. The anthocyanins are all based chemically on a single aromatic structure, that of cyanidin, and all are derived from this pigment by addition or subtraction of hydroxyl groups or by methylation or by glycosylation (see Fig. 7.1).

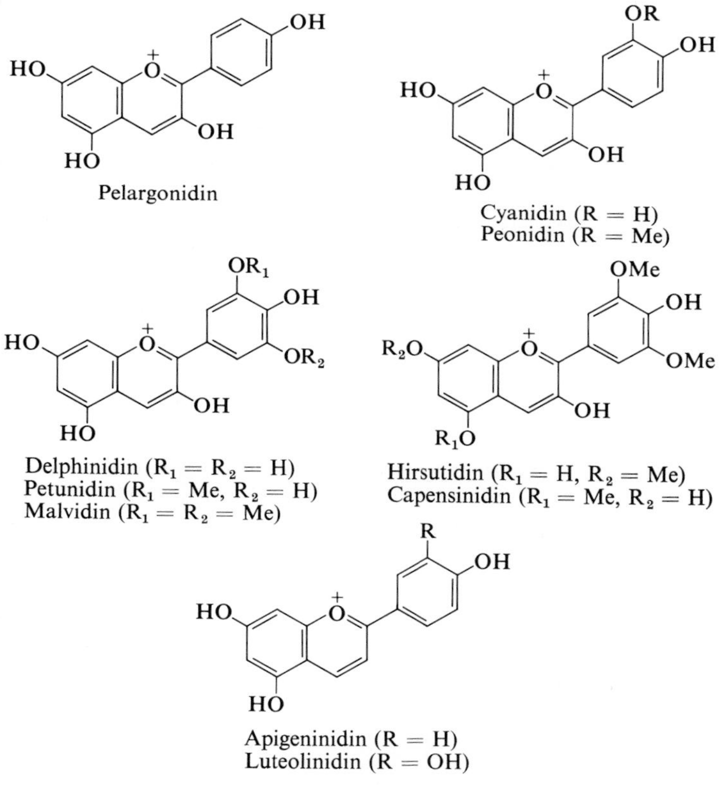

Pelargonidin

Cyanidin (R = H)
Peonidin (R = Me)

Delphinidin ($R_1 = R_2 = H$)
Petunidin (R_1 = Me, R_2 = H)
Malvidin ($R_1 = R_2$ = Me)

Hirsutidin (R_1 = H, R_2 = Me)
Capensinidin (R_1 = Me, R_2 = H)

Apigeninidin (R = H)
Luteolinidin (R = OH)

Fig. 7.1 The structures of plant anthocyanidins.

There are six common anthocyanidins (anthocyanin aglycones formed when anthocyanins are hydrolysed with acid), the magenta-coloured cyanidin being by far the most common. Orange–red colours are due to pelargonidin with one *less* hydroxyl group than cyanidin, whereas mauve, purple and blue colours are generally due to delphinidin, which has one *more* hydroxyl group than cyanidin. Three anthocyanidin methyl ethers are also quite common: peonidin derived from cyanidin; and petunidin and malvidin, based on delphinidin. Among the rarer anthocyanidins in plants are those with

O-methylation at the 5- or 7-hydroxyl groups (e.g. capensinidin from *Plumbago capensis*, hirsutidin from *Primula hirsuta*) and those that lack a hydroxyl residue at the 3-position and which are called 3-desoxyanthocyanidins (e.g. apigeninidin and luteolinidin; see Fig. 7.1).

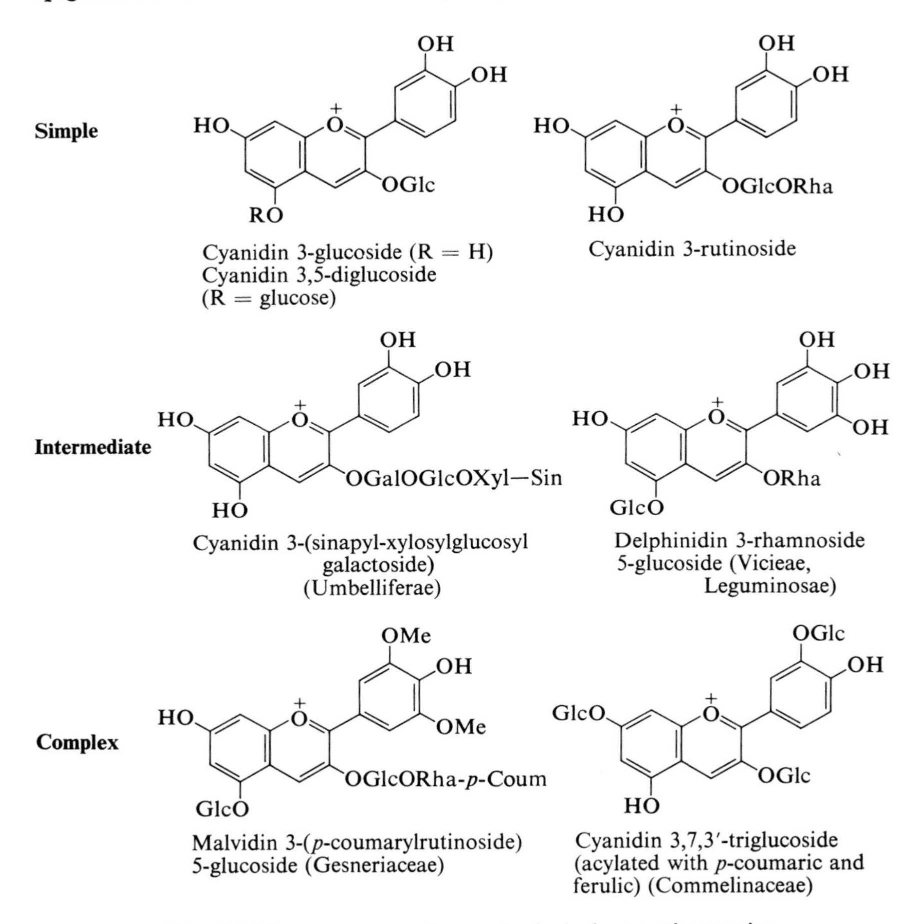

Fig. 7.2 The structures of some typical plant anthocyanins.

Each of the above anthocyanidins occurs with various sugars attached as a range of glycosides (i.e. as anthocyanins). The main variation is in the nature of the sugar (often glucose, but may also be galactose, rhamnose, xylose or arabinose), the number of sugar units (mono-, di- or triglycosides) and the position of attachment of sugar (usually to the 3-hydroxyl, or to the 3- and 5-hydroxyl groups; see Fig. 7.2). In a few rare cases, sugars may be attached to the 7- or 3′-hydroxyl group. A further structural complication in some anthocyanins is the presence of aliphatic or aromatic acyl groups. These acylated anthocyanins generally have the acyl group or groups attached through one of the sugar hydroxyl groups. Some typical anthocyanin structures are illustrated in Fig. 7.2.

Delphinidin, cyanidin and, to a much lesser extent, pelargonidin may also be formed in acid-hydrolysed plant tissue from colourless polymeric tannins present, from compounds originally known as leucoanthocyanidins and now known as proanthocyanidins or flavolans. Production of anthocyanidin in this way constitutes the main method for detecting these colourless substances in plants. They are regularly present in wood and leaf of woody plants, but very often appear in the flowers of these plants. In such cases, if the tissue is already coloured, it is not always clear in preliminary analysis whether the anthocyanidin detected is derived from leucoanthocyanidins or from anthocyanin pigment already present.

Although anthocyanins are almost universal in vascular plants (they have been detected in a few mosses, in young fern fronds as well as in gymnosperms and angiosperms), they are replaced by a superficially similar group of pigments, the betacyanins, in one order of higher plants, the Centrospermae. The replacement of anthocyanin by betacyanin in this order is not complete, since two families, the Caryophyllaceae and Molluginaceae, retain cyanic pigmentation. The taxonomic significance of this replacement of anthocyanin is discussed more fully in section V of this chapter.

Although anthocyanin pigments may not be readily detectable in every species of flowering plant, the ability to synthesize anthocyanin is undoubtedly universally present. Thus anthocyanin can often be induced in otherwise anthocyanin-free tissue by physiological means. Also, anthocyanin synthesis may develop in tissue culture even in cases when the whole plant from which the culture is derived lacks visible pigmentation, for example, *Machaeranthera gracilis* (Compositae) (see Harborne, 1979b).

Anthocyanins are formed biosynthetically, as are other flavonoids, by the condensation of three malonyl-CoA units with *p*-coumaryl-CoA. The first product of this reaction is a chalcone and this intermediate is isomerized to a flavanone, further oxidized to a dihydroflavonol and this is modified to yield anthocyanidin. The final stages in synthesis include *O*-methylation, glycosylation and acylation. An outline of the pathway to cyanidin and to other structurally related flavonoids is given in Fig. 7.3. Most of the enzymes of the pathway have now been characterized, although the exact mechanism by which the dihydroflavonol is converted into anthocyanidin remains obscure.

The main function of anthocyanins is to provide different flower colours for purposes of attracting various pollinators to the plant. In fact, flower

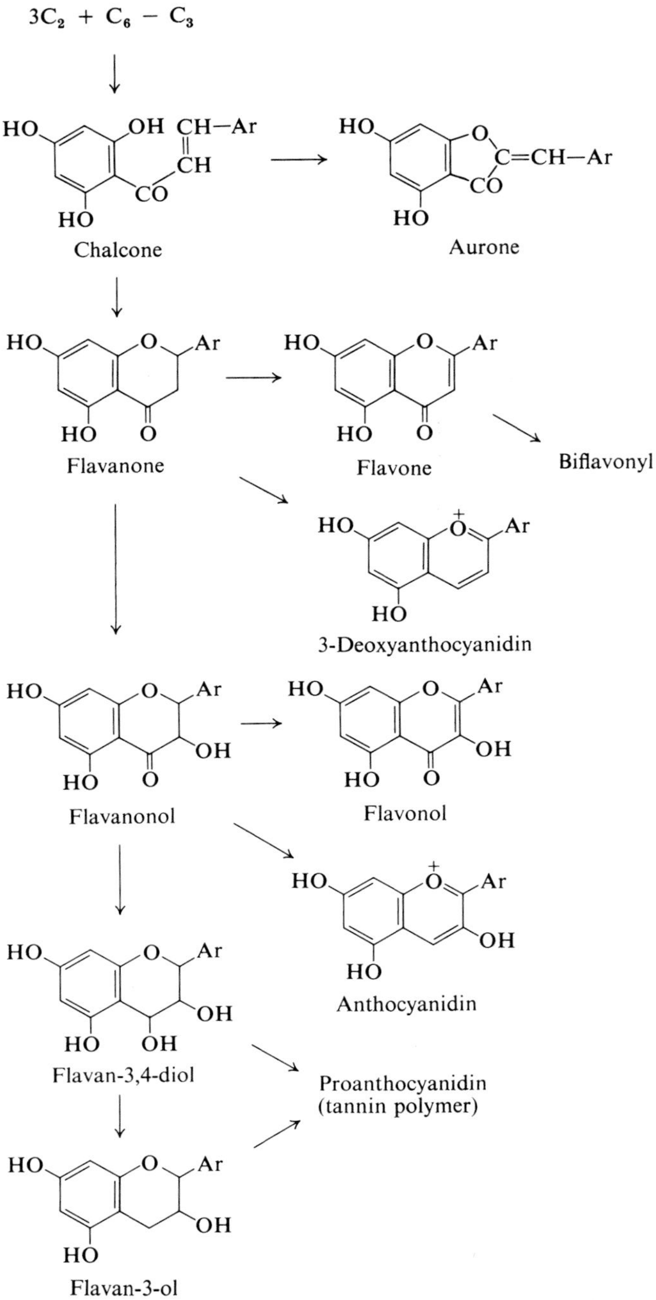

$3C_2 + C_6 - C_3$
HO
OH
CH—Ar
CH
CO
HO
Chalcone
HO
O
C=CH—Ar
CO
HO
Aurone
HO
O
Ar
HO
O
Flavanone
HO
O
Ar
HO
O
Flavone
Biflavonyl
HO
O
Ar
HO
3-Deoxyanthocyanidin
HO
O
Ar
OH
HO
O
Flavanonol
HO
O
Ar
OH
HO
O
Flavonol
HO
O
Ar
OH
HO
Anthocyanidin
HO
O
Ar
OH
HO
OH
Flavan-3,4-diol
Proanthocyanidin
(tannin polymer)
HO
O
Ar
OH
HO
Flavan-3-ol

Fig. 7.3 Biosynthetic pathway to the flavonoid pigments.

colours and anthocyanidin type (i.e. pelargonidin, cyanidin or delphinidin) are correlated with the colour preferences of different pollinating vectors. In Polemoniaceae, for example, species pollinated by humming birds contain pelargonidin with some cyanidin, lepidopteran-pollinated species contain mainly cyanidin and bee-pollinated species mainly delphinidin in the flowers (Harborne and Smith, 1978a). This function has to be borne in mind when considering the taxonomic utility of floral anthocyanins. In fact, although the anthocyanidin type may vary with pollinating vectors within a given family, the nature of the glycosidic attachment is usually much more constant and may well be of systematic significance.

A more detailed account of the plant anthocyanins may be found in Harborne (1967a). The distribution of anthocyanins in plants has been reviewed by Harborne (1963a, 1967a), Timberlake and Bridle (1975) and Hrazdina (1982).

B. Methods of detection

Anthocyanins are water-soluble pigments, best extracted from fresh plant tissue, although recently dried plants may yield satisfactory extracts. Pigment solutions are readily obtained by macerating flower or fruit tissue with methanol containing 1% (v/v) concentrated HCl. Leaf tissue is preferably extracted with aqueous HCl so as to avoid contamination with chlorophylls. In larger scale separations, crude extracts are conveniently purified by ion-exchange chromatography.

Anthocyanins are readily differentiated from betacyanins by their greater stability to hot acid and their different behaviour on electrophoresis (see section V, p. 161). Known anthocyanins can be identified by direct comparison with authentic markers (see Harborne, 1967a, 1973b). Different pigments have characteristically different chromatographic properties and absorption spectral characteristics. Further identification depends on hydrolytic studies and the detection of the respective anthocyanidin, sugars and acyl groups (if present). NMR and MS procedures have not been widely applied to anthocyanins because of low volatility and insolubility in preferred solvents, but newer techniques (e.g. fast atom bombardment MS) are becoming available which work well with these substances.

Earlier surveys of anthocyanins in plants were carried out using simple colour solubility tests on crude extracts (Lawrence *et al.*, 1939). More accurate surveys have subsequently been based on paper chromatographic procedures (e.g. Hayashi and Abe, 1956). In general, there are limits to the information available from such surveys, particularly regarding the glycosidic patterns. Since the combined sugars are usually of taxonomic interest, it is essential wherever possible to identify these rigorously in every taxon after isolating the anthocyanins in the pure state.

C. Taxonomic utility

1. *General*

In an earlier review of the distribution of anthocyanin pigments in higher plants, it was suggested (Harborne, 1963a) that the glycosidic pattern of anthocyanins 'was a more useful character than that of anthocyanidin type, because it is much less variable genetically'. Subsequent work has substantiated this view. Anthocyanin characters continue to be most useful at the lower levels of classification and it is the variation in glycosidic complexity which is of most taxonomic significance. In addition, there are a few rare anthocyanidins known in plants and these are of taxonomic interest at higher levels of classification.

A major problem with using anthocyanin characters is the limited information available on their natural occurrence. In fact, the anthocyanins have only been adequately characterized in some 200–300 species of angiosperms (Harborne, 1967a; Timberlake and Bridle, 1975; Hrazdina, 1982). Many families have still not been studied at all. The fact that living plants in flower are needed for anthocyanin analysis is also a limitation. This can be circumvented by extracting flowers in the field (even in the tropical jungle), storing the extracts on chromatograms and then developing the chromatograms later in the laboratory (Lowry, 1972). One advantage of the newer chromatographic methods is that it is possible to identify anthocyanins using very limited amounts of tissue and, in favourable cases, the extract from a single flower.

2. *Anthocyanidin type*

Certain plant families (e.g. the Plumbaginaceae, Primulaceae and Gesneriaceae) are distinguished by having present anthocyanidins of unusual structure; in such cases, the distribution of the rare pigments appears to be of taxonomic interest. In the Plumbaginaceae, the unusual structural feature is the presence of a 5-*O*-methyl group and the 5-*O*-methyl ethers of delphinidin, petunidin and malvidin have been reported variously from *Plumbago* (in four of five cyanic species) and *Ceratostigma* (in both species examined) (Harborne, 1967b). These three rare anthocyanidins are accompanied by the structurally related flavonols (e.g. azaleatin, 5-*O*-methylquercetin) in the same and related plants. The taxonomic interest is that these flavonoids are only found in members of the other tribe in the family, the Staticeae. On its own, this unusual 5-methylation would not be particularly significant. However, there are many biological features and also other chemical characters (e.g. plumbagin, see section VI, p. 164) which separate the Plumbagineae from the Staticeae. So much so that Linczevski (1968) has proposed that the *combination* of biological and chemical differences between the tribes justifies raising them to familial rank.

It is interesting that the Primulaceae, a family placed close to the Plumbaginaceae in most systems, is also distinguished by having unusual methylated anthocyanidins. Here, methylation is at the 7-hydroxyl residue and two novel pigments characterize the genus *Primula*: hirsutidin (7-methylmalvidin) and rosinidin (7-methylpeonidin). 7-Methylation of anthocyanidins is not unique to Primulaceae, since hirsutidin has been detected in flowers of the periwinkle, *Catharanthus roseus* (Apocynaceae) (Forsyth and Simmonds, 1957). A report by Thakur and Ibrahim (1974) of hirsutidin in *Linum usitatissimum* could not be substantiated in a more rigorous investigation (Dubois and Harborne, 1975).

Within Primulaceae, hirsutidin and rosinidin have so far only been found in the genus *Primula* and the closely related genus *Dionysia* (Harborne, 1968, 1969b). The two pigments were found in 30 of 41 cyanic *Primula* species and are so widely present that it is their absence from the subgenus *Auganthus* (inc. *P. obconica*, *P. sinensis*) that is noteworthy. Here they are replaced by the more common malvidin derivatives. These chemical results support Wendelbo's (1961) reconstruction of this subgenus and also fit in with cytological data, which are distinctive in these plants (Bruun, 1932).

Finally, in the case of the Gesneriaceae, the unusual anthocyanins are 3-desoxyanthocyanins. Three pigment aglycones are present: apigeninidin, luteolinidin and columnidin, the third compound being specific to the family and as yet not fully characterized. Apigeninidin and luteolinidin (for structures, see Fig. 7.1) are known elsewhere in plants in isolated instances (e.g. in the genus *Sorghum*, Gramineae), but their major occurrence is with columnidin in the Gesneriaceae, but only in the New World subfamily Gesnerioideae (Table 7.1). Here, they are almost universally present, occurring in 15 of 19 genera studied, and they largely replace the common anthocyanidins in these taxa. By contrast, 3-desoxyanthocyanidins have never been detected in the Cyrtandroideae, in spite of pigment investigations in 40 species of 15 genera.

The disparate distribution of these rare pigments in Gesneriaceae provides a positive contribution to the systematics of the family, since it confirms a revision of Von Fritsch's earlier subfamily classification, proposed by Burtt (1962) on the basis of geography (New World/Old World) and anisocotyly. In particular, *Columnea* was originally placed in the Cyrtandroideae because of the superior ovary. The presence of columnidin in all ten species studied fits in with its realignment with the Gesnerioideae because of its New World origin and its isocotyledonous seedling habit. There are also differences in yellow pigmentation between the two subfamilies (see section III, p. 143) which can be used to assign genera to one or other groups (Harborne, 1967b).

Table 7.1 Distinction between subfamilies in the Gesneriaceae in their anthocyanidin patterns

Genus (species surveyed)	Common anthocyanidins	Desoxyanthocyanidins
Subfamily Gesnerioideae		
Achimenes (1)	Pg, Mv	—
Alloplectus (1)	—	Lt, Co
Chrysothemis (1)	Cy	Ap, Lt
Columnea (10)	—	Co
Episcia (3)	Pg, Cy	Co
Gesneria (2)	Pg	Ap, Lt
Hypocyrta (1)	—	Lt
Koellekeria (1)	—	Co
Kohleria (2)	Pg	Ap, Lt, Co
Nautilocalyx (1)	—	Co
Rechsteineria (3)	—	Ap, Lt
Rhabdothamnus (1)	Pg, Cy	—
Rhytidophyllum (1)	Cy, Dp	—
Sarmienta (1)	—	Lt
Sinningia (3)	Pg, Cy, Mv	Co
Smithiantha (1)	Pg	Ap
Titanotrichum (1)	Cy	—
Trichantha (1)	—	Co
Subfamily Cyrtandriodeae		
Aeschynanthus (6)	Pg, Cy	—
Agalmyla (1)	Pg	—
Boea (1)	Cy, Dp	—
Chirita (2)	Mv	—
Cyrtandra (1)	Cy	—
Dichiloboea (1)	Dp, Pt, Mv	—
Dichotrichum (1)	Pg	—
Didissandra (1)	Mv	—
Didymocarpus (4)	Cy, Dp, Mv	—
Jerdonia (1)	Cy	—
Ornithoboea (1)	Pt, Mv	—
Paraboea (1)	Mv	—
Rhynchotechum (1)	Pt	—
Saintpaulia (1)	Cy, Mv	—
Streptocarpus (17)	Cy, Mv	—

Key: Common anthocyanidins, Pg = pelargonidin, Cy = cyanidin, Dp = delphinidin, Pt = petunidin, Mv = malvidin : desoxyanthocyanidins, Ap = apigeninidin, Lt = luteolinidin, Co = columnidin. Data from Harborne (1966, 1967a) and Lowry (1972); for simplicity, glycosidic variation has been omitted.

Variation in the type of common anthocyanidin is occasionally systematically interesting. This is true of pigmentation in the fruits of the holly genus *Ilex* (Aquifoliaceae). Four anthocyanins are present—the 3-glucosides and 3-sambubiosides of pelargonidin and cyanidin—but they do not occur together. Thus, in the subgenus *Aquifolium*, the two pelargonidin glycosides are the pigments in 10 of 11 species in section *Aquifolium*, whereas the two cyanidin glycosides provide colour in all nine species studied in the section *Lioprinus* (Santamour, 1973). A similar dichotomy between anthocyanidin type in fruit and sectional classification is also apparent in the subgenus *Prinos*. Although surveys have so far been limited, these data do suggest that the presence of cyanidin or pelargonidin in the fruit may be a distinguishing feature of value in future taxonomic revision of what is a large (450–500 species) and complex genus.

3. *Glycosidic patterns*

Although certain glycosidic patterns of the anthocyanins are common (e.g. 3-glucoside, 3,5-diglucoside) there are many more complex patterns with a variety of other sugars or acylated sugars which are of more restricted occurrence (Fig. 7.2). Such glycosidic patterns may show correlations with taxonomy, particularly when contrasted with the distribution of the more common patterns. In general, where sufficient species have been surveyed, it is apparent that a given genus is likely to be characterized by a particular glycosidic pattern (see Harborne, 1963a). In only a few cases (e.g. *Antirrhinum*, *Rhododendron*) are contrasting patterns present within the same genus. Not infrequently, a particular pattern may extend from a genus and typify a whole family. For example, a unique glycosidic pattern based on cyanidin 3-xylosylglucosylgalactoside and its acyl derivatives occurs in *Daucus* but also extends throughout the Umbelliferae (Harborne, 1976). Another type, 3-lathyroside (3-rhamnosylgalactoside) is also present in *Daucus* and this one even extends to a neighbouring family, the Araliaceae (Ishikura, 1975).

An example of glycosidic patterns being employed for taxonomic purposes is in the flower pigments of the tribe Vicieae, family Leguminosae. A rare type of glycoside, in which the 3-sugar is rhamnose instead of glucose, occurs regularly and consistently in species of *Vicia*, *Lathyrus*, *Pisum* and *Cicer*. Until recently, these genera, together with *Lens*, constituted the tribe Vicieae and it appeared that the presence of these rare anthocyanins was absolutely characteristic of the one tribe (Harborne, 1971). Very recently, however, the taxonomy has been revised (Kupicha, 1977) and *Cicer* has been separated into its own tribe Cicereae, with the comment that it represents a bridge-taxon between Vicieae and Trifolieae. In this context, it was clearly of interest to examine further members of the tribe Trifolieae and, as will be seen from Table 7.2, 3-rhamnoside-5-glucosides were found in two of five genera surveyed. It now appears, therefore, that the 3-rhamnoside character does have a distribution outside the Vicieae and, in fact, it provides a character that divides the genera of the Trifolieae into two groups (Table

7.2). Finally, it is worth noting that *Abrus*, a small atypical genus once included in the Vicieae, is clearly different in its anthocyanins (Table 7.2) and this confirms its present position in its own tribe, Abreae.

Table 7.2 Distribution of different anthocyanidin glycosides in genera of the Vicieae and Trifolieae

Tribe	Genus	Anthocyanidin 3-sugar[a] rhamnose	glucose
Vicieae	*Vicia*	+	—
	Lathyrus	+	—
	Pisum	+	—
Cicereae	*Cicer*	+	—
Trifolieae	*Parochetus*	+	—
	Trigonella	+	—
	Medicago	—	+
	Trifolium	—	+
	Ononis	—	+
Abreae	*Abrus*	—	+

[a]Anthocyanidins are commonly present with the sugar glucose in the 5-position. Data refer essentially to examination of petal pigments. Leaf pigments are occasionally of a different glycosidic type. Data on *Abrus* refer to seed coat pigmentation. (For reference, see Harborne, 1971.)

III. Yellow flavonoids

A. Chemistry and distribution

The term yellow flavonoid is used here to refer to those water-soluble compounds of the flavonoid type which contribute to yellow flower colour in plants. The name 'anthoxanthin' might be more appropriate for such substances, but this term, as far as it is still used today, has a wider connotation and refers principally to the colourless flavonoids, which are dealt with separately in section IV (p. 146).

Although yellow lipid-soluble carotenoids (see section VII, p. 172) are the most widespread yellow flower pigments, they can be completely replaced by yellow flavonoids in plants such as cotton, corn marigold, primrose and carnation. Not infrequently, yellow flavonoids co-occur with carotenoids, reinforcing the yellow impact of the lipid-soluble pigments. Where the two types of yellow pigment occur together, the yellow flavonoids may have a special role due to their absorption properties in the ultraviolet region, and they can act as honey guides to pollinating insects. This happens, for example, in *Rudbeckia hirta* flowers (Thompson *et al.*, 1972). In such cases, the flavonoid may be distributed only in the epidermal cells of the inner part of the corolla or ray, whereas the carotenoids are diffused throughout the petal. Like the carotenoids, yellow flavonoids also occur, often hidden by chlorophyll, in other parts of the plant and indeed, in general, they have a wider distribution in leaves than in flowers.

Yellow flavonoids are conveniently divided into four groups: chalcones, aurones, yellow flavonols and yellow flavones. The chalcones and aurones are together known as 'anthochlors', because they can be detected *in vivo* by means of the alkaline vapour of a cigar or a bottle of ammonia; when the yellow petal containing an anthochlor is so treated, there is a dramatic colour change to orange or red. Yellow-flowered angiosperm species were in fact surveyed for anthochlors by Gertz (1938) some time before the actual chemical structures were determined (Geissman, 1941). A typical chalcone is butein, the related aurone being sulphuretin; both classes of compound occur in petals as glycosides (Fig. 7.4). Chalcones and aurones are closely related chemically, aurones being formed from chalcones both in the laboratory and biosynthetically by an oxidative process. In fact, chalcone–aurone pairs are commonly, but not invariably, found in petal tissues. This is true particularly of their occurrence in the Compositae, the only family where they can be said to occur widely. Chalcones and aurones have been found in about 20 other plant families, mostly in isolated genera but occasionally (Onagraceae, Leguminosae) as a regular feature (Bohm, 1975).

Two series of yellow flavonols are known: those based on gossypetin (8-hydroxyquercetin) and those based on quercetagetin (6-hydroxyquercetin) (Fig. 7.4). Biosynthetically, they are formed by pathways similar to those leading to the common flavonols (see Fig. 7.3). Indeed, both gossypetin and quercetagetin are conceivably formed simply by the oxygenation in the 8- and 6-position respectively of quercetin as the precursor molecule. Both gossypetin and quercetagetin were originally isolated as yellow flower pigments from *Gossypium* and *Tagetes* spp. respectively and are present in glycosidic form. Methylated derivatives are also known, e.g. 8-methylgossypetin from *Lotus corniculatus* and patuletin (6-methylquercetagetin) from *Tagetes patula*. Gossypetin and its derivatives occur in at least 12 angiosperm families and they are also present in other phyla (e.g. in *Equisetum* spores). Quercetagetin, by contrast, has a lesser distribution in about eight families. The two types rarely co-occur, but both types are present in the Leguminosae and the Compositae (Harborne, 1975a).

Flavones related in structure to gossypetin and quercetin are known. 8-Hydroxyluteolin (hypolaetin) and 6-hydroxyluteolin are both yellow and contribute to the yellow colour in a few plants. 6-Hydroxyflavones are, however, more common as leaf than as flower pigments (Harborne, 1975a).

Another yellow flavone of a different structure is isoetin (2′-hydroxyluteolin) which provides yellow flower colour in *Hieracium pillosella* and several related Cichorieae (Harborne, 1978).

Chalcones

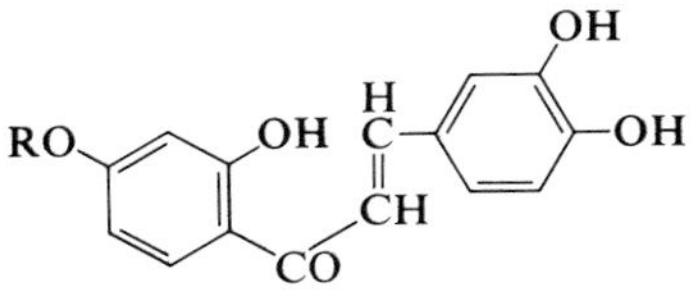

Butein (R = H)
Coreopin (R = Glc)

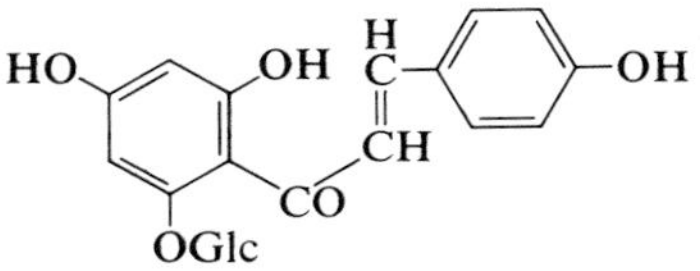

Isosalipurposide

Aurones

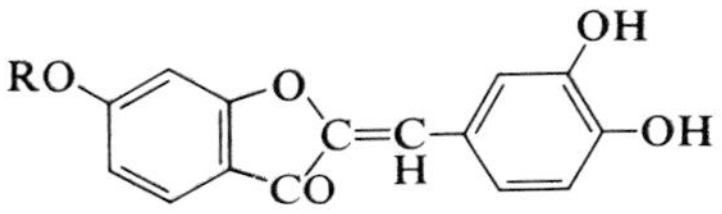

Sulphuretin (R = H)
Sulphurein (R = Glc)

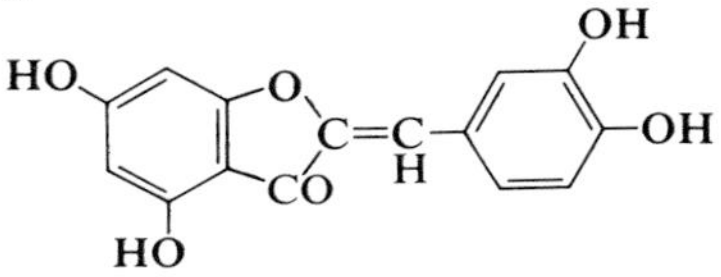

Aureusidin

Yellow Flavonols

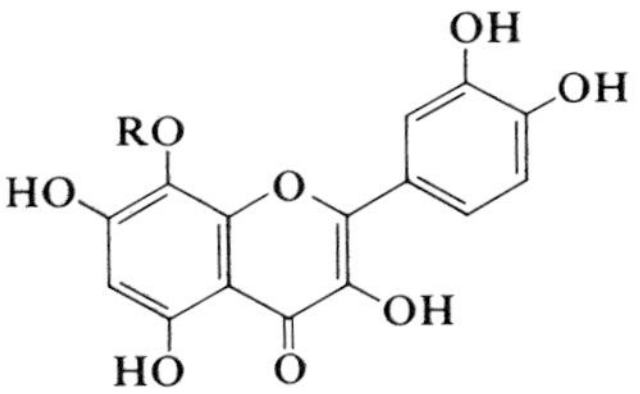

Gossypetin (R = H)
8-Methylgossypetin (R = Me)

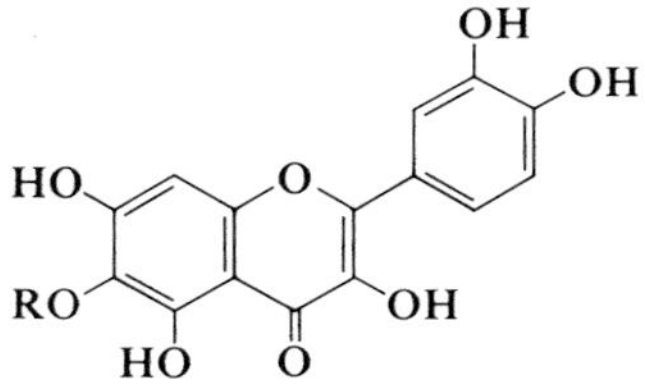

Quercetagetin (R = H)
Patuletin (R = Me)

Yellow Flavones

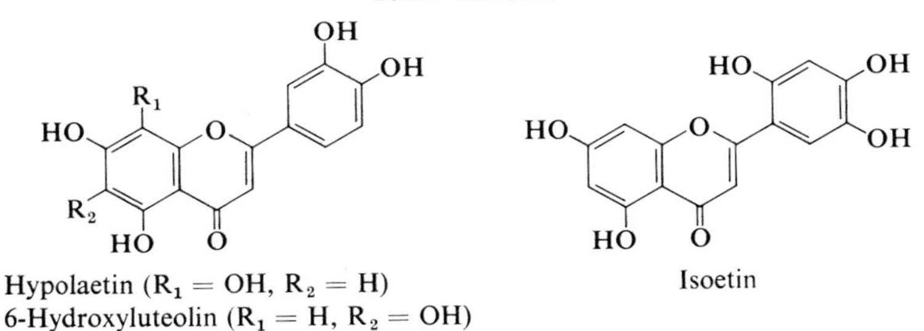

Hypolaetin (R_1 = OH, R_2 = H)
6-Hydroxyluteolin (R_1 = H, R_2 = OH)

Isoetin

Fig. 7.4 Some yellow flavonoids in plants.

B. Detection

Chalcones and aurones are detected by their characteristic visual colour change to orange–red in the presence of ammonia. Although yellow-flowered species can be screened for anthochlors with an ammonia bottle in the field, such tests are not infallible, and confirmatory tests must be carried out subsequently in the laboratory. The pigments should be separated chromatographically and their characteristic spectra determined. Chalcones and aurones can be distinguished from each other on paper chromatograms when examined by ultraviolet irradiation. The former appear as deep-brown absorbing spots, changing to deep red with ammonia, whereas aurones are bright yellow, changing to intense orange–red. Also, chalcones tend to lose their yellow visible colour when heated in acid (due to isomerization to flavanone), whereas aurones remain yellow.

Yellow flavonols and flavones can be recognized by their visual colours, coupled with their appearance on paper under ultraviolet irradiation as black absorbing spots that do not change when fumed with ammonia. They are distinguished from the much more common colourless flavonols by their spectral maxima which extend into the visible region. They tend to be unusually unstable in alkaline solution. Gossypetin can be identified during chromatographic surveys by the fact that it gives a blue colour on paper with alcoholic sodium acetate, whereas quercetagetin is unaffected (for details, see Harborne, 1973b).

As with other flavonoids (see section IV, p. 146), NMR and mass-spectral studies are valuable for confirming identifications, once the pigments have been obtained in quantity in the pure state.

C. Chemotaxonomic utility

1. *Chalcones and aurones*

Wherever these pigments occur regularly in a given plant family, there appears to be some correlation in their distribution with systematics. Thus, the chalcone isosalipurposide has been found in flowers of about one-third of 40 species surveyed in the Onagraceae. It is present in two tribes, the Onagreae (*Camissona*, *Gaura*, *Oenothera*, *Xylonagra*) and the Jussiaeeae (*Ludwigia*), being apparently absent from members of the other five tribes (Dement and Raven, 1973). It is correlated with the occurrence of yellow flower colour in *Oenothera* species, only being absent from white-flowered species (three of the eleven studied). Similarly, various chalcones and aurones have been recorded in yellow-flowered species of the Gesneriaceae, but only in members of the subfamily Cyrtandroideae (in seven genera); yellow-flowered members of the subfamily Gesnerioideae have carotenoids instead (Harborne, 1967c). Again, the aurone aureusidin has been identified as an inflorescence or fruit pigment in Cyperaceae, but only in Australian taxa (Clifford and Harborne, 1969), in spite of a wide survey of sedges from other geographical areas (Williams and Harborne, 1977a).

The most important occurrence of anthochlors from the taxonomic view point is in the family Compositae, where these substances, as far as present surveys have extended, have a significant distribution pattern (Harborne, 1977a). The greatest concentration of these compounds is in the subtribe Coreopsidineae of the tribe Heliantheae, for example, in *Bidens*, *Coreopsis*, *Cosmos* and *Dahlia* spp. (Crawford and Stuessy, 1981). Another major occurrence is in the genera *Baeria*, *Lasthenia* and *Syntrichopappus*, normally classified in the tribe Helenieae (Bohm, 1977). In a recent revision of the Compositae, however, the tribe Helenieae has been disbanded and its members distributed among other tribes (Turner and Powell, 1977). In this revision, it is highly significant that the three genera with anthochlors mentioned above have been re-allocated to the Coreopsidineae on morphological grounds. Clearly, the anthochlor chemistry supports this reassignment.

At and below the generic level, too, anthochlors are of interest in the Compositae. Their distribution in *Coreopsis* spp. has been extensively scrutinized and they are widespread. Among other interesting facets of the *Coreopsis* anthochlors, is the fact that they occur as very complex glycosides in the primitive sections of the genus, but as simple glucosides in the more advanced groups (Crawford, 1978). Advancement is thus by reduction in the complexity of the sugar components. Species of the section *Coreopsis* are marked off by their containing both marein and lanceolin, two chalcone glycosides otherwise found separately in other sections.

Finally, one might refer to the occurrence of coreopsin, the most common chalcone of *Coreopsis* spp., in all known species of the genus *Pyrrhopappus*. This is quite an unrelated source, since *Pyrrhopappus* belongs to the subtribe Microseridinae of the tribe Cichorieae (Northington, 1974; Harborne, 1977b). Within this subtribe, however, *Pyrrhopappus* is the only genus with a basic chromosome number of six and which grows in the Eastern United States. Pigment chemistry is thus closely correlated with chromosome number and geography, since the other genera of the subtribe (*Agoseris*, *Microseris*, *Krigia*) have yellow carotenoids, a chromosome number of five or nine and a habitat in California.

2. *Yellow flavonols and flavones*

Since it is relatively rare in its occurrence, gossypetin represents an interesting taxonomic marker at several levels of classification. It has a major occurrence in the family Ericaceae but here it is almost exclusive to the tribes Rhodoreae (four of five genera positive) and Phyllodoceae (seven of eight genera positive) (Harborne and Williams, 1973a). The relationships of the various tribes of the subfamily Rhododendroideae are still debated (see, e.g., Watson, 1965) and the chemical evidence would suggest that there is a special affinity between these two particular tribes. Gossypetin also occurs throughout the related Empetraceae (Moore *et al.*, 1970) but is not present in other associated families, namely Epacridaceae, Clethraceae and Diapensiaceae. It thus chemically links Empetraceae with Ericaceae, a link already apparent in the morphology, embryology and general habit of the two families.

Gossypetin is clearly a polyphyletic character within the angiosperms generally, since it occurs in a number of unrelated sources. It is present, for example, in the monocotyledonous family, Restionaceae. It is accompanied here by the 7-methyl ether, and by the related 8-hydroxyflavone, hypolaetin. It occurs widely in Australasian taxa (in nine of fourteen spp. surveyed) but is generally absent from South African species (33 spp. surveyed; Harborne, 1979b). In fact, geographical differences in the family are closely correlated with anatomical differences (Cutler, 1969), in spite of the fact that taxonomists have grouped Australian and South African species into the same genera. The gossypetin data thus reinforce the anatomical findings, which show that the family needs considerable taxonomic revision at the generic level.

A fourth family in which gossypetin occurs is the Primulaceae, and here it is restricted to three genera, *Primula*, *Dionysia* and *Douglasia*, which all belong to the Primuleae (Harborne, 1968). Within *Primula* spp., its occurrence follows sectional classification. It is uniformly present as a flower pigment in sections *Vernales* and *Sikkimenses* but is absent from yellow-flowered members of section *Candelabra* which have carotenoids instead. Its occurrence thus supports existing views of classification in *Primula*, without adding a new dimension.

Quercetagetin has a more limited distribution in plants than gossypetin and is mainly of interest as a taxonomic marker in the Compositae, where it occurs in several tribes (Harborne, 1975a). One significant occurrence is in the genus *Eriophyllum*, the flowers of which are pigmented by both quercetagetin and patuletin (Harborne and Smith, 1978b). This genus is traditionally placed in the Helenieae, but has recently been moved to the Senecioneae (Turner and Powell, 1977), although this reassignment has been disputed (Nordenstam, 1977). The chemical results support an association with the Heliantheae, since 6-hydroxyflavonols have not been found in the Senecioneae, but do occur regularly in *Iva*, *Parthenium* and *Xanthium* genera and in other Heliantheae (Swain and Williams, 1977). It is highly significant in this context that Baagøe (1977) on the basis of ligule microcharacters, also favours placing *Eriophyllum* in the Heliantheae.

A yellow flavone which has an interesting distribution is 6-hydroxyluteolin. It occurs widely in various sympetalous families (Harborne, 1975a). It is present, for example, in leaves of *Valerianella*, but only in six of the 40 species surveyed (Greger and Ernet, 1973). Its exclusive occurrence in these six Mediterranean species acts as a marker for this group, since all the remaining taxa have common flavones or flavonols. Cytological and morphological characters also separate these six species and the authors propose a revision of Boissier's original sectional classification in the light of the new findings. A number of other examples where 6-hydroxyflavones are of taxonomic importance at generic level could also be quoted. Their overall distribution in the angiosperms is, however, still far from clear and more surveys are needed to determine whether they have a significant pattern of occurrence at the higher level of classification.

IV. Colourless flavonoids

A. Chemistry and distribution

The majority of plant flavonoids are not directly coloured although they often contribute to flower and fruit colours. Thus flavonols and flavones are important co-pigments to the anthocyanins and are required in most flowers for the full expression of anthocyanin colour at the slightly acid pH of the cell sap of corolla tissues. Also, when present in sufficient concentration, they may shift purple or red anthocyanin towards the blue region so that blue flower colour is often due to a combination of anthocyanin and flavonol in the petals.

Flavonols and flavones also occur universally in white flowers, where their presence in the cell vacuole provides 'body' to the petals, so that they in fact appear cream, buff or ivory rather than dead white. Although practically colourless, these and other flavonoids considered in this section absorb strongly in the ultraviolet region of the spectrum. This means that they may be important functionally in flowers for attracting those pollinators (e.g. bees) whose vision extends into the ultraviolet region (Harborne and Smith, 1978a). It also means that such compounds, whether they occur in flowers or in leaves, are readily detected on chromatograms by the range of absorbing or reflecting colours they display on paper in ultraviolet irradiation.

The most common colourless flavonoids are the flavonols and flavones. Of minor importance, in terms of natural distribution, are the flavanones, dihydroflavonols, biflavonyls, dihydrochalcones, isoflavones and proanthocyanidins. The biosynthetic interrelationships of these various flavonoid classes are indicated in Fig. 7.3. Although not flavonoids *per se*, the plant xanthones will be included in this section, since they resemble flavonoids in their colour reactions and may also act as co-pigments to anthocyanins in certain flowers (e.g. *Iris* sp.).

Flavonols (Fig. 7.5) are very widely distributed in plants, both in flowers and in leaves. Like the anthocyanidins (see p. 131), they occur most frequently in combination with sugar as glycosides. Although some 150 flavonols are known, only three are at all common. Kaempferol (corresponding in structure to the anthocyanidin pelargonidin); quercetin (cf. cyanidin); and myricetin (cf. delphinidin). Like the corresponding anthocyanidins, these three flavonols are well separated by simple one-dimensional paper chromatography or TLC and their presence or absence in plants is easily screened. The other known flavonols are mainly simple structural variants of one or other of the three common flavonols and are generally of restricted occurrence. In the case of quercetin, a range of *O*-methyl ethers are known, the 3′-methyl ether (isorhamnetin) and the 5-methyl ether (azaleatin) being but two examples. The latter compound is distinctive in having an intense yellow fluorescence, particularly when chromatographed, so that it is unusually easy to recognize in plant extracts even when present only as a trace component.

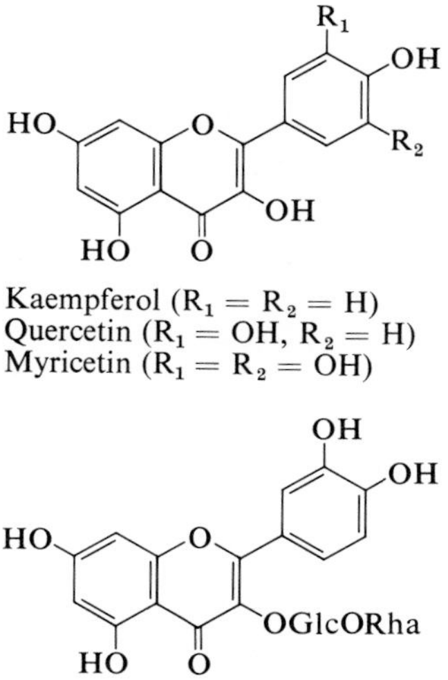

Kaempferol ($R_1 = R_2 = H$)
Quercetin ($R_1 = OH$, $R_2 = H$)
Myricetin ($R_1 = R_2 = OH$)

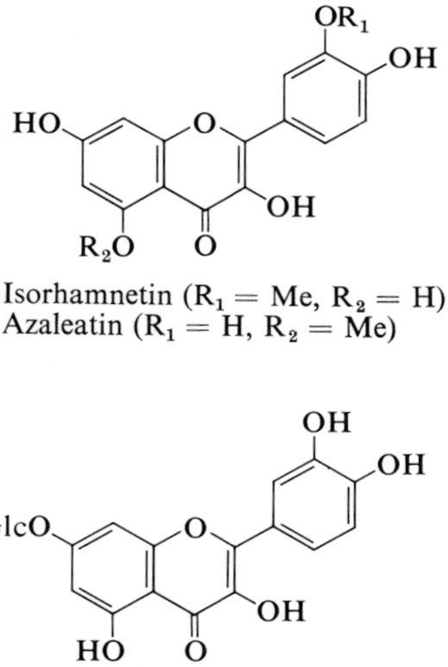

Isorhamnetin ($R_1 = Me$, $R_2 = H$)
Azaleatin ($R_1 = H$, $R_2 = Me$)

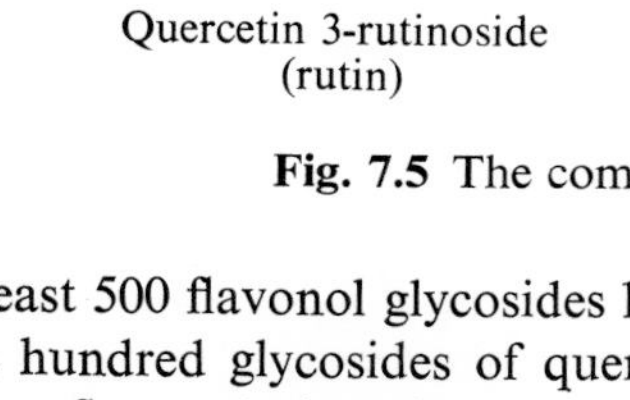

Quercetin 3-rutinoside
(rutin)

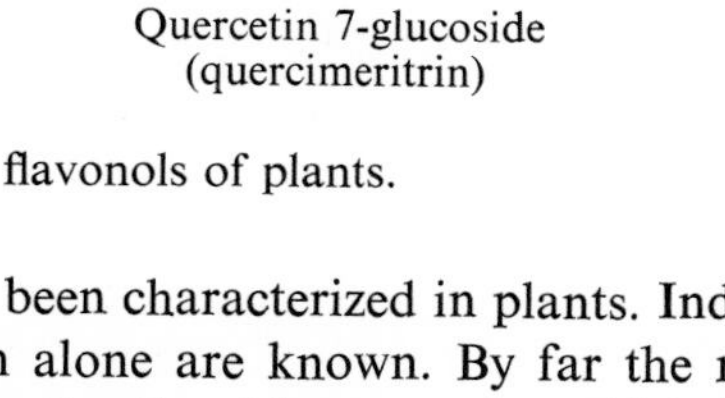

Quercetin 7-glucoside
(quercimeritrin)

Fig. 7.5 The common flavonols of plants.

At least 500 flavonol glycosides have been characterized in plants. Indeed, over a hundred glycosides of quercetin alone are known. By far the most common flavonol glycoside is quercetin 3-rutinoside, or rutin, which is so often found that it rarely provides a useful taxonomic indicator. Other well known glycosides are the 3-glucoside (isoquercitrin), the 3-rhamnoside (quercitrin) and the 7-glucoside (quercimeritrin). More complex flavonol glycosides include those with three or four different sugars attached and those with acyl substituents in addition.

Flavones (Fig. 7.6) differ from flavonols in lacking a 3-hydroxyl substitution; this affects their ultraviolet absorption, chromatographic mobility and colour reactions so that simple flavones can be distinguished from flavonols on these bases. Flavones tend to replace flavonols as the major leaf flavonoids in herbaceous plants, so that the ability to distinguish flavones from flavonols is often of systematic importance. There are only two common flavones, apigenin and luteolin; these correspond in hydroxylation pattern to kaempferol and quercetin. The flavone tricetin, the myricetin analogue, is known but it is of rare occurrence. The 3′,5′-dimethyl ether of tricetin, namely tricin, is more common, since it occurs widely in grasses, palms and sedges. Indeed, it is so regularly present in these three families that it acts here as a taxonomic character linking these groups.

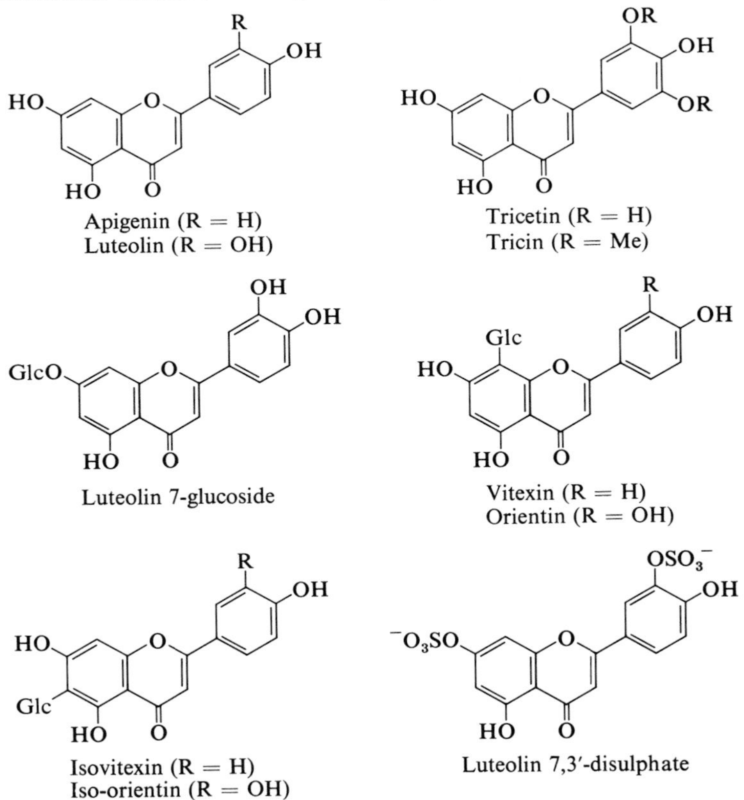

Fig. 7.6 Some common flavones and glycosylflavones of plants.

Flavones occur with sugar attached but the range of *O*-glycosides is less than in the case of the flavonols. A common type is the 7-glucoside, exemplified by luteolin 7-glucoside (Fig. 7.6). Flavones, unlike flavonols, occur remarkably with sugar bound by a carbon–carbon bond to the aromatic nucleus. A series of such glycosylflavones have been described, two examples being vitexin, the 8-*C*-glucoside of apigenin and orientin, the 8-*C*-glucoside of luteolin. 6-*C*-Glucosides are also known, as are glycosylflavones that have *O*-sugars attached through hydroxyl groups. The carbon–carbon bond in such substances is extremely resistant to acid hydrolysis, so it is possible to distinguish these *C*-glycosides from the *O*-glycosides which are more readily hydrolysed.

Flavonols, flavones and glycosylflavones occasionally occur naturally in plants in covalent combination with inorganic sulphate. Such sulphate conjugates carry a negative charge and are distinguished from other combined forms by their electrophoretic mobility in acid solution. Sulphated flavonoids have an interesting taxonomic distribution in both monocotyledonous and dicotyledonous families (Harborne, 1977c).

Among the minor flavonoids of plants are the biflavonyls, of which some 70 are known. These dimeric compounds are formed by carbon–carbon or carbon–oxygen coupling between two flavone (usually apigenin) units. Most also carry *O*-methyl substituents. A typical biflavonyl is ginkgetin (see Fig. 7.7), which occurs in the autumnal leaves of the Maidenhair tree, *Gingko biloba*. Biflavonyls occur characteristically in the gymnosperms, although they are notably absent from the Pinaceae and Gnetaceae. They also have a very limited distribution in a few angiosperms (especially in some Guttiferae) and in one or two non-seed plants (e.g. *Selaginella*).

Flavanones are isomeric with the chalcones (see p. 142) and the two classes are interconvertible *in vitro* and *in vivo*. Not infrequently, they are found together, although flavanones do occur sporadically in plants without their chalcone analogues. From the taxonomic viewpoint, the only important occurrence of flavanones is in the genus *Citrus*, where they occur throughout in a range of glycosidic forms. One such glycoside, naringin, is a soluble bitter principle of the Seville orange. Dihydroflavonols are related to the flavanones as flavonols are to flavones and they, like the flavanones, are of restricted occurrence. They are mostly known from heartwood tissues of trees, where they occur together with the related flavonols. Another reduced form of flavonoid are the dihydrochalcones, such as phloridzin (see Fig. 7.7). These are of very limited distribution, mainly being reported in Rosaceae, especially *Malus* (see p. 156) and in certain Ericaceae.

Isoflavones, of which nearly 200 are known, are isomeric with the flavones but are of much rarer occurrence. They are found almost entirely in one subfamily, the Papilionoideae, of the Leguminosae. There are also disparate reports of isoflavones in Amaranthaceae, Chenopodiaceae, Compositae, Moraceae, Myristicaceae and Rosaceae (Dicotyledoneae), in Iridaceae and Stemonaceae (Monocotyledoneae) and in Cupressaceae and Podocarpaceae (Gymnospermeae). Isoflavonoids can be divided into three groups according to their physiological activities. Compounds such as 7,4′-dihydroxyisoflavone (daidzein) and 5,7,4′-trihydroxyisoflavone (genistein) and their methyl ethers are weak natural oestrogens found in forage clovers. Complex isoflavans such as rotenone from derris root are powerful natural insecticides and piscicidal agents. Finally, there are the related coumestans such as pisatin, from pea leaves, which are phytoalexins, protective substances formed in plants in response to fungal disease attack (see chapter 15).

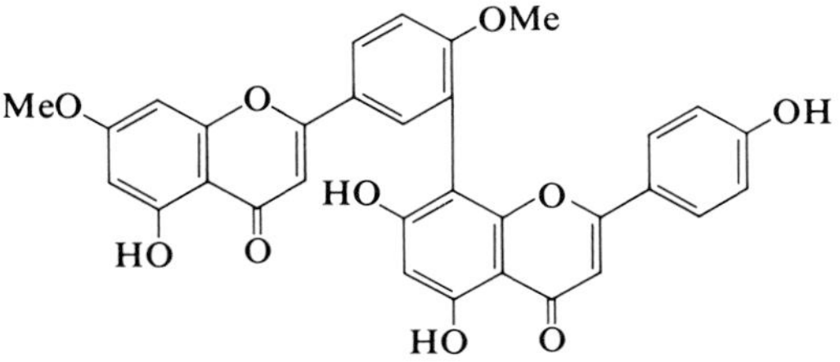

Biflavonyl: Ginkgetin

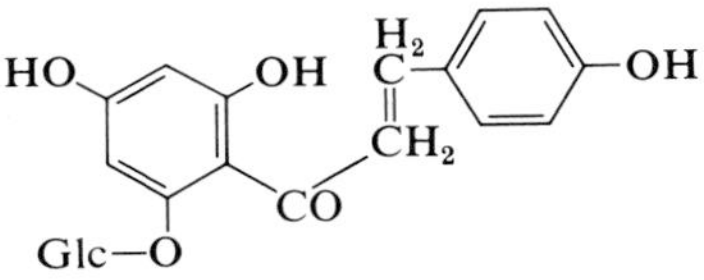

Dihydrochalcone: Phloridzin

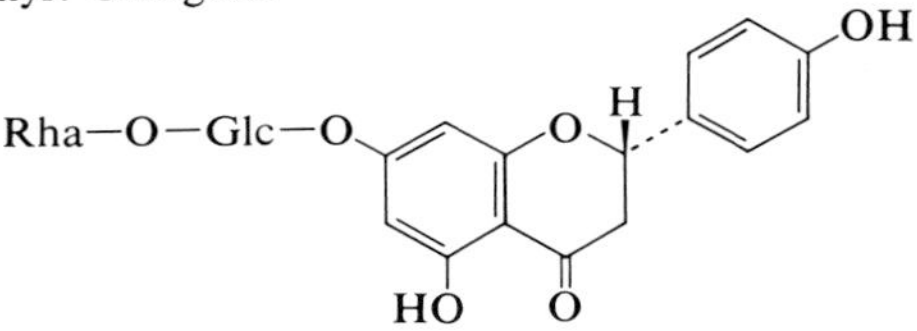

Flavanone: Naringin

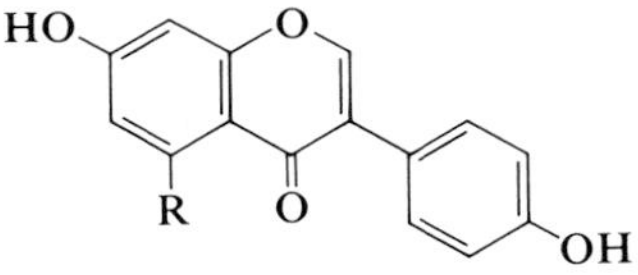

Isoflavones: Daidzein (R = H)
Genistein (R = OH)

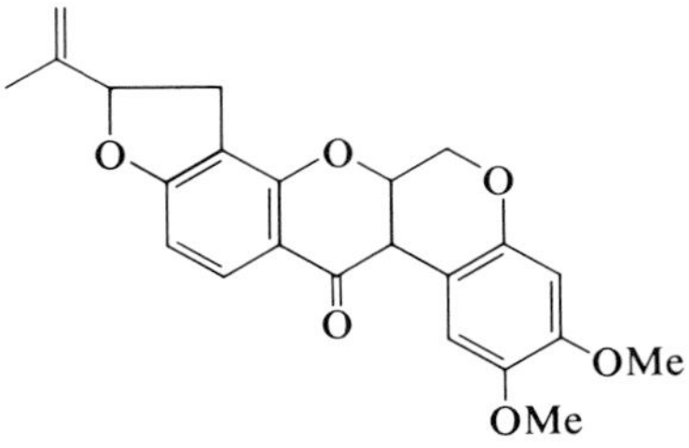

Isoflavanoid: Rotenone

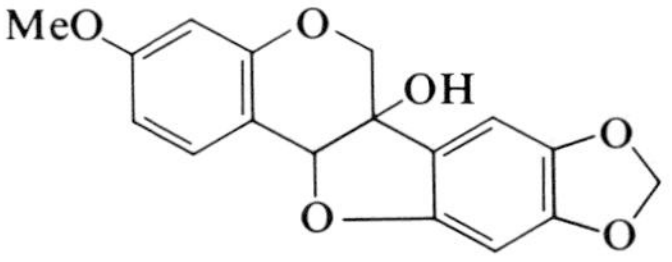

Coumestan: Pisatin

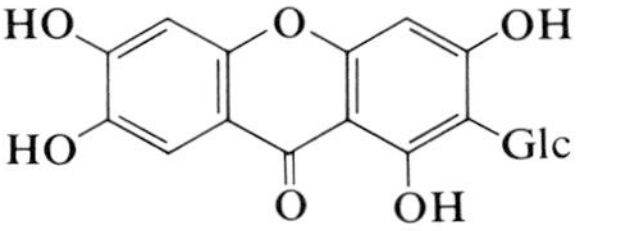

Xanthone: Mangiferin

Fig. 7.7 Minor flavonoids in plants.

The most widespread and abundant of the minor flavonoid classes are the proanthocyanidins, which occur in leaves of most woody plants and in ferns. They are oligomers and polymers, formed by carbon–carbon links between catechin monomers, which usually accompany them in the same plants. Proanthocyanidins are characterized by the identification of the coloured anthocyanidin liberated on treatment with hot acid. Procyanidins are most common, prodelphinidins occasionally occur and propelargonidins are relatively rare.

Xanthones are phenolic derivatives similar in colour reactions to flavones and flavonols. Chemically, they differ in that they are formed biosynthetically by the condensation of a phenylpropanoid (C_6–C_3) unit with two (not three) malonyl-CoA moieties. They are distinguished from flavonoids by their distinctive spectral properties. Almost all known xanthones are confined to four plant families: the Gentianaceae, Guttiferae, Moraceae and Polygalaceae. One particular xanthone, mangiferin, which is *C*-glycosylated, is much more widespread and it frequently occurs in conjunction with *C*-glycosylflavones or in place of them. According to Bate-Smith (1965), mangiferin has a vicarious distribution in plants. Although other xanthones often occur in the free state, a number of their *O*-glucosides have been reported in recent years (Hostettmann and Wagner, 1977). They are of chemotaxonomic importance in relation to the classification within the genus *Gentiana*.

A comprehensive and up-to-date account of the naturally occurring flavonoids is provided in the text of Harborne *et al.* (1975), with its later supplement (Harborne and Mabry, 1982). The distribution of xanthones in plants is reviewed by Carpenter *et al.* (1969) and by Rezende and Gottlieb (1973).

B. Detection

Colourless flavonoids are most conveniently detected in the first instance by two-dimensional paper chromatography of a direct alcoholic extract of a plant (for details, see Harborne, 1973b). Similar results can be achieved by TLC on microcrystalline cellulose. The positions the various flavonoids occupy on a two-dimensional chromatogram and their different colours under ultraviolet irradiation, with and without ammonia vapour, are indicated in Fig. 7.8. The basic flavonoid pattern present is then determined by acid hydrolysis of the crude extract or by direct hot-acid treatment of leaves, when the flavonoid aglycones are released and can be extracted into an organic solvent. These aglycones can then be identified by one-dimensional paper chromatography by direct comparison with markers. By this means, it is possible to say whether flavonols, flavones or glycosylflavones are present and which of the common representatives of these types occur. Further confirmation is provided by spectral studies; ultraviolet, mass and nuclear magnetic resonance spectroscopy are useful in that order.

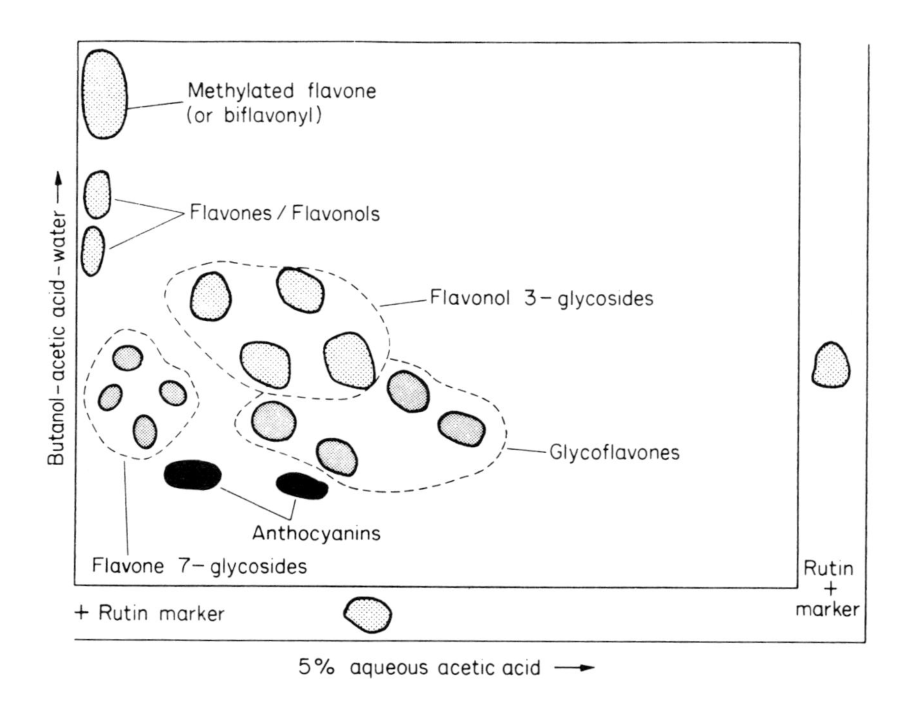

Fig. 7.8 Two-dimensional chromatogram of leaf flavonoids.

The identification of minor flavonoids is sometimes more difficult, since they may not be so readily apparent on paper chromatograms. Isoflavones are difficult to characterize in plant-screening programmes, since they do not respond to any one specific colour reaction. They are best separated by TLC and then detected as dull-purple absorbing spots, which respond positively to a general phenolic spray reagent. Flavanones can be detected as a class by the red colour developed in solution when they are reduced in alcoholic solution with magnesium and concentrated HCl; the reaction can be applied as a chromatographic spray.

General procedures for flavonoid identification can be found in Harborne (1967a, 1973b). The systematic identification of flavonoids by ultraviolet and NMR procedures is detailed in the texts of Mabry *et al.* (1970) and of Markham (1982).

C. Chemotaxonomic utility

1. *General*

The flavonoids are the most widely used of all secondary constituents in taxonomic studies. This is partly because of their ubiquitous occurrence in higher plants and of their great structural diversity. Also, they are stable chemicals, readily studied in small plant samples or even in tiny fragments of herbarium specimens. An even more compelling reason for their popularity with plant taxonomists is that they can be detected by paper chromatography without using chromagenic sprays and can be identified by a few simple procedures. Another advantage is that many plants have been surveyed for flavonoids and related phenolics (e.g. Bate-Smith, 1962) and much background information is already available on the flavonoid profiles of most plant families (see Hegnauer, 1962–1973). The distribution of different flavonoid classes at the ordinal level among angiosperms has recently been reviewed (Gornall *et al.*, 1979).

The ease of detecting flavonoids and other phenolics on two-dimensional chromatograms has led to their misuse as 'spot pattern' data, i.e. as a series of unknown chemical components. One of the dangers inherent in such an approach is that the same chemical compound may be 'counted' more than once (see Harborne, 1975b). In order to evaluate the significance of flavonoid occurrences, it is essential to at least partly identify the various substances present. In some cases, it may be sufficient to establish which *classes* of flavonoid are represented, i.e. the presence or absence of flavonols, flavones or glycoflavones. This is a minimal requirement in such studies.

Flavonoids have been employed as taxonomic characters at almost every level of systematic enquiry. It is thus not possible here to mention more than a few selected examples. The taxonomic value of flavonoids has been discussed elsewhere by a number of authors (see especially Alston and Turner, 1963a; Erdtman, 1963; Alston, 1967; Swain, 1975; Harborne, 1975b).

2. *Flavonoid patterns in Ulmaceae*

A simple but striking example where the analysis of flavonoid type within a family has led to a taxonomic reassessment is the recent survey of elm leaves by Giannasi (1978). Eighty species were examined and each of 15 genera was characterized by the presence of flavonols *or* glycosylflavones but not both (Table 7.3). This chemical dichotomy is remarkably compatible with the generally accepted division of the genera into the subfamilies Ulmoideae or Celtidoideae (see also Lebreton, 1965). The only exceptions are *Ampelocera*, *Aphananthe* and *Gironniera* (in part), which are usually regarded as celtoid but which possess the ulmoid flavonols. However, recent anatomical and morphological studies indicate that these three genera may not be as unambiguously celtoid as originally thought. *Ampelocera*, for example, although it possesses celtoid fruits, has ulmoid leaf venations and in floral anatomy is intermediate. Indeed, the chemical data suggest that all these genera should be regarded as intermediate between the two groups, until further biological studies have been accomplished.

Table 7.3 Flavonoid types identified in genera of the Ulmaceae

Subfamily[b]	and Genera (number of species studied)	Presence or absence[a] of:	
		Flavonols	Glycoflavones
Ulmoideae	*Hemiptelea* (1)	+	–
	Holoptelea (1)	+	–
	Phyllostylon (2)	+	–
	Planera (1)	+	–
	Ulmus (9)	+	–
	Zelkova (15)	+	–
Celtidoideae	*Ampelocera*	+	–
	Aphananthe	+	–
	Gironniera subg. *galumpita* (3)	+	–
	Gironniera subg. *Gironniera* (5)	–	+
	Celtis (2)	–	+
	Chaetachme (1)	–	+
	Lozanella (1)	–	+
	Parasponia (3)	–	+
	Pteroceltis (1)	–	+
	Trema (12)	–	+

[a]Flavonols identified: Kaempferol, quercetin and myricetin.
[b]Giannasi (1978) also looked at a further four genera, but these were all taxa of unclear affinities and they have been omitted for the sake of simplicity

3. *Heartwood flavonoids in pines*

One of the most thorough pioneering studies of flavonoids in relation to taxonomy has been that of Erdtman and his co-workers in Sweden, who have established a clear-cut correlation between the heartwood constituents of these trees and the subgeneric classification of *Pinus* into *Haploxylon* and *Diploxylon* species (Table 7.4). The survey extended to 52 of the known 90 species and thus represented an ascertainment of well over 50% (Erdtman, 1963). Chemically, the work was simplified by the absence of glycosylation, since the compounds occur in the heartwood in the free state. Nevertheless, a mixture of up to eight closely related flavonoids was variously present (Table 7.4).

Table 7.4 Distribution of flavonoids in *Pinus* heartwoods

Subgenus and subsection (no. of species surveyed)	Flavones			Flavanones			Flavanonols	
	1	2	3	4	5	6	7	8
Haploxylon								
Cembrae (3)	+	+	–	+	+	–	+	–
Flexiles (1)	+	+	–	+	+	–	+	–
Strobi (8)	+	(+)	(+)	+	(+)	(+)	+	(+)
Cembroides (1)	+	+	–	+	+	–	+	–
Gerardianae (2)	–	–	–	+	+	(+)	+	+
Balfourianae (2)	+	+	–	+	+	–	+	–
Diploxylon								
Leiophyllae (2)	–	–	–	+	–	–	+	–
Longifoliae (2)	–	–	–	+	–	–	+	–
Pineae (1)	–	–	–	+	–	–	+	–
Lariciones (7)	–	–	–	+	–	–	(+)	–
Australes (9)	–	–	–	+	–	–	+	–
Insignes (12)	–	–	–	+	–	–	+	–
Macropae (2)	–	–	–	+	–	–	+	–

Key: 1, chrysin; 2, tectochrysin; 3, strobochrysin; 4, pinocembrin; 5, pinostrobin; 6, strobopinin; 7, pinobanksin; 8, strobobanksin. +, uniformly present; (+) not present in all species; –, absent. Data adapted from Erdtman (1963).

The above finding has proved to be useful in relation to the classification of an unusual species of pine, *P. krempfii*, which some taxonomists consider to belong to the subgenus *Haploxylon* in spite of the fact that it has its leaves in pairs. Because of this anomalous morphology, it has also been put into its own genus *Ducampopinus*. Erdtman *et al.* (1966) looked at its heartwood compounds and found all the characteristic flavones, flavanones and flavanonols of *Haploxylon* pines. This firmly places it chemically here, so that there is no support for the creation of a new genus for it. True, it does contain one new flavanone, 6,8-dimethylpinocembrin; however, the occurrence of such a compound is completely expectable, since the closely similar 6- and 8-monomethyl flavanones are widely present in *Haploxylon* pines.

4. *Leaf flavonoids of apples and pears*

Although apple and pear species are usually recognized today as belonging to separate genera in the Rosaceae, this has not always been so (see e.g. Willis, 1960); indeed Linnaeus himself treated them as one genus, *Pyrus*. Taxonomic characters for separating them are minor and relatively few in number (see Table 7.5). Their closeness is reflected in the fact that a pear–apple hybrid has been achieved, although hybrid plants have not so far survived beyond the seedling stage. In this situation, it is interesting to find that from the view point of flavonoid chemistry in the leaves, the two groups are quite distinctive.

Table 7.5 Biological and chemical features separating *Pyrus* from *Malus*

Genus	Taxonomic characters	Major leaf phenolic	Minor leaf flavonoids[a]
Pyrus (30 spp.)	Styles free; fruit pyriform, turbinate or globose, flesh with stone cells; flowers in corymbs	Arbutin	Chrysoeriol 7-glucoside, luteolin and apigenin 4′-glucosides
Malus (25 spp.)	Styles united at base; fruit globose, flesh usually without stone cells; flowers in umbels	Dihydrochalcones: phloridzin or hydroxyphloridzin	Chrysin 7-glucoside, flavanone glucosides, flavonol 4′-glucosides and azaleatin

[a]Chemical data summarized by Challice (1974).

All known taxa have been surveyed and there is a major difference in the presence in *Malus*, but not *Pyrus*, of dihydrochalcones (Table 7.5). In fact, the main leaf phenolic in *Pyrus* is not of the flavonoid class, but is the simple phenol, arbutin. There are also quite a number of minor flavonoid differences between the two groups (Table 7.5). Chemically, therefore, the two groups are different and their taxonomic recognition as distinct genera would seem to be fully justified. Chemical surveys have been extended to other Rosaceae and a number of different flavonoid profiles are apparent. The results have illuminated generic and tribal interrelationships elsewhere in the family (see Challice, 1974).

5. *Isoflavone patterns in the Leguminosae*

Isoflavones occur in rich profusion in only one group of plants, the subfamily Papilionoideae of the Leguminosae. They are found indiscriminately in flower, leaf, seed, root and heartwood. As taxonomic markers, they are highly characteristic of the group, but their distribution within the subfamily at present seems to be erratic. Although they are reported from nearly all the tribes, they may be found in all species of some genera (e.g. *Cytisus*, *Ulex*), in most species of others (e.g. *Trifolium*) and in rather few species in yet others (e.g. *Lathyrus*). One apparent correlation with tribal classification is the accumulation of 5-methylgenistein in leaves of just one tribe, the Genisteae. However, it is not always present, having been detected in only about two-thirds of the 93 species surveyed (Harborne, 1969a).

The isoflavone character is thus one that should be useful to legume taxonomists. Unfortunately, it is not yet of much significance, largely because of the incomplete information on its natural distribution. Recently, an attempt has been made to analyse the existing isoflavone data from a biosynthetic and structural view point. The results are presented in a three-dimensional diagram, where the three axes represent additional hydroxylation over that present in genistein, the North–South axis referring to A-ring oxygenation and the East–West axis to B-ring oxygenation. Figure 7.9 shows some typical 'biogenetic maps' produced in this way (Cagnin *et al.*, 1977). The nine genera illustrated belong to three tribes and no striking relationship is apparent with tribal classification. However, *Machaerium* and *Dalbergia* genera are clearly similar and both belong to the Dalbergieae. *Pterodon*, also Dalbergieae, is superficially similar to these but does differ (note extension on horizontal axis) and in fact represents a different phyletic line. *Millettia* and *Piscidia* (both Tephrosieae) have similar diagrams, which reflects these similarities in containing rotenoids as part of the isoflavonoid complement. This development is important in that these diagrams present a 'translation' of flavonoid structures in a form that can be readily appreciated by systematists, who do not need to have any knowledge of the chemistry involved.

6. *Flavonoid patterns in* Baptisia

In order to make full use of flavonoid data for taxonomic purposes, all known taxa within a given group need to be screened and all substances have to be completely identified. In practice, such an ideal has rarely been achieved. One of the few groups where this has been done is the legume genus *Baptisia*, which contains 18 species of perennial herbs native to North America. Flavonoids in both leaves and flowers have been characterized and a rich array of substances discovered: nine flavone, 16 flavonol and 18 isoflavone glycosides (Markham *et al.*, 1970). The systematic significance of these data has been analysed in part (Harborne *et al.*, 1971) and is discussed elsewhere in this volume (chapter 11). Here, brief mention may be made of a comparison of these data on *Baptisia* with similar results on a closely related genus *Thermopsis* (Dement and Mabry, 1972). By contrast with *Baptisia* where the species fall into four main flavonoid groups, all 13 *Thermopsis* species are identical in profile. In reviewing the biological implications of this comparison, Dement and Mabry (1975) were able to conclude that: (1) *Baptisia megacarpa* represents an ancestral complex, chemically similar to extant *Thermopsis*, which probably gave rise to several *Baptisia* lines; (2) *Baptisia* is more advanced than *Thermopsis*, since there are more complex flavonoid substitutions in *Baptisia*; and (3) *Baptisia* exhibits more evolutionary vigour than *Thermopsis*, in view of the variety of patterns apparent in the former genus.

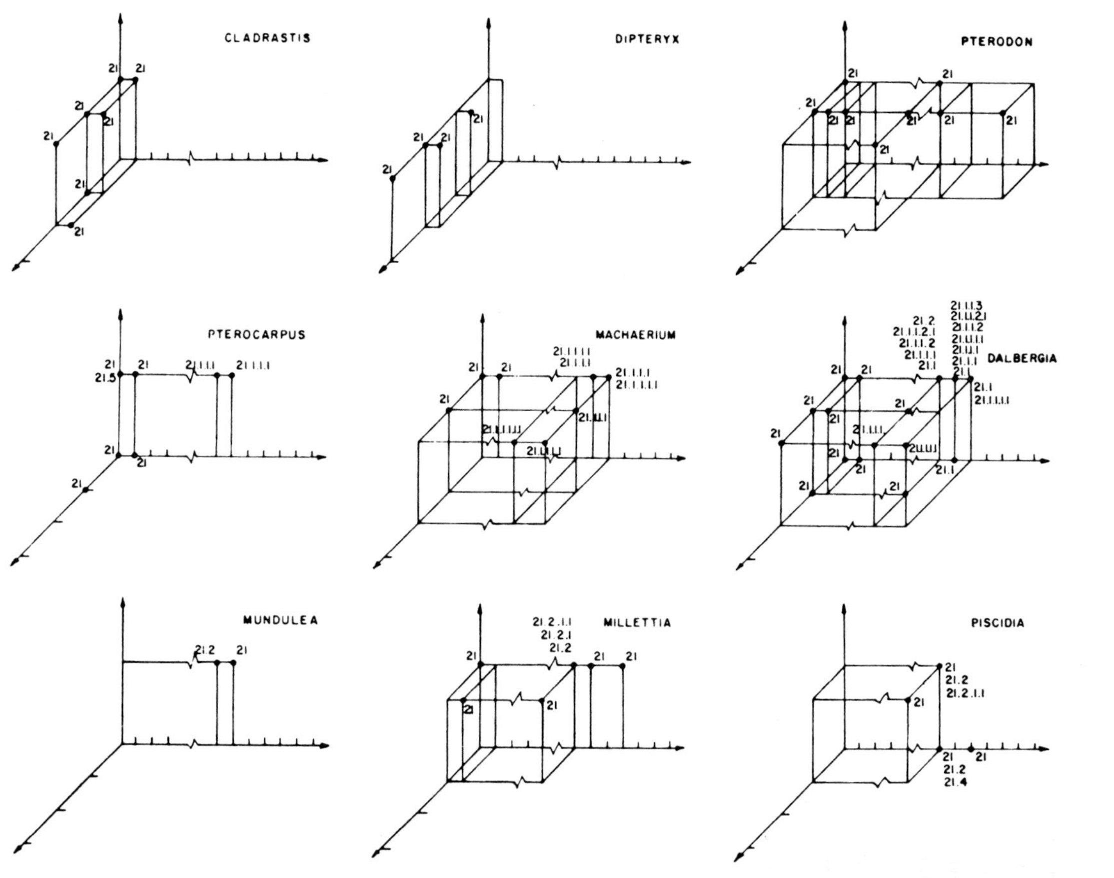

Fig. 7.9 Isoflavone biogenetic maps of some legume genera. (From Cagnin *et al.*, 1977.)

V. Betalain pigments

A. Chemistry and distribution

Apart from the ubiquitous green photosynthetic catalysts chlorophylls *a* and *b*, there are surprisingly few nitrogenous pigments known in higher plants. Indeed, the only significant group are the sap-soluble betalains, which are widespread in just one order of higher plants, in members of the Centrospermae. The collective term betalain encompasses the purple betacyanins, together with the yellow betaxanthins. Both groups of pigment have a common betalamic acid chromophore and only differ in that in betacyanins betalamic acid is linked to cyclodopa, an aromatic dopa-derived moiety, whereas in the betaxanthins it is linked to one or other of a number of protein amino acids (Fig. 7.10).

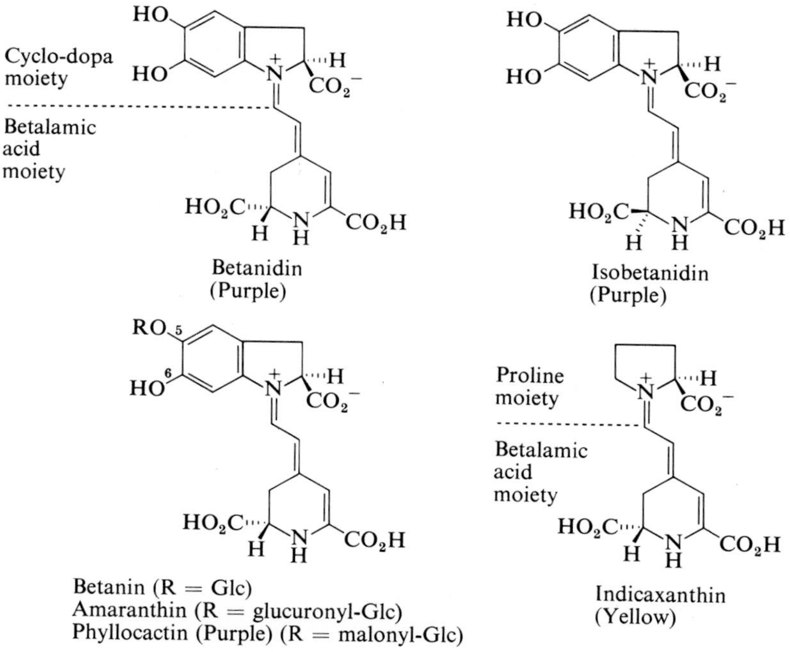

Cyclo-dopa moiety
Betalamic acid moiety
Betanidin (Purple)
Isobetanidin (Purple)
Betanin (R = Glc)
Amaranthin (R = glucuronyl-Glc)
Phyllocactin (Purple) (R = malonyl-Glc)
Proline moiety
Betalamic acid moiety
Indicaxanthin (Yellow)

Fig. 7.10 Structures of betacyanins and betaxanthins.

The betacyanins are all derived from two aglycones, betanidin and isobetanidin and are variously substituted derivatives with sugars or acylated sugars at either the 5- or 6-hydroxyl group. Betanidin and isobetanidin were first isolated by the hydrolysis of betanin, the major pigment of beetroot, *Beta vulgaris*; the two isomers (Fig. 7.10) exist in equilibrium with each other and are interconvertible. Common betacyanins are the 5-glucoside betanin, the 6-glucoside gomphrenin 1 and the 5-glucuronosylglucoside amaranthin. A number of aromatic and aliphatic acylated derivatives of these glycosides are known, phyllocactin the malonyl ester of betanin from *Phyllocactus hybridus* being one example.

The betaxanthins are typified by indicaxanthin, a proline-substituted betalamic acid, which was isolated as the yellow pigment of *Opuntia ficus-indica* fruits. Structural variation in this series consists simply in variation in the nature of the amino acid substitution. The other eight known betaxanthins are thus similar in colour and structure to indicaxanthin; proline can be replaced in turn by 4-hydroxyproline, glutamate, glutamine, dopa, dopamine, tyramine, aspartate and methionine sulphoxide.

The biosynthesis of the betalains involves the condensation together of two amino acids, one of which is always dopa (3,4-dihydroxyphenylalanine). This molecule of dopa undergoes a series of rearrangements and oxidation to give betalamic acid as key intermediate (Fig. 7.11). The involvement of betalamic acid in the pathway is supported by its detection as a natural constituent in a number of Centrospermae. Furthermore, betalains are readily synthesized in the laboratory by condensing betalamic acid with the appropriate amino acid moiety. The isomeric isobetanidin derivatives often accompany betanidin derivatives in plants but betanidin glycosides also occur free of the corresponding isomers. It is thus likely that betanidin pigments are formed first in the biosynthetic pathway and that isobetanidin derivatives are subsequently produced from them by spontaneous epimerization.

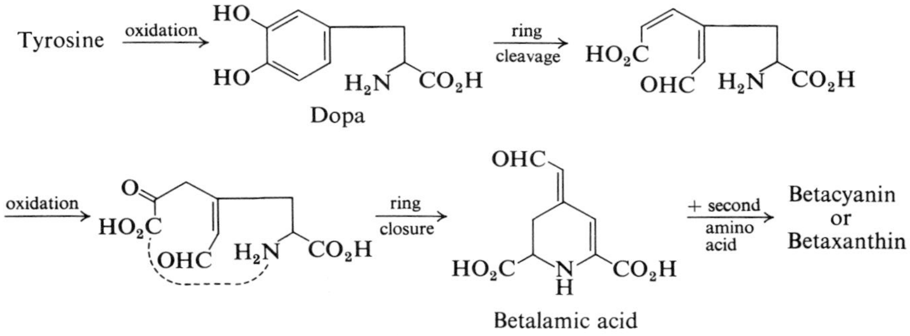

Tyrosine $\xrightarrow{\text{oxidation}}$ Dopa $\xrightarrow{\text{ring cleavage}}$ $\xrightarrow{\text{oxidation}}$ $\xrightarrow{\text{ring closure}}$ Betalamic acid $\xrightarrow{\text{+ second amino acid}}$ Betacyanin or Betaxanthin

Fig. 7.11 Biosynthetic origin of betalain pigments.

Betalains occur consistently in nine families of the Centrospermae (Table 7.6) where they replace anthocyanin pigments. They are found mainly in flowers and fruits, but are also present in leaves, stems and roots. Functionally, they are similar to the anthocyanins. They occur in flowers and fruits to attract animals for purposes of pollination and seed dispersal respectively. Intriguing unanswered questions are whether animals can 'see' betacyanins differently and whether betacyanin synthesis represents a significant ecological adaptation in habitats containing a special range of animal pollinators.

Table 7.6 Plant families containing betalains

Family	Main betacyanin type[a]
Aizoaceae	Betanin
Amaranthaceae	Amaranthin, betanin
Baselliaceae	Unidentified
Cactaceae	Betanin, phyllocactin
Chenopodiaceae	Amaranthin, phyllocactin
Didiereaceae	Unidentified
Nyctaginaceae Phytolaccaceae Portulaccaceae	Betanin

[a]Excludes a number of pigments, which are mainly reported so far in one or two species. If betanin is reported, betanidin is also often found as well.

True betalains are only known in higher plants, but several similar purple and yellow pigments based on betalamic acid have been reported in the fungus *Amanita muscaria* (Döpp and Musso, 1973). Whether such pigments occur regularly in the Basidiomycetes is not yet known. It is, however, apparent that quite a range of other types of nitrogenous pigment are present in fungi and bacteria (Thomson, 1976).

An excellent review of the chemistry and distribution of betalain pigments in plants is that of Piattelli (1976). Other important reviews are those of Mabry and Dreiding (1968) and Reznik (1980).

B. Detection

Betalains are labile substances, subject to oxidation and fading in solution, and are of necessity isolated from fresh plant tissue by extraction into alcoholic or aqueous acid. The most reliable method of distinguishing between betacyanins and anthocyanins is paper electrophoresis in acid buffer, pH 2.8, when betacyanins migrate as anions and anthocyanins as cations. Betaxanthins are similarly recognized by their anionic mobility, whereas water-soluble yellow flavonoids are immobile under these conditions. Paper chromatography is also a useful screening procedure, since betalains have higher R_F

values in aqueous solvents and lower R_F values in butanolic solvents than anthocyanins. There are also differences in the visible spectrum; betacyanins have maxima between 534 and 554 nm and betaxanthins at about 480 nm.

For their isolation, betalains are subjected to preliminary purification on ion-exchange resin, followed by separation on a polyamide column, eluted with methanol in citric acid (Piattelli and Minale, 1964). For structural identification, NMR spectroscopy is the most valuable technique (Wyler *et al.*, 1963).

C. Taxonomic utility

Much has been written on the chemotaxonomic and phyletic importance of the betalain character (see also chapter 12, p. 306). The subject is controversial (Bendz and Santesson, 1974) and unresolved in that there is no general agreement as yet as to the degree of weighting that should be given to this highly distinctive chemical feature which clearly divides one of the most natural orders of plants into two groups. Thus, there are nine families in the Centrospermae with betalain pigmentation (Table 7.6) and two (Caryophyllaceae, Molluginaceae) without. In these last two families, common anthocyanins replace betacyanins as the purple colouring of flowers and fruits; betaxanthins are also absent, and are replaced in part by yellow flavonoids.

The taxonomic importance of the betalain/anthocyanin dichotomy is strengthened by the fact that the two types of water-soluble pigment have never been found to co-occur in the same plant. It should, however, be pointed out that, although anthocyanins have not yet been found in any of the betalain families, only some 300 species of a possible total of 8000 have actually been screened for their pigments. It is still conceivable that anthocyanin might be found in one of the more primitive members of the betalain families not yet adequately surveyed.

If one accepts the existing data on betalain occurrences as being representative, the results may be considered taxonomically useful in two ways. They can be employed to interrelate the different betalain families or to determine the position of 'borderline' members of these families. In terms of family interrelationships, more chemical studies are still needed, since the actual betalains present in some families such as the Didiereaceae have not yet been fully characterized. Nevertheless, it is apparent from the existing data (Table 7.6) that Amaranthaceae and Chenopodiaceae are at present linked through the common occurrence of amaranthin, whereas the Chenopodiaceae and Cactaceae are linked by the pigment phyllocactin. The full significance of such links are not as yet clear.

There are several examples where studying the presence or absence of betalain has helped in dealing with anomalous taxa; two may be quoted as typical. Thus *Hectorella* is a monotypic New Zealand endemic, which was earlier placed in the Caryophyllaceae. Recently, it proved possible to induce pigmentation, not normally present, in the roots of this plant and this turned out to be betalain (Mabry *et al.*, 1978). This result clinched a decision to move *Hectorella* to the Chenopodiaceae on the basis of certain morphological and anatomical features which also did not agree with its association with the Caryophyllaceae.

By contrast, the Gyrostemonaceae, an Australian group of 17 species and five genera, is an example of a taxon once placed in or near the Phytolaccaceae, which now appears to belong outside the order. No betalains have ever been detected in these plants; furthermore, a characteristic centrospermous anatomical feature (a P-type plastid) is also not present (Goldblatt *et al.*, 1976). Even more weighty phytochemical evidence for the removal of these plants from the Centrospermae is the presence of glucosinolates (mustard oil glycosides) in *Codonocarpus* and *Tersonia* (Kjaer and Malver, 1979). This and chromosomal data indicate that the Gyrostemonaceae have affinities with families of the Capparales.

VI. Quinone pigments

A. Chemistry and distribution

The natural quinone pigments are usually yellow, orange or red in colour; occasionally, they may be blue or green or even black. Well over 450 structures are known (Thomson, 1971, 1976). Although quinones are widely distributed in nature and exhibit great structural variation, they make no major contribution to visual colour in higher plants. Thus, they are most often present in bark, latex, heartwood or root or may occur as a pigment deposit on the undersurface of the leaves. Also, they frequently occur in a colourless reduced hydroquinone form, the coloured pigment only being released after hydrolysis of a bound glycoside and oxidation.

In higher plants, quinone colours seem to be incidental to their more important functions as herbivore deterrents. Many quinones are biologically active, being toxic to some degree and sometimes having vesicant properties. Their defensive role in plants is substantiated by the fact that a number of simple benzoquinones are also active in the defence secretions of a range of insects, including that of the bombardier beetle.

In the case of lower plants, quinones have a scattered distribution, but here they occur in the free state and do make a useful contribution to the colour of the organisms. They are found in the fruiting bodies of many basidiomycetes and are also regular constituents of lichens.

Benzoquinones

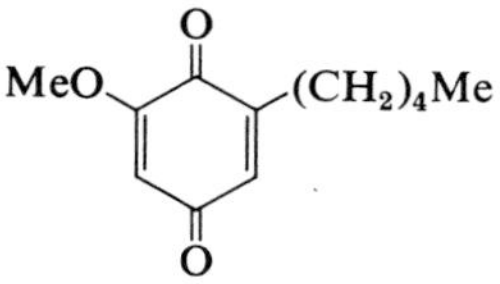

Primin
glandular hairs
Primula obconica

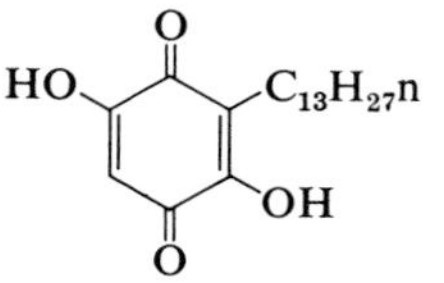

Embelin
roots, Myrsinaceae

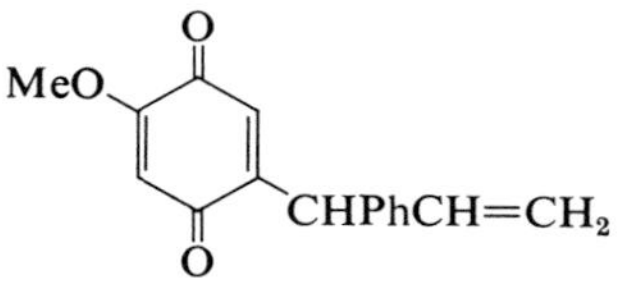

Dalbergione
Dalbergia heartwood

Naphthoquinones

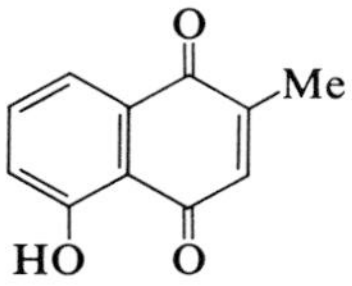

Plumbagin
Plumbago root

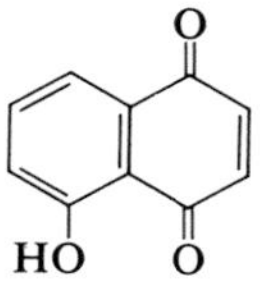

Juglone
Juglans

Anthraquinones

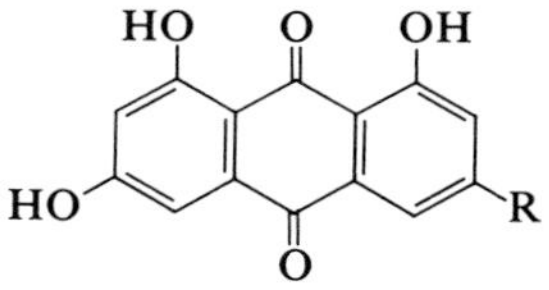

Emodin (R = Me)
Aloe-emodin (R = CH_2OH)

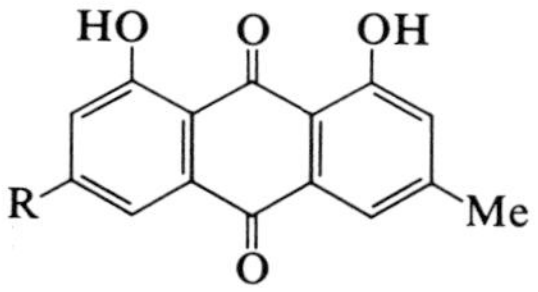

Physcion (R = OMe)
Chrysophanol (R = H)

Extended Quinones

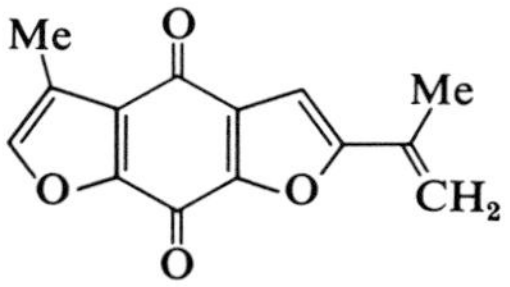

Cyperaquinone

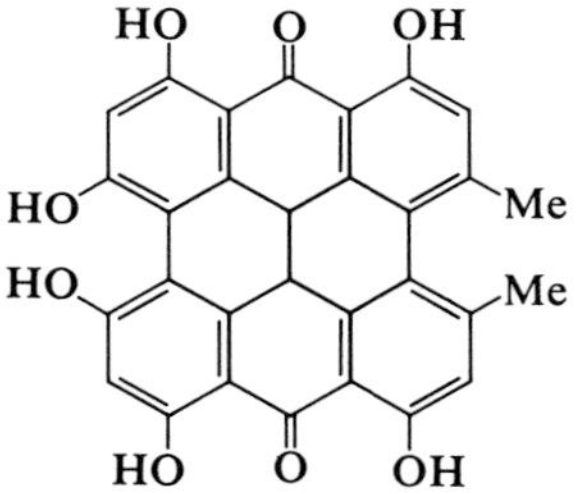

Hypericin
Hypericum

Fig. 7.12 Structures of some typical plant quinones.

Quinones (Fig. 7.12) all contain the same basic pigment chromophore, that of benzoquinone itself, which consists of two carbonyl groups joined in conjugation with two carbon–carbon double bonds. The yellow quinone colour intensifies and shifts towards the red when further double bonds are present in a conjugated system (as in anthraquinones). For convenience, quinones can be divided into four groups of increasing complexity: benzoquinones, naphthoquinones, anthraquinones and extended quinones (Fig. 7.12). Most quinones are hydroxylated and have phenolic properties. They may occur *in vivo* combined with sugar; this is particularly true of anthraquinones. Anthraquinones are also distinctive in occurring in dimeric quinol form, as dianthrones.

Benzoquinone itself occurs in insects (e.g. in the bombardier beetle) but is only known in plants in the reduced form as hydroquinone. Arbutin, the monoglucoside of hydroquinone, is frequently present in members of Ericaceae and Rosaceae and also occurs sporadically in several other families. Primin, a substituted benzoquinone, is the pigment present on the surface of *Primula obconica* leaves which is responsible for the allergenic rash produced occasionally in those handling this plant. Two further groups of benzoquinone should be mentioned, although they only occur in trace amounts in plants and have a universal distribution. These are the ubiquinones, involved in cellular respiration, and the plastoquinones, substances with a key role in photosynthesis in green plants.

Of the many known naphthoquinones, juglone and plumbagin are two simple representatives (Fig. 7.12). Each characterizes the genus from which they were first isolated, *Juglans* and *Plumbago* respectively, but both have a wider distribution in addition. They are toxic compounds with vesicant properties, staining the skin brown or purple. Although most of the known naphthoquinones are *p*-quinones, a limited number of *o*-quinones have also been described. One notable *o*-naphthoquinone is dunnione, the orange-red pigment from the under surface of the leaves of *Streptocarpus dunnii* (Gesneriaceae).

Anthraquinones comprise the largest group of natural quinones (some 200 pigments) and several structures have a fairly regular distribution among higher plants and among micro-organisms. Emodin, aloe-emodin, physcion and chrysophanol are found together as mixtures in several sources. Emodin, in particular, is known from the Polygonaceae, Iridaceae, Liliaceae and Lythraceae. Emodin is also known in glycosidic form and may be found as the corresponding anthrone or dianthrone.

Surveys of plant tissues for quinones have been linked to their practical use as natural dyestuffs (e.g. the quinones of madder, *Rubia tinctorum*) or their medicinal use as purgatives. The cathartic action in Man of extracts of senna pods and cascara bark are due to the anthraquinone derivatives present. Surveys have indicated that anthraquinones are present in some dozen plant families, of which the Rubiaceae is the most consistent source. Other anthraquinone-containing families include Rhamnaceae (*Rhamnus*), Leguminosae (mainly *Cassia*), Polygonaceae (*Rheum*, *Rumex*), Bignoniaceae, Verbenaceae, Scrophulariaceae and Liliaceae (*Aloe*). Naphthoquinones are, by contrast, more sporadic in distribution. They occur occasionally with anthraquinones (e.g. in Bignoniaceae and Verbenaceae) but are otherwise found in more diverse groups; the Ebenaceae (*Diospyros*) is a particularly rich source. Finally, the benzoquinones, a much smaller group of pigments, have a number of unrelated family occurrences. A most consistent source are the heartwoods of the two legume genera, *Dalbergia* and *Machaerium*, which contain a series of modified isoflavanoid quinones, such as dalbergione.

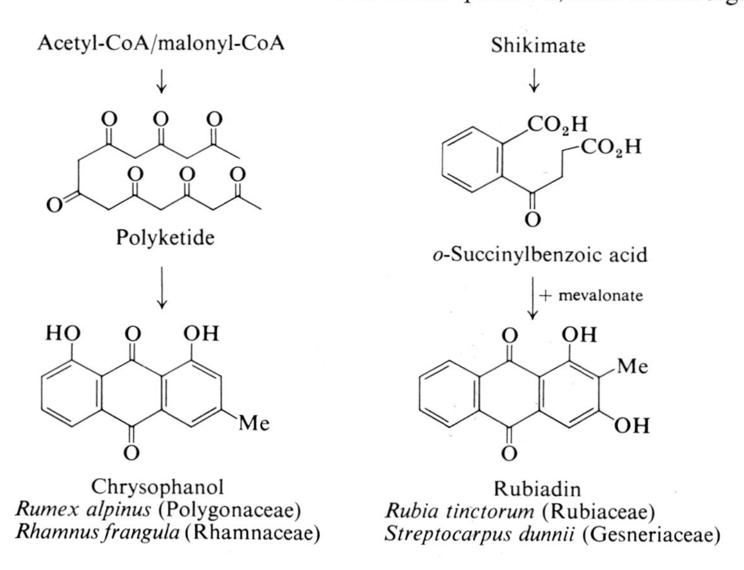

Fig. 7.13 Alternative biosynthetic pathways to plant anthraquinones.

Biosynthetic diversity has to be considered when considering the taxonomic significance of quinones. Although two pathways are known to higher plant anthraquinones, no less than four pathways are known for the various naphthoquinones (Thomson, 1971; Bentley, 1975). Anthraquinones may be formed by the acetate–malonate or the *o*-succinylbenzoic acid routes (Fig. 7.13). The first pathway is followed in Polygonaceae and Rhamnaceae and the second pathway operates in Rubiaceae and Gesneriaceae and also probably in Bignoniaceae and Verbenaceae. These two pathways are also used for the synthesis of naphthoquinones, but two further alternative routes to naphthoquinones have also been delineated (Fig. 7.14). What is most striking is the fact that three almost identically substituted simple naphthoquinones—plumbagin, juglone and chimaphilin—are made by three quite distinct routes in Plumbaginaceae, Juglandaceae and Pyrolaceae, respectively. The presence of structurally similar quinones in these three families cannot, therefore, be used to relate them to each other chemically, since the biosynthetic origins are different in each case.

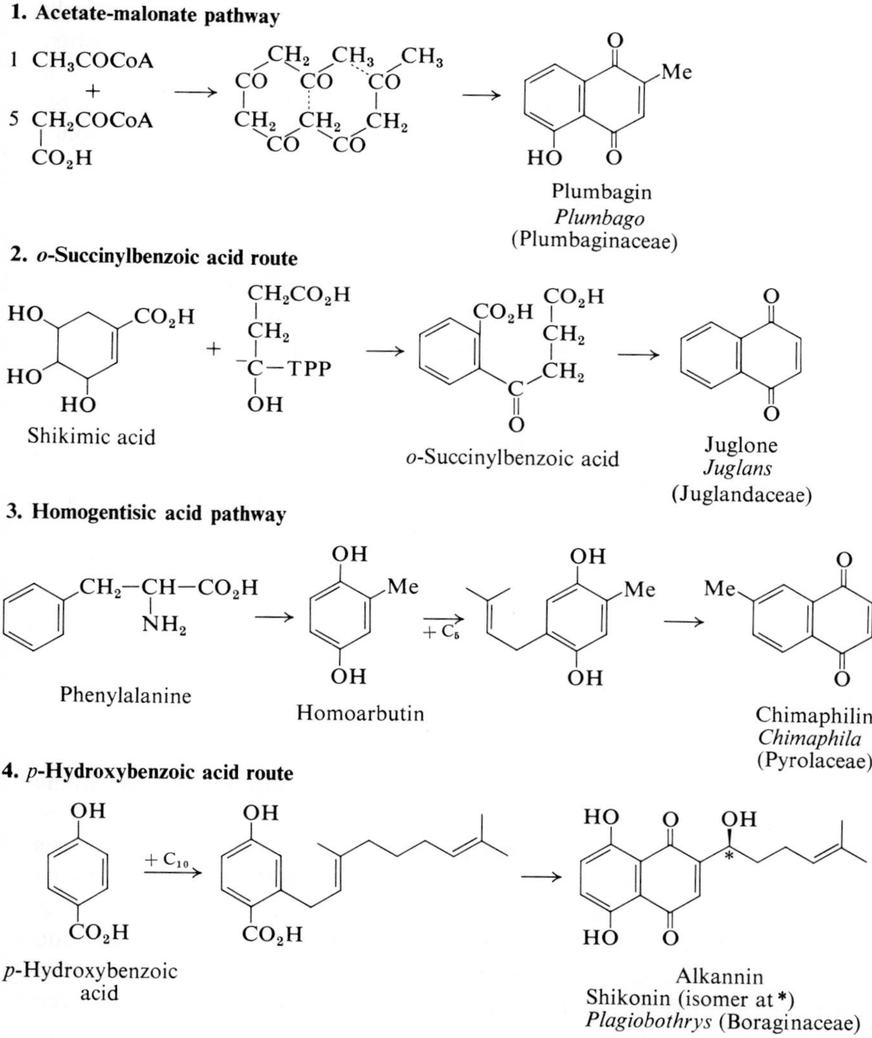

1. Acetate-malonate pathway

1 CH_3COCoA + 5 $CH_2(CO_2H)COCoA$ → → Plumbagin
Plumbago
(Plumbaginaceae)

2. *o*-Succinylbenzoic acid route

Shikimic acid + → *o*-Succinylbenzoic acid → Juglone
Juglans
(Juglandaceae)

3. Homogentisic acid pathway

Phenylalanine → Homoarbutin → ($+C_5$) → Chimaphilin
Chimaphila
(Pyrolaceae)

4. *p*-Hydroxybenzoic acid route

p-Hydroxybenzoic acid → ($+C_{10}$) → Alkannin
Shikonin (isomer at *)
Plagiobothrys (Boraginaceae)

Fig. 7.14 Biosynthetic routes to naphthoquinones.

The biosynthetic versatility of secondary metabolism in plants also leads to the formation of *o*- and *p*-quinone systems in compounds that are entirely of terpenoid origin (see chapter 6). For example, a taxonomically interesting series of diterpenoid quinones called the coleones have been identified in glandular secretions on leaves of various Labiatae species (Eugster, 1980). The reddish pigments appear to be highly characteristic of *Coleus*, *Fuerstia*, *Plectranthus* and *Solenostemon* species (all in the Plectranthinae of the subfamily Ocimoideae) but their absence from other labiate groups has yet to be firmly established.

B. Detection

An important preliminary in detecting quinones in higher plants is to ensure that, if the pigment is likely to occur in colourless bound form, it is released in the free oxidized state before tests are applied. As a routine, plant tissues may be immersed in 2 M HCl and heated for 20 minutes in the presence of air. This will liberate the free pigment, which can then be extracted into an organic solvent. Simple colour reactions are still useful for confirming that a pigment is a quinone. Reversible reduction to a colourless form and restoration of colour by aerial oxidation is diagnostic. Quinones also give intense bathochromic shifts in colour with alkali, but such colour changes are also given by other types of phenolic pigment, such as the flavonoids.

TLC on silica gel is a general procedure for detecting and separating quinones and will clearly distinguish quinones from other phenolic substances. Simple benzoquinones and naphthoquinones are very lipid-soluble and will separate in benzene, chloroform or petroleum or their mixtures. By contrast, the highly polar anthraquinones require more complex solvent mixtures (e.g. benzene–ethyl acetate–acetic acid) for their separation. Quinones are readily detected by their natural colours on TLC plates; examination by ultraviolet irradiation may provide a more sensitive means of detecting pigments present in trace amounts. Those quinones that have a labile hydrogen radical adjacent to the carbonyl group can be specifically detected on TLC plates by the blue or green colour which develops in the Craven test (see Hausen, 1978).

For quinone identification, spectral measurements are mandatory (Thomson, 1971). The ultraviolet and visible spectrum indicates the class of quinone present, since the number and position of absorption bands increase with the complexity of the structure. Quinones give characteristically intense bands of carbonyl absorption in the infrared spectrum and in the mass spectrum can be recognized by the ready loss of one and two molecules of carbon monoxide from the parent ion peak.

Because of the small amounts of plant available with lichens, special micro-methods have been developed for detecting anthraquinones in them. Lichen fragments are introduced directly into a combined gas chromatography–mass spectrometry apparatus at 120–160°C and the anthraquinones are sufficiently volatile to be detected and identified (Santesson, 1970). A micro-method based on electron-spin-resonance spectroscopy has also been proposed for detecting quinone pigments in higher plants (Pedersen, 1978). This procedure can be used to detect, identify and quantify quinones and quinols directly in crude plant extracts.

C. Chemotaxonomic utility

1. *General comments*

So far, relatively little use has been made of quinones as chemotaxonomic markers in plants. A major problem at the higher levels of classification is their apparently sporadic occurrence in mainly unrelated family clusters. The biosynthetic diversity in pathways to similar structures has already been mentioned as a difficulty in interpretation. A serious practical problem is that quinones tend to occur most consistently in inaccessible parts of the plant, for example, the roots. This is true of the naphthoquinone plumbagin in its occurrence in Plumbaginaceae. Nevertheless, a number of widely differing plant genera are variously characterized by the presence of specific quinones, so that these pigments do have potential at the lower levels of classification.

2. *Benzoquinones*

A series of hydroxybenzoquinones with long alkyl side chains (e.g. embelin, Fig. 7.12) occur regularly in the Myrsinaceae. Ogawa and Natori (1968) surveyed roots, rhizomes, barks and fruits of eleven Japanese species of the family and found such compounds in eight of them. Embelin and its relatives have also been detected in at least eleven other species from other geographical areas. There is thus at least the suggestion of an interesting chemotaxonomy, but more surveys are needed, since the Myrsinaceae has about a thousand species.

Primin is a benzoquinone with possible chemotaxonomic value in a family not unrelated to Myrsinaceae, namely the Primulaceae. Primin was first isolated from glandular hairs of *Primula obconica* leaves and characterized as the substance causing skin dermatitis to those handling this plant. Primin also occurs in bound form within the leaf and a screening of Primulaceae has revealed its presence in 20 of 82 species studied (Hausen, 1978). It occurs, not surprisingly, mainly in the genus *Primula*, the pattern of its distribution being related to some extent with sectional classification.

Finally, an example of a benzoquinone of phylogenetic interest may be mentioned. This is dalbergione, a pigment responsible in part for the heartwood colour of trees of the genus *Dalbergia*. Dalbergione or related structures have been found in all 17 *Dalbergia* species surveyed and similar compounds also occur in the related *Machaerium* (11 species studied). These two legume genera are found both in Africa and in South America. On the basis of this chemistry, and also of geography, Braga de Oliviera *et al.* (1971) have proposed that *Machaerium* evolved from *Dalbergia* in South America, but only after the separation of the continents and the formation of the Atlantic ocean. This seems to be a plausible hypothesis, although one rather difficult to substantiate in any other way.

3. *Naphthoquinones*

Several simple naphthoquinones are useful taxonomic markers at generic or family level. These include plumbagin, juglone and 7-methyljuglone (for structures, see Figs 7.12 and 7.14). Plumbagin, for example, occurs in plants of the Plumbaginaceae, chiefly in the roots but also in other organs. It is uniformly present in all four genera (and all species surveyed) of one tribe, the Plumbagineae (Harborne, 1967b; van der Vijver, 1972) but has never been detected in the other tribe, the Staticeae. Its distribution is correlated with differences in pollen morphology and geography at the tribal level. Plumbagin is also absent from *Aegialitis annulata*, a curious relict member of the family from Queensland, Australia which represents a 'bridge' species between the two tribes (Harborne, 1967b).

Plumbagin also occurs in the unrelated Droseraceae, together with 7-methyljuglone. Here, all 20 species of sundews investigated (representing four genera) contain one or other or both of these quinones in the roots (Zenk *et al.*, 1969). These two quinones, however, were found to be absent from the monotypic *Byblis* (Byblidaceae) and *Roridula* (Roridulaceae), two taxa at one time included in the Droseraceae. This chemical difference bolsters the recent separation from Droseraceae of these two groups, which has been proposed on morphological grounds. Finally, there is juglone itself, which occurs as a consistent marker in the genus *Juglans* and is also present in *Carya* and *Pterocarya* (all Juglandaceae) (see Thomson, 1971).

4. *Anthraquinones*

Several attempts have been made to relate anthraquinone distribution to taxonomy at the generic level in such groups as *Aloe*, *Cassia* and *Rumex*, but the data available are conflicting and incomplete and it is difficult to assess at the present time the taxonomic utility of anthraquinones. For example, several earlier surveys were carried out on anthraquinones in the genus *Rumex* (Polygonaceae) (e.g. Jaretzsky, 1926; for summary, see Alston and Turner, 1963a) which suggested that the pigments were present in primitive members of the genus but were absent from the more advanced species such as *R. acetosa*. More recently, Fairbairn and El-Muhtadi (1972) surveyed 19 species chosen to represent all three subgenera and found anthraquinone mixtures in all plants, including *R. acetosa*. Emodin, chrysophanol and physcion were always present and there were, in addition, a variable number of other partly identified constituents also in the tissues. Anthraquinones would thus seem to be universal in the genus, any chemotaxonomy being dependent on variations in the rather complex mixtures of compounds present.

Other genera in Polygonaceae, especially *Rheum* and *Emex*, also contain anthraquinones similar to *Rumex*. It appears that more recent studies have not upset Jaretzsky's (1926) earlier observation that the anthraquinone-containing subfamily Polygonoideae can be separated from the pigment-free Eriogonoideae on the basis of this character.

Information on the distribution of anthraquinones in *Cassia* and other legumes has been summarized by Harborne (1971) and that on the occurrence of aloe-emodin and related structures in *Aloe* and other Aspheloideae by van Rheede van Outshoorn (1964) and by Williams (1975). Further surveys are needed in both of these groups to establish fully their taxonomic value.

5. *Extended quinones*

Roots and rhizomes of members of the Cyperaceae are often coloured and recent studies have revealed the presence of a unique series of conjugated benzoquinone derivatives, based in part on cyperaquinone (see Fig. 7.12), in these tissues. A survey of some 100 species representing 29 genera of the family has also revealed an interesting chemotaxonomy for these pigments (Allen *et al.*, 1977). In general, cyperaquinone and related structures occur almost exclusively in *Cyperus*, and here only in subgenera *Cyperus* (in 13 of 26 species) and *Mariscus* (in 12 of 14 species). There were isolated occurrences in *Scirpus articulatus* and *Remirea maritima*. The latter is the only species of the subfamily Rhynchosporoideae reported as positive and is thus anomalous. This taxon, although placed in the Rhynchosporoideae because of its single-flowered spikelet, has been shown to have a reduced form of the *Cyperus* spikelet and hence Kern (1962) and others have advocated its transfer to *Cyperus*, as *C. pedunculatus*. The chemical data in this instance appears to be decisive, since the presence in *Remirea* of cyperaquinone, dihydrocyperaquinone and tetrahydrocyperaquinone would seem to anchor it firmly into the genus *Cyperus*. Flavonoid data on these plants is also in keeping with the transfer (Harborne *et al.*, 1982). Within *Cyperus*, the quinone data have been analysed systematically and although the results are consistent with Kukenthal's sectional classification, they do indicate that some rearrangement of sections within the subgenera might be needed.

One final example of a systematically interesting quinone is hypericin, a substance regularly present in petals, stems and leaves of *Hypericum* species in the family Hypericaceae. It is an extended quinone which is very stable, since it can be readily detected in herbarium tissues up to 120 years old. In studying its distribution in over 200 species of *Hypericum*, Mathis and Ourisson (1963) developed a system of 'printing' fragments of herbarium tissue directly on to the origin of a paper chromatogram in order to detect it. In fact, with a few scattered exceptions, it occurs exclusively in species belonging to the sections *Hypericum* and *Campylosporus*. Its more detailed occurrence within these sections fits in best with the classification of Keller (1925). It is an interesting marker, since it is specifically localized in particular anatomical glands within the plant and its distribution at the subsectional level varies according to the part of the plant where these glands are located.

VII. Carotenoid pigments

A. Chemistry and distribution

Carotenoids, which are C_{40} tetraterpenoids, are an extremely widely distributed group of lipid-soluble pigments, found in all kinds of plants from simple bacteria to yellow-flowered composites. In animals, one particular carotenoid, β-carotene, is an essential dietary requirement, since it provides a source, through hydration and splitting of the molecule, of vitamin A, a C_{20} isoprenoid alcohol. Carotenoids, through dietary intake, also provide many brilliant animal colours, as in the flamingo, starfish, lobster and sea urchin. In plants, carotenoids have two principal functions: as accessory pigments in photosynthesis and as colouring matters in flowers and fruits. In flowers, they mostly appear as yellow colours (e.g. daffodil, pansy and marigold), whereas in fruits, they may also be orange or red (e.g. rose hip, tomato and paprika). Taxonomically, it is their presence in flowers and fruits that is of most interest.

Although there are now over 300 known carotenoids (Isler, 1971), only a few are common in higher plants. Well known carotenoids are either simple unsaturated hydrocarbons based on lycopene or their oxygenated derivatives, known as xanthophylls. The chemical structure of lycopene (see Fig. 7.15) consists of a long chain of eight isoprene units joined head to tail, giving a completely conjugated system of alternate double and single bonds, which is the chromophore giving it colour. Cyclization of lycopene at one end gives γ-carotene whereas cyclization at both ends provides the bicyclic hydrocarbon β-carotene. β-Carotene isomers (e.g. α- and ε-carotene) only differ in the positions of the double bonds in the cyclic end units. The common xanthophylls are either monohydroxy- (e.g. lutein and rubixanthin), dihydroxy- (zeaxanthin) or dihydroxyepoxy- (violaxanthin) carotenes.

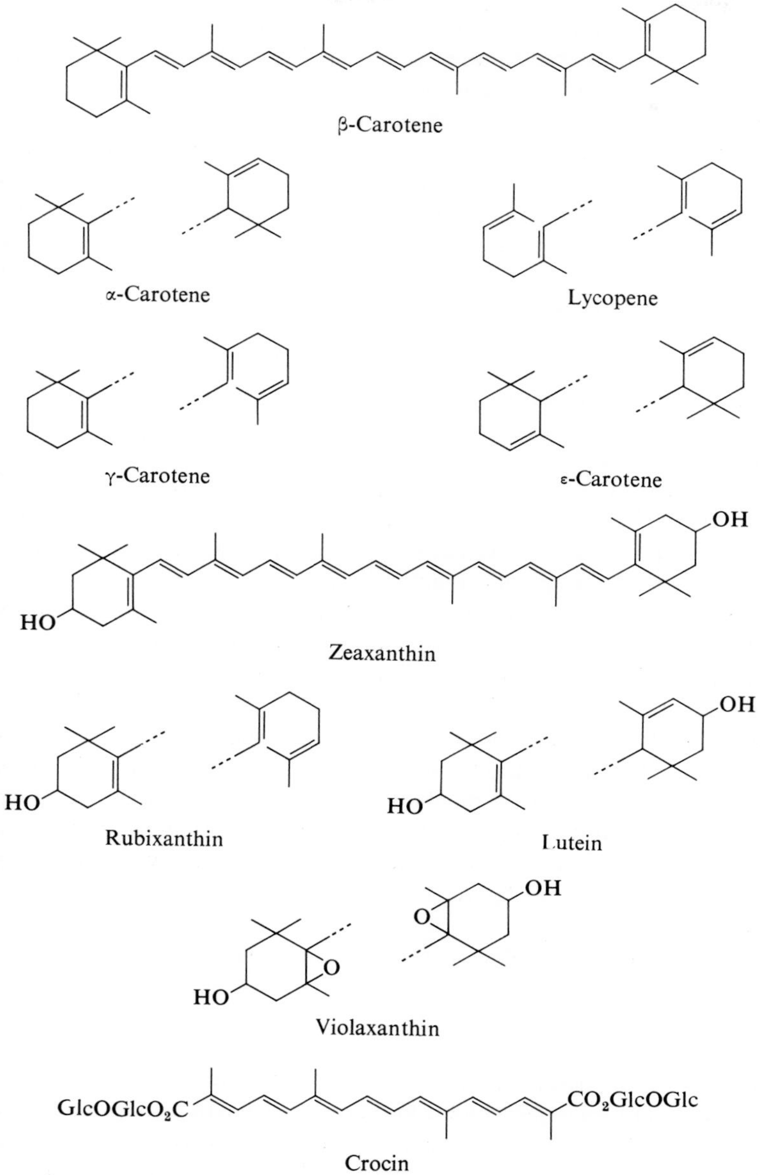

β-Carotene

α-Carotene

Lycopene

γ-Carotene

ε-Carotene

HO OH

Zeaxanthin

HO

Rubixanthin

HO OH

Lutein

O OH

HO O

Violaxanthin

$GlcOGlcO_2C$ $CO_2GlcOGlc$

Crocin

Fig. 7.15 Structures of some common carotenoids.

Most of the rarer carotenoids have more complicated structures and may be more highly hydroxylated (e.g. with keto groups), more highly unsaturated (e.g. with allenic or acetylenic groups) or more extended with additional isoprene residues (giving C_{45} or C_{50} carotenoids). Just one example of a rarer carotenoid is rubixanthin (see Fig. 7.15) which is a characteristic major pigment in the fruit of *Rosa* species.

Combined forms of carotenoid occur, especially in flowers and fruits of higher plants, and they are usually xanthophylls esterified with fatty acid residue, for example, palmitic, oleic or linoleic acids. Glycosides are normally very rare; in higher plants, the best known is the water-soluble crocin, the gentiobiose derivative of an unusual C_{20}-carotenoid crocetin, the yellow pigment of meadow saffron, *Crocus sativa*. An unrelated occurrence of crocin is in the Scrophulariaceae, where it occurs in flowers of *Verbascum* and *Nemesia* (Harborne, 1966).

When isolating a carotenoid from a new higher plant source, the chances are fairly high that it will be β-carotene, since this is by far the most common of these pigments. In quantitative terms, however, it is not as important as certain xanthophylls. The annual production of carotenoids in plants has been estimated at 108 tons a year and this refers mainly to the synthesis of fucoxanthin (widespread in marine algae) and of lutein, violaxanthin and neoxanthin. These three latter compounds occur, with β-carotene, universally in the leaves of higher plants. Traces of α-carotene, cryptoxanthin or zeaxanthin may also occur in the lipid-soluble fraction of leaf extracts. In flowers too, carotenoid mixtures are the rule rather than the exception. The pigments are often highly oxidized with epoxides being common and carotenes only being present in traces. The amount of oxidation, however, varies from species to species and, in the corona of certain narcissi, the red colour is due almost entirely to high concentrations (up to 15% of the dry weight) of β-carotene. Finally, when studying fruit pigments, it should be remembered that acyclic carotenoids tend to accumulate in some cases (e.g. lycopene in tomato) and that rather specific compounds may be synthesized by particular plant groups (e.g. rubixanthin by *Rosa* species, capsanthin by *Capsicum* species and so on).

Reviews of the known distribution of carotenoids in plants are contained in Isler (1971) and Goodwin (1976a, 1980).

B. Detection

Carotenoids are unstable pigments and undergo oxidation, particularly on TLC plates, and *trans–cis* isomerism. Pigments have to be isolated from fresh tissue with diethyl ether and it is advantageous to carry out identifications rapidly, storing solutions in the dark under nitrogen if necessary. Since carotenoids may occur in flowers and fruits in esterified form, it is necessary to saponify crude extracts with aqueous alkali. After recovering the pigments in the ether, they can then be identified following separation by TLC on silica gel. Column chromatography on alumina or magnesium oxide is useful for preparative separations.

Identification of carotenoids is based on co-chromatography with authentic markers, together with spectral comparisons. The visible spectra of carotenoids are highly characteristic between 400 and 500 nm, with a major peak at about 450 nm and usually two minor peaks on either side. The exact positions of the maxima vary from pigment to pigment and can be used for identification purposes (see Davies, 1976).

Nuclear magnetic resonance and mass spectrometry are important methods in the structural analysis of new pigments, although, because of low volatility, the latter procedure may be difficult to apply.

C. Chemotaxonomic utility

1. *General*

Undoubtedly, the most important application of carotenoid patterns to systematics lies not in higher plants, but in algae. Here, there is a much wider variation in carotenoid types and each algal class generally has a characteristic pattern of pigments (see Goodwin, 1976b, 1980). The patterns found can also be related to the biosynthetic development of structural complexity in carotenoids, so that the data have phyletic implications. The pattern fits in, apart from a few exceptions, with the views of algal phylogeny derived from morphological considerations. Fungi are also fairly rich in carotenoids and taxonomic conclusions have been drawn from some of the more significant variations in pigment patterns (see Goodwin, 1976b; Fiasson *et al.*, 1968).

By contrast with the situation in the above phyla, the distribution of carotenoids in flowers and fruits of higher plants has been described as 'capricious'. Repeated attempts to discern taxonomic significance in the existing data have failed. However, this is largely due to the fact that most of the results have been obtained incidentally to other types of investigation. Indeed, it is difficult to point to any surveys that have been sufficiently thorough from the taxonomic viewpoint. Most of the data available represents very poor sampling of a given genus or family. In the belief that carotenoids may prove to be of systematic interest in higher plants, a summary of the present information on flower and fruit pigments will now be given.

2. *Flower carotenoids*

Most yellow flower colour is based on carotenoid synthesis, some orange and red shades are also derived from these plastid pigments, rather than from anthocyanin. Carotenoids are thus relatively widespread in angiosperm flowers. Goodwin (1976b) divides the plants that have been analysed for carotenoids into three main groups. There are those with highly oxidized pigments, such as *Chrysanthemum coronarium* with mutatochrome (β-carotene-5,8-epoxide), flavoxanthin and β-carotene-5,6-epoxide. Then there are those with simple hydrocarbons, such as *Mimulus cupreus* with β-carotene. Thirdly, there are those with species-specific pigments, such as *Gazania rigens* with gazanixanthin. Plants falling into all three categories can, however, be found in the same family.

Some idea of the limited results available on flower carotenoids can be seen in Table 7.7, which assembles most of the data currently available on carotenoids in legume flowers. Although there is some correlation with generic difference, the data are so incomplete that it is difficult to draw any useful conclusions.

Table 7.7 Carotenoids present in yellow-flowered legumes

	Carotenes						
Species	α-Carotene	β-Carotene	Lutein	Lutein 5,6-epoxide	Flavoxanthin	Violaxanthin	Chrysanthemaxanthin
Acacia decurrens var. *mollis*	−	+	+	−	−	−	−
Acacia discolor	−	+	+	−	−	−	−
Acacia linifolia	−	+	+	−	−	−	−
Cytisus scoparius	+	+	+	+	+	−	+
Delonix regia	−	+	+	+	+	+	+
Genista tridentata	+	+	+	−	−	−	−
Laburnum anagyroides	−	+	−	+	−	+	−
Lotus corniculatus	+	+	+	+	−	+	−
Ulex europeaus	+	+	−	+	−	−	−
Ulex galli	+	+	−	+	+	−	−

One of the most extensive surveys to date at family level is that of Valadon and Mummery (1971) on composite flowers. A study of 26 species representing 18 genera revealed complex mixtures of up to 24 pigments in the petals. As the authors put it, the survey 'failed to demonstrate any carotenoid that may have been used as a taxonomic marker' in the family. One rare pigment, *cis*-

taraxanthin, was noted but this occurred in two unrelated sources, in *Helianthus decapitatis* (Heliantheae) and in *Taraxacum kok-saghyz* (Cichorieae). Considering that the Compositae contain about 1300 genera and 23 000 species, it is hardly surprising that such a limited sampling should have been fruitless in terms of revealing taxonomic characters. It may be suggested from these and other results discussed by Goodwin (1976b) the case for flower carotenoids as taxonomic markers is as yet non-proven and has still to be properly tested.

Fruit carotenoids Carotenoids make a major contribution to fruit colour in angiosperms and these pigments have been studied in relation to the ripening process in both edible and ornamental fruits (Goodwin and Goad, 1970). No less than eight different patterns can be discerned (Goodwin, 1976b). The fruit pigments may be chloroplast carotenoids (cucumber), lycopene-based (*Diospyros kaki*), β-carotene derivatives (*Physalis alkekengi*), epoxides (hawthorn), species-specific (red pepper), poly-*cis*-derivatives (*Pyracantha angustifolia*), seco- and apocarotenoids (*Murraya exotica*) or present in traces (*P. rogersiana*).

In some genera, fairly consistent patterns have been noted. In *Rosa*, rubixanthin is characteristic, often as the major pigment. Again in *Cotoneaster*, all three species studied have β-carotene, mutatochrome and cryptoxanthin in common. In other cases, there is much diversity. Thus in *Pyracantha*, *P. rogersiana* has only traces of fruit carotenoids, *P. flava* has β-carotene derivatives and *P. angustifolia* has poly-*cis*-carotenes. In general, it is difficult to make much sense of the available data. However, as with the floral carotenoids, the results are still very sparse and more information is needed before it is possible to assess the taxonomic value of carotenoid patterns in fruits.

VIII. Conclusion

Six major classes of plant pigment are here recognized as being of chemotaxonomic interest (Table 7.8). Three classes are flavonoid, the others being nitrogenous, quinonoid or terpenoid in biosynthetic origin. Numerically, the flavonoids (some 1000 structures) predominate but there are also large numbers of quinonoids and carotenoids. The leaf carotenoids and quinones and also the two chlorophylls are of no taxonomic interest, since they are always present in the chloroplast and are essential for photosynthesis. Apart from these, the only ubiquitous pigments are the flavonoids. The remaining pigment types are more limited in occurrence and thus have little value as chemical markers at the higher levels of classification.

Two major problems in using many of these pigments for taxonomic purposes is their instability (so that they can only be looked at in living plants) and their rather sporadic distribution. The latter problem can be overcome in the case of anthocyanins and betalains in that otherwise colourless tissue can often be induced to produce red or purple pigmentation by changing the physiological status. Nevertheless, the only class of pigment that does not have either of these drawbacks is that of the colourless flavonoid. It is not surprising therefore to find that these substances have already been extensively employed in systematic investigations. They are of proven convenience and usefulness and will undoubtedly continue to be widely studied by systematists.

Table 7.8 The major classes of plant pigment

Pigment class	Number of known structures	Distribution in organs	Natural distribution
Anthocyanins	300	Especially flowers; also fruits, leaves	Universal, except the Centrospermae
Yellow flavonoids	100	Flowers, but often also in leaves	Sporadic, in some 20 families
Colourless flavonoids	600	All parts of the plant	Ubiquitous; always available for comparison
Betalains	50	Flowers, fruits and other tissues	Only in the order Centrospermae
Quinonoids	300	Often hidden in root, bark, leaf	Sporadic, in some 30 families
Carotenoids	400	Flowers, fruits and leaves	Same set in leaves; capricious and variable in flowers and fruits

Of the other classes of pigment, the betalains have commanded the most attention. Their singular restriction in occurrence to one order of higher plant families is their most remarkable feature, but there are also many other interesting facets to these amino acid-based pigments. Certain basic information about them is still lacking and their functional role in pollination ecology deserves to be further explored. Quinonoids are only found in a few plant families but, where they do occur, they show considerable promise as taxonomic markers. Finally, there are the carotenoids of higher plants. Although the present evidence indicates they are very capricious in their natural distribution in flowers and fruits, they should not be completely dismissed from consideration. Few proper surveys have been attempted and future work may well show them to be of some significance and worthy of systematic consideration. Even leaf carotenoids may be taxonomically interesting, following the intriguing discovery of a characteristic ε,ε-carotene-3,3′-diol, called lactucaxanthin, in chloroplasts of lettuce *Lactuca sativa* (Siefermann-Harms *et al.*, 1981). In a survey of species in the tribe Cichorieae of the Compositae, lactucaxanthin was found in six other related species but was absent from 16 other, more distant, taxa. It will be interesting to see if other taxonomic markers are present in the leaf carotenoid fraction of plants.

8
Hidden Metabolites

I.	Introduction	180
II.	Fatty acids and lipids	183
III.	Alkanes and related hydrocarbons	191
IV.	Polyacetylenes	198
V.	Sugars and sugar derivatives	204
VI.	Conclusion	215

I. Introduction

According to recent estimates, as many as 30 000 organic molecules have been characterized from plants. There is thus a veritable cornucopia of low molecular-weight constituents in plants which are of potential interest to chemosystematists. Some of these compounds are volatile and advertise their presence by their odour (chapter 5); others, such as alkaloids, can be recognized by their bitter taste (chapter 6); and yet others are identifiable because of their colours (chapter 7). In addition to these apparent substances, there are more compounds hidden away within the plant which are non-apparent to the casual observer. They are further hidden in that they are not readily detected in the field by means of simple colour tests; sophisticated techniques such as gas chromatography are required for their analysis. This present chapter is concerned with these hidden metabolites and their possible utility as taxonomic markers in plants.

The term metabolite is used deliberately, since the substances considered here are either lipids or sugars and they have a primary role in plant metabolism, particularly in relation to the storage of energy. Lipids and sugars are closely connected since, in spite of their chemical differences, they are metabolically interconvertible. For example, lipid accumulated and laid down in the seed can be changed to free sugar in the germinating seedling, via the glyoxylate cycle and a reversal of the glycolytic pathway (see, e.g. Goodwin and Mercer, 1982). It is perhaps surprising, in view of their close involvement in primary metabolism, that lipids and sugars should be of taxonomic interest. Nevertheless, there is considerable chemical variation within these two groups of metabolite, particularly in the way they are stored in plant seeds. The fact that they are metabolically important does affect their synthesis, accumulation and turnover and it is essential when comparing the seed fatty acids of different species that the effects of physiological factors should be eliminated as far as possible.

Our first group of substances, the fatty acids, occur mainly in bound form as fat (or oil or triglyceride) and can only be examined after saponification of the fat and removal of the glycerol to which they are bound. Fatty acids occur in high concentrations principally in plant seeds and fruit coats, where they may constitute up to 50% of the dry weight. Variation in the fatty acids of seed fats has long been a point of interest to chemotaxonomists. As early as 1936, Hilditch and Lovern in a paper entitled 'The Evolution of Natural Fats' were able to point to a number of correlations between fatty acid composition and plant phylogeny. More recently, Shorland (1963) has shown that plant families can be placed into a variety of classes and subclasses according to the fatty acid profiles of their seed lipids.

The next class of substance to be considered are the alkanes, hydrocarbons formed directly or indirectly from fatty acids by decarboxylation. After synthesis within the leaf, these substances are deposited on the leaf surface and, although not always apparent to the naked eye, accumulate here as crystalline-like deposits of wax which protect the leaf from desiccation. Such external leaf lipids are readily collected by washing the surface for a few seconds with an organic solvent. Again, when analysed by modern chromatographic techniques (Eglinton *et al.*, 1962), these hydrocarbons show a variety of patterns which are related to the taxonomy of the plant from which the washings are collected.

A third class of fatty acid-derived compounds of taxonomic importance are also hydrocarbons, but differ from the alkanes of the leaf cuticle in having a high degree of unsaturation in the form of acetylenic bonds. These hydrocarbons have several acetylenic groups and are conveniently described as polyacetylenes. Over 650 such acetylenic compounds have been described (Bohlmann *et al.*, 1973). Found especially in roots of plants, they are very labile and readily polymerize after isolation. Although only present in concentration in five highly advanced angiosperm families, their presence has considerable phyletic import. Their widespread occurrence in the Compositae and the Umbelliferae argues for a close chemical relationship between these two taxa.

Although the three sugars glucose, fructose and sucrose are universally distributed as the main forms of low molecular-weight carbohydrate, there are other plant sugars, particularly oligosaccharides, which are of more limited occurrence. These may replace sucrose to various degrees as the major storage form of carbohydrate in seeds, roots and other underground storage organs. Legume seeds contain considerable quantities of the oligosaccharides raffinose and stachyose, while roots of umbellifers contain a characteristic trisaccharide, called umbelliferose after the family name. There is thus an interesting chemotaxonomy in the storage carbohydrate in plants, which is separate from the storage of high molecular-weight sugars as polysaccharides, which will be considered later in chapter 19.

Within the low molecular-weight sugar fraction of plant cells are a number of other sugar derivatives which tend to vary in their occurrence from one plant to another. There are sugar alcohols (or polyols), such as sorbitol, which is present characteristically in fruits of *Sorbus* and related Rosaceae. There are also cyclitols, such as pinitol, which occurs regularly in pine species. The chemotaxonomy of both these classes of sugar derivatives will be considered later in this chapter.

Before concluding this introduction, a word needs to be said about two other common classes of simple metabolite, the protein amino acids and the organic acids. Although attempts have been made to relate the pattern of common amino acids to taxonomy (cf. Alston and Turner, 1963a), it is now generally accepted that the protein amino acids do not offer characters for taxonomic consideration. Qualitatively, all 20 common amino acids must be present in plant cells, to provide all the necessary components for protein synthesis. Even where there are quantitative variations in the amino acid pool, these are so directly related to the growth and development of the plant or to physiological stresses that they cannot be considered to constitute reliable characters. It is only the non-protein amino acids of plants that can be considered of taxonomic significance (see chapter 6).

Protein amino acids should not, however, be completely ruled out from consideration. Recently, for example, it has been discovered that plant nectars contain trace amounts of these essential amino acids and they provide an important source of nitrogen for butterflies and birds feeding on the nectars. A study of the amino acids of nectars has revealed differences at the species level which seem to be consistent. Indeed, hybrid plants between species with different amino acid profiles show additive inheritance and contain the amino acids of both parents (Baker and Baker, 1976).

Plant organic acids are another group of primary metabolites, eight of which are universally distributed as trace constituents since they are key intermediates of the Krebs tricarboxylic acid cycle. Occasionally, plant acids may also accumulate in the leaves. A special feature of species with crassulacean acid metabolism (so-called CAM plants) is their ability to store three organic acids—malic, citric and isocitric—in the leaves during the night and then use them up during the day. This represents a physiological adaptation in succulent plants to dry climates, the diurnal variation in organic acid content being due to the fact that photosynthesis occurs mostly at night. There are occasional reports of other organic acids accumulating (e.g. malonic acid in leaves of legumes) without any obvious physiological explanation. In some cases, the accumulation of organic acid in the leaf cells is such that it crystallizes out as a salt. This is characteristic of oxalic acid, which forms a calcium salt. The crystals are known as raphides and, since they are often sharp and pointed, they appear to have some value to the plant in that they lower the palatability of the leaf to certain herbivores. Taxonomically, raphides have an interesting distribution (Gibbs, 1963). However, since raphides are usually considered as anatomical rather than chemical features of plants, they will not be further discussed here. Their occurrence in the plant kingdom is dealt with in anatomical texts (see e.g. Metcalfe and Chalk, 1950).

Finally, another situation where organic acids accumulate is in fruits. This accumulation is associated with acidity and repellency, a characteristic of unripe fruit. Thus, the concentration of acid decreases during ripening or else the acidity is balanced by an increase in sugar content. There is an element of chemotaxonomy in the acids that accumulate, at least among cultivated fruits. Thus citrus fruits accumulate citric, apple fruits malic, grape fruits tartaric and so on (Ulrich, 1970). The situation has, however, not been explored extensively among wild species. It is possible that organic acid content of the fruit might eventually provide a character for taxonomic assessment. At present, however, there is no clear evidence indicating that organic acids of either leaves or fruits are useful taxonomic markers in plants.

II. Fatty acids and lipids

A. Chemistry and distribution

Fatty acids occur mainly in plants in bound form, esterified to glycerol, as fats or lipids. Lipids comprise up to 7% of the dry weight of leaves of higher plants but occur in considerably greater amounts in the seeds or fruits of many, if not most, plants. In the seeds, they provide an important storage form of energy for use during subsequent germination; the concentration of lipid may be as much as 50% of the dry weight of the seed. Seed oils from plants such as the olive, palm, coconut and peanut are used as food fats, for soap manufacture and in the paint industry. Plant fats, unlike animal fats, are often rich in unsaturated fatty acids. The variation from plant to plant in the extent of this unsaturation has long been recognized, since the degree of unsaturation of the fat often determines what use is made of a particular plant oil.

Chemically, lipids fall into three main classes; the neutral fats or triglycerides; the polar phospholipids, with phosphate and a base (choline, ethanolamine or serine) substituent; and the glycolipids, with a sugar (usually galactose) substituent (Fig. 8.1). The latter two groups have important functions as structural elements in cell membranes and do not show much chemical variation. From the chemotaxonomic viewpoint, almost all attention has been given to the neutral lipids.

Structure	*Name*	*Symbolic structure*
Saturated Acids		
$CH_3(CH_2)_{10}CO_2H$	Lauric	12 : 0
$CH_3(CH_2)_{12}CO_2H$	Myristic	14 : 0
$CH_3(CH_2)_{14}CO_2H$	Palmitic	16 : 0
$CH_3(CH_2)_{16}CO_2H$	Stearic	18 : 0
$CH_3(CH_2)_{18}CO_2H$	Arachidic	20 : 0
Unsaturated Acids		
$CH_3(CH_2)_5CH{=}CH(CH_2)_7CO_2H$	Palmitoleic	16 : 1 (9c)
$CH_3(CH_2)_7CH{=}CH(CH_2)_7CO_2H$	Oleic	18 : 1 (9c)
$CH_3(CH_2)_4CH{=}CHCH_2CH{=}CH(CH_2)_7CO_2H$	Linoleic	18 : 2 (9c12c)
$CH_3CH_2CH{=}CHCH_2CH{=}CHCH_2CH{=}CH(CH_2)_7CO_2H$	Linolenic	18 : 3 (9c12c15c)
Unusual Acids		
$CH_3(CH_2)_7CH{=}CH(CH_2)_5CO_2H$	Petroselinic	16 : 1 (6c)
$CH_3(CH_2)_7CH{=}CH(CH_2)_{11}CO_2H$	Erucic	22 : 1 (13c)
$CH_3(CH_2)_7C{=}C(CH_2)_7CO_2H$ (ring bridged by CH_2)	Sterculic	9,10-methylene-18 : 1
$CH_3(CH_2)_4CH{-}CH{-}CH_2CH{=}CH(CH_2)_7CO_2H$ (epoxide ring, O)	Vernolic	12,13-epoxy : 18 : 1(9

Lipid structures

Neutral triglycerides	Phospholipids	Glycolipids
CH_2OCOR_1	CH_2OCOR_1	CH_2OCOR_1
R_2COOCH	R_2COOCH	R_2COOCH
CH_2OCOR_3	CH_2OPO-base	CH_2O-sugar

R_1, R_2, R_3 = hydrocarbon sidechains of different fatty acids

Fig. 8.1 Structures of representative plant fatty acids and of lipids.

Structural variation in fats is due to the differences in the fatty acid residues which substitute the three positions in the basic glycerol structure. Triglycerides are simple if the same fatty acid is present in all three positions, but they are usually mixed and have different acid components. In fact, any given plant fat is likely to be a complex mixture of many triglycerides. Detailed analysis of the individual lipids is, therefore, rarely attempted.

Instead, chemical surveys have concentrated on isolating the neutral lipid fraction and then identifying the fatty acids released during saponification.

Although numerous fatty acids are known in plants, most lipids have the same fatty acid residues. In fact, there are about 20 common acids and it is the quantitative variation in the relative amounts of these acids that usually provides characters of taxonomic significance. The common acids have a similar chain length centred around C_{16} and C_{18} and are either saturated or partly unsaturated (Fig. 8.1). The most common saturated acids are palmitic (16 : 0) and stearic acids (18 : 0), but lauric (12 : 0), myristic (14 : 0) and arachidic acid (20 : 0) are also fairly common.

The major unsaturated acids are all based on C_{18}. The mono-unsaturated oleic acid (18 : 1) comprises 80% of the fatty acid content of olive oil, 59% in peanut oil and is often accompanied by the di-unsaturated linoleic acid (18 : 2). The tri-unsaturated relative is linolenic acid, which occurs in linseed oil to the extent of 52% of the total acid, with 15% linoleic and 15% oleic. Palmitoleic acid (16 : 1) is widespread as a minor unsaturated acid but occurs in unusual abundance (64% of total) in the seed oil of *Doxantha unguis-cati* (Bignoniaceae). Unsaturated acids can exist in both *cis* and *trans* forms but most natural acids have the *cis* configuration.

Of the more than 100 rarer fatty acids known as lipid components (see Hitchcock and Nichols, 1971), only a few can be mentioned here. Most of the rarer acids are structural analogues of one or other of the common acids and can be related to them just as the non-protein amino acids can be related to the protein amino acids (see p. 98). There is some evidence that the rarer fatty acids, like the non-protein amino acids, may be harmful to animals eating the seeds. Erucic acid, a characteristic acid of the Cruciferae, has been described as having toxic effects in mammals, if ingested in sufficient amounts. It occurs in high levels in rape seed oil, *Brassica napus*, which is used in animal feeds and, because of the possible harmful effects, rape varieties with low erucic acid content have been developed. Erucic acid (22 : 1) has a longer chain length than most common fatty acids. Although occurring characteristically in members of the Cruciferae, it has several unrelated occurrences in *Tropaeolum* (Tropaeoleaceae) and in cereals (Gramineae).

Three other taxonomically interesting rarer fatty acids are petroselinic, sterculic and vernolic acids (for structures, see Fig. 8.1). Petroselinic acid is a positional isomer of the common acid, palmitoleic, and was first reported (to the extent of 76% of the total acids) in the seed oil of parsley, *Petroselinum sativum*. It is now known to be widely present in the Umbelliferae, the family to which parsley belongs, and also occurs in *Aralia* spp., in the closely allied Araliaceae.

Sterculic acid is structurally unusual in the presence of a unique cyclopropene ring and it occurs extensively in seed oils of Sterculiaceae and the related Malvaceae. Vernolic acid is structurally related to oleic acid but differs in having an epoxy substituent at positions 12 and 13. It occurs especially in the Compositae, in the genus *Vernonia* and other members of the tribe Vernonieae, but has also been detected in individual members of Dipsacaceae, Euphorbiaceae, Onagraceae and Valerianaceae.

Examples of rarer fatty acids with additional functional groups other than those already mentioned are: stearolic, an acetylenic acid from Santalaceae; ricinoleic, a hydroxy acid from castor bean; and chaulmoogric, a cyclopentene-substituted acid from Flacourtiaceae. Yet other unusual fatty acids have been obtained from lower plants; a considerable number of these are specifically present in certain bacteria (especially branched chain acids) or in particular algae.

The common saturated fatty acids are formed biosynthetically by a basic process of carbon chain elongation. A starter unit, acetyl-CoA, is linked with a repeating unit, malonyl-CoA in a regular sequence by which the carbon chain is extended two carbons at a time, the process being catalysed by a multienzyme complex, fatty acid synthetase. The process is thus geared to produce fatty acids with an even number of carbon atoms and, in fact, acids that do not have this basic pattern are extremely rare. Unsaturation is introduced as a final step in synthesis, but there is still some controversy about the exact details of the mechanism by which double bonds are formed. Other functional groups are also presumably introduced in the final stages of synthesis. The processes by which fatty acids are synthesized are common to all organisms, although there may be a few minor differences in the properties of the synthetase enzyme complex.

There is a considerable literature on the chemistry and biochemistry of lipids, from which further details can be obtained. The best general reference to plant lipids is the book of Hitchcock and Nichols (1971). The chemistry of fatty acids is covered in the monograph of Gunstone (1967). More recent developments in the biochemistry of plant lipids are discussed in Galliard and Mercer (1975).

B. Detection

Since lipids are ubiquitously present in plants, in all parts, there is rarely any need for detecting the presence or absence of this character. Nevertheless, it may be useful to determine the concentration of lipid in a given tissue, prior to hydrolysis and separation of the fatty acid components. This can be done by extracting a given weight of dried powdered seed or other tissue with light petroleum (b.p. 40–60°C). After filtering from the residue and removing the solvent, the resulting crude oil can then be weighed, to give a direct measurement of total oil. The degree of unsaturation in the oil is often determined directly, by measuring the iodine number (the amount of iodine taken up) or by argentation TLC. In this latter procedure, the oil is chromatographed on a silica-gel plate, previously treated with silver nitrate, whereby the saturated lipids move to the front while the unsaturated lipids are progressively held back. Lipids can be detected on the plate by spraying with a fluorescent dye, which binds specifically with them.

Although the fatty acids produced by saponification (or hydrolysis) of lipids can be separated and estimated by TLC (Stahl, 1969), gas chromatographic techniques are always employed because of the excellent sensitivity and the great accuracy in determining the relative amounts. Typically, fruits may be depulped and the seed coats washed, dried and crushed. The lipids are then extracted into benzene–methanol (2 : 1, v/v) and the samples saponified by refluxing with added sulphuric acid (10%) for 90 minutes. By this procedure, the fatty acids released undergo transmethylation and are converted into their methyl esters. These are more volatile than the free acids and are more readily separated by GLC on columns coated with polyethylene glycol adipate or silicone oil. Typical procedures are described by Hitchcock and Nichols (1971). If rarer acids are present, these may need more careful identification, using standard spectral and chemical procedures of organic chemistry (Gunstone, 1967).

C. Taxonomic utility

1. *General comments*

Only fats present in seed, fruit coat or leaf have been investigated to any extent, so that it is probably wise to concentrate in surveys on these tissues. In fact, unusual fatty acids are normally present in the seed fats; furthermore, most of the available data (Shorland, 1963) refers to analyses of seeds. In most cases where patterns in the seed and leaf have been compared, they are found to be different and quite unrelated.

Advantages of concentrating on seed fats are the relative richness in triglyceride content and the usually conservative nature of chemical constituents present in seeds. Nevertheless, considerable care is necessary in selecting seeds for fatty acid analyses. Significant changes in lipid content take place during seed maturation so that it is essential to restrict analyses to mature seeds. An even greater source of variation is due to the largely uncontrollable influences of different environments on plant growth and ultimately on seed development. This problem of variability, which is particularly acute when collecting seed in the wild, can only really be solved by wide sampling of seeds of a given species. In many recent studies of seed fatty acids, up to 30 individual samples may be examined in order to obtain representative values (Stone *et al.*, 1969).

2. *Family patterns in seed fats*

In 1963, Shorland comprehensively reviewed the data available at that time on fatty acid patterns in plant lipids; more recent results are available in Hitchcock and Nichols (1971). Two main points emerge from these reviews. First, a particular fatty acid pattern is characteristic of a given plant family and plant families can be placed into four main classes (and various subclasses) according to their fatty acid constituents. These classes may refer simply to varying concentrations of the common acids or to the presence of a particular type of rare fatty acid (Table 8.1). Secondly, the fatty acid patterns so found in various families bear no relationship to the systematic classification of these families. Closely related families have dissimilar patterns (compare Fumariaceae and Papaveraceae or Scrophulariaceae and Bignoniaceae in Table 8.1). Unrelated families may have almost identical patterns; for example, the Lauraceae in the dicotyledons and Palmae in the monocotyledons are both rich in myristic and lauric acids. Occasionally, two related families may have an unusual acid in common (e.g. Umbelliferae and Araliaceae with petroselinic acid); the value of such an acid as a marker is usually diminished, however, by its presence in several unrelated sources as well.

Since the major review of Shorland (1963), the pace of research on seed fatty acids has quickened and more data have come to hand. Most surveys are carried out for economic reasons and taxonomic aspects are still sometimes neglected. However, the main result of more recent investigations has been to cast doubt on the view that all species within a given family will have the same fatty acid profile.

This is clearly not so in a very large family, such as the Compositae, where a number of rarer fatty acids are found variously in different tribes of the family (Heywood *et al.*, 1977). Even in the Cruciferae, which at one time appeared to be marked off by the presence of erucic acid, there seems to be considerable variation. Appelqvist (1976), in reviewing fatty acid patterns in this family, pointed out that erucic acid has only been found to be present in 75% of the total number of species surveyed and the levels present may vary from 0 to 60% of the total fatty acids. There are also large intrageneric and intraspecific variations in erucic acid content among the positive species. The character cannot therefore be considered to characterize the family chemically as clearly as can the more consistent glucosinolate feature (see chapter 5).

It is possible that the distribution of petroselinic acid in seed oils of the Umbelliferae may similarly be found to be limited when further investigations are carried out. Analyses so far have been restricted to about 80 taxa (Hegnauer, 1973), which is a poor sample in a family that contains nearly 3000 species. Furthermore, petroselinic acid is not restricted to the Umbelliferae. In one survey by Kartha and Khan (1969), ten out of ten species

Table 8.1 Classification of seed fats in angiosperm families

Class (subclass)	Major fatty acids	Families[a]
I	Linoleic and/or linolenic with oleic	
	(a) linolenic-rich	15 families including Actidiniaceae, Boraginaceae, Ericaceae, Labiatae
	(b) linoleic-rich	35 families including Gramineae, Liliaceae, Papaveraceae, Scrophulariaceae, Solanaceae
II	Linoleic, oleic with linolenic or other polyunsaturated acid	7 families including Cucurbitaceae, Euphorbiaceae, Rosaceae
III	Palmitic, oleic and linoleic	28 families including Acanthaceae Anacardiaceae, Capparaceae, Fumariaceae
IV	Palmitic, oleic, linoleic and unusual fatty acid(s)	
	(a) with unsaturated acids e.g. petroselinic	18 families including Umbelliferae, Araliaceae
	(b) with saturated acids e.g. myristic, lauric	24 families 6 families including Lauraceae, Palmae, Myristicaceae

[a]Adapted from Shorland (1963) where complete lists of relevant families are given.

tested were positive. Unfortunately, however, one of the species tested was incorrectly ascribed to the Umbelliferae due to its confusing common name of black cumin; the plant in fact is *Nigella sativum*, which belongs to the Ranunculaceae! In fact, there are records of petroselinic acid having been detected also in Simaroubaceae, Theaceae and Euphorbiaceae.

3. *Fatty acids below the family level*

Alston and Turner (1963), in commenting on the situation regarding fatty acid patterns in the angiosperms, emphasized that little attempt had been made to use fatty acids directly to solve systematic problems. The situation is still largely the same today, although the picture is not quite as negative as it then appeared. Several useful surveys have been conducted within particular families and some tentative correlations with taxonomy have been noted. Vickery (1971) surveyed seed fats in 26 species of the Proteaceae, determined the relative proportions of some 17 standard acids and found that more acids were present in one subfamily, the Grevilleoideae, than in another, the Proteoideae. Again, Hopkins *et al.* (1969) determined the concentrations of six acetylenic fatty acids in 22 species (from 13 genera) of the Santalaceae and noted some incomplete correlations with generic classification. Yet again, Morice (1975a,b) examined 33 species of Cyperaceae, recorded relative amounts of five saturated and five unsaturated acids and found some correlations with tribal classification. In all these and other cases, however, the number of species sampled was really too small to draw any definite taxonomic conclusions.

Table 8.2 Fatty acid profiles in the Nyssaceae and Cornaceae

	Fatty acid content (%)[a]				
Family, genus and species	16:0	18:0	18:1	18:2	18:3
Nyssaceae					
Nyssa 3 spp. (average)	8.1	3.1	15.5	32.5	40.8
Davidia involucrata[b]	5.1	1.3	20.2	31.7	41.2
Camptotheca acuminata	8.3	4.5	11.3	12.7	63.3
Cornaceae					
Cornus 11 spp. (average)	6.8	2.0	19.1	71.5	0.9
Mastixia philippensis	33.5	6.0	8.5	38.0	14.0
Aucuba japonica	13.3	6.9	62.8	8.5	14.7

[a]For key to acids, see Fig 8.1. These are average values from several samples; actual values range about ± 2–5% around the average.
[b]Comparative data are available (see text) for *Actinidia arguta* (Actinidiaceae) which has a higher linolenic acid content (Earle *et al.*, 1960) than *Davidia* spp.

More successful results have been obtained when fatty acid analyses have been restricted to smaller plant groups, as in the investigations of Stone *et al.* (1969) on *Carya* (Juglandaceae), of Nordby and Nagy (1974) on *Citrus* (Rutaceae) and of Hohn and Meinschein (1976) on *Nyssa* and *Cornus* species. Some of the seed data obtained by the last workers is summarized in Table 8.2. The results are reliable in that, on average, five samples were analysed for each species and the data subjected to extensive statistical analyses. The survey also covered most known taxa at both the generic and species level. Furthermore, the results have some bearing on taxonomic problems within this group of related plants. To mention just one, there is the problem of the handkerchief tree, *Davidia involucrata*, which has been variously placed, (a) in *Nyssa*, (b) as a separate genus in the Nyssaceae, (c) in its own family Davidiaceae and (d) in the Actinidiaceae. The fatty acid data (Table 8.2) argue rather convincingly for a close relationship between *Davidia* and *Nyssa*, which is also reflected in wood, fruit and inflorescence characters. Thus, the chemical evidence favours (a) or (b) rather than (c) or (d). In this way, the fatty acid results seem to have made a really positive contribution to the taxonomy of these plants. It suggests that, in future, fatty acid patterns are going to be of most interest to systematists at the lower rather than at the higher levels of plant classification.

III. Alkanes and related hydrocarbons

A. Chemistry and distribution

Since all plants have a similar looking waxy coating on leaves and other organs, one would not immediately suspect much variation in this chemical feature. Indeed, early studies of the chemistry of leaf waxes revealed the presence of a few simple saturated hydrocarbons, called alkanes, in most species. It was not until the advent of thin-layer and gas chromatography and their application to leaf waxes that the full extent of both qualitative and quantitative variation in these waxes was actually realized. Eglinton *et al.* (1962), in a GLC survey of waxes from crassulacean plants, were among the first to show that species could be distinguished by the alkane patterns of their leaves. Although the qualitative pattern is relatively similar from plant to plant, there are considerable quantitative variations. Alkanes also occur in fungi and other lower plant groups, and the pattern is generally like that found in higher plants (Weete, 1972). The function of alkanes in the cuticular waxes of plants is a protective one, the water-repellent properties providing a means of controlling water balance in the leaf and stem. Their universal presence in leaf coatings may also be to provide a measure of disease resistance to the plant.

Alkanes are saturated long-chain hydrocarbons with the general formula $CH_3(CH_2)_nCH_3$. They are usually present in the range C_{25} to C_{35} carbon atoms, i.e. general formula with $n = 23$ to 33. Odd-numbered members of the series, C_{25}, C_{35}, etc. predominate over the even-numbered members of the series, often to the extent of 10 : 1. The major constituents of waxes are thus C_{27}–C_{31} alkanes. Examples are *n*-nonacosane, $C_{29}H_{60}$ and *n*-hentriacontane, $C_{31}H_{64}$, the major constituent of candelilla wax from *Euphorbia* species. Biosynthetically, these hydrocarbons are related to the fatty acids (see section II, p. 184 and, in the simplest instances, are formed from them by chain elongation and subsequent decarboxylation. Their origin by loss of one carbon (through decarboxylation) from fatty acids with even carbon numbers explains why the predominant alkanes have an odd number of carbons.

There are also a considerable number of alkane derivatives in waxes, formed by the introduction of unsaturation, branching of the chain, or oxidation to alcohol, aldehyde or ketone. Branching most commonly occurs near the end of the carbon chain. Two types may be mentioned; isoalkanes, with general formula $(CH_3)_2CH\text{-}(CH_2)_n\text{-}CH_3$; and anteisoalkanes, general formula $(CH_3)(C_2H_5)CH(CH_2)_n\text{-}CH_3$. Branched alkanes are by no means universally present and rarely occur in any quantity. Olefinic alkanes, or alkenes, have a similar distribution in that they occur fairly frequently but in relatively low amounts. Exceptionally high amounts of alkenes have been detected in rye pollen, rose petals and sugar cane.

Hydrocarbon alcohols are fairly common in plants, ceryl alcohol $CH_3(CH_2)_{24}CH_2OH$ being a regular constituent in many cuticular waxes. By contrast, aldehydes and ketones are rare. One taxonomically interesting and rare type of ketone in waxes are long chain β-diketones. An example is the C_{29} compound $CH_3(CH_2)_{10}COCH_2CO(CH_2)_{14}CH_3$ found in waxes of certain *Eucalyptus* species (Horn *et al.*, 1964). Such β-diketones have also been found in *Buxus* (Buxaceae), several Gramineae and *Rhododendron* (Ericaceae) waxes.

Finally, one other group of wax constituents should be mentioned; these are the cuticular fatty acids which are generally of a much longer chain length than the lipid fatty acids (see section II, p. 184). Indeed, they are biosynthetically related to the co-occurring alkanes and have chain lengths ranging from C_{24} to C_{32}. Lignoceric acid is the saturated C_{24} acid and cerotic the C_{26} acid. In addition, the cutin may contain a series of hydroxy acids of the *same* chain length as the lipid fatty acids. Thus 10,16-dihydroxypalmitic acid is a major component in many plant cutins. In gymnosperms, 9,16-dihydroxypalmitic acid is found instead, and in ferns and lycopods the major acid is 16-monohydroxypalmitic acid (Hunneman and Eglinton, 1972).

B. Detection

One of the major problems in analysing plant alkanes is contamination, since these hydrocarbons are widely present in petroleum products and thus may be present in traces in laboratory solvents and glassware. It is essential to employ clean glassware and redistilled solvents and also to avoid contact with stopcock grease or plastic tubing. Another source of possible contamination is contact with 'Parafilm', a thermoplastic sealing material widely employed in phytochemical laboratories. Indeed, this material is very useful since, when washed with benzene or hexane, it will give a solution containing the standard range of *n*-alkanes, which can then be used for GLC comparison with alkanes from a plant extract (Gaskin *et al.*, 1971).

Extraction of plant waxes is simply carried out by dipping unbroken leaves or stems into diethyl ether or chloroform for about 30 seconds; this removes the surface alkanes without attacking cytoplasmic constituents. An alternative procedure is to Soxhlet extract dried powdered leaf for several hours in hexane. Such an extract will be greatly contaminated with leaf lipids and some fractionation will be necessary before pure hydrocarbons are obtained. For example, steam-distillation may be desirable in order to remove any essential oils from such an extract.

It is common practice to fractionate the direct wax extract either in order to remove undesirable components (such as lipids) or to separate the hydrocarbon classes according to polarity. In many cases, all that is necessary is to pass the crude extract through an alumina column and elute with light petroleum (b.p. 40–60°C). The first fraction contains the alkanes, the later fractions containing the alcohols, aldehydes or ketones. The purity of the alkane fraction can be checked by infrared spectroscopy. Gas chromatography is then carried out by standard procedures, a variety of column coatings having been employed from Apiezon L grease to silicone oil (Eglinton *et al.*, 1962). There is a linear relationship between the logarithm of the relative retention time on the column and the alkane carbon atom. It is then possible to use this relationship to identify individual alkanes. When genuine alkanes of known carbon atoms are added separately to plant extracts, there will be an intensification of appropriate peaks when the extract is rerun through the column. Further tests have to be made in order to detect branched alkanes and unsaturated derivatives (e.g. alkenes). Identifications are conveniently confirmed by use of combined gas chromatography–mass spectrometry (GC–MS). For further practical details, the text of Martin and Juniper (1970) may be consulted.

C. Taxonomic utility

Ever since alkane patterns were first related to plant taxonomy in the early 1960s, phytochemists have stressed the need for caution in interpreting the results. The identification of hydrocarbons in leaf washings requires considerable skill and the novice may well obtain misleading data due to contamination of samples (see preceding section). Experimental conditions used for comparative analyses of alkanes therefore need to be carefully controlled and regularly monitored. In addition, the study of plant alkanes is subject to all the standard biological variables. Patterns are influenced to a greater or lesser extent by environment, age of the plant, part of the plant sampled and so on. Most of these variables are documented and discussed in the several reviews already available on alkane chemotaxonomy (Eglinton and Hamilton, 1963; Douglas and Eglinton, 1967; Herbin and Robins, 1969; Martin and Juniper, 1970). Here it seems most useful to consider briefly what is known of alkane patterns in turn at the family, generic and species level.

The possibility of using alkane patterns at the family level was discarded by Eglinton and Hamilton (1963), on the basis of the patterns in leaves that they produced from 13 families and 21 genera. The diversity was such that it was not possible to draw any taxonomic conclusions. Furthermore, the nature of the data did not appear to permit one to be able to assign a species to a given family from the alkane contents.

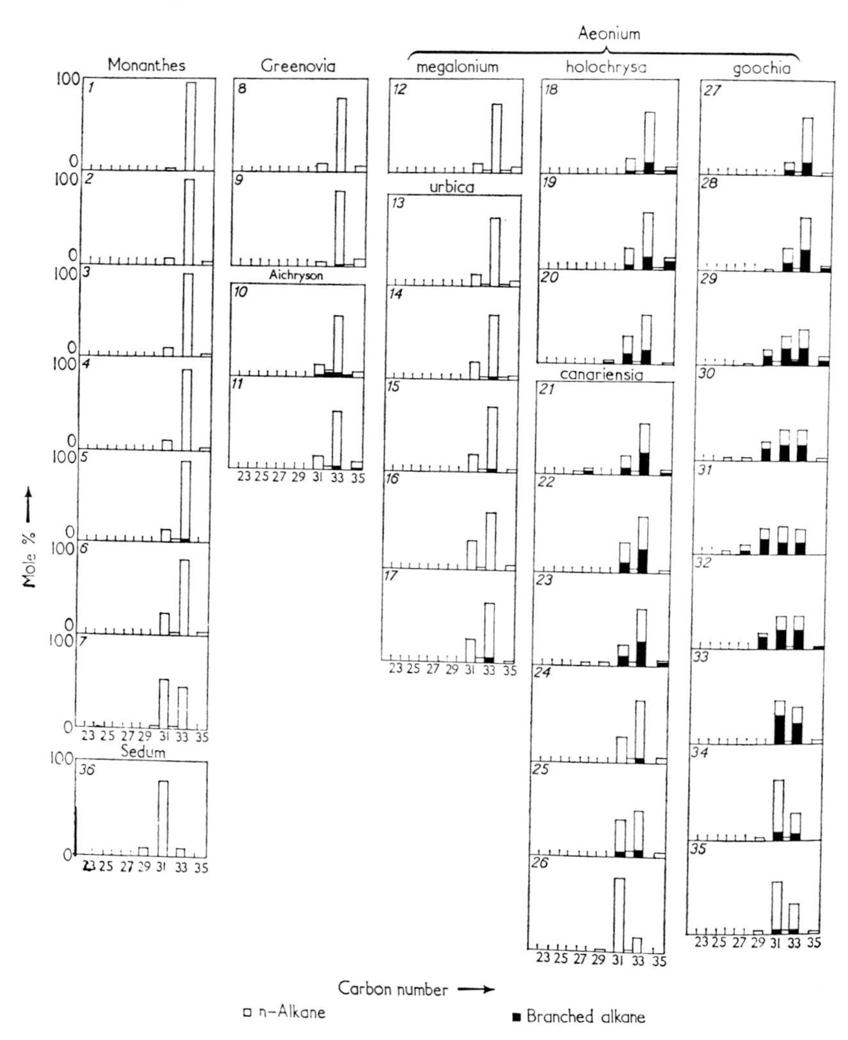

Fig. 8.2 Distribution (mol %) of normal (□) and branched (■) alkanes C_{25} to C_{35} in surface waxes from the leaves of *Monanthes*, *Greenovia*, *Aichryson* and *Aeonium* species (from Eglinton and Hamilton, 1963).

How far alkane patterns are related to taxonomy at the generic level is still not entirely clear, in spite of the fact that a variety of genera have now been explored in some depth. Some of the early data of Eglinton *et al.* (1962) on genera within the subfamily Sempervivoideae of the Crassulaceae is illustrated in Fig. 8.2. Correlations are apparent (e.g. *Sedum* is distinctive in having C_{33} alkane dominant) but there is no sharp cut off between most of the genera included. In *Aeonium*, patterns are shown for species within five sections, but here again it is not possible to discern clear-cut correlations with sectional classifications. Indeed, on the basis of these alkane profiles, a quite different arrangement of species was suggested, which would put together species from different sections.

Quite recently, Scora *et al.* (1975) compared alkane patterns in 22 taxa of *Persea* with that of *Beilschmiedia meirsii*, a closely related genus in the same family, Lauraceae. *Beilschmiedia meirsii* has, in fact, an alkane distribution quite different from those in *Persea* spp. In *B. meirsii* $C_{27}H_{56}$ is the dominant alkane, whereas in *Persea* spp. $C_{33}H_{68}$ is dominant; there are also other lesser differences in the alkane pattern. Thus, in favourable circumstances, it is possible to use alkane patterns to draw distinctions between genera. However, even in this case, a wider sampling is still needed to confirm the differences, since up to 40 species of *Beilschmiedia* have been described.

At the species level, there are considerable data indicating that alkane pattern is usually distinctive, although not invariably so. Also, it appears that alkane patterns in other parts of the plant besides the leaf may be more reliable for differentiating between species. Herbin and Robins (1968) examined alkanes of 63 species of *Aloe* (Liliaceae) and confirmed the species specificity that was apparent in the earlier experiments of Eglinton *et al.* (1962) on *Aeonium* (Crassulaceae) (see Fig. 8.2). In *Aloe*, however, rather better results emerged from a study of alkanes of the perianth than from those of the leaf. In a similar survey of species in the tuberous *Solanum* (Solanaceae), Mecklenburg (1966) found that alkane patterns in the inflorescence rather than the leaf enabled one to separate species. The type of GLC profile obtained with *Solanum* spp. is illustrated in Fig. 8.3. This shows the 'reproducibility' of two samples from the same species (*S. polyadenium*) and the differences that can exist between different species (*S. polyadenium* and *S. hougasii*). Here it should be pointed out that these inflorescence waxes in *Solanum* differ from most leaf waxes in the relatively high proportion of branched alkanes present. Finally, in an examination of *Citrus* alkanes, Nordby and Nagy (1977) found that those of the peel were most distinctive. These authors were able to distinguish between mandarins, oranges, lemons, grapefruits and limes on the basis of the peel alkane patterns.

Although much data have accumulated on plant alkanes, these patterns have rarely been applied to the actual solution of taxonomic problems. An interesting example in which alkanes have made a positive contribution to

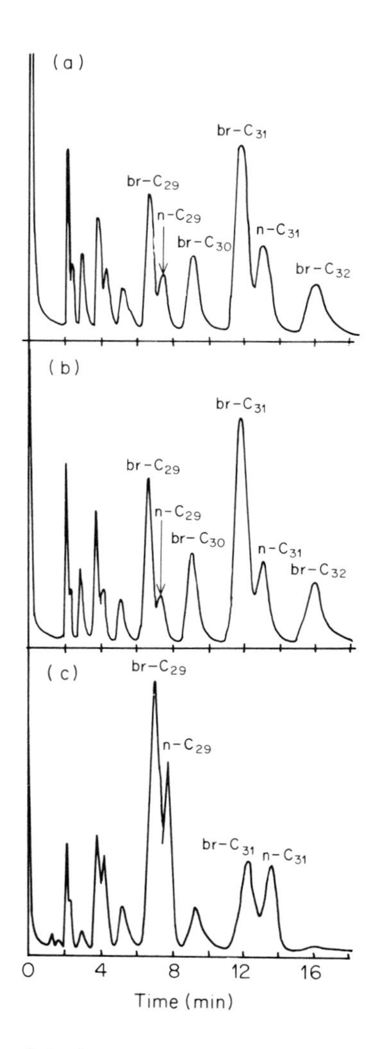

Fig. 8.3 GLC profiles of hydrocarbons from the inflorescences of two clones of *S. polyadenium* (a and b) and one of *S. hougasii* (c) (from Mecklenburg, 1966).

species identification is in *Arbutus* (Ericaceae) (Sorensen *et al.*, 1978). Three species have been described as native to the United States, *A. arizonica*, *A. texana* and *A. menziesii*. The last two, although allopatric, overlap in morphological characters. Furthermore, they cannot be distinguished by looking at herbarium specimens; only field observations allow them to be separated morphologically. The results of alkane analysis on 16 samples from three species are given in Table 8.3. Although there is infraspecific variation, all three taxa can be distinguished by differences in alkane proportions.

Table 8.3 Alkanes of *Arbutus* leaves expressed as a percentage of the total alkane content.

Collection no.[a]	Collecting locality	C_{17}	C_{18}	C_{19}	C_{20}	C_{22}	C_{23}	C_{24}	C_{25}	C_{26}	C_{27}	C_{28}	C_{29}	C_{30}	C_{31}	C_{32}	C_{33}	C_{35}
A. menziesi																		
7332A	Olympic Peninsula	1.6									3.5		32.5		45.3		16.7	
7331C	Roseburg, Oregon	1.3		1.1							2.6	1.0	29.7	1.1	42.4	1.2	18.7	
7330A	Redwood L., Calif.										1.4	1.7	22.5	1.1	42.2	1.8	25.1	
7338C	Santa Cruz Mountains, Calif.	1.2									2.6	16.4	20.5		35.5	1.1	19.6	
A. arizonica																		
7329A	Herb Martyr Dam Arizona	5.4	1.0	1.1							3.6	1.4	11.0	1.0	30.3	3.0	38.8	
7329B		2.7	1.5	1.8								1.3	15.5	1.1	31.9	3.0	31.5	7.7
7328A	near Portal, Arizona	1.3	1.6								3.1	11.2	11.2	3.1	26.7	2.8	37.4	
7326A	South Fork Camp Portal, Arizona										2.8	1.3	20.6	1.9	41.2	3.3	27.9	
7326X											3.6	1.3	22.0	1.3	40.3	3.0	25.9	
Arbutius texana																		
7298A	Guadalupe Mountains New Mexico	24.4	1.4								2.7	3.3	13.5	1.7	29.2	17.7	1.5	
7298C		16.3	1.1						1.0	1.0	3.0	2.0	22.7	1.3	28.1	17.9	1.6	
7298D		22.5	1.0	1.0					1.1	1.1	2.7	3.8	33.0		28.4	5.1		
7310A	Davis Mountains Texas	23.5	1.5	1.4	1.0	1.3	1.6	1.5	1.1	1.0	2.7	3.4	32.5		25.1			
7317A	Big Bend National Park, Texas	12.5	1.0					2.6	1.6	1.0	2.5	8.0	15.1	1.3	29.6		21.3	
7317B		20.1	1.0	1.1							2.2	13.6	9.0	1.5	24.8	1.0	21.8	
7317C		11.8				1.5	1.8	18.5	2.3	3.3	1.7	7.6	15.5		22.8		9.5	

[a]Collection numbers refer to populations, letters to individual trees within a population
Values <1.0 not shown.

Arbutus texana is particularly well separated, with its high content of C_{17} alkane. *Arbutus arizonica* and *A. menziesii* differ mainly in their content of C_{33} hydrocarbon. It is interesting that the two species that are furthest apart geographically (*A. menziesii* occurs in Oregon) are closest together in alkane patterns. In conclusion, the *n*-alkane biochemistry of *Arbutus* spp. usefully combines with morphological data in supporting the continued recognition of three species as occurring in the western United States.

Finally, it has been established by phytochemists working with leaf waxes that more valuable chemotaxonomic insights may be obtained in some plants by concentrating on other components besides the simple alkanes. Thus, in a recent study of *Rhododendron* waxes, four long-chain β-diketones were found in considerable quantity in a number of species (Evans *et al.*, 1975a). Taxonomically, their occurrence correlated with subsectional classification and also they were more frequent in lepidote than in elepidote species. Steryl acetates were also discovered in the waxes of certain other species (Evans *et al.*, 1975b). Clearly, these characters will be of some value for taxonomic discrimination in what is a very large and complex genus.

IV. Polyacetylenes

A. Chemistry and distribution

Polyacetylenes are a very unusual group of natural substances because of their considerable unsaturation and high reactivity. They are thermally unstable and may even explode during distillation. On exposure to daylight, they may decompose with the production of red or blue colours. What is so remarkable is that most of them are sufficiently stable to be isolated and characterized by standard phytochemical techniques. Indeed, well over 650 polyacetylenes are now known as plant products (Bohlmann *et al.*, 1973). A few are simple hydrocarbon derivatives, such as penta-yne, which is widely present in members of the Compositae. Most have additional functional groups (the presence of which tends to stabilize the reactive acetylenic bonds) and are either alcohols, ketones, acids, esters, aromatics, furans, pyrans or thiophenes. Some typical structures are illustrated in Fig. 8.4.

Polyacetylenes have a taxonomically restricted distribution, since they occur regularly in only five angiosperm families, namely the Campanulaceae, Compositae, Araliaceae, Pittosporaceae and Umbelliferae. The Compositae is by far the richest source and it contains many acetylenic structures which are unique to the family. Very rarely, polyacetylenes may be formed elsewhere in plants in response to fungal invasion as phytoalexins. Indeed, wyerone acid (see Fig. 8.4) is formed this way in *Lens* and *Vicia* species in the Leguminosae, and falcarindiol in *Lycopersicon* Solanaceae) (see chapter 15).

Straight chain hydrocarbon	$CH_3-(C\equiv C)_5-CH=CH_2$ Penta-yne
Alcohol	$CH_2=CH-CHOH-(C\equiv C)_2CH_2CH=CH(CH_2)_7H$ Falcarinol
Ketone	$CH_2=CH-CO(C\equiv C)_2CH_2CH=CH(CH_2)_7H$ Falcarinone
Ester	$CH_3-(C\equiv C)_3-CH=CH-CO_2Me$ Dehydromatricaria ester
Aromatic	$Ph-CH_2-C\equiv C-C\equiv CMe$ Capillen
Furan	$CH_3CH_2CH=CH-C\equiv C-CO-$(furan ring, O)$-CH=CHCO_2H$ Wyerone acid
Tetrahydropyran	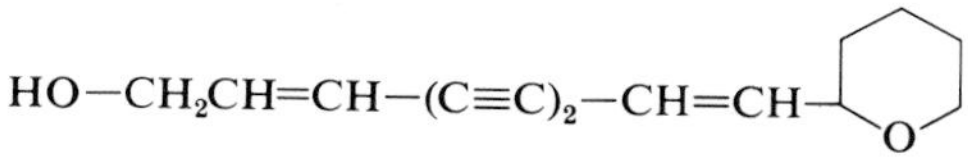 $HO-CH_2CH=CH-(C\equiv C)_2-CH=CH-$(tetrahydropyran ring, O)
Thiophen	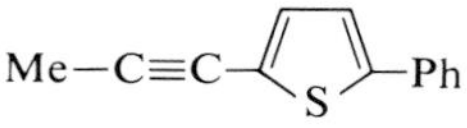 $Me-C\equiv C-$(thiophene ring, S)$-Ph$ (three linked thiophene rings, S S S) Terthienyl

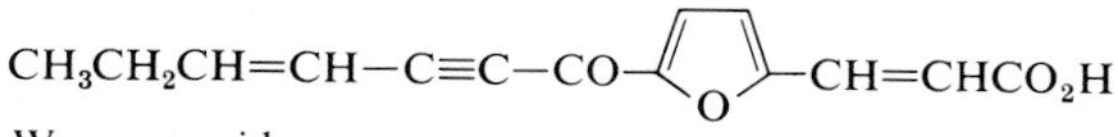

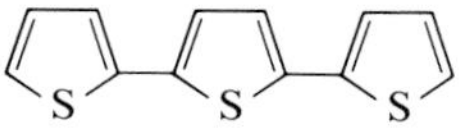

Fig. 8.4 Structures of some representative polyacetylenes.

Acetylenic fatty acids, which have already been mentioned briefly in section II of this chapter, may accompany polyacetylenes in the above five families and one substance in particular, crepenynic acid, is a key precursor of polyacetylenes (Fig. 8.5). Related fatty acids such as stearolic are found in a number of other families, without being accompanied by polyacetylenes; such sources include Santalaceae, Malvaceae and their related taxa. The only other major source of polyacetylenes in the plant kingdom is in the higher fungi, where they occur in two basidiomycete families, the Agaricaceae and Polyporaceae. The fungal compounds differ in having a chain length mainly between C_8 and C_{14}, whereas the higher plant acetylenes are mostly C_{14} to C_{18} compounds.

In their biosynthesis, polyacetylenes are probably mostly formed from oleic acid, via the key acetylenic intermediate crepenynic acid (see Fig. 8.5). This intermediate undergoes a variety of modifications, including successive dehydrogenations to yield the range of polyacetylenes that are known. Simple chain shortening may be by decarboxylation to give a C_{17} polyacetylene such as falcarinone. More drastic chain shortening may occur to give a C_{13}

hydrocarbon such as penta-yne or the C_{10} ester, dehydromatricaria ester. Reaction of two adjacent acetylenic groups with reduced sulphur may occur in some plants, giving rise to thiophene derivatives. In certain cases, all of the acetylenic bonds may be used up in this way, to give ultimately a substance such as α-methylterthienyl (Fig. 8.5). Although clearly lacking any acetylenic substitution, this latter substance is regarded as an acetylenic derivative by Sorensen (1977), because of its biosynthetic origin.

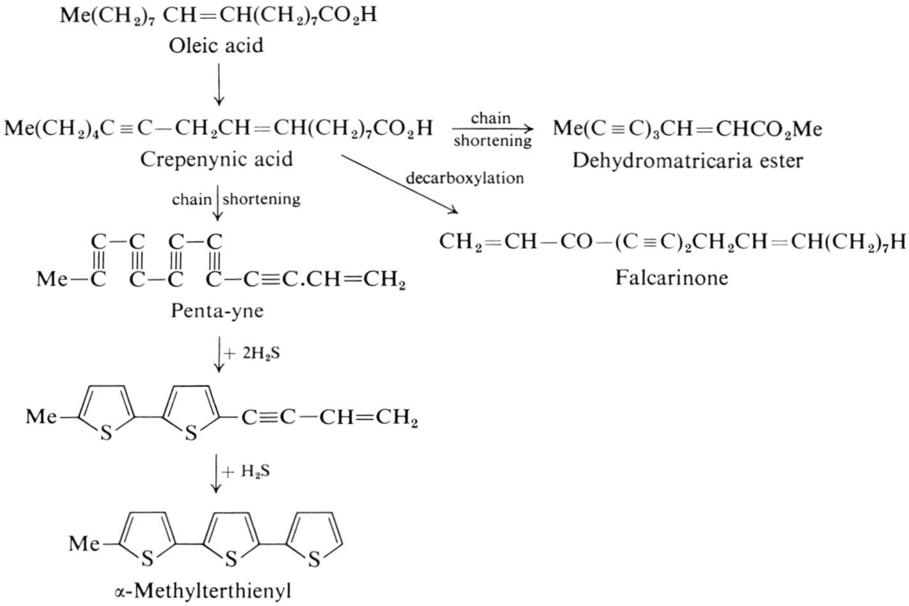

$Me(CH_2)_7CH{=}CH(CH_2)_7CO_2H$
Oleic acid

↓

$Me(CH_2)_4C{\equiv}C{-}CH_2CH{=}CH(CH_2)_7CO_2H$ —chain shortening→ $Me(C{\equiv}C)_3CH{=}CHCO_2Me$
Crepenynic acid — Dehydromatricaria ester

Crepenynic acid —decarboxylation→ $CH_2{=}CH{-}CO{-}(C{\equiv}C)_2CH_2CH{=}CH(CH_2)_7H$
Falcarinone

chain ↓ shortening

$Me{-}C{\equiv}C{-}C{\equiv}C{-}C{\equiv}C{-}C{\equiv}C.CH{=}CH_2$
Penta-yne

↓ $+ 2H_2S$

Me–(thienyl)–(thienyl)–$C{\equiv}C{-}CH{=}CH_2$

↓ $+ H_2S$

Me–(thienyl)–(thienyl)–(thienyl)
α-Methylterthienyl

Fig. 8.5 Biogenetic origin of polyacetylenes.

The function of polyacetylenes in plants has largely been neglected, but there is some evidence that they are involved in plant–animal or plant–plant interactions. Some are highly poisonous to mammals, for example those found in the roots of the water dropwort *Oenanthe crocata*, whereas others are piscicidal or insecticidal. Many also display antibiotic activity. Recent evidence has shown that the toxic and antibiotic effects of polyacetylenes are enhanced by ultraviolet irradiation, i.e. they have a photosensitizing capacity (Camm *et al.*, 1975).

B. Detection

Polyacetylenes are readily extracted from fresh root or leaf tissue by maceration in the presence of diethyl ether. This ether extract, after drying and concentrating, can then be developed on thin layers of silica gel using any lipid-type solvent system (e.g. pentane–ether; 9 : 1, v/v). On spraying the plate with isatin in concentrated sulphuric acid and then heating, acetylenics, if present, will appear as brown or green spots. Such preliminary detection methods need to be confirmed by ultraviolet spectroscopy, since almost all polyacetylenes show intense and characteristic peaks in the 200–320 nm region. Crude ether extracts can also be directly screened in the ultraviolet spectrophotometer for polyacetylenes (see, e.g. Bentley *et al.*, 1969).

Unfortunately, ultraviolet spectroscopy is not an infallible guide to the presence or absence of polyacetylenes, since just a few compounds fail to give a series of intense peaks. In such cases, infrared spectra may be measured, since there is a characteristic band for the acetylenic triple bond. For complete identification, other spectral procedures (NMR or MS) are invaluable (Sorensen, 1968; Bohlmann *et al.*, 1973).

C. Taxonomic utility

Since over 1100 species of Compositae and 200 species of Umbelliferae have been surveyed for polyacetylenes, much is known about their natural distribution. In the majority of these species, the acetylenic components have been isolated and identified; further, the results have been collected together in monographic form (Bohlmann *et al.*, 1973). Thus the chemistry is secure; unfortunately, taxonomic aspects have not been so fully developed. With the results available, it is possible to perceive certain correlations with taxonomy. In general, however, the acetylenes have not been interpreted in the detail they deserve. Also, they have rarely been employed in a decisive way for evaluating taxonomic relationships. It is only possible here to comment briefly on their value in interrelating those families that contain them and their use in relating taxa, primarily at the generic level, within the Compositae.

The five dicotyledonous families regularly containing polyacetylenes in their tissues fall into two morphologically distinct groups, the Araliaceae, Pittosporaceae and Umbelliferae and the Campanulaceae and Compositae. The pattern in the first three families is a uniformly simple one, based on aliphatic hydrocarbons, alcohols and ketones. One particular marker, falcarinone, is widespread in Araliaceae and Umbelliferae and the related alcohol falcarinol has been detected in the one species of Pittosporaceae which has been fully analysed, namely *Pittosporum buchananii.* The chemical data thus support the unity of this grouping of families, which most taxonomists recognize today as comprising the same single order, the Araliales.

The fairly recent discovery of acetylenics in *Pittosporum* has been a valuable aid, together with various other chemical markers, in confirming the re-assignment of Pittosporaceae to the Araliaceae–Umbelliferae alliance.

Previously, it had been placed near the Saxifragaceae because of the floral morphology. However, this new alliance is also indicated by a series of other biological features in Pittosporaceae, which distinguish it from Saxifragaceae, including root and wood anatomy, trinucleate pollen type and presence of schizogenous oil ducts (Hegnauer, 1969).

The chemical link between Campanulaceae and Compositae through polyacetylenes is a more tenuous one. Thus, the Compositae contain an enormous range of substances, and many complex and highly derived structures. By contrast, the Campanulaceae, as far as it has been studied (some 32 species), has only one main type based on terminal substitution in the hydrocarbon chain by a tetrahydropyran group (see Fig. 8.4). This is a type of compound that is not known in the Compositae. Nevertheless, the polyacetylene character itself does link the two families and also confirms the significance of other chemical links, for example the presence of a distinctive type of storage polysaccharide (see chapter 19) between the two groups.

Within the Compositae, much information is available at the tribal, subtribal and generic level (Bohlmann *et al.*, 1973; Sorensen, 1977), but it is not possible to draw too many conclusions from these data. At one extreme, the tribes Heliantheae, Anthemideae and Cynareae are particularly rich in complex structures and at the other there are the Cichorieae and Senecioneae which are poor in acetylenes. The Anthemideae and Astereae are chemically distinct from other tribes by the common occurrence of C_{10} acetylenes. However, these two tribes are not particularly close to each other in overall morphology so that the acetylenic link is an enigmatic one.

The general poverty of acetylenics in the Senecioneae has some taxonomic interest in that it has been argued that the few taxa that are positive, and especially the genera *Arnica* and *Doronicum*, might be considered anomalous and thus removed from the tribe. Although there are other more compelling reasons for removing *Arnica* from the tribe, *Doronicum* is quite characteristically senecionoid in its floral morphology and cytology and could not thus be dislodged by the acetylenic data (Nordenstam, 1977).

Greger (1977) has given detailed consideration to the polyacetylene types found in the Anthemideae and has summarized the available data in diagrammatic form (Fig. 8.6). The main type of acetylene in this tribe is based on C_{14} hydrocarbons, whereas C_{13} compounds are considered to be derived forms. Acetylenes based on dehydromatricaria ester (DME) and with amide substituents are also biosynthetically derived. By contrast, the most primitive feature in the tribe is the dihydrofalcarinone type, present in *Cotula* spp. and related taxa. In summary, there are a range of structural types, which can be placed in order of biosynthetic complexity. It will be seen (Fig. 8.6) that these different types, with some overlapping, are discretely distributed among the different generic groupings within the tribe. This pattern (Fig. 8.6) fits in reasonably well with what is known of generic relationships from morphological, karyological and embryological findings. There is also a striking correlation between acetylenic structures and the geographical separation of the tribe between the Northern and Southern hemispheres. Finally, the absence of acetylenics from the *Ursinia* group may represent the loss of metabolic activity during evolution in a habitat where ecological pressures differed from those present where the original anthemoid stock evolved.

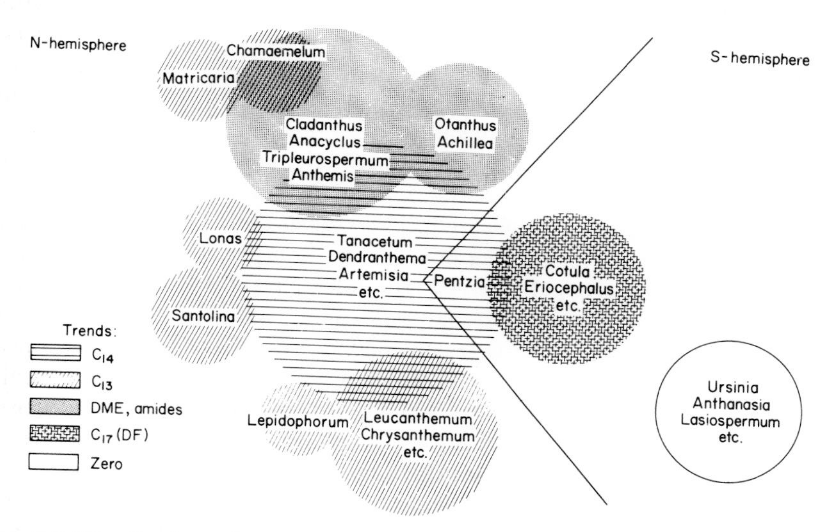

Fig. 8.6 Distribution of acetylenic structural types among genera of the tribe Anthemideae, family Compositae.

The significance of acetylenic data at the species level in the Compositae has occasionally been considered and there is some evidence that the patterns are usually invariant within a given species. In the case of the Anthemideae, the root acetylenes of seven taxa in the *Artemisia dracunculus* complex have been analysed in some detail (Greger, 1979). This group includes tarragon, a plant used for several centuries both medicinally and for flavouring food. Chemically, there is a clear separation of the three species, *A. dracunculiformis*, *A. glauca* and *A. pamirica*, which have sometimes been subsumed into *A. dracunculus*, from *A. dracunculus* s.str. The former group has aromatic acetylenes of a 'primitive' type (e.g. capillen), whereas the latter (four samples studied) has isocoumarin-substituted acetylenics, which are biosynthetically advanced. These differences in acetylenic patterns are reflected in the respective chromosome numbers: the first three species are diploid, whereas

A. dracunculus s.str. may be hexa-, octa- or decaploid. Thus both the cytology and chemistry indicate that the direction of evolution within the group has been from the first three species towards the more highly developed spice plant *A. dracunculus*.

V. Sugars and sugar derivatives

A. Chemistry and distribution

Sugars occupy a central position in plant metabolism; not only are they the first organic compounds produced in the photosynthetic carbon cycle but they also provide the major source of respiratory energy to the cell. Such sugars of intermediary metabolism (e.g. ribulose bisphosphate, glucose 1-phosphate and fructose 1,6-bisphosphate) are common to all photosynthetic organisms. Sugars also provide a major means of storing energy, either in low molecular-weight form as oligosaccharide (sucrose) or in high molecular-weight form as starch. Sucrose has a special role, too, as a transport form of energy. Again, glucose and other monosaccharides are required as precursors of cell wall synthesis. In addition, many other classes of plant constituent from nucleic acids to plant glycosides contain sugars as essential features of their structures. Finally, sugars play a number of ecological roles, for example in plant–animal interactions, since flower nectars are mainly sugar and water and in the detoxification of foreign substances entering the plant cell.

In many of these functions, the same sugars are involved in all plants and there is no significant variation. The pool of free sugar in higher plants invariably contains glucose, fructose, sucrose and traces of other sugars such as xylose, galactose and rhamnose. Variation does occur in the low molecular-weight storage of sugar and various oligosaccharides, sugar alcohols and cyclitols have been found to replace sucrose in this capacity in many plants. There are also variations in the ratios of sugar content in nectars. Such variations will be the main topic of discussion in this section. Variation in the sugar content at the polysaccharide level will be discussed later, in chapter 19.

Three classes of low molecular-weight sugar are thus of potential interest to the chemotaxonomist: the oligosaccharides, the sugar alcohols and the cyclitols (Fig. 8.7). Oligosaccharides are formed from the linking together of the same or different monosaccharides and are described as di-, tri- or tetrasaccharides, according to the number (2, 3 or 4) of monosaccharide units present. Structural variation is considerable, since even when considering the known disaccharides based on two glucose units, ten isomeric structures are possible. Variation is according to whether the linkage from the reducing end of one glucose unit to the other has an α- or β-configuration and to which of the five hydroxyl groups of the second glucose unit the linkage is made (1→1, 1→2, 1→3, 1→4 or 1→6). The possibilities of structural variation is enhanced when different monosaccharides are linked together to form a tetra- or pentasaccharide. It is not surprising, therefore, that Bailey in 1965 was able to list as many as 500 oligosaccharides as occurring in nature, in free or bound form; today, a similar list would probably extend to at least 700 sugar derivatives.

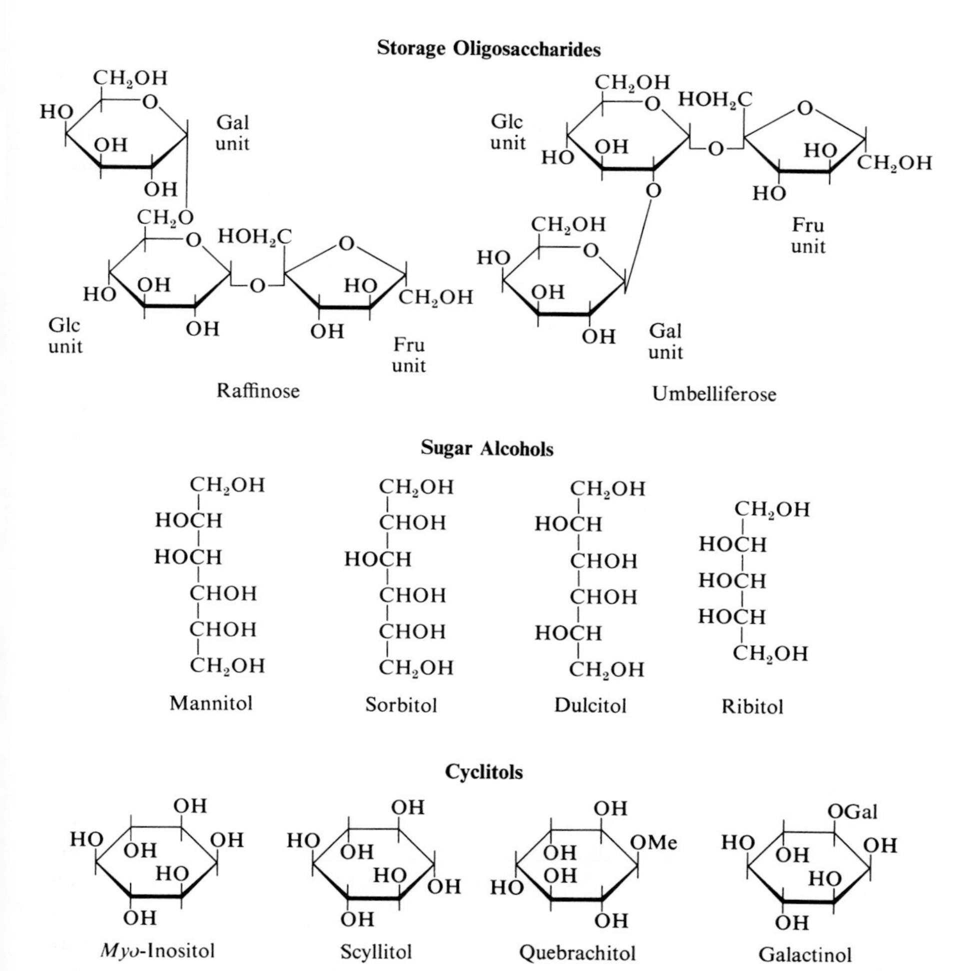

Fig. 8.7 Structures of plant storage sugars. Sugar abbreviations: Gal, galactose Glc, glucose; Fru, fructose.

Table 8.4 Storage oligosaccharides of taxonomic interest in plants

Oligosaccharide	Distribution
Disaccharide	
Trehalose (α-glucosyl-(1→1)-α-glucose)	Widespread in Selaginaceae; not known in higher plants.[a]
Trisaccharides	
Raffinose (6G-α-galactosylsucrose)	Widespread in seeds of higher plants, especially common in Leguminosae.
Planteose (6F-α-galactosylsucrose)	Characteristic in *Plantago* (Plantaginaceae); also elsewhere in seeds
Gentianose (6G-β-glucosylsucrose)	Restricted to rhizomes of *Gentiana* spp. (Gentianaceae).
Manninotriose (6-α-digalactosylglucose)	In *Fraxinus* manna, in cocoa beans and hazel nuts, *Corylus avellana*.
Umbelliferose (2G-α-galactosylsucrose)	Widespread in Umbelliferae, in all tribes of the Apioideae; also in Araliaceae and Pittosporaceae.
Tetrasaccharides	
Stachyose (6G-α-digalactosylsucrose)	Rhizomes of *Stachys* (Labiatae), seed of *Glycine max;* generally widespread.
Lychnose (6G,1F-α-digalactosylsucrose)	Roots of *Lychnis dioica* (Caryophyllaceae); restricted to tribe Sileneae.
Pentasaccharide	
Verbascose (6G-α-trigalactosylsucrose)	Roots of *Verbascum* (Scrophulariaceae), seeds of legumes; fairly widespread.
Hexasaccharide	
Ajugose (6G-α-tetragalactosylsucrose)	Roots of *Ajuga nipponensis* (Labiatae), also in *Verbascum*, *Salvia* and *Vicia*.

[a]Earlier reports of trehalose in beech, *Fagus sylvatica* and in cabbage *Brassica oleracea* could not be corroborated in later studies (Gussin, 1972).

Fortunately, the number of storage oligosaccharides so far found to accumulate in plant tissues is very much less than this. In fact, only ten structures in addition to sucrose (α-glucosylfructose) need to be considered (Table 8.4). These are all, like sucrose, non-reducing oligosaccharides, i.e. the reducing end of the hexose unit (usually glucose or galactose) is directly linked to a keto sugar (usually fructose). The most important series of storage sugar is that in which galactose units are added in turn to sucrose, through the 6-hydroxyl position of the glucose unit. The simplest member of the series is raffinose, with one galactose unit (see Fig. 8.7); other members are stachyose (two galactoses), verbascose (three galactoses) and ajugose (four galactoses). These substances are widespread in plants, particularly in seeds, tubers, rhizomes and roots, but also in fruits, barks and leaves. They are also, like sucrose, translocated in plant phloem.

In addition to the one major series of galactosyl-substituted sucroses, other structures occur which appear to characterize a particular genus or family. For example, the trisaccharide umbelliferose is an isomer of raffinose in which the galactose is linked to the 2- rather than the 6-position of the glucose in sucrose. This has been found characteristically in roots and other storage tissues of members of the Umbelliferae. Again, the tetrasaccharide lychnose, first isolated from *Lychnis dioica* (Caryophyllaceae), is only known to occur in this species and related taxa in the same tribe (Sileneae) of this family.

Most of the taxonomically interesting oligosaccharides of plants are shown in Table 8.4. In addition, it should be mentioned that a number of reducing oligosaccharides (e.g. maltose, primeverose) have been detected free in plants, but their occurrences at present seem to be quite limited. Even lactose (4-β-galactosylglucose), the sugar of cow's milk, has been characterized in plants but principally in one main source, the chicle plant, *Manilkara zapota* (Sapotaceae).

The sugar alcohols, the next class of storage sugar of taxonomic importance, are produced *in vitro* by reduction of the aldehyde group of monosaccharides. Thus, sorbitol can be obtained from glucose, dulcitol from galactose, mannitol from mannose and ribitol from ribose (Fig. 8.7). Curiously, the hexitol, mannitol is by far the most common of the sugar alcohols in higher plants, although the related hexose mannose cannot be detected in any quantity as occurring free in plants (Herold and Lewis, 1977). Most of the known sugar alcohols of higher plants (Table 8.5) are hexitols, with six carbon atoms, but those with fewer (e.g. ribitol) or more (e.g. volemitol) are also known.

The natural distribution of the plant alcohols falls into a similar pattern to that of the storage oligosaccharides. A few are common, mannitol, dulcitol, sorbitol and (probably) ribitol. The rest are known from only a single or a few plant sources and may be of interest as taxonomic markers. About half the sugar alcohols shown in Table 8.5 have been found also in algae, fungi or lichens. However, these lower plants have also yielded a number of sugar alcohols not so far described in higher plants (see Lewis and Smith, 1967a). Since the common sugar alcohols are formed as their phosphates in the photosynthetic carbon cycle, their biosynthetic origin is fairly clear. They are also known to exist in dynamic equilibrium with the corresponding keto sugars.

The third class of sugars, the cyclitols, are also sugar alcohols and are related to the alicyclic sugar alcohols already mentioned, but they differ in being carbocyclic. Thus, they are alcohols based on the six-membered carbon ring of cyclohexane. They usually have six hydroxyl groups, one or more of which may also be methylated (Fig. 8.8). Optical isomerism, as with other sugars, is a structural feature. Four isomeric hexahydroxy inositols are known, *myo*-inositol (or *meso*-inositol), L-inositol, D-inositol and scyllitol (or *scyllo*-inositol). Of these four, *myo*-inositol is of universal occurrence, best known as the lipid constituent of the calcium reservoir of seeds, phytic acid. Of the other three, scyllitol is the most widespread, being recorded in over seven plant families.

Table 8.5 Distribution of sugar alcohols in higher plants

Sugar alcohol	Distribution[a]
Pentitol	
Ribitol[b]	In *Adonis vernalis* (Ranunculaceae) and *Bupleurum falcatum* (Umbelliferae); probably also elsewhere
Hexitols	
Sorbitol[b]	In fruits and leaves of members of Rosaceae
D-Mannitol[b]	Common in at least 50 plant families
Dulcitol[b]	Characteristic of Celastraceae, but also in three other families
Allitol	Only in two spp. of *Itea* (Saxifragaceae)
L-Iditol	Only in fruits of *Sorbus aucuparia* (Rosaceae)
Polygalitol	In two families: Polygalaceae and Aceraceae
Styracitol	Only in fruits of *Styrax obassia* (Styracaceae)
Hamamalitol	In *Primula* spp. (Primulaceae)
Heptitols	
Volemitol[b]	In *Primula* spp., e.g. roots of *P. elatior*
Perseitol	In avocado, *Persea americana* (Lauraceae)
Octitol	
D-Erythro-D-galacto-octitol	In avocado, *P. americana* (Lauraceae)

[a]For references, see Plouvier (1963)
[b]These polyols also occur in various algae, lichens or fungi

Of the methylated inositols, pinitol, the 3-methyl ether of D-inositol, is quite common, being present in at least 13 angiosperm families. Quebrachitol, the 2-methyl ether of L-inositol, is also well known, being recorded in some 11 families. Two monomethyl inositols, sequoyitol and 1-methylmucoinositol, are confined to gymnosperms and are not known to any extent in angiosperm families. Yet other inositol derivatives such as 2-*C*-methylscyllitol (or mytilitol) are only known in other plant phyla, such as the red algae. The natural distribution of cyclitols is reviewed by Plouvier (1963).

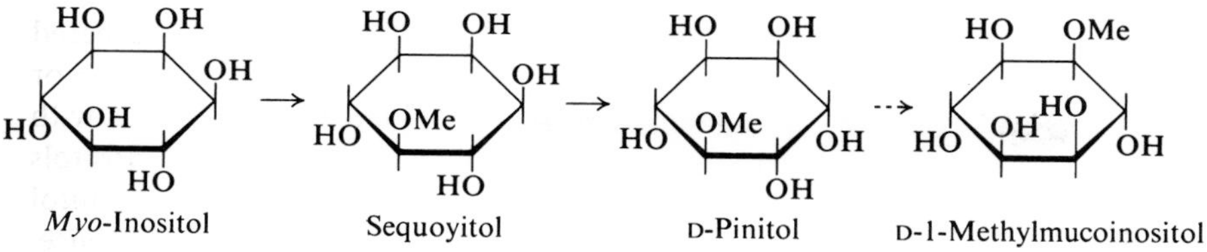

Myo-Inositol → Sequoyitol → D-Pinitol ⇢ D-1-Methylmucoinositol

Fig. 8.8 Biosynthesis of 1-methylmucoinositol in gymnosperm leaves.

Inositols occur occasionally linked to hexoses. One such derivative is galactinol (1-*O*-α-D-galactosyl-*myo*-inositol), first isolated from sugar beet. It is probably widespread in plants, since it has an intermediate role in the transfer of galactose units in the biosynthesis of storage oligosaccharides such as raffinose and stachyose (Senser and Kandler, 1967). It is not, however, involved in the biosynthesis of all galactosyl-containing oligosaccharides, since the galactose donor molecule in the synthesis of umbelliferose in umbellifer roots is uridine diphosphogalactose and *not* galactinol (Hopf and Kandler, 1974). Inositols themselves are synthesized by cyclization of the corresponding hexose such as glucose. Thus, *myo*-inositol is formed from glucose 6-phosphate, via *myo*-inositol 1-phosphate, by an enzyme-catalysed cyclization, the enzyme being called glucose 6-phosphate cycloaldolase.

B. Detection

Sugars can be screened in crude plant extracts by one-dimensional paper chromatography (PC) in one or other of the main solvent systems devised for them. The sugars are developed by dipping the paper in a chromogenic reagent (e.g. aniline hydrogen phthalate) and heating at 100°C for 10 minutes. Because of the relatively low mobilities of higher oligosaccharides (trisaccharides and above), the paper may require extended development beyond the end of the paper. The three nectar sugars—glucose, fructose and sucrose—can be separated by PC in butanol–acetone–water (4 : 1 : 5, by vol.). Sugars are well resolved in the same solvents when separated on thin layers of microcrystalline cellulose. Although gas chromatography is less convenient than PC for sugar detection, it is an ideal procedure for their quantitative determination (as in sugar nectars). Sugars need to be converted into their trimethylsilyl ethers prior to GLC. The classic paper on the GLC of sugars is that of Sweeley *et al.* (1963); the topic is also reviewed by Holligan and Drew (1971).

Sugar alcohols can be separated on paper in a system such as propanol–ethyl acetate–water (7 : 1 : 2, by vol.) and detected by spraying with alkaline silver nitrate. For both polyols and cyclitols, electrophoretic separation is a recommended additional procedure. The analysis of polyols in plants is reviewed by Lewis and Smith (1967b). A guide to techniques for identifying sugars and their derivatives is provided in Harborne (1973b). One further procedure, developed by Kandler (1964) for comparative studies of sugar production in plants, should be mentioned. This involves exposing a small leaf sample, set up in a special chamber, to $^{14}CO_2$ for about half-an-hour. After extraction and two-dimensional PC of the extract, the sugars that are labelled with ^{14}C are detected by radioautography. It is an accurate and sensitive technique and has been extensively used by Kandler and his group in chemotaxonomic studies of sugar alcohols (e.g. Sellmair *et al.*, 1977).

C. Taxonomic utility

1. *Nectar sugars*

Indications of different patterns in the glucose (Glc), fructose (Fru) and sucrose (Suc) contents in nectars came during a chromatographic survey of nectars collected from over 900 angiosperm species (Percival, 1961). Three sugar patterns were readily distinguished: Glc and Fru dominant (e.g. Cruciferae, Umbelliferae); Suc dominant (e.g. *Berberis*, *Helleborus*); and equal amounts of all three sugars (*Abutilon*). Other variations were also apparent; occasionally, oligosaccharides in addition to sucrose (e.g. raffinose, melibiose) were detected. A particular pattern was found to be consistent in a given species and this was unaffected by diurnal or seasonal variations, water content of the nectar or ageing in the flower. Although traces of invertase, an enzyme that hydrolyses sucrose to glucose and fructose, are present in nectars, there was little evidence that it was active in affecting the sugar ratios.

Subsequent studies have confirmed the view that nectar sugars do vary consistently in different plants. Baskin and Bliss (1969) surveyed extrafloral nectaries in 30 species of Orchidaceae and found the main pattern (Suc dominant) to resemble that of intrafloral nectaries previously studied by Percival (1961). Again, a survey of nectars in *Rhododendron* flowers indicated the presence of five rather than three sugar patterns, all within the same genus. Each pattern was fairly equally distributed among the 55 species surveyed, the two extra patterns being Suc only (as distinct from Suc with traces of Fru and Glc) and equal amounts of three sugars with a second oligosaccharide. The patterns of distribution, however, were not consistent enough with sectional classification in the genus to warrant extending the survey to further species (cf. Harborne, 1982).

Finally, Rix and Rast (1975) have tried to use nectar sugars as an aid in reclassification in the monocotyledonous genus, *Fritillaria*. A range of patterns, obtained by GLC quantification, was observed among the 30 species studied and the results were chemotaxonomically useful. *Fritillaria imperialis*, well known as the most distinctive fritillary, was the only species that lacked sucrose completely in its nectar. Also, two groups of species recently recognized within the section *Trichostylae* had different patterns, the Glc/Fru ratio was 1 : 1 in one and nearer 10 : 1 in the other. There is thus some indication that nectar sugars may be employed usefully in chemotaxonomic studies at the lower levels of classification.

2. *Storage oligosaccharides*

An early survey of oligosaccharides in leaves of the Gramineae suggested that the commonly occurring raffinose was replaced in at least one genus (*Bromus*) by simple fructans and in others (*Festuca*, *Lolium*) by a raffinose isomer (MacLeod and McCorquodale, 1958). Unfortunately, the sampling of the grasses was very limited (only 22 species to represent 11 tribes) and the work did no more than suggest the chemotaxonomic possibilities of storage sugars. In a wider survey of angiosperm storage tissues, Jeremias (1962) established that the raffinose series of oligosaccharides are very commonly present. Indeed, raffinose itself was detected in 220 spp. from 35 families, stachyose in 165 spp. of 46 families, verbascose in 24 spp. of seven families and ajugose in six spp. from three families. The decreasing frequency of reports with increasing molecular size probably reflects in part the greater difficulty there is in detecting verbascose and ajugose, particularly when they are only present in small amounts. It appears, therefore, that these oligosaccharides are widely present in plants and are unlikely in general to be of much value as taxonomic markers. Consistent quantitative variations in the amounts of the different members of the series may occur, as in legume seeds (see Courtois and Percheron, 1971), and they may be of some taxonomic significance at generic or species levels.

By contrast with the raffinose series, most of the other oligosaccharides reported (see Table 8.4) are of more limited occurrence. They may be of interest taxonomically although present information is too limited to be certain whether their distribution patterns are significant or not (cf. Meier and Reid, 1982). In the case of umbelliferose, the storage sugar of the Umbelliferae, it is not yet clear how characteristic it is of this family. Although it has been detected in representative members of all tribes in the subfamily Apioideae (Crowden *et al.*, 1969), it is not clear whether it occurs universally in other subfamilies. In fact, other storage sugars have been detected in members of the subfamily Saniculoideae, namely sucrose in *Sanicula* and isoketose in *Eryngium* (see Hegnauer, 1971). The presence of umbelliferose in Araliaceae and Pittosporaceae (Meier and Reid, 1982) adds yet one more piece of chemical evidence linking these families with the Umbelliferae (cf. Hegnauer, 1969).

Table 8.6 Distribution of cyclitols in higher plants

Cyclitol	Distribution
Pentols	
D-Quercitol	In leaves of all *Quercus* spp. (Fagaceae), in roots and bark of Menispermaceae, variously in seven other families
L-Viburnitol	In Asclepiadaceae, Caprifoliaceae (only in *Viburnum tinus*), Compositae and Menispermaceae
Hexols	
myo-Inositol	Universal
L-Inositol	Euphorbiaceae (incl. rubber latex) and Compositae
D-Inositol	Only in wood of *Pinus lambertiana* (Pinaceae)
DL-Inositol	In fruits of *Viscum album* (Loranthaceae)
Scyllitol	In Calycanthaceae and six other families
Hexol methyl ethers	
Sequoyitol	Only in gymnosperms, but widely present
D-Bornesitol	In Apocynaceae and the related Rubiaceae
L-Bornesitol	In Boraginaceae, Rhamnaceae, Apocynaceae, Leguminosae and Proteaceae
D-Ononitol	In several Leguminosae (e.g. *Ononis*) and once in Flacourtiaceae (*Kigelia africana*)
Dambonitol	In latex of *Castilloa* (Moraceae) and of *Nerium* and *Vinca* (Apocynaceae)
D-Pinitol	In gymnosperms (six families) and angiosperms (13 families)
L-Pinitol	Only in *Artemisia dracunculus* (Compositae)
L-Quebrachitol	In Ulmaceae (in *Celtis* but not *Ulmus*) and in ten other families
D-1-Methylmucoinositol	Widely in the gymnosperms, but not in Pinaceae

3. *Sugar alcohols*

Most of the distributional data on polyols in green plants has been derived from surveys directed towards finding new structures, so that the taxonomic implications of these occurrences are still far from clear. The results up to 1963 have been summarized by Plouvier (1963) and are indicated in brief in Table 8.6. Mannitol is so widespread that its occurrence has little taxonomic input as such. By contrast, sorbitol and dulcitol do seem to be potential markers for the families Rosaceae and Celastraceae respectively. Only the distribution of sorbitol has been examined in some detail. It is mainly confined to subfamilies Spiraeoideae, Pomoideae and Prunoideae. In the largest subfamily, the Rosoideae, it is generally absent, but does occur exceptionally in *Rhodotypos*, *Kerria* and *Neviusia*, three genera which form a highly distinctive group, placed in a separate tribe, the Kerrieae, and which show affinities with *Spiraea* (Spiraeoideae). A recent critical re-examination of sorbitol occurrence in Rosaceae (Wallaart, 1980) has confirmed the above findings and also shown a clear cut correlation between sorbitol accumulation and chromosome number within the family.

The only other family where characteristic polyols occur and where proper surveys have been initiated is the Primulaceae. Here, there are three interesting polyols: volemitol; hamamelitol, the alcohol corresponding to hamamelose, 2-hydroxymethyl-D-ribose; and clusianose, the α-galactoside of hamamelitol. All three compounds are confined to the genus *Primula* and to certain sections and subsections within the genus (Sellmair *et al.*, 1977; Kremer, 1978) so that they are of some interest in terms of the still rather confused sectional and subsectional classification within this large genus.

4. *Cyclitols*

The taxonomic position regarding plant cyclitols is similar to that of the sugar alcohols. From Plouvier's (1963) review of the earlier literature, it is apparent that cyclitols are widely distributed in higher plants, but often a particular structure may appear capriciously in a number of unrelated sources. More information has become available since 1963 but it has not yet established cyclitols as important taxonomic markers in plants. To indicate the systematic potentialities of these substances, it is worth mentioning briefly their occurrence within the gymnosperms and also within a single angiosperm family, the Compositae.

Representative surveys throughout the Gymnospermae have indicated that three cyclitols, sequoyitol, D-pinitol and D-1-methylmucoinositol are widespread (Plouvier, 1963; Dittrich *et al.*, 1971). These three substances fall into a biosynthetic series, all being derived originally from *myo*-inositol (Fig. 8.7). A key step in the pathway is the epimerization of sequoyitol to D-pinitol, a reaction that seems to be specific to gymnosperm tissues. D-Pinitol is then by further modification converted into 1-methylmucoinositol, the end product of the sequence (Fig. 8.7).

The distribution of the above cyclitols in the Gymnospermae is shown in Table 8.7. Several interesting points may be made about these results. The general absence of methylmucoinositol from the Cycadopsida and from the Chlamydospermae confirm their often-recognized separation from the main group of conifers, the Coniferopsida. The exceptional occurrence of this marker in *Welwitschia mirabilis* of one of the outer groups is not entirely unexpected, since the Welwitschiaceae do show some morphological characteristics (e.g. formation of cones) that link it with the Coniferopsida proper.

Another interesting point in Table 8.7 is the unique absence of methylmuco-inositol within the Coniferopsida from just one family, the Pinaceae. This ties in with other chemical features which distinguish the Pinaceae from the rest of the conifers, notably the absence of biflavonyls, otherwise prevalent (see chapter 7) and the presence of distinctive leaf lipids (Jamieson and Reid, 1971).

Table 8.7 Distribution of cyclitols in gymnosperm leaves

Class and family	Presence or absence of[a] Sequoyitol	D-pinitol	1-methylmucoinositol
Cycadopsida			
Cycadaceae	+	+	−
Zamiaceae	+	(+)	−
Ginkgoaceae	+	+	−
Coniferopsida			
Araucariaceae	+	−	+
Cephalotaxaceae	+	+	+
Cupressaceae	(+)	+	+
Pinaceae	(+)	+	−
Podocarpaceae	(+)	(+)	+
Taxodiaceae	+	+	+
Taxopsida			
Taxaceae	+	+	+
Chlamydospermae			
Ephedraceae	−	−	−
Gnetaceae	−	−	−
Welwitschiaceae	−	−	+

[a] + uniformly present, (+) variably present, − absent. Data from Dittrich *et al.* (1971). Cyclitols detected in leaves or needles.

Although a larger number of inositol derivatives occur in angiosperms than in the gymnosperms, no clear correlations emerge between cyclitol patterns and family classification. However, some families are quite rich in structures. The Compositae, for example, is distinctive in containing L-inositol quite widely; this isomer is otherwise rare in angiosperms (see Table 8.7). Although there are no less than six other cyclitols reported in the family, these are all apparently restricted to a single tribe, genus or species. Thus, L-viburnitol occurs in several genera of the Anthemideae, L-quebrachitol in *Artemisia* spp. (except *A. dracunculus*), L-pinitol in *A. dracunculus*, scyllitol in *Vernonia altissima* and leucanthemitol, D- and L-pinitol all in *Leucanthemum vulgare*. There is thus chemical variation in the cyclitol series worthy of further systematic exploration in this family.

VI. Conclusion

In this chapter, several classes of primary metabolite (Table 8.8) have been considered from a chemotaxonomic viewpoint. Their utility as taxonomic markers is profoundly affected by their metabolic importance. Inevitably, this reduces their attractiveness in chemotaxonomic investigations, since much sampling is necessary in order to eliminate physiological variables. This may explain why less attention has been given to fatty acids, alkanes or sugars in recent chemosystematic studies. Nevertheless, there are sufficient indications from the literature that all these substances may be of interest when seeking chemical markers for solving particular systematic problems.

Table 8.8 Hidden metabolites in plants

Class	Number of known structures	Organs found in	Natural distribution
Fatty acids	150	Seeds, fruit coats: also leaves	Universal, but much variation at family level.
Alkanes	50	Leaf waxes; also coatings on other organs	Universal, but much quantitative variation
Polyacetylenes	750	Chiefly in roots; also leaves	In Compositae, Umbelliferae and three other families
Storage sugars	50	Seeds, roots, rhizomes	Widespread

One advantage is that fatty acids can always be found in seeds and alkanes in surfaces waxes; they are characters always at hand for systematic consideration. Analyses can be carried out on relatively small amounts of plant tissue. The position regarding storage sugars is less clear, since it is possible that many plants rely simply on sucrose as the major low molecular-weight form of stored energy. However, a range of other oligosaccharides and of sugar alcohols may partly or completely replace sucrose in yet other plants. Certainly, in a number of advanced angiosperm groups, they are known to accumulate and appear to offer interesting chemotaxonomic criteria.

The polyacetylenes have been included in this chapter because of their biosynthetic origin from fatty acids, but in some ways they fit in better in an earlier chapter on plant toxins. Their function, as far as it is known, seems to lie in the area of plant and animal interactions. Unlike the other metabolites considered here, they have a very limited natural occurrence. They are thus only of taxonomic value in the few families where they accumulate. Their chemical instability poses certain difficulties in their analysis. Within these limitations, however, they are potentially a most important group of plant substances for chemotaxonomic consideration.

9

Environmental and Genetic Variability

I.	Variation in secondary chemistry	216
II.	Flavonoids	219
III.	Terpenoids	224
IV.	Alkaloids	234
V.	Conclusion	236

I. Variation in secondary chemistry

Variation in taxonomic characters can be both the bane and joy of the systematist, depending upon its extent and nature. If the variation observed is solely environmentally induced, this is mostly ignored as a classificatory character. Such variation can be determined by careful experimental garden studies (transplants, etc.) or, indeed, in the case of exomorphic features, can be inferred from observations in the field (e.g. a given population might contain plants that show variation in height, leaf size, degree of pubescence, etc., all or some of which might be correlated with one or more purely local situations of shade, soil or microhabitat). As noted in chapter 4 the capacity of the field systematist to assess the likely significance of such variability in natural populations, by visually examining the characters and making *in situ* correlative judgments, constitutes a major advantage of the morphological approach over that of the chemical.

Many examples of environmentally produced variations in both morphological and chemical features are known, although this is better documented for morphological features. In general, it can be said that the usual environmental variables encountered in nature affect relatively trivial characters, taxonomically speaking. At least the variation in the characters concerned is usually one of quantity and not quality. This is true for both morphological and chemical characters.

Environmentally induced, *quantitative* variations among selected chemical constituents are well known, mainly for the so called 'secondary compounds' of plants which include flavonoids, terpenoids and alkaloids. And, just as with morphological features, the extent and cause of such variation within a given taxon must be evaluated anew on a character-to-character basis in each instance, *if certainty* as to the cause of such variation is desired.

Secondary compounds, in that they reflect the expression of enzymes, are likely to vary from organ to organ, at various developmental stages in those organs and from plant to plant within populations. Because of this the taxonomist must know 'how to measure an avocado', to borrow a phrase from that excellent layman's text, *Plants and Man*, written by the late Edgar Anderson. That is, good taxonomic treatments can only come out of studies in which multiple comparisons are made of characters taken from comparable organs at comparable stages of development. The chemist's frequent failure to do this has prompted Erdtman (1973) to remark, rightly so, that 'Unfortunately the parts of plants investigated [by chemists] is often not mentioned in the literature, thus making the [chemical] information of little value'.

On the basis of research and observations available to date, however, certain generalizations can be made with respect to the extent of variation anticipated for a given character. For example, plant height, flowering time, and size of vegetative features are more likely to vary within a given population, because of genetic and environmental factors, than are floral parts. The same can be said for selected chemical characters. Thus the quantitative variability of flavonoids appears to be more stable over a range of environmental conditions than are terpenoids, with alkaloids perhaps running a poor third. But the metabolically important enzymes are believed to be relatively invariant within populations of closely related taxa, except for isoenzymic variations, which may be looked upon as analogous to a kind of floral variation, and even then certain enzymes, such as cytochrome *c* are not known to exist in isoenzymic form, at least in higher organisms.

Nevertheless, one of the most frequently asked questions of the 'outsider' contemplating the results or use of chemical data is, 'how do you know that the compounds concerned are not environmentally induced?' In most cases, the chemosystematist using such characters will not know this. However, just as in the case of morphological characters, he can assume that their production has a genetic base, much as the morphological worker must assume that the characters with which he is concerned have a genetic base for, more frequently than not, the systematist simply has no experimental, or often times observational, data to distinguish between environmental or genetic variation within a particular group.

It should be clear, then, that the variability of chemical characters found among individuals in a given chemosystematic study will most likely reflect at least these several variables:

1. developmental differences among the organs or organisms being studied;

2. environmental conditions prevailing at the site or sites from which the samples are taken;
3. genetic differences among the individuals and populations studied;
4. inconsistency of sampling techniques (which may be viewed as a restatement of them in part); and
5. inconsistent techniques of extraction and analysis.

If the worker is exceedingly careful with respect to items 1, 4 and 5, most of the *qualitative* variability encountered will probably be due to genetic variation while the *quantitative* variability will probably reflect environmental factors, as will be discussed in more detail below.

One would like to think that relatively few chemosystematic studies will reflect flaws in sampling or technique, at least to any high degree, but such is often not the case. The novice with exceptional expertise in chemistry but with little, if any, training in plant systematics is likely to incorporate considerable variability into his data as a result of items 1 and 4, whereas the systematically apt, but chemically inept (i.e. one who is just learning technique or who does not appreciate the fragile nature of molecules or is ignorant of the operation of his particular analytical apparatus) is likely to incorporate much spurious variability into his study as a result of item 5.

The likelihood of incorporating variability into one's data as a result of inconsistent technique, to say nothing of faulty technique, can be considerable. For example, R. H. Flake, E. von Rudloff and B. L. Turner (unpublished observations) gathered terpenoid data in successive years from natural populations of *Juniperus virginiana* using a gas chromatographic apparatus. Their data were collected on the same instrument, using the same column in the same laboratory with the same, highly qualified, organic chemist in charge. And essentially the same results with respect to the interpretation of these two sets of data were obtained (i.e. the results evaluated separately reaffirmed the existence of clinal gradation in terpenoids over a 1500 mile transect; Flake *et al.*, 1973). But when the *two* sets of data were treated together, numerically, the populations clustered by the year of sampling (not by populational terpenoid quantitation), *suggesting* that regional climatic differences in the 2 years were responsible for this remarkable result.

After much initial excitement, disbelief took over and, after careful scrutiny of all the possible variables, it was discovered that the differences (all quantitative) in the sets of data gathered were due to a relatively slight modification in the speed of delivery of the carrier gas to the apparatus concerned during the second year. The chemist could hardly believe that this simple change in schedule might affect the relative quantities among the approximately 40 compounds studied. Test runs subsequently proved the fact, and the point made here is that what seems a trifle in the manipulation of technique can often introduce unanticipated results.

Even in such a simple procedure as leaf extraction of flavonoids for paper chromatography, all kinds of spurious variation can be encountered. These may be due to poorly measured fluids, excessive extraction, varying amounts and ages of plant material used in the extraction and so on. There is yet another hazard: Coradin and Giannasi (1980) report that leaf analysis for flavonoids, in otherwise perfectly preserved tropical plant specimens, yielded no flavonoids since they were first soaked in formaldehyde or similar solutions to protect them from tropical decay. Thus one must exercise care in the selection of specimens to be examined; this might include knowledge of collecting technique.

Adams (1972) has called to the fore some of the more common errors in chemosystematics and has analysed these numerically to ascertain how they might affect classifications. He concluded that the most serious errors are those relating to seasonal variation, at least for terpenoids in *Juniperus* spp.

In the sections that follow we will review briefly some of the pertinent research bearing on the problems raised here, mainly with respect to flavonoids, terpenoids and alkaloids, since examples among these generally cover the range of variation expected. The occurrence of natural variation among macromolecules, especially protein bands and isoenzymes, will be discussed in chapter 17.

II. Flavonoids

As a group, the flavonoids are relatively uniform structurally. Because of this, their stability in dried plant tissue, and the fact that they make up most of the visual pigments in vascular plants, flavonoids, especially anthocyanins, have been much studied by biochemical or physiological geneticists (e.g. Harborne, 1967a; Kirby and Styles, 1970). Nevertheless, except for anthocyanins, there have been relatively few, carefully controlled, experimental studies conducted as concerns environmental effects on the production of this class of compounds.

One of the best documented experimental studies on the effect of controlled environmental variables on the flavonoid chemistry of plants has been that of McClure and Alston (1964). Working with controls and *in vitro* cultures on defined media of the small aquatic angiosperm, *Spirodela oligorhiza* (duck weed), they report on 52 different, often extreme, treatments, carried out in quadruplicate, including variation in light intensity, temperature, composition and concentration of nutrients, along with the addition of growth factors such as auxin and gibberellic acid. In almost every instance the pattern of approximately 15 flavonoids, which was resolved by two-dimensional paper chromatography, remained essentially unchanged.

Differences, when detected, were nearly always quantitative; rare qualitative differences usually involved only minor components of the profile, and even then such 'absence' was thought to be quantitative, i.e. below the level of chromatographic detection.

This work was only part of a much larger study of the family Lemnaceae, to which the species belongs. McClure and Alston (1966) subsequently published similar research on 186 clones representing a world-wide collection of most of the species recognized within the four genera of this family. In all, 47 flavonoids were studied and the clones that contained them were grown under equivalent controlled conditions. Infraspecific variation in the qualitative production of these flavonoids was found in only one flavone glycoside of the species *Lemna perpusilla*. Further, as noted, in part, above, they exposed the two species *Spirodela oligorhiza* and *S. polyrhiza* to 62 different environmental regimes and, among the 15 or more flavonoids making up the chromatographic pattern of each, only 'minor variations' in their flavonoid glycosides could be found. They noted that, under the conditions of their experiment, the environmental variables employed induced morphological changes in certain species such that identification of these, using the routine criteria, would have proved difficult, if not impossible; however, in their words, 'Each could be conclusively identified by its flavonoid chemistry'.

It proves instructive to compare McClure and Alston's carefully documented study of these two species with the study of these same two species by Ball *et al.* (1967). These workers concluded after culturing clones of *S. polyrhiza* in media deficient in phosphorus and nitrogen, one on an 8-hour light/day cycle and another on a 16-hour light/day cycle, and observing that four spots were found in clones cultured under the latter cycle and only one or two spots in clones of the former, that 'it becomes obvious that chromatographic comparisons of plants taken from populations may provide a better record of environmental conditions than genetic differences'. They further concluded that the results of their study indicate that 'for taxonomic use, chromatographic analyses should be attempted (1) only on plants which have been maintained under uniform conditions or (2) on plants in which prior trials, under controlled conditions, show the chromatographic complement to be unaffected by the environment'.

It should be noted that McClure and Alston, put together a carefully devised research programme (or series of experiments) to test for possible variability of flavonoids among clones grown under a wide set of environmental regimes. Characteristically, they made no broad general claims as to the efficacy of the biochemical approach, in spite of results which might have been seized upon for this purpose. According to Parks *et al.* (1972), the duckweed experiments of Ball *et al.* (1967) were repeated and this work revealed many more constituents in the two organisms than was reported by these latter workers, much as McClure and Alston found. This suggests that the 'absent' compounds in the work of Ball and his co-workers were most likely due to their inability to detect these compounds, presumably because of their weak extractions (obtained from 0.1 g wet weight per extraction). However, we do not fault their research so much as we fault their generalizations from the sparse data generated.

Actually, exposing these *Spirodela* spp., or almost any other organism, to conditions of drastic phosphorus deficiency, as is well known, can produce considerable stress on a plant, including death. For example, Bieleski and Johnson (1972), working with phosphorus-deficient *oligorhiza*, showed that such deficiencies occasioned a 10–20-fold increase in phosphatase activity, this being centred in the epidermal tissues of root and lower frond surfaces, instead of the cells surrounding the internal vascular tissue where such activity normally resides. In addition, these workers discerned a 'new' phosphatase isoenzyme in the deficient plants that was not detected in the control plants. Certainly, such dramatic 'variation' will not lessen the interest of biochemical geneticists interested in such characters for systematic purposes (cf. chapter 17). Anyway, most plants growing on soils highly deficient in phosphorus in nature would probably appear 'sick' and hence would not be collected; just as the exomorphic worker will ignore such plants in his sampling, so should the chemosystematist.

Alston himself, being a physiological geneticist by training, was acutely aware of both the genetic control of flavonoids and their susceptibility to environmental influences. Thus he states (Alston, 1967) 'it is probable that the anthocyanins, which have been so valuable in the study of biochemical genetics, are among the most variable of all the many flavonoids. For example, anthocyanins are dependent upon light for their synthesis while other flavonoids are not entirely dependent upon light, although they may be affected by light. Also, certain specific inhibitors, such as benzimidazole, affect specifically anthocyanins and do not greatly affect other flavonoids'.

This kind of thinking (undoubtedly influenced by his experience and familiarity of the relatively stable leaf flavonoids in *Baptisia* spp.), along with pragmatic considerations (i.e. you can sample leaves for flavonoids long before flowering is achieved), led Alston to do most of his chemosystematic work on foliar tissue. In fact, he often expressed himself to the effect that leaves appeared to have more consistent flavonoid patterns than floral parts, or at least they were more readily interpretable.

That this need not be so seems clear from the observations of Harborne (1967c) and the research of Parks *et al.* (1972). The latter workers in particular, made comparative studies of experimentally grown, inbred lines of four species of *Gossypium* and found that the flavonoid variation among floral parts was less than that found among leaves on these same plants. Reasons for the greater variability among foliar extracts were suggested as possibly due to the difficulty in selecting comparable leaves at the same stage of development, or perhaps due to analytical problems connected with the detection of low concentrations. It follows that flavonoid complements in floral parts might be more consistent in that they can be more readily collected at a given developmental stage. (This probably accounts in part for the remarkable flavonoid consistency found in the studies of McClure and Alston (1966): they were using *whole* plants synchronized as to date of culture, etc. Thus, developmental stages for a given organ *per se* never entered the picture.)

Parks *et al.* (1972) conclude from their studies that the degree of environmentally induced flavonoid variation should be determined for the particular plant organ to be analysed before a chemosystematic investigation using flavonoids is begun. Which, of course, is good advice, but this need not be determined by *experimental* studies in each instance, anymore than it is necessary to do this for morphological characters. As already noted, one should attempt to work with chemical constituents from organs at comparable stages of development selected from a wide range of 'normal' looking plants in natural populations. If in the preliminary stages of such sampling, relatively high consistency in the flavonoids concerned is found, fine; if there is discovered a meaningless array of variability among the flavonoids, then something is either wrong with the techniques (either sampling or chemical), or variation disclosed is real. If the latter, then the worker will have hit upon a remarkable situation which should be studied yet further and reported upon, for relatively few, if any, carefully documented studies with such findings have been reported using flavonoids.

It is generally believed that the continuous variation in quantitatively inherited characters (e.g. terpenes) is the result of the cumulative effects of several-to-many genes (polygenic inheritance). Environmental influences on the quantitative expression of such characters is believed to be large. Flavonoids, on the other hand, in that they are controlled by the so-called major genes, are generally believed to show little, if any, quantitative variation. That this need not be so is clear from the carefully controlled and exceedingly well-documented studies of Jana and Seyffert (1971) on the inheritance of anthocyanins in *Matthiola incana*. Using isogenic lines and two alleles at each of two loci with a common genetic background, they found that the phenotypes of the nine possible genotypes were *qualitatively* distinguishable but that quantitative variation among each might be considerable. They conclude that the latter may be greatly influenced by a slight change in the genetic background, but is less susceptible to the variation in environmental conditions. Dominance was found to be the most important factor in the concentration of anthocyanin in floral tissues followed by such nonallelic interactions as epistasis and pleiotropy. Nevertheless, there was detected 'marked variations' between years, or growing conditions, on some of these genetic expressions; they surmise therefore that the influence of environment on a given genetic expression may be considerable. In short, they conclude that environmental influences are *not* limited to those characters which are controlled by polygenes and that one's inability to detect recognizable effects of environment on the qualitatively inherited traits may be merely their inability to detect intangible differences in the intensity or quality of such phenotypes in different environments.

The experiments of Jana and Seyffert (1971), and the conclusions therefrom, are reported upon in some detail because of the recurring belief that much of the flavonoid variation found in nature is excessively influenced by environmental factors. Thus Abrahamson and Solbrig (1970), working with transplants from 19 natural populations of *Aster* spp., conclude that from a taxonomic point of view the secondary compounds (none of which was identified but were most probably phenolics and flavonoids) were not particularly stable, noting that 'The seasonal and environmental variation found should be a fair warning to those attempting to use similar methods of chromatography of secondary compounds for taxonomic purposes using random field samples or herbarium samples. At least in the heterophylli *Aster* secondary compounds vary greatly during the growing season, and are also affected by the particular environment of the plant. The environment affects both the number as well as the intensity of spots produced'.

Well, almost certainly the spots under investigation by these workers varied for two major reasons: their use of 'mature basal leaves' which were not said to have been collected at any particular nodal point, and by their techniques, which probably did not permit the recognition of diminished amounts of the compounds concerned. Besides, *critical* studies of this type ought to be performed by a reasonably well trained chemosystematist as noted by Adams (1974), so that identification of the compounds might be made; in fact, many of their variable spots may simply have been artifacts of technique. Identifications would have helped clarify this possibility. It should be emphasized that their results reflect differences between two clones from each of ten or more plants from single populations, all of which were randomized and grown in a transplant garden. Unfortunately, it is not possible to ascertain from their published data whether or not interclonal variation contributed significantly to the results obtained. The fact that spots from individual plants vary in the natural populations is not surprising; indeed, it is expected. And that 'seasonal' variation for a given clone-pair might be found is also expected, simply because the organs sampled will be at different developmental stages.

Fahselt (1971), working with the flavonoid constituents of *Dicentra canadensis*, noted that the compounds occurred with less than 100% frequency, reasonably concluding that 'This variability may reflect some genetic variability between populations, and is almost certainly due in part to the difficulty in detection of weaker spots'. Chemical variability and its cause has been commented upon by Murray *et al.* (1972a) who studied the even more variable secondary compounds (terpenes) of *Mentha*, concluding that 'Oil differences due to the time of harvest are minor compared to those due to inherited differences'. In fact, as will be noted below, the expressions of terpenoids in plants is under fairly rigid genetic control, in spite of their considerable quantitative variation among individuals in natural populations.

In short, what Abrahamson and Solbrig (1970) have succeeded in showing, as far as their study of the secondary chemistry is concerned, is that seasonal variation in populations will occur, much as has been done by Asker and Fröst (1972a,b) for *Potentilla* and Akabori (1978) for *Adiantum*, as have other workers (e.g. Voirin and Lebreton, 1972) for yet other genera and compounds (cf. below). And that is why chemosystematists should study individuals in populational form, emphasizing those aspects of the variation that prove taxonomically significant, much as the morphological worker must do if his work at the specific level is to be meaningful.

Variation upon a theme, whatever its cause, may be expected, both within and between populations; indeed this has been found in nearly all studies of a geographical nature (cf. chapter 10). However, in the words of Alston (1967):

> If chemical variation in plants is so excessive, so capricious, and so generally immune to analytical interpretation that predictability is essentially lacking, then such data hardly can be utilized scientifically at all, certainly not by systematists. But, if variation in secondary products is generally responsive to various factors similar to those factors which govern the morphology of a species, then chemical variation can be described in such a way as to introduce limits to the variation and to discern its pattern and thus, of course, to allow predictability. Since the latter alternative is nearly axiomatic, in my opinion, then the problem is merely to become familiar with chemical variation, its origin and factors affecting it, and its limits and its meaning, if possible, just as any good systematist would attempt to study morphological or cytological variation.

III. Terpenoids

The number of volatiles, especially terpenes, known to occur in plants has increased dramatically over the past decade, largely because of the development of chromatographic techniques and sophisticated methods for their detection and identification. For example, the number of sesquiterpene structures alone has increased from about 200 in 1964 to over 1000 in 1971 and some of the lower terpenes such as hemiterpenes and monoterpenes, once thought to be relatively rare and restricted amongst plant groups generally, are now also known to number nearly 1000 (Loomis and Croteau, 1973; Nicholas, 1973). In fact, it appears obvious that all living things are capable of synthesizing some form of terpene. For example, algae contain the common monoterpenes pinene, limonene, cineole and carvone (Katayama, 1955), and even animals, especially insects, accumulate these, either biosynthetically or as excretory products from plant diets (Nicholas, 1973).

A. Variability in herbaceous plants

1. *Genetic control*

In spite of their wide occurrence, however, unlike the flavonoids, relatively little experimental work has been carried out on the genetic and environmental factors that serve to regulate the production of terpenoids. Some of the better documented genetic studies have been those of Hefendehl and Murray (1972) and Murray *et al.* (1972b) on species belonging to the herbaceous genus *Mentha*. The expression of terpenes in this genus are clearly controlled by relatively simple genetic systems. Especially interesting have been the studies of Murray and Lincoln (1970) and Lincoln *et al.* (1971) in which it was shown that in *M. citrata* and *M. aquatica* (Murray and Hefendehl, 1973), there is a dominant *Lm* gene which prevents the conversion of limonene into more biosynthetically advanced compounds such as carvone, pulegone, menthofuran, menthone, menthol and menthyl acetate, compounds that are normally developed in those strains of *Mentha* having the recessive *lm* gene. Plants with the dominant *Lm* gene normally accumulate limonene and cineole. Further, these workers detected a dominant *I* gene in *M. citrata* that interrupts terpene synthesis at an even earlier stage than the *Lm* gene. The dominant *I* gene largely prevents the synthesis of limonene and cineole so that linalol and linalyl acetate are accumulated. Thus it is possible for a single gene to block the appearance of an array of compounds belonging to the same class. Fortunately, for the chemosystematist at least, such 'blocking' genes do not appear to occur as common segregates in natural populations; rather, their detection has resulted from the unexpected segregation (as a result of cross-overs, etc.) of such genetic factors following interspecific hybridizations. Because of the possibly pronounced effects of such genes early in a biosynthetic pathway, it is important that the chemosystematist become familiar, not only with the structural complexity of the compounds under investigation, but also with the metabolic pathways leading to their formation.

Irving and Adams (1973) have discussed the genetic and biosynthetic relationships of monoterpenes in the *Hedeoma drummondii* complex, an herbaceous group also belonging to the mint family. This complex is made

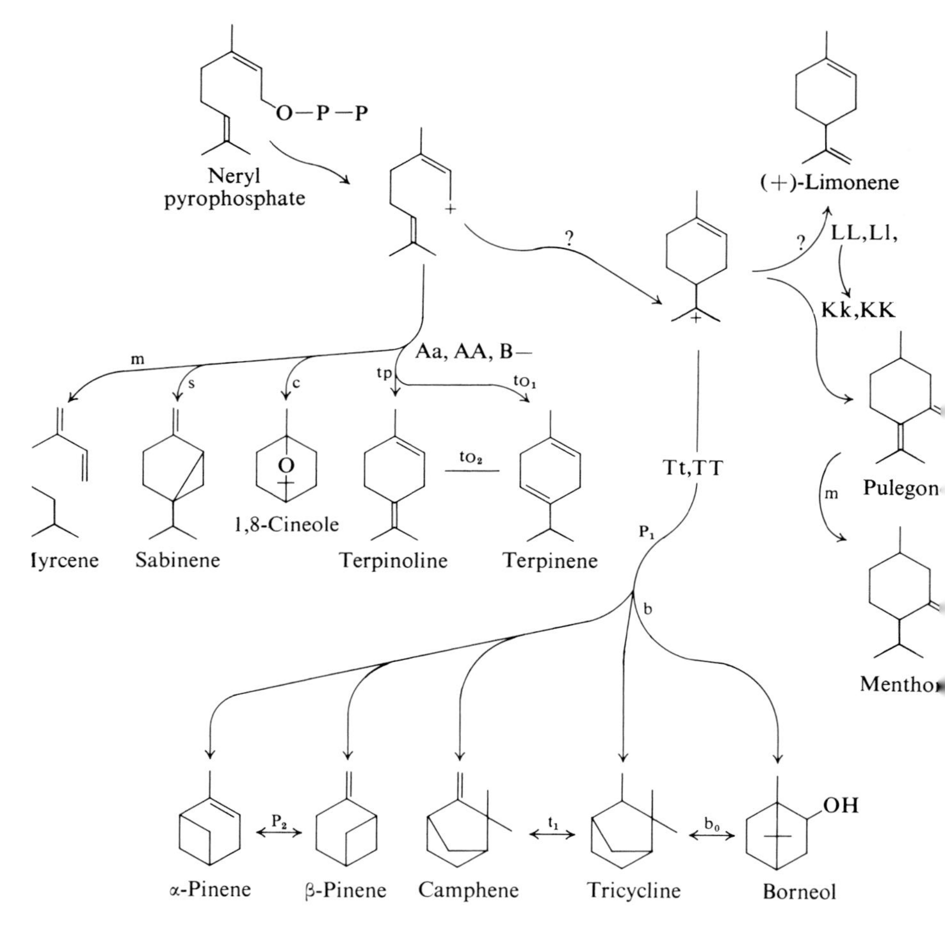

Fig. 9.1 Biosynthetic and genetic synopsis of the monoterpenes in the *Hedeoma drummondii* complex. Lower case letters represent alternative sites of gene and/or enzymic action. (From Irving and Adams, 1973.)

up of three taxa, each with its own array of monoterpenoid types: *H. drummondii* possessing mostly monocyclic compounds; *H. reverchonii* mostly acyclic; and *H. serpyllifolium* mostly bicyclic compounds. Each of the approximately 20 terpenes studied were estimated to be controlled by relatively few genes, generally ranging from one to four. Their conclusion as to the possible biosynthetic pathways leading to the production of this group of compounds is shown in Fig. 9.1. It seems clear that 'blocking' genes interposed between one or more of the critical intermediates shown in their diagram may serve to channel production into one or more alternate pathways. This is consistent with the genetic inferences made by Lincoln and Murray (1971) for *Mentha*; in fact, the terpenoid types which characterize the three closely related taxa of the *Hedeoma drummondii* complex probably had their origin through relatively simple gene localizations in the natural populations or breeding pools of the taxa concerned. How they arose and under what selection pressures, if any, is another question.

2. *Environmental control*

Although the research published to date suggests strongly that the qualitative production of terpenes is genetically determined, there is much evidence to suggest that, at least among rapidly differentiating herbaceous plants such as *Mentha* spp. and seedling trees (Rockwood, 1973), as well as developing organs on woody plants (Adams and Hagerman, 1976), such compounds are readily affected by environmental factors. Loomis and Croteau (1973) discuss this problem and have presented compelling evidence to suggest that environmental factors can affect relative terpenoid yields. Using *Mentha* spp., they found that periodic analysis of comparable organs at comparable stages of development grown under controlled environmental conditions such that contrasting warm-day, cool-night data might be compared, resulted in quite significant differences in the amount of terpenes produced, as shown in Fig. 9.2. These authors conclude that monoterpenes and sesquiterpenes are metabolically quite active in the growing plant 'and are not, as was once supposed, inert metabolic end products'.

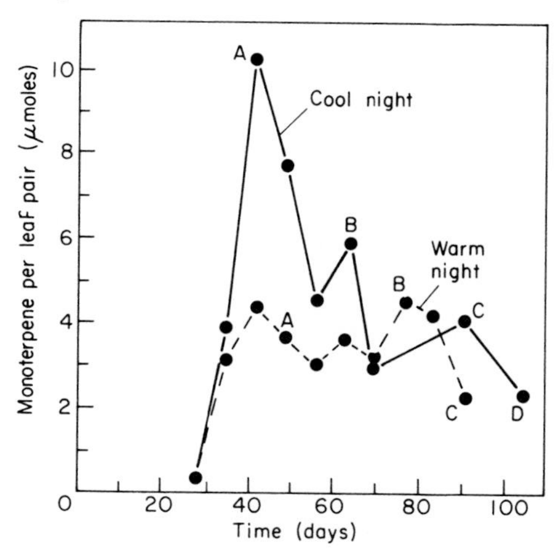

Fig. 9.2 Periodic analysis of total monoterpenes of leaf pair number 9 from peppermint plants grown under controlled environmental conditions. 'Cool night': 16 hour day, 25°C; night 8°C. 'Warm night': 14 hour day, 25°C; night 25°C. A, Time at which flower initiation could be recognized macroscopically; B, time at which first flowers appeared; C, full bloom; D, end of bloom. (From Loomis and Croteau, 1973.)

A number of other workers (e.g. Sprecher, 1958) have also reported periodic variation in terpenes from a number of mostly herbaceous plants. Thus, Banthorpe and Wirz-Justice (1969) reported that the terpenes of *Tanacetum vulgare* varied such that the same plants might yield 20% more oil at night than by day.

With the possible exception of *Mentha* spp. and selected genera within the mint family (e.g. *Satureja*; Lincoln and Langenheim, 1976), most of the published work on the genetic versus environmental control of terpenoid production has been in the odoriferous economically important families Rutaceae and Pinaceae. The latter family in particular has been much-studied by forest geneticists, mostly in connection with efforts to produce better and more resistant (to predation) timber plantings. Since most of this work is relatively recent and has been performed by research units keenly aware of the need of extensive population and provenance studies, along with thorough chemical analyses, it will be reviewed in some detail here.

B. Variability among woody plants, especially gymnosperms

1. *Developmental and environmental variation*

As already noted, during early developmental stages plants and their organs are likely to show considerable variation in their secondary compounds, especially terpenes. Mature woody plants, however, have been found to be relatively stable as regards their 'stored' terpenoids, both in their wood and in mature vegetative tissues. Thus Rockwood (1973) found that the monoterpene concentrations in 14-month-old seedlings of *Pinus taeda* were more similar to those of mature trees than to the monoterpene concentrations found in 9-month-old seedlings of the same species.

Zavarin *et al.* (1971a), after a careful study of the volatile oils in *Pinus ponderosa*, report that 'The most important step towards decreasing the seasonal age and variability of needle samples in ecological or chemotaxonomic studies was found to involve exclusion of sample collections during a period between the initial appearance of young needles and the time that they reach maximum length'. Such sampling precautions do not preclude significant variations in at least a few chemical characters, for these authors note that methylchavicol (or esdragole) varies significantly, not only during periods of active growth of the organs in which they occur, but also may vary significantly from month to month and year to year, although in a somewhat predictable fashion. Concentrations of methylchavicol were found to be lowest in the developing needles and in the oldest needles (Fig. 9.3). (It is interesting that the production of the phenylpropanoid methylchavicol in *Pinus ponderosa* is reported to be drastically affected by air pollutants; but this was not found to be so for the common monoterpenes of that species

(Cobb *et al.*, 1972).) Less regularity was observed in the periodic changes of the monoterpene, 3-carene, with growth and senescence of needles and no regularity could be detected in the changes associated with α-pinene. They conclude that 'The total variance associated with changes in individual oil constituents due to season and needle age can be regarded as composed of two components, a systematic component, involving the same changes in foliage oil of the trees examined, and a random component. Both components can adversely affect any correlative studies based on these oils (e.g. correlations with susceptibility of *P. ponderosa* to fungal, beetle, or smog damage, genetic or chemotaxonomic studies) by obscuring any existing correlations through addition of extraneous variance (random changes) and by introducing non-existing correlations through addition of extraneous covariance (systematic changes). Thus, some standardization of sampling involving season and needle age is indicated'. The highest systematic and random changes detected in this study were those associated with methylchavicol and summer sampling respectively. By omitting such data from their computations and by omitting juvenile leaves, they were able to minimize, to some extent, the variance component of the data concerned.

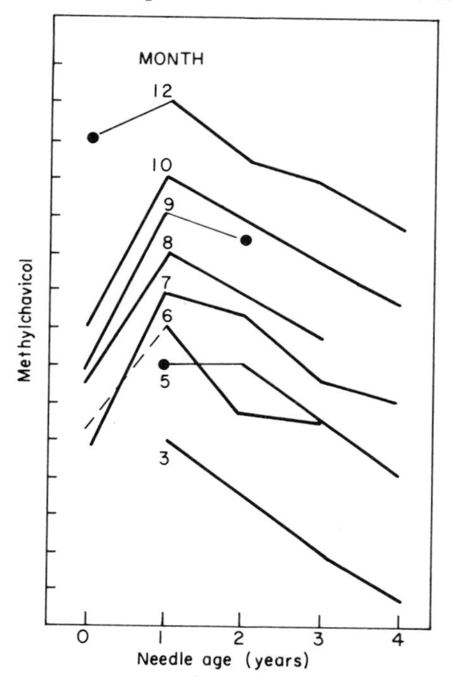

Fig. 9.3 Yearly changes in methylchavicol content relative to data for year old needles of *Pinus ponderosa* taken as = 1.0, ordinate divisions = 0.1. ——, 0.01 level; - - - -, 0.1 level; ——, not significant.

In a similar study on *Picea glauca*, von Rudloff (1972a) concluded that the major changes in volatile terpenes of the leaves, buds and twigs of this species 'take place only in the new shoots during the early part of summer' and that the relative amounts of β-pinene, limonene and myrcene in the buds may change significantly in the fall and winter but that 'the volatile oil of the leaves and twigs remain constant in quantitative composition during late summer, fall and winter.' This latter finding was consistent with his previous conclusions from similar work on *Pinus contorta* (Pauly and von Rudloff, 1971), *Pseudotsuga menziesii* (von Rudloff, 1972b), and several species of *Picea* (von Rudloff, 1962, 1966, 1967) and *Juniperus* (von Rudloff, 1968; Vinutha and von Rudloff, 1968). Juvonen (1966) also reports similar findings for *Pinus sylvestris*, as does Zavarin *et al.* (1971b), as already noted.

Adams (1970a,b), however, does show that the distillates of 'bulk' foliage (intact, terminal branches) may show measurable quantitative variation in the winter months, but not so noticeable as that reported for the summer. Adams' study, and perhaps those of others who also report similar variations (e.g. Poltartchenko *et al.*, 1968), has not been so meticulously partitioned as to the likely sources of variability for, as noted by von Rudloff (1972a) and Adams and Hagerman (1976), failure to separate bud material and developing needles from mature foliage may incorporate considerable variability into one's sample. However, studies by Adams *et al.* (1981) reveal no significant differences among 39 terpenoids between young and adult foliage of *Juniperus horizontalis*. They relate this anomaly to 'indeterminate growth' patterns in the latter species.

More recently, Powell and Adams (1974) and Adams and Powell (1976) have made an in-depth study of the variability of terpenes in natural populations of *Juniperus scopulorum*, attempting to correlate terpenoid changes in the intact shoot with climatic data gathered at the actual site being sampled. They were unable to find ready correlations between periodic terpenoid changes among the trees analysed and the climatic factors measured, suggesting perhaps that most of the differences were simply due to sporadic metabolic activity or developmental changes in the growth of needles and formation of buds, for it is difficult to collect a terminal branch and estimate, without detailed examination, what the relative ratio of mature leaves to immature leaves and buds might be. Nevertheless, they did find that the winter foliage was least variable from month to month and from year to year, presumably because, during this period, the metabolic activity of the plants is at a minimum, relatively speaking.

Diurnal variation in terpenoids of *Juniperus scopulorum* has also been studied (Adams and Hagerman, 1977). Using three-way analysis of variance of 37 compounds obtained from four trees during summer growth (temperature range: 18–32°C (64–89°F) daily) they noted that 36 of these showed significant differences among trees, 11 with significant differences between days, 13 with significant diurnal variation and nine which showed 'some significant interaction term differences'. Most of the variation occurred among trees. Further, they showed that 'oxygenated terpenes and sesquiterpenes tended to increase during the day, while sabinene decreased until late evening and increased during the early morning'. Nevertheless, diurnal effects were not found to be important sources of error if 'character weighting were used to maximize the genotypic differences'. This study reaffirmed the rationale of sampling *J. scopulorum* during periods of inactive growth (Powell and Adams, 1973) as has been noted by several other authors for conifer trees.

2. *Genetic variability*

Most of the literature relating to the genetic control of terpenoids among plants has appeared relatively recently, primarily because prior to the development of gas chromatography identification and quantification of these compounds was difficult. Mirov (1956) was one of the early pre-chromatographic workers in this area, having published an excellent population study of *Pinus contorta* and *P. banksiana* in which it was concluded, mainly as a result of studies on natural hybrids, but also with data from at least one synthetic cross, that the individual expression of terpenes in these two taxa seemed to be controlled by relatively few genes, much as in the herbaceous species of *Mentha* and *Hedeoma*, discussed above. For example, he noted that in hybrids bicyclic terpenes tended to dominate over monocyclic terpenes and that quantitative variation in these compounds seemed also to be genetically controlled, presumably by relatively few genes. Subsequent work by Hanover (1966) and numerous other investigators (e.g. Squillace and Fisher, 1966; Wilkinson *et al.*, 1971; Rockwood, 1973), mostly using data obtained from provenance studies, have tended to agree with these conclusions. That is, the presence or absence of some terpenes seem to be under simple gene control, while the expression of others, perhaps a majority, seem to be largely polygenically controlled, albeit by relatively few gene loci. Thus Squillace (1971), as a result of studies on 811 individuals of *Pinus elliotii*, from 16 clones and 45 control-pollinated 'families' produced by matings among 21 individuals, concluded that, although composition varied greatly among individual trees, this was mostly due to genetic factors and that all of the major terpene constituents (α-pinene, β-pinene, myrcene and β-phellandrene) showed very broad-sense heritabilities and at least moderately strong, narrow-sense heritabilities'. Franklin and Snyder (1971) studying phenotypic variation of monoterpenes in trees of *Pinus palustris*, derived from controlled experimental crosses, found 'that over half of the total genetic variance and slightly less than half of the phenotypic variance was of

Table 9.1 Percentage composition of the leaf oil of *Picea glauca* from samples collected on different sides of a tree[a] (From von Rudloff, 1967)

Compound	Tree 1				Tree 2			
	East	South	West	North	East	South	West	North
	(0.55)[b]	(0.34)	(0.39)	(0.45)	(0.31)	(0.40)	(0.42)	(0.24)
Tricyclene	0.4	0.2	0.3	0.2	0.3	0.2	0.3	0.2
α-Pinene	4.0	3.1	5.3	3.0	4.0	3.5	5.2	5.2
Camphene	4.1	4.0	5.0	3.2	6.7	5.9	7.0	7.3
β-Pinene	1.3	0.4	1.9	0.5	0.4	0.2	1.5	1.7
Myrcene + 3-Carene	4.9	3.6	5.1	2.6	9.3	8.4	10.0	10.0
Limonene + β-Phellandrene	16.8	13.6	15.3	12.4	14.8	14.5	14.7	14.5
Terpinolene	0.1	trace	trace	0.1	trace	trace	0.2	0.3
1,8-Cineole	0.2	0.1	trace	0.2	trace	0.1	0.1	trace
Camphor	57.4	62.0	55.5	64.8	52.5	56.3	47.5	47.0
4-Terpinenol	0.3	0.2	0.3	0.2	0.4	0.3	0.2	0.3
Borneol	0.2	0.2	0.2	0.2	0.5	0.3	1.5	1.8
α-Terpineol	0.2	0.2	0.2	0.2	0.1	0.1	0.4	0.4
Bornyl acetate	13.0	12.0	10.5	11.5	10.5	9.5	9.7	10.0
Piperitone	0.1	0.1	0.1	0.2	trace	trace	0.1	0.2
Cadinenes	trace	trace	trace	0.1	0.2	0.2	0.6	0.5
Sequiterpenes	0.2	0.2	0.1	0.2	0.1	0.2	0.2	0.4

[a]From two old trees (50–55 years) at the same location, September 17, 1966.
[b]Figures in parentheses are percentage yields obtained with the circulatory distillation apparatus.

the additive genetic type'. It appears fairly certain, then, that the qualitative expression of volatiles in these studies is nearly always under rigid genetic control; indeed, simple inheritance (dominance versus recessive genetic factors) has been established for several terpenes. Environmentally induced quantitative variation in these characters does occur, but most of this variation is also under fairly strong genetic control. Further, compared to herbaceous groups, the terpenes of gymnospermous trees appear to be much less prone to vary in mature organs. They also appear to have a lower degree of metabolic turnover of these compounds in mature structures, to judge from the evidence available to date.

Using consistent steam distillation procedures and gas chromatographic units with automated recorders, von Rudloff (1972a) reckons that the analytical errors incorporated into his numerous studies as a result of such technique and instrumentation averages about 2–3% for the major terpenoid components, the latter ranging between 2 and 90% in a given oil sample. As indicated earlier, however, if different chromatographic columns are used, or if the programmed run or rate of gas flow is not kept the same, the introduction of such seemingly comparable data may incorporate considerable 'analytical error' into a given study. In addition, von Rudloff estimates that sampling errors of a biological nature for the major components examined in his studies may range from 5 to 10%; for minor compounds (those making up 0.5–2.0% of the volatile oil) errors will be twice as high or more. He concludes, therefore, that for his particular studies 'variation above 10% (relative) of the major and 25% of the minor constituents were considered as significant'.

This is probably a fairly conservative estimate of such variability if careful sampling procedures are followed, but for the inexperienced or ignorant field collector, taking samples from natural populations without regard to plant condition, time of season and developmental stage of the plants concerned, variance in one's data due to errors in sampling may be much higher. Some early workers using terpenes were apparently unaware of the need of such sampling precautions; consequently they tended to attribute greater effect of environmental factors upon these compounds than seems to have been warranted.

If the investigator knows his plants, however, and carefully designs his collecting techniques, variability in one's data due to maturational differences among the plants or organs being studied is not likely to be more than 5% or so. For example, Table 9.1 shows the variation in terpenes encountered by von Rudloff (1967) in sampling four sides of a single tree of *Picea glauca*. Similar, mostly trivial, variations were also encountered by Scora *et al.* (1966) for the terpenes from the rinds of the citrus plant *Poncirus trifoliata*.

Further, von Rudloff (1972a) collected foliage from the same tree during the same month over a 5-year period (Table 9.2) and did not detect significant variation in the relative concentration among terpenes. He remarks upon this as follows: 'the foliage oil from a single white spruce tree did not change significantly over a 5-year period when the foliage was collected during the fall . . . . This confirms the previous conclusion that the type and relative amount of terpene appear to be under strict genetic control'. He notes, however, that annual variations in the yield of such oils may vary for environmental reasons, but the *percentage yield* of each compound remains about the same.

Table 9.2 Percentage composition of the volatile oil of the foliage of a single white spruce tree[a] collected each fall during 5 years. (From von Rudloff, 1972a.)

	Year				
	1967	1968	1969	1970	1971
Compound[b]	0.19[c]	0.21	0.24	0.23	0.28
α-Pinene	5.5	5.0	5.8	6.2	6.0
Camphene	5.8	5.8	7.8	8.8	8.1
β-Pinene	2.5	2.8	2.5	2.6	2.8
Myrcene	8.1	9.4	9.6	10.2	8.7
Limonene	15.0	14.5	14.8	15.3	14.5
Camphor	40.6	41.7	39.6	38.0	40.3
Borneol	0.8	12.3	1.0	1.1	1.2
Bornyl acetate	12.1	12.3	11.4	10.5	11.8

[a]Fifty-year-old tree, Sandy Lake, Sask. Foliage collected in September and early October each year.
[b]Major components only.
[c]Percentage yield of oil.

IV. Alkaloids

In spite of an extensive body of literature, alkaloids have not proven especially useful in systematic studies, except as broad general markers at the tribal level or higher (Gottlieb, 1982). Alkaloids are exceedingly variable at the generic level and lower; indeed, it is difficult to obtain consistent results with repeated sampling even at the cultivar level.

The reasons for this are not entirely clear but presumably a large factor has to do with the fact that most alkaloids arise out of rather ubiquitous basal metabolic pathways; i.e. the biosynthetically 'primitive' precursors leading to their formation are generally believed to be relatively simple amino acids and terpenes (Robinson, 1968, 1981). This explains, in part, the seemingly independent origin in remote phyletic groups of the structurally simple alkaloids such as caffeine, harmine, and nicotine.

As an example, the relatively common alkaloid anabasine is believed to arise out of pathways leading from the ubiquitous amino acid lysine or its decarboxylated product cadaverine. Robinson (1968), citing the work of others, points out that anabasine can be synthesized by crude enzyme extracts of the legume *Pisum sativum* if cadaverine is added to the extracts as a substrate. This species, however, does not produce anabasine *in vivo*. Crude enzyme extracts from species belonging to the legume genus *Lupinus* can also form anabasine if tetrahydroanabasine is added as a substrate under *in vitro* conditions. Because of these results, Robinson rightly remarks:

> Since neither compound is normally present in pea or lupine plants, their formation in extracts is obviously an *in vitro* artifact which may serve as a warning against indiscriminate extrapolation of *in vitro* findings to proposals concerning normal pathways of alkaloid biosynthesis. On the other hand, such results may point out that the enzymes needed for alkaloid biosynthesis are widespread and that only differences in cellular organization or degradative ability prevent all plants from accumulating alkaloids.

Another example of this sort is provided by experiments on *Vicia faba*. This species and those related to it, including most of the tribe Viceae, lack the so-called lupine alkaloids. It is interesting to note that *V. faba*, while not containing compounds of this type, will nevertheless synthesize lupanine and hydroxylupanine *if* sparteine is injected into the intact seedlings of this plant. One can conclude from this work that either the oxidizing enzymes are not specific to the alkaloid biosynthesis pathway or, possibly, *V. faba* represents an evolutionary line whose ancestors contained lupine-type alkaloids but which through mutations has lost certain enzymes of the biosynthetic pathway without losing the oxidases.

Robinson (1974) emphasizes the dynamic metabolic state of alkaloids in plants, commenting upon their fluctuating concentration and rate of turnover, both as a result of development and ageing. This is exemplified by *Catharanthus roseus* (periwinkle) from which several dozen or more indole alkaloids have been isolated. Virtually no alkaloids can be detected in seeds but they appear during germination and at 3 weeks of age are found throughout the plant. They then disappear almost completely only to reappear again at about 8 weeks. Similar anomalous shifts and changes are reported for yet other plants (Robinson, 1974), including significant diurnal fluctuations (Fairbairn and Wassel, 1964; Vagujfalvi, 1973).

The considerable variation encountered in alkaloids from population to population, plant to plant, and organ to organ within the same plant is quite remarkable; considering the fact that their production is almost certainly governed by genetic factors (Legg and Collins, 1971; Barker and Hovin, 1974; Waller and Nowacki, 1978). However, genetic control is presumably under weak metabolic restraints since the quantitative and qualitative variation of alkaloids within a given plant is highly influenced by environmental factors.

V. Conclusion

As noted in the several chapters to follow, flavonoids, terpenoids and alkaloids have been the favourite compounds of plant investigators. No doubt *all* primary and secondary metabolites are under varying degrees of genetic and environmental control. Thus it behoves the phytochemist to be careful in his selection of material, aware of the experimental errors involved in their preservation and the artifacts to be expected in extraction of materials. But above all, he should be specific about the origin of his material, the plant parts investigated, their stage of development, and for goodness sake he should provide good herbarium material for voucher specimens!

10

Application of Chemistry at the Infraspecific Level

I.	Introduction	237
II.	Polymorphism	238
III.	Infraspecific structure of plant populations	241
IV.	Chemical investigations at the infraspecific level	243
V.	Conclusion	265

I. Introduction

Numerous workers have contributed to the ever mounting list of chemical components found among higher plant species. For any given plant, phytochemical investigation is usually limited to relatively few samples, usually one or two individuals from a single population, and even then little effort has been made to compare whole plants at comparable stages of development (to say nothing of the failure to document investigations through the preservation of voucher specimens). Alston (1967) reviewed critically many of these studies and predicted that future chemosystematic work would see 'a gradual shift in emphasis from the problem-exposing or the data-exposing phase to the problem-attacking phase'. As will be indicated in the several case studies considered in some detail below, this observation has come true. Indeed, since some of the more intriguing evolutionary problems exist at the infraspecific or population level and, because of the difficulty in perceiving and/or gathering morphological data for their resolution, it appears that such problems have proven especially attractive to the chemosystematist.

Although a broad conceptual framework exists for the evaluation of chemical data at the specific level or lower (Alston, 1967; Turner, 1967, 1970; Flake and Turner, 1968; Mabry, 1973a,b), relatively few studies have attempted to accumulate sufficient chemical data for meaningful systematic interpretations. Plant systematists in general, accustomed as they are to working with numerous, easily observed, usually quite variable morphological characters, are rightfully suspicious of the sweeping generalizations and extrapolations often made by chemosystematists using blatantly inadequate population data. Indeed, until the comparative phytochemist became as much aware of inter- and intrapopulation variability in chemical characters as the classically oriented systematist had been to morphological characters, most plant systematists tended to play down the significance of chemical data.

A few early workers in the area of chemosystematics, notably Mirov (1956) working with *Pinus*, were very much aware of the kinds of data and sampling needed for the adequate study of populations but, largely because of inadequate instrumentation, the time involved in merely accumulating relatively simple listings of data precluded the intensive survey for numerous compounds from a large assemblage of individuals and populations. In addition, numerical methods using high-speed computers, essential to the evaluation of such large massses of data, have been relatively recent developments (Flake and Turner, 1968; Adams, 1972; Sneath and Sokal, 1973).

II. Polymorphism

Chemical differences among individuals, especially those in continuously varying outcrossing populations, is the rule in nature. Detection of clear-cut character differences, be these morphological or chemical, between individuals in a given population is often referred to as polymorphism. Those characters or character states which can be said to be present or absent are usually spoken of as being polymorphic. As noted, however, by Salthe (1972, p. 427) 'at the level of genes and gene products all variability is polymorphic. One either has allele *x* or does not have it: there are no degrees of having it'. At the phenotypic level, however, when a single gene has an overriding effect on a trait, one monitors the alleles predominantly present at that locus in examining the trait. This is the situation normally encountered under field conditions when working with secondary compounds such as flavonoids where single-gene control of trait expression is the rule.

Studies of single trait polymorphisms have been well-established in Man; they are also well described in plants, especially among cultivars. Polymorphic traits may vary indiscriminately throughout the range of a taxon (i.e. without meaningful geographical correlation as regards expression of the trait in the population) or these may be expressed clinally. Because of polymorphic expression in one or more characters it is difficult, if not impossible, to cluster *individuals* in a meaningful way so as to reflect infraspecific population structure; rather, population statistics must be used in such studies. In short, what one must do is to estimate population variance within and between populations so that the populations themselves might be clustered. This simulates, in principle, the classical morphological methods, as noted by Flake and Turner (1968).

A number of interesting studies of character polymorphisms have been made, one of the more notable being that of D. A. Jones and co-workers (1978) on cyanogenic and acyanogenic forms of *Lotus corniculatus* and *Trifolium repens*. These workers attribute, in part, at least some of the polymorphism to predators, namely herbivores. Other workers have also reported that differential herbivory may be responsible for the maintenance of polymorphism. Thus Rice *et al.* (1978) account for the infraspecific variability of monoterpenes among *Satureja douglasii* (Labiatae) as due to palatability selection from the generalist Molluscan herbivore, *Ariolimax dolichophallus* (the Banana Slug).

In a genetic sense, then, polymorphism is universal. It has been estimated that 40% or more of all gene loci may be polymorphic in animals and this presumably holds for plants, although estimates of the latter are few and highly heterogeneous (Selander, 1976). Numerous reasons have been put forward to account for phenotypic polymorphism in nature (S. J. Jones, 1973), varying from recurrent mutation to genetic drift. To rule out mutation as a complicating consideration, it is perhaps appropriate to refer to a population as polymorphic for a given character when two or more distinct inherited traits co-exist at frequencies too great to be attributed to mutation. In common parlance, a population is considered polymorphic for a given trait only if the rarest form has a population frequency greater than 1%. This, for example, would rule out most albino traits, and perhaps many other of the occasional individual variant traits reported for plants.

Tétényi (1968, 1970) in reference to polymorphic chemical characters, has coined the term 'polychemism'. He champions the idea that detection of such phenomena should be dignified by some formal nomenclature, although it is not clear whether he conceives of chemical taxa as comprising population or geographic breeding units comparable to species and varieties, or to the mere clustering of chemical components, irrespective of their existence as elements within individuals of a population or breeding unit. It is unlikely that significant suites of chemical characters will exist independently of morphologically discernible taxa (in a population sense) so that the desirability of an independent chemical nomenclature is moot. Of course, cryptic taxa or sibling species, perhaps largely based on chemical characters, are certain to come to light as noted by the work on lichens, discussed below, but formal recognition in such instances can be readily accomplished by existing nomenclatural methods. Justification of such recognition, however, should ascribe to the notion that taxa exist (theoretically) as interbreeding or phyletic population units which show some meaningful discordance with yet other population units.

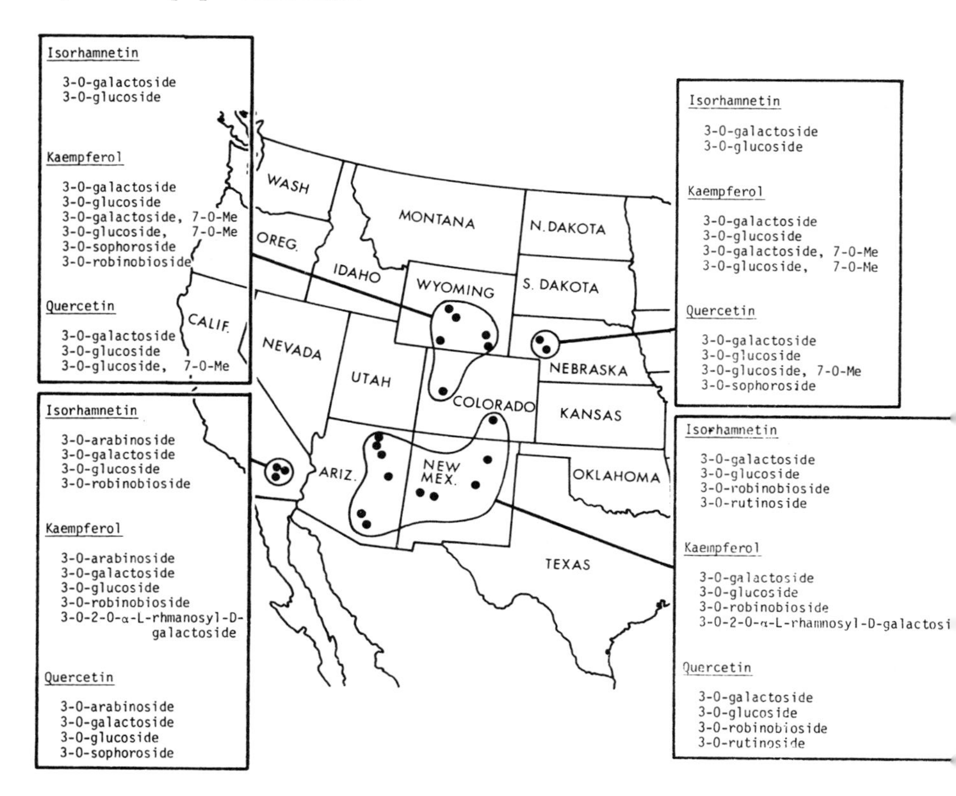

Fig. 10.1 The geographic distribution of flavonoids among 22 populations of *Chenopodium fremontii* in the western United States. (From Crawford and Mabry, 1978.)

In any event, the need for an independent taxonomic nomenclature to provide for the recognition of 'chemical races' as proposed by Tétényi, as noted by Alston and Turner (1963a,b), is perhaps ill advised in that chemical races can be readily acknowledged in an informal sense; these need only be recognized in a formal sense where they make sense! In which case, nomenclature can follow the present Code of Botanical Nomenclature, using chemical appellations, if this is deemed necessary. Crawford and Mabry (1978), working with infraspecific chemical variation among 22 populations of *Chenopodium fremontii*, make this point neatly. At least four chemical races within this taxon were detected, as noted in Fig. 10.1. It should be emphasized that these are *population* chemotypes and not individual chemotypes. Suites of characters characterize the races which, for the most part, are indistinguishable morphologically. The complexity of applying a nomenclature to the races so as to communicate a meaningful chemical concept can only be imagined. The method of communication of these data by the authors for the taxon concerned is simple and unequivocal. There was no obvious need for an independent system of nomenclature.

III. Infraspecific structure of plant populations

Ehrendorfer (1968) has summarized and evaluated much of the early work in the area of infraspecific differentiation, showing that variability in morphological, chromosomal, as well as in chemical characters in a population can be exceedingly complex. Most patterns of variation are correlated with ecological or geographical patterns or gradients. Factors affecting such patterning include (1) size of population, (2) environmental position, (3) migration routes, (4) variation and selection and (5) reproductive mechanisms. He recognized two extreme types of geographical and ecological differentiation in populations that he referred to as allopatric and partially sympatric. Allopatric differentiation occurs when subpopulations become divergent at the periphery of an already variable gene pool so that differentiation takes place without the build-up of strong barriers to gene exchange among individuals of the parental population. Divergence of subpopulations usually depends on isolation of the latter from the parental gene pool. Any new contact of isolated peripheral populations with the parental gene pool, if gene flow is relatively unrestricted, will produce clinal or stepwise intergradation (depending on ecogeographical factors) or at least regional differentiation. Partially sympatric differentiation occurs when inbreeding subpopulations are formed within the geographic confines of the parental population, usually because of abrupt intrinsic barriers to gene exchange among at least some of the subpopulations. This produces a population structure of pattern which is essentially a mosaic, the barriers to gene exchange having produced a sorting of variation and divergence among the populations.

These two phyletic models for population differentiation are illustrated in Fig. 10.2. They are at best highly simplified and all manner of variations on and between the models can be anticipated. What matters is that the systematist be made aware of the kind and extent of chemical variation likely to be encountered in natural populations, always keeping in mind that the variation is likely to reflect, to some extent, the influence of all the factors mentioned above.

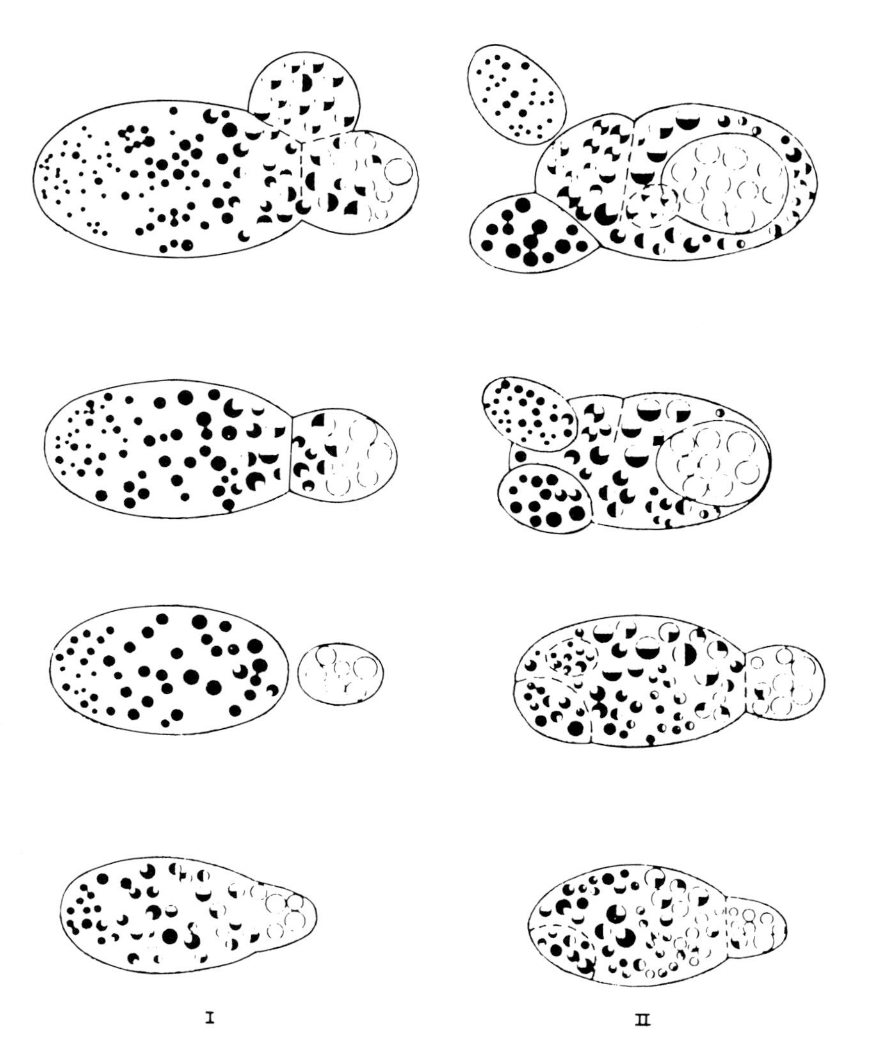

Fig. 10.2 Evolutionary scheme in two idealized population groups over four time levels: I (allopatric); II, partly sympatric differentiation. Further explanation in the text. (From Ehrendorfer, 1968.)

Most of our knowledge of differentiation in natural plant populations has come out of the study of morphological or chromosomal characters. Chemosystematic investigations, especially those involving the use of micromolecular data, have largely been concerned with taxonomic problems at the specific level, particularly where hybridization among the taxa concerned has been a factor in the variation (see chapter 13). As noted, however, there has been a recent trend toward more intensive studies at the infraspecific level using chemical data and, as will be indicated below, these have paid off handsomely.

IV. Chemical investigations at the infraspecific level

Several interesting publications have appeared on the significance of variation in chemical constituents in and among infraspecific populations. Many of the earlier studies have been reviewed by Alston (1967), including those of his colleagues and students working with *Baptisia* spp. Turner (1969) has also reviewed several interesting papers in this area and Mabry (1970) has given a detailed account of his very extensive population studies in *Ambrosia*, and has reviewed (Mabry, 1973a,b) yet other studies of a similar nature.

In the present review of infraspecific variation in chemical constituents, we have selected five studies that seem to us especially instructive. Each of these adds a significant dimension to the field and taken together they provide a historical perspective which, we hope, will compensate for our inability, because of space limitations, to cover all those published reports relating to the subject.

The alkaloid investigations on *Cinchona* spp. by W. H. Camp (1949) is important because, to our knowledge, it is one of the earliest population studies in which chemical data have been brought to bear on a systematic problem. The study would not have been possible at the time (undertaken during World War II, 1942–1945) were it not for government support of the project, for it engaged the services of several first-rate taxonomists and organic chemists, the purpose being to locate high-yielding quinine trees. To judge from Camp's paper, very little planning went into the systematic side of the 'team effort', especially as concerns population sampling and what might be needed for adequate statistical treatment. Still, the study is important because it showed the existence of chemical clines across a broad region, and this only a few years after the term cline was proposed for such character gradients (Huxley, 1942).

The several reports of chemical variation in populations of lichens by Culberson and co-workers is reviewed because it illustrates the value of intensive infrapopulation studies at the local level, even when only a few compounds are available for investigation.

The work of Flake and co-workers and that of Adams, using computer 'read-outs' in the form of automatic data processing, is significant in that it points to the direction that future studies in this area might take. It also provides a very convenient method for the presentation of chemical data since variation-trends across broad geographical regions for single compounds or those in combination can be visualized at a glance.

The work of Mabry and workers on species of *Ambrosia confertiflora* and that of Vernet and yet other European workers on *Thymus vulgaris* are selected for review since these are herbaceous genera which have been intensively investigated by a number of workers, using morphological, chemical and cytological methods. *Ambrosia confertiflora* is quite complex, complicated by polyploidy and high variability in both chemical and morphological characters, whereas *Thymus vulgaris*, to judge from the published studies is relatively simple, both chemically and morphologically.

Finally, a number of miscellaneous studies will be reviewed, especially those by Zavarin and co-workers and Von Rudloff and co-workers on conifer trees. Temperate conifers are especially amenable to combined biochemical and population studies since the compounds involved (terpenes) are easily quantified and therefore readily used in a comparative sense. Also, of course, population size and samples are readily taken during periods (winter months) of relatively stable biochemical activity (cf. chapter 9).

A. Clinal variation in *Cinchona*

Camp (1949), in an often overlooked but significant chemosystematic paper (modestly entitled '*Cinchona* at High Elevations in Ecuador'), presented population data for four alkaloids (quinidine, cinchonidine, quinine and cinchonine) found in the bark of a group of taxa belonging to the genus *Cinchona*. Quantitative variation in these compounds was obtained from over 120 trees representing 11 populations, the latter samples along a north–south transect of 500 miles from Colombia to Peru (see Fig. 6.4 in chapter 6, p. 86).

The study suffers in that the population samples are of different sizes (from 3 to 30 individuals) and statistical data are not given for all of the populations studied. Also, little information is presented as to the age of the trees, season collected, or how the bark samples were subsequently stored or treated. Although not evaluated statistically, the data available to him suggested that the taxon under investigation was composed of a chemical cline divisible into 'steps'.

In a complementary paper on *Cinchona* spp. by Schramm and Schwarting (1961), which is essentially an extension of Camp's early work, it was shown that the cline in total alkaloids shown in Fig. 6.4 extended to the 10° north latitude. The percentage of total crystallizable alkaloids (based upon 652 individual tree samples) remained at approximately 5% through the 5° north latitude and then progressively declined northward until they were essentially absent at 10° latitude.

The study on *Cinchona* spp. is significant in that it is one of the very earliest attempts to apply chemical data to population problems, concluding, in this

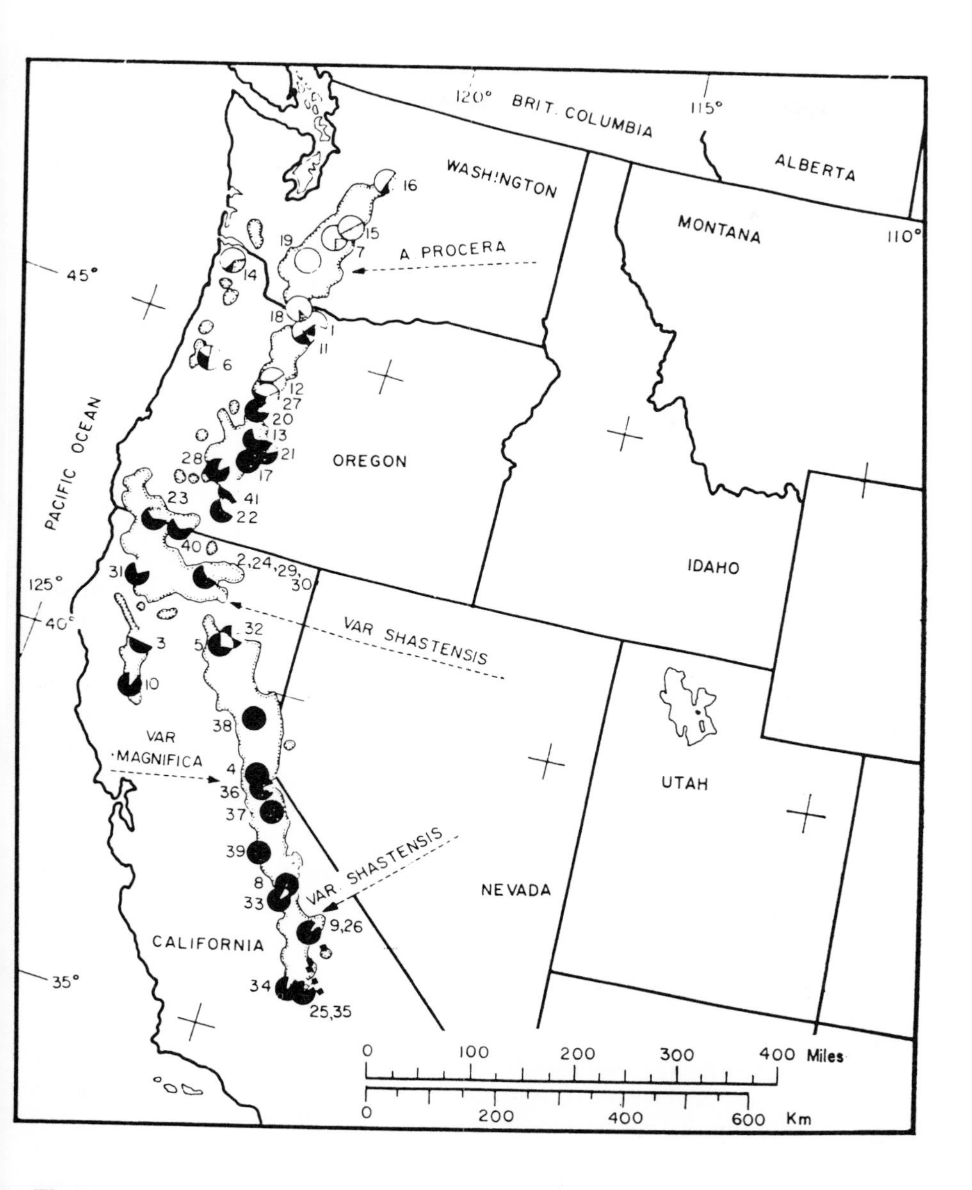

Fig. 10.3 A simplified geographic distribution of *Abies magnifica* and *A. procera* and location of the populations sampled (large circles). The black, stippled and white areas within each ring represent the percentage of low, intermediate and high limonene trees within a population. Values refer to population numbers. (From Zavarin *et al.*, 1978.)

instance, that what had been treated variously as three or more specific taxa was in reality a clinal complex composed of three intergrading population units. Camp (1949) also concluded that the cline was largely a result of gene-flow into the taxon from adjacent species and proposed the term *exogenous cline* to describe such intergradation. He distinguished this from the usual situation in which genetic differentiation *within* the population (i.e. mutation and selection among local or intergrading habitats) accounts for the pattern of variation disclosed. He termed the latter an *endogenous cline*. These are important concepts since both types undoubtedly occur in nature, both singly and in combination.

Endogenous chemical clines have been reported for several taxa, the most notable being that of Wilkinson *et al.* (1971) for White Spruce (*Picea glauca*); von Rudloff (1972b) for Douglas Fir (*Pseudotsuga menziesii*); and Zavarin *et al.* (1978) for two species belonging to the genus *Abies* (*A. procera* and *A. magnifica*). This last study is somewhat equivocal in that some workers interpret the two taxa as intergrading as a result of hybridization, although the carefully argued study of Zavarin *et al.* (1978) suggests that two taxa evolved out of ancestral elements which occupied the region in Eocene time (over 30 million years ago) and that the two taxa have 'undergone little morphological change in the past ten million years'. Under this view the transitional elements of southern Oregon and northern California (Fig. 10.3) are ancestral relicts that gave rise to the two elements and constitute the midportion of an endogenous cline.

In the *Juniperus* studies outlined below, it will be shown that what was previously thought to be an exogenous cline in southern populations of *J. virginiana* is in fact an endogenous cline. In other accounts of variation in *J. virginiana* (e.g. Flake *et al.*, 1978; cf. chapter 13) what was taken to be an endogenous cline among northern populations proved, upon closer chemical examination, to be an exogenous cline.

B. Variability within species of lichens

Most of the population studies of plant taxa published to date have been relatively broad regional studies. A notable exception has been that of Culberson and Culberson (1967) who studied the variation in lichen acids among a group of lichens belonging to the genus *Ramalina*. Their study of 980 plants (10 from each of 98 quadrats) is effectively summarized in Figs 10.4 and 10.5; briefly, it shows how much information can be obtained from detailed population work even when only a few compounds in a very local area are examined.

Zonation of chemotypes within the *Ramalina siliquosa* complex similar to that reported for the cliff face in Wales was subsequently reported for a maritime cliff in Portugal (Culberson, 1969). Because each of the six chemotypes had different but overlapping geographic distributions, and because they appeared to occupy different niches where they occurred in close proximity (Fig. 10.5), Culberson (1967) recognized the chemotypes as six sibling species.

Subsequent work by Culberson *et al.* (1977) extended the above studies and they report that an increasing degree of oxidation of the major secondary products characterized the more northern populations of the six sibling species in Europe and that on the 'telescoped environments' or maritime cliffs, the zonation of sympatric chemotypes (or sibling species) showed similarities to those same correlations. For example, the chemotype *R. crassa* is most abundant in southern Europe and it also occurs in the more sheltered habitat on maritime cliffs. In the words of the authors (Culberson *et al.*, 1977):

> 'The ranking of chemotypes by the northern limits of their ranges shows a previously unsuspected correlation with the chemical structure of the secondary products involved. If the products are ranked by increasing numbers of oxidation steps in their biosynthesis [Table 10], the order of the compounds correlates exactly with the order of the corresponding chemotypes by northern limits of ranges. It also approximates the order by local ecology.'

They conclude that their evidence concerning chemotypes in the *Ramalina siliquosa* complex leads to the 'inescapable conclusion to us that chemotypes, not plastic morphotypes, have been the basic evolutionary units in this group'.

Sheard (1978) has reviewed much of the controversy surrounding the work of Culberson and colleagues on *Ramalina siliquosa* and its segregates. He concludes, after numerical analysis of both chemical and morphological characters, that the zonation of chemotypes described by Culberson and Culberson (1967) can be attributed in part to the presence of two species occupying different niches.

In addition to their *Ramalina* studies, Culberson and Culberson (1976) have reported an equally interesting chemosystematic phenomenon in the lichen genus *Cetrelia*. They found that the 15 species of this genus synthesize natural products in characteristic biogenetically meaningful sets which they term chemosyndromes. The chemosyndromes (and therefore the taxa) can be ordinated by the length of the side chains of the constituent compounds (Fig. 10.6), which they interpret as indicative of chemical evolution towards shorter side chains.

There are reportedly six morphologically well-defined species groups in the 15 taxa of *Cetrelia* which they examined. The chemosyndromes that characterize the 15 taxa could be grouped about these six morphological norms in a relatively meaningful way. Because of their weak morphological differentiation among several of these syndromic types, many, if not most,

Fig. 10.4 The promontory in North Wales studied for the behaviour of the chemical races of the *Ramalina siliquosa* lichens. All of the cliff faces support a *Ramalina* vegetation. To the south and west, the *Ramalina* zone faces the sea and is directly above a *Verrucaria maura* zone (V), which in turn is above the algae of the *Fucus-Ascophyllum* zone (F), here seen exposed at low tide. Toward progressively more sheltered conditions around the headland, the *Ramalina* zone on the northwest side faces a rocky beach at the left of the photograph; on the northeast, the most protected place of all, it faces a grassy slope (G). The location of the six line-transects, indicated by numbers, is approximate and the distance between transects 1 and 2, 2 and 3, and 3 and 4 (actually 2.5, 3.4 and 4.1 m respectively) is distorted by perspective. Fig. 10.5 shows a breakdown of the chemical types found in the six transects. (From Culberson and Culberson, 1967.)

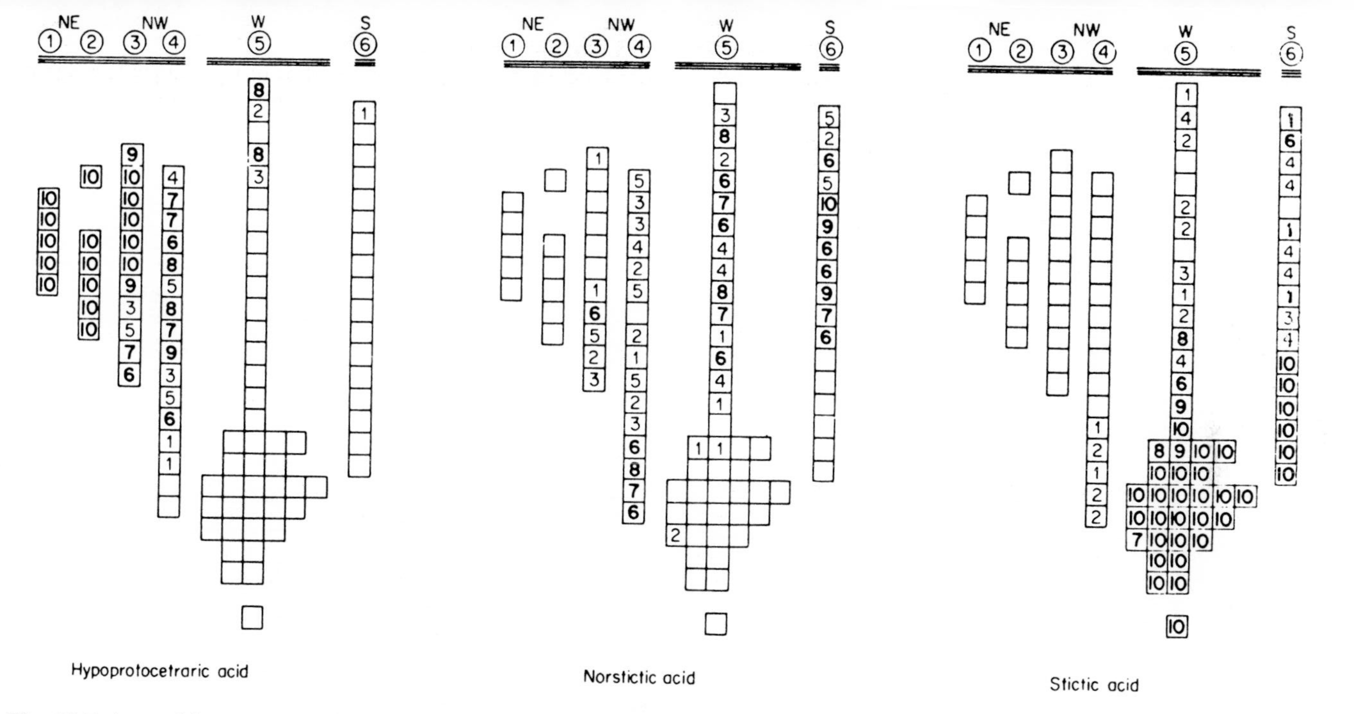

Fig. 10.5 A graphic summary of the results, showing the distribution of the three most common chemical races in the six transects indicated in Fig. 10.4: the hypoprotocetraric acid type (left), the norstictic acid type (centre) and the stictic acid type (right). The values 1 to 6 in circles refer to the six vertical transects, the position of which on the cliff is indicated by corresponding values in Fig. 10.4. The blocks (35 cm × 35 cm) of the transects are shown in their topographic position with respect to each other from transect to transect. Ten plants were taken from each of the 98 blocks, the numbers indicating how many plants per block were of each type. Some corresponding blocks have values totalling less than 10 because data for 42 individuals of three additional but rare chemical types are not given here. Where blocks are missing altogether, there were no *Ramalina* lichens. (From Culberson and Culberson, 1967.)

workers would recognize these as sibling species. Culberson and Culberson (1976) have noted, however, that 'confirmed [lichen] morphologists argue that chemical changes lack extensive genetic bases, are of questionable adaptive value, and have been superficially veneered upon morphology'. Several attempts have been made to reconcile the views of morphological and chemical proponents, but none has proven satisfactory. It is likely, as concluded by the Culbersons, that differentiation in *Cetrelia* occurred through a combination of parallel morphological and chemical changes.

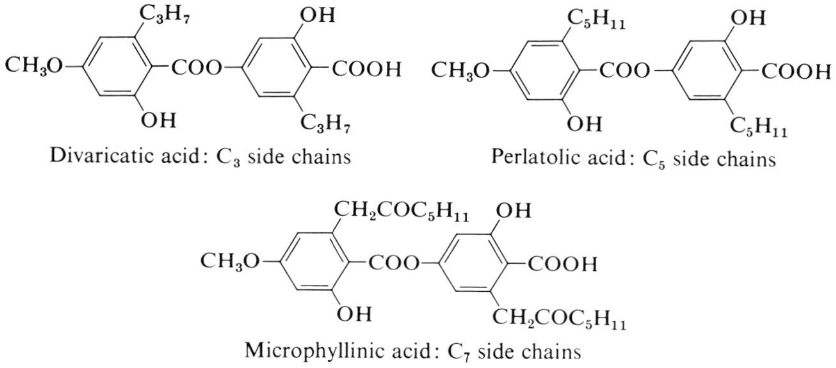

C_3H_7 OH C_5H_{11} OH
CH_3O COO COOH CH_3O COO COOH
OH C_3H_7 OH C_5H_{11}

Divaricatic acid: C_3 side chains

Perlatolic acid: C_5 side chains

$CH_2COC_5H_{11}$ OH
CH_3O COO COOH
OH $CH_2COC_5H_{11}$

Microphyllinic acid: C_7 side chains

Fig. 10.6 Some typical depsides varying in their side chain substitution in the lichen genus *Cetrelia*.

Culberson *et al.* (1977) reported an additional set of chemosyndromes in the *Parmelia pulla* group of lichens. This group comprises 18 species or races which can be ordinated into six chemosyndromic types, based upon six orcinol-type depsides that occur as constant major constituents in at least one taxon and whose minor constituents are all biogenetically close to the major constituents. Again, as in *Cetrelia*, these could be arranged, for the most part, so as to show a progression towards shorter side chains, this presumably relating to phyletic lineages.

The popularity (or unpopularity, whichever) of chemosystematic work with lichens undoubtedly rests upon the absoluteness of the results obtained. In a given plant, compounds either exist or do not. This makes for the ready construction of keys and distribution maps. Since lichens, for the most part, reproduce vegetatively, it can be argued that what one is circumscribing, using only chemical characters, are chemical groupings that may or may not coincide with the more conventional 'biological species' of the more advanced plant groups. Thus, in *Ambrosia confertiflora* (discussed below) it would be possible to recognize several sibling species using one or a few sesquiterpene lactones, to say nothing of the presumably independently varying flavonoid types.

There appears to be no easy resolution to the systematic problems posed by lichens, as noted by Robinson (1975) who states, 'The present view [of lichen speciation] would not require much less stability of lichen 'species' in nature than is generally suspected, but it would suggest that when changes do occur they are mostly from hybridization than from mutation, and that some 'species *may* have originated more than once'. In other words, given the long evolutionary history of lichens, he supposes that occasional hybridization among the infrequent sexual morphological forms spawns much of the chemical variation which is then maintained by clonal propagation and that what is circumscribed chemically may or may not coincide with meaningful phyletic lineages.

Other than lichens, extensive chemical studies have not been made of plant groups which are largely composed of asexual reproducing populations. However, it would appear that the situation among lichens is not too different from that found in populations of the phytoplankton genus *Daphnia*, which is composed of species complexes or populations, many of which are maintained by clonal reproduction. Thus in *D. magna*, genetic studies have shown that large numbers of chemical clones co-exist; but, since this species is capable of occasional sexual reproduction, it has been argued that clonal co-existence is short term, for new clones may enter the population from sexual eggs much as suggested for lichens by Robinson (see above). But, even in obligate parthenogenetic species such as *D. pulex*, allozymic variation is apparently common in natural populations. Thus Hebert and Crease (1980) report that chemical clones could be recognized in 11 populations which they surveyed 'of which as many as seven were found in a single habitat'.

C. Variability within selected perennial herbs

1. Ambrosia confertiflora (*Compositae*)

Detailed surveys for sesquiterpene lactones both between and among populations of this perennial, herbaceous, weedy species were made by Renold (1970) in a doctoral study. Portions of this work have been often quoted (Mabry, 1973a,b; Yoshioka *et al.*, 1973; Payne, 1976), but the study remains unpublished. It is an excellent piece of work in which considerable care went into the sampling procedures and much effort was expended in an effort to reconcile chemical variability with morphological and cytological variables.

Renold examined about 250 populations of this taxon and these could be roughly classified into leaf forms as shown in Fig. 10.7. Altogether, 187 of these were examined for sesquiterpene lactones and the results of this study are shown in Fig. 10.8. Because of time limitations, Renold was unable to examine individually each of the 10 plants that he collected at a given site.

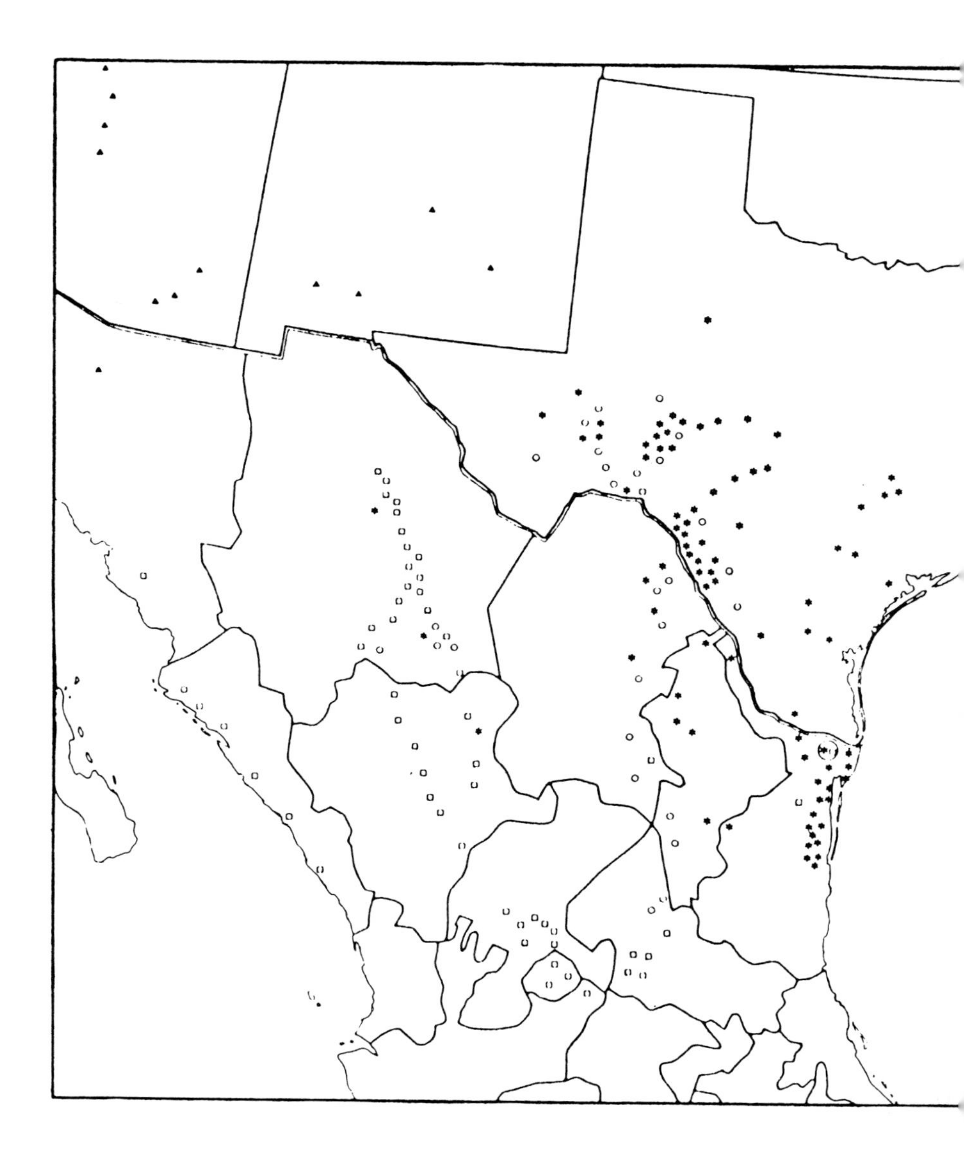

Fig. 10.7 Distribution of morphological leaf forms in *Ambrosia confertiflora* populations in the southern United States and northern Mexico. Symbols: *, alternate-leaved form; □, lacy-leaved form; ▲, hispid-leaved form; o, intermediate form between the 'attenuate-leaved form' and the 'lacy-leaved form'; (*□), occurrence of two 'forms' within the same population.

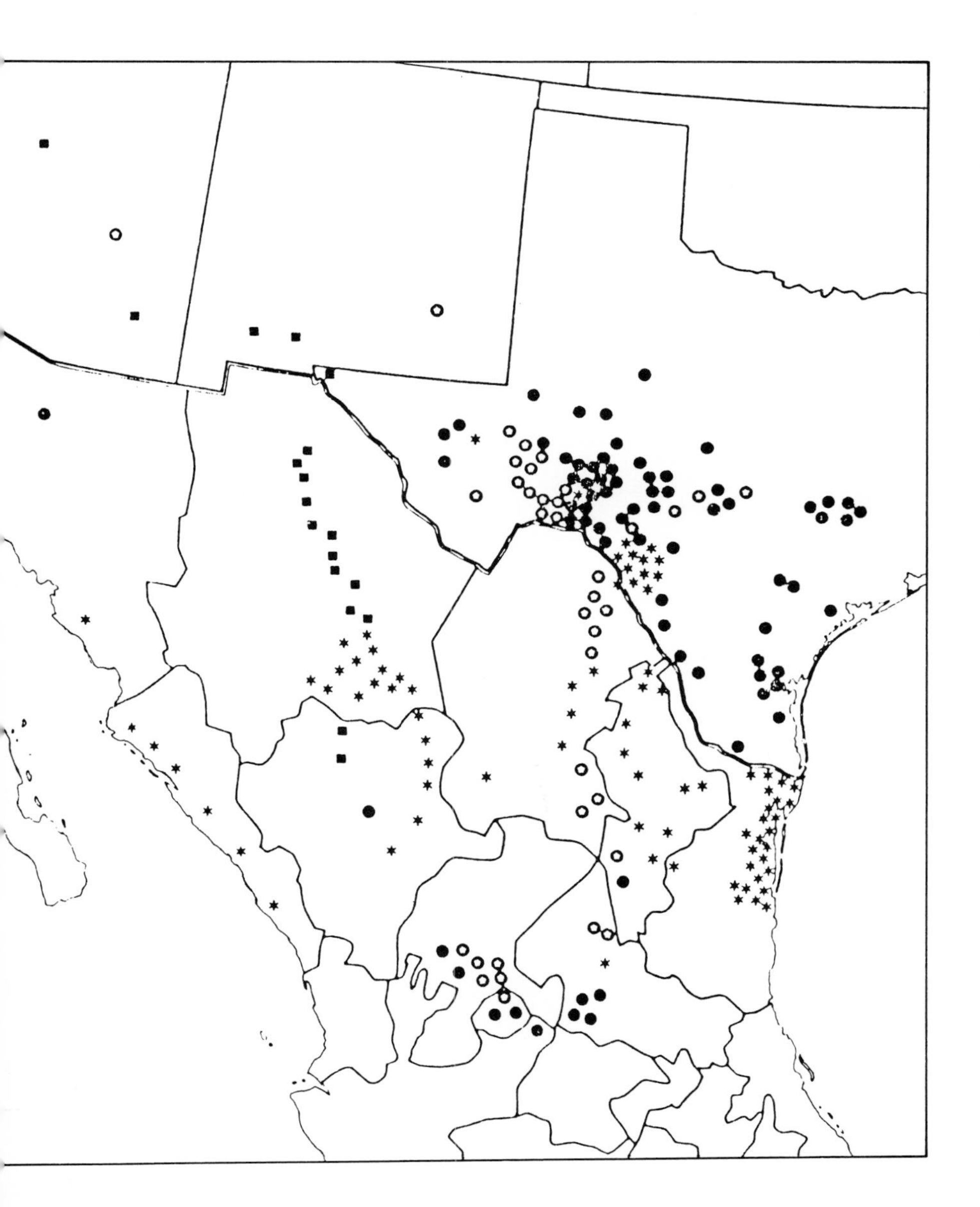

Fig. 10.8 Distribution of sesquiterpene lactones in *Ambrosia confertiflora* populations. Key: ⋆ germacranolide—eudesmanolide race; ■ germacranolide race; ● pseudoguaianolide—monolactone race; ✪ pseudoguaianolide—dilactone race.

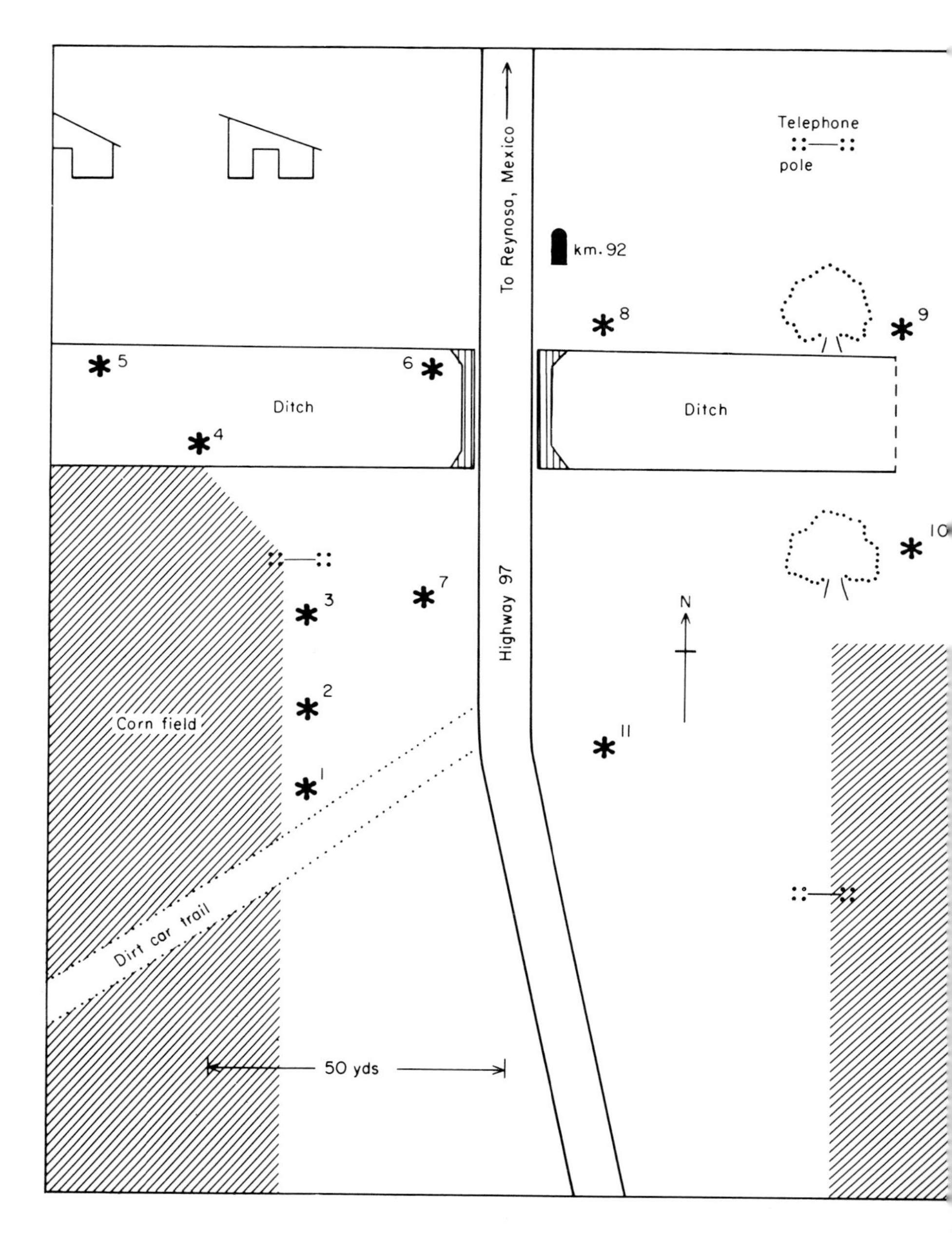

Fig. 10.9 Distribution of 11 individuals of *Ambrosia confertiflora* at a single site in Reynosa, Mexico. Relative percentages of six sesquiterpene lactones found to occur in these are shown in Fig. 10.10. (From Renold, 1970.)

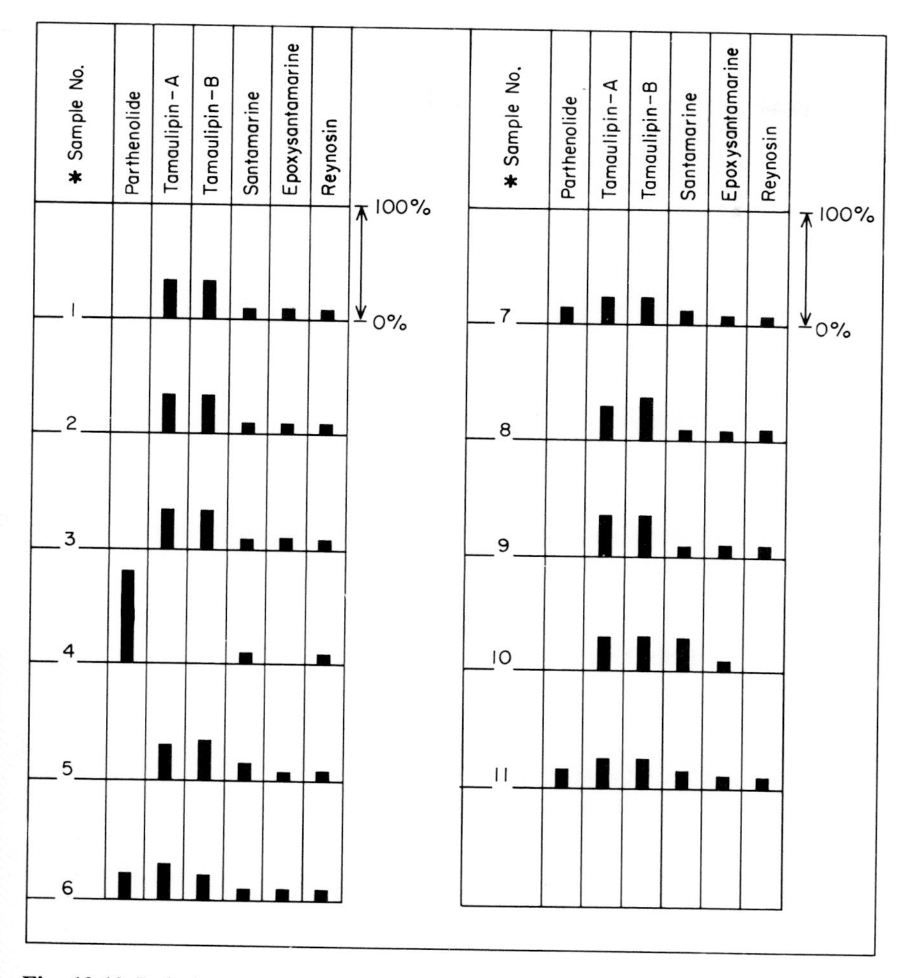

Fig. 10.10 Relative percentages of sesquiterpene lactones found in the individuals of *Ambrosia confertiflora* indicated in Fig. 10.9. (From Renold, 1970.)

Instead, these were bundled together as representative of the population concerned and extracted for their sesquiterpene lactones. Nevertheless, within a given sample site, he never encountered more than one of the four chemical types recognized. In other words, they were mutually exclusive. Transplant and developmental studies showed that environmental and ageing variables did not produce qualitative effects, although quantitative variation was noted.

In those several natural populations in which *individuals* were examined, only one population was found in which individuals representing two of his chemical 'races' were observed. An example of the kind of qualitative and quantitative variation found among individuals of a given population is shown in Figs 10.9 and 10.10. In fact, only six populations of the approximately 187 examined, were found to contain compounds from two different chemical races. Renold rightly concludes that the rarity of such 'hybrid' populations 'implies the presence of unique and genetically quite stable populations, at least with regard to their sesquiterpene lactone constituents'. An attempt was made by Renold to relate the four chemical types to leaf types in the species. He concluded that 'hispid-leaved' populations were characterized by Chihuahuin chemotypes with an absence of Tamaulipin chemotypes; that 'attenuate-leaved' populations were characterized by Tamaulipin and Confertiflorin chemotypes with a near absence of Chihuahuin chemotypes; and that 'lacy-leaved' populations were characterized by a predominance of Chihuahuin and Tamaulipin chemotypes (but with a significant set of yet other types). No particular chemotype was found to dominate exclusively a particular morphological type but, in general, broad geographical patterns could be recognized, in spite of intergradation in the types of leaf form that occurred sporadically in regions of allopatry, if not sympatry.

Attempts to relate the chemotypes or, indeed, morphotypes to the several polyploid levels ($n = 36$, 45 and 54) within *Ambrosia confertiflora* failed, because polyploidy has been sporadically superimposed upon the various races, presumably in relatively recent times. However, only 20 populations were examined for chromosomes and more extensive sampling may yet reveal strong correlations.

In spite of the cytological obfuscation, Renold was able to postulate, on chemical grounds alone, that the species, as it is known today, radiated out of an ancestral complex in central Mexico. On biogenetic grounds the Tamaulipin chemotype was considered to be primitive and the more highly oxidized Chihuahuin and Artemisiifolin types to be derived or arising from it. The more northern Confertiflorin and Psilostachyin races are believed to have arisen out of the latter, having spread subsequently into Texas and adjacent areas.

2. Thymus vulgaris (*Labiatae*)

In contrast to the sesquiterpenoid studies on *Ambrosia confertiflora*, several phytochemical workers, using volatile oils, report rather remarkable correlations of chemical characters with geographical or associational types even over very short distances. Thus, Vernet (1976) was able to relate chemotypes of *Thymus vulgaris* to a mosaic of ecological or vegetational units over an 80 km^2 area (Fig. 10.11). He found that, when using two of the more abundant

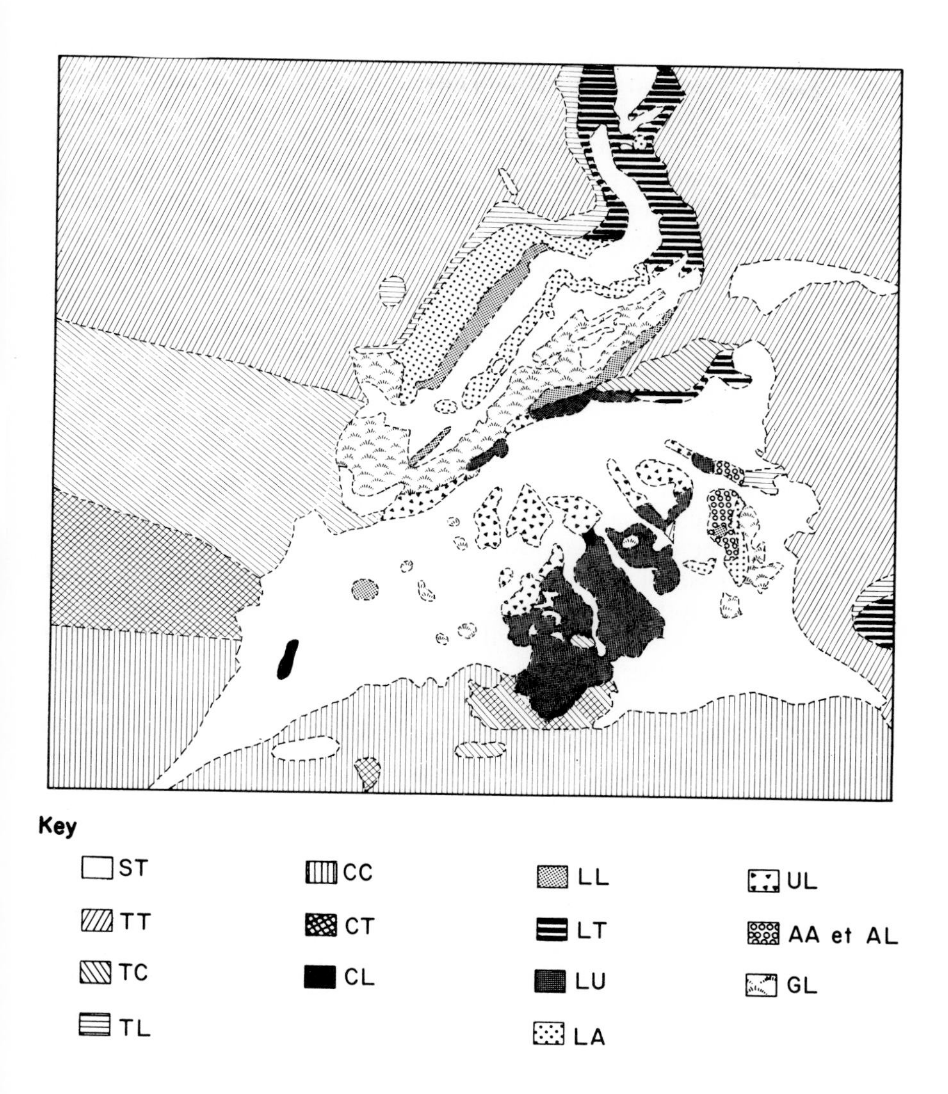

Fig. 10.11 Distribution of chemotypes of *Thymus* in a 10 km² tract in northcentral France. Populations are identifiable by the combination of two terpenoids as follows: T, thymol; C, carvacrol; L, linalol; U, thuyanol; A, α-terpineol; G, geraniol; ST, without *Thymus* populations. Chemotypes are said to be correlated with specific soil types in the region concerned. (Redrawn from Vernet, 1976.)

monoterpenoids, very precise correlations of chemotypes and habitat could be made. Changes in such correlations could be detected within distances as short as 10 metres, and nearly always linked to ecological shifts. Proceeding from dry to relatively humid habitats, he reports clinal shifts from populations with phenolic terpenoids (carvacrol or thymol plants), then linalol, thuyanol-4 and finally to α-terpineol populations. This chemical shift in a population's make up was found to be independent of the successional state of the vegetation. He notes that a two-chemotype delineation is superior to delineation attempts using only the presence or absence of a single chemotype.

Strong correlation of chemical constituents with geographical regions and ecological factors are reported for yet other species of *Thymus* (Adzet *et al.*, 1977) and to a lesser extent for the legume species *Anthyllis vulneraria* (Gonnet, 1980).

It is difficult to believe that such precision of habitat delineation can be obtained by the methods presented by Vernet (1976), especially knowing the complex genetic make up of most plant populations, their propensity for widespread sporadic dispersal, and especially the individualistic way that species tend to be distributed within and between associated vegetational types. Nevertheless, considerable experimental work, especially genetic, went into Vernet's study and the results (Fig. 10.11) are especially impressive. Additional work is needed, for it does appear that similar studies of yet other taxa, using either volatile compounds, flavonoids or sesquiterpene lactones, are not so promising for such infraspecific delineation (e.g. Bragg *et al.*, 1978, in *Prosopis juliflora*; Crawford and Mabry, 1978, in *Chenopodium fremontii*; Kelsey *et al.*, 1975, in *Artemisia tridentata*; Levy and Fujii, 1978, in *Phlox carolina*; Sutherland and Park, 1967, in *Myoporum deserti*; etc.).

3. Satureja douglasii (*Labiatae*)

Jean Langenheim and her co-workers in California have attempted studies similar to those of Vernet (discussed above) for the perennial evergreen mint *Satureja douglasii* (Rhoades *et al.*, 1976). They analysed ten populations of this species. These were scattered over a 250 hectare tract which varied from dense redwood forest, through a chaparral–redwood border to a meadow–chaparral border. From this ecologically diverse area they were able to classify population types as either A or B, the former containing large amounts of limonene and carvone, but small amounts of piperitenone, piperitone, pulegone and isomenthone; B-type populations were essentially the reverse of A-type populations.

Although cognizant of the genetic base for such compounds, they nevertheless performed reciprocal clonal-transplant studies in the different habitats (which differed mainly in exposure to sun) and performed experiments in growth chambers. It was noted that the B-type plants showed considerable variation in the more reduced compounds (isomenthone and piperitone) and that increase in the latter was correlated with increasing shade. Further, B-type plants that had very different compositions in the field, produced similar compositions when grown together under controlled conditions. In addition to light intensity, they found that developmental processes also affected intrapopulation variability, noting that much of the observed variation among the 3-oxymonoterpenoids was 'probably an artifact of sampling shoots of slightly different ages'. Large differences in monoterpene yield in *S. douglasii* can also be affected by moisture stress, as shown by Gershenzon *et al.* (1978). Nonetheless, genetic control of terpenoid compounds is largely the rule, at least as far as yield is concerned under field conditions, high-yielding genotypes being correlated with low-light high-herbivory conditions and vice versa (Lincoln and Langenheim, 1979).

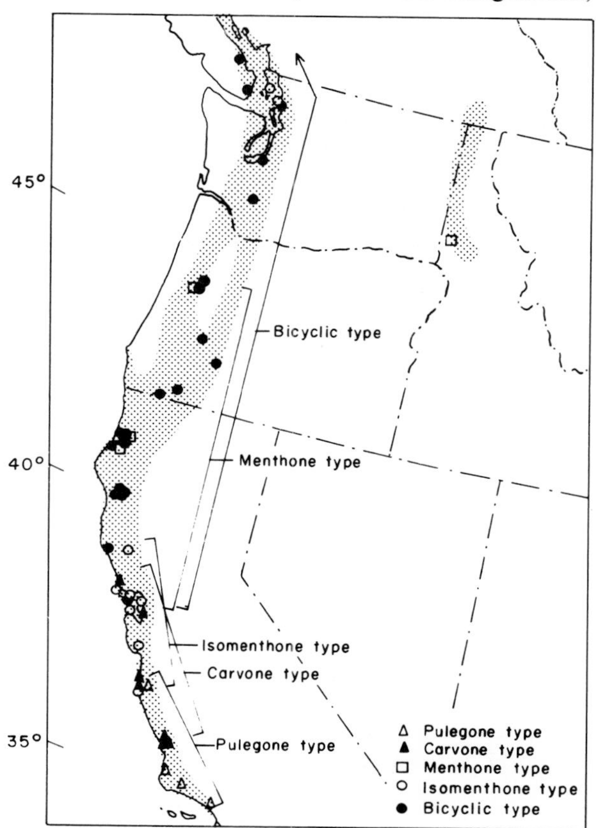

Fig. 10.12 Geographic range of *Satureja douglasii* and the five compositional types. (From Lincoln and Langerheim, 1976.)

Lincoln and Langenheim (1976) also analysed the monoterpenes of *S. douglasii* throughout its range. They were able to recognize five compositional types over a broad region as shown in Fig. 10.12. Using the dominant species-components of the vegetation at any given site as an indicator of habitat-type, they attempted to obtain correlation of the genetically determined variation in composition (i.e. chemotypes) with environmental conditions. It was concluded that although chemical composition types could be partially parcelled by such correlations, this did not sort out possible environmental conditions which determine the patterns of composition. They contrast their work on *S. douglasii* with that accomplished on *Thymus* spp. (discussed above) and conclude that, unlike the latter, *S. douglasii* shows polygenic chemical clines which occur 'irrespective of discrete compositional type variation'.

D. Variability within evergreen temperate trees

Species of evergreen woody plants which exist in large numbers over a wide region are especially amenable to chemosystematic study. Their considerable size permit ready recognition and, more important, population sizes throughout the distribution area can be readily ascertained. In short, a total census is possible. This permits systematic sampling *and* resampling if necessary. This contrasts with annual plants, which vary considerably from year to year, depending on climatic events; perennial, non-woody species, even if readily located, may also vary in number from year-to-year in population sites, either through predation, or simply by failure of shoot regeneration during periods of atypical climatic conditions, mainly drought, not to mention the greater susceptibility to stress that most such plants show, as noted for *Satureja douglasii* (above) and yet other perennial forms (cf. chapter 9). In addition, woody evergreen plants of temperate and cold regions, if sampled for their chemical components during the winter months, tend to show less diurnal, weekly or monthly variation in the quantitation expression of secondary components, mainly terpenes, than they do during summer months, a period of more active metabolic turnover of such compounds (Adams, 1970a,b; see also chapter 9).

E. Numerical analysis of terpenoid data in juniperus

One of the best documented infraspecific studies using terpenoids has been that of Adams (1977) on the gymnosperm *Juniperus ashei*. His study is one of the few that has integrated both chemical and morphological data in the examination of infraspecific variation and, coupled with multivariate analysis and computer graphic methods, serves as a model for what good infraspecific analysis might involve.

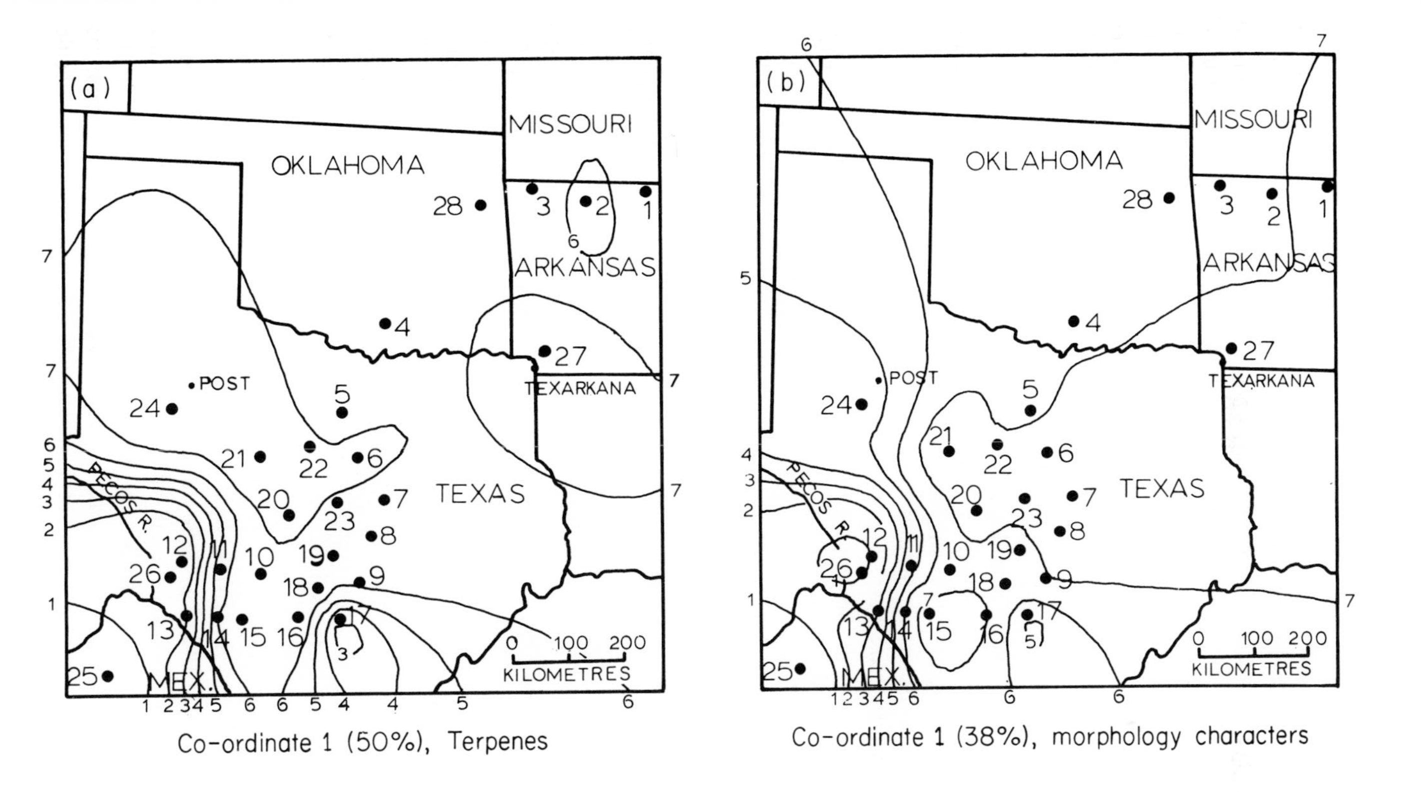

Fig. 10.13 Contour analysis of (a) terpenoid data and (b) morphological data from 28 populations (15 plants per population) of *Juniperus ashei*. Additional explanation in the text. (From Adams, 1977.)

Juniperus ashei is a widely distributed, conspicuous, weedy tree that Adams was able to sample (15 trees per population) uniformly and at regular intervals throughout the range of the species (Adams, 1977). Using 59 terpenoids (terpenes and sesquiterpenes) from the foliage, similarities were computed between populations using characters weighted by their variance among and within populations. The similarity matrix (between 28 populations) was then factored by principal co-ordinate analysis (PCO) and the resulting principal axes individually contour-mapped onto a geographical base map (Fig. 10.13a).

The first co-ordinate axis of the terpenoid similarities accounted for 50% of the variation between populations and clearly separates populations from the trans-Pecos Texas/northern Mexico region (nos. 12, 13, 25 and 26, Fig. 10.13) from the central Texas and northern populations. However, one exception was observed in that population 17 was more similar to the divergent trans-Pecos Texas/northern Mexico populations than to adjacent central Texas populations. A comparable analysis based on 15 morphological characters resulted in a major trend (38% of the variation) that is essentially identical to the terpenoid pattern (Fig. 10.13b), even including the similarity of population 17 to the trans-Pecos Texas/northern Mexico populations.

The above results have been confirmed using another multivariate statistical method (canonical variate analysis; R. P. Adams, personal communication) which resolves the correlations between variables. With this information, Adams examined past patterns of distribution and migration during the Pleistocene (with possible dating to much earlier times) to explain the extant distribution data. The most likely hypothesis to account for the latter appears to be that widespread extinctions occurred during the Wisconsin glacial advance (or in previous glaciations). This, along with the predilection of the species to prosper on calcareous soils, permitted relatively recent and rapid (re-)colonization of post-glacial habitats (Fig. 10.14). This led to very uniform populations in spite of disjunctions of 200 to 300 kilometres between populations (cf. Figs 10.13 and 10.14). Population 17 appears to be a relict of the ancestral type (lower camphor type, Adams *et al.*, 1980) that is at present restricted to the semi-arid desert margins in the trans-Pecos Texas/northern Mexico region (Fig. 10.14). Additional evidence on these relicts was obtained by computing the similarity between the most similar extant species to *J. ashei* (*J. saltillensis* from Northern Mexico) and contour mapping these similarities (Fig. 10.15a,b). Both the terpenoids and morphology showed a similar trend of increased similarity of the divergent populations to *J. saltillensis*. Although hybridization of the relict population could explain this trend in northern Mexico and the trans-Pecos Texas region (however, it is improbable that stabilized hybrids would persist over such a large geographical area), it could not explain the co-differentiation of population 17, which must surely be a relict.

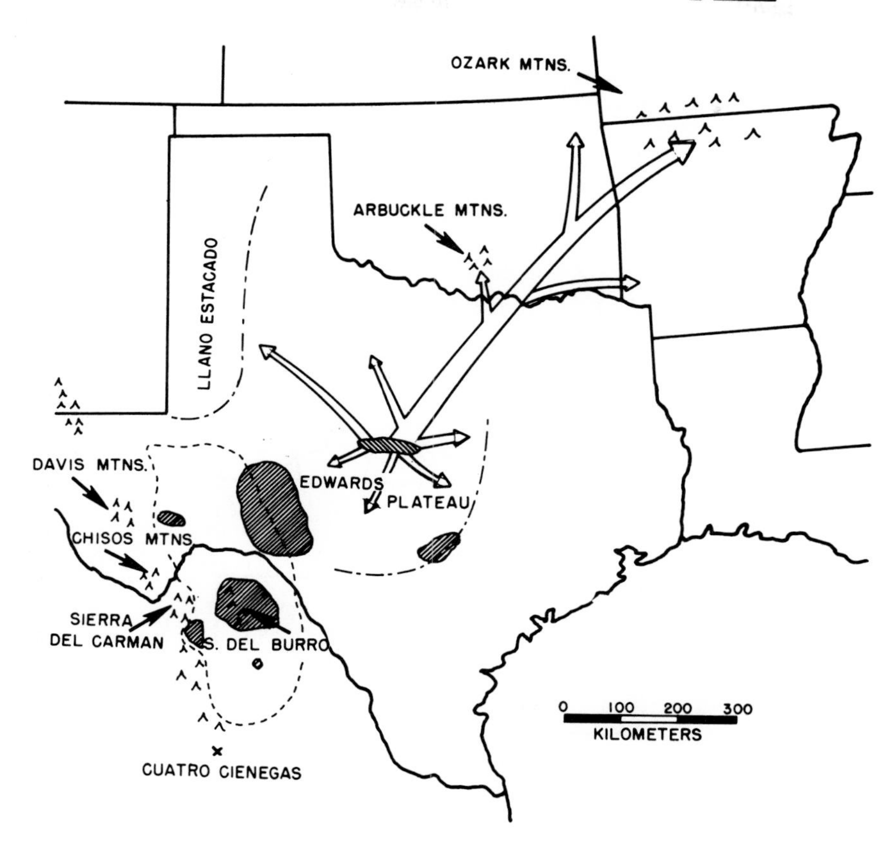

Fig. 10.14 Hypothetical post-glacial migrations of *Juniperus ashei.* Additional explanation in text. ▧, Remnant adapted to mesic environment during the pluval period; ▨ present distribution of relict populations. (From Adams, 1977.)

The above studies by Adams and co-workers was important for several reasons. It clearly demonstrates the advantages of terpenoid data for population work, especially in combination with morphological data. The study revealed that ancestral populations persist in nature for many generations and can be discovered through the analysis of extant data. More importantly, it combines numerous chemical characters by the use of multivariate methods and computer graphics into patterns of variation that could be used and/or tested to explain several biological phenomena.

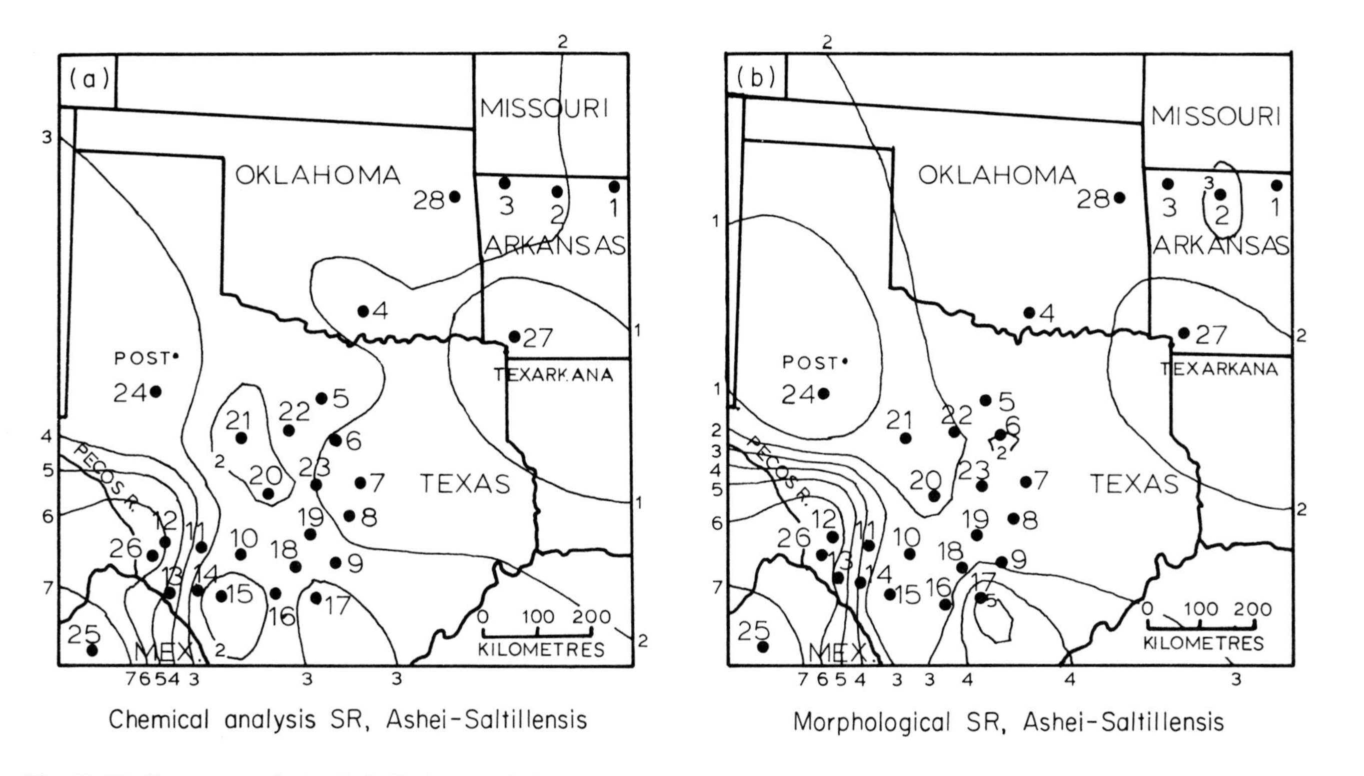

Fig. 10.15 Contour analysis of similarity trends between *Juniperus ashei* and its most closely related species, *J. saltillensis*; (a) terpenoid data, using the same populations examined in Fig. 10.12 and (b) morphological data. Additional explanation in text. (From Adams, 1977.)

As a footnote to the above discussion, it should be noted that Adams (1977) extracted several other trends from the similarity matrices and found that the morphological data proved to be more difficult to reproduce when changing technicians, etc. than the chemical data. That is to say, the morphological data contained a larger error variance than the chemical data.

Zavarin *et al.* (1971b) studied the leaf monoterpenes of several species of *Cupressus* along the California coast, including *C. sargentii*. The latter has the greatest distributional range of the several coastal species, occurring from latitude 35° to 40° north. Of the 136 trees examined across the range of this species, it was noted that the more southern populations (Chorro Creek, 35° 29′ and Zaca Peak, Lat. 34° 46′) appeared to be considerably different from the more northern populations, especially in the percentage composition of limonene and sabinene. It is clear from visual examination of these data that the Chorro Creek, Zaca Peak and the more northern plants have quite different average values and might be readily recognized by population samplings using the two terpenoid characters concerned, to say nothing of yet other characters. Morphological workers have not singled out the two more southern populations as being especially different, but the chemical data suggests otherwise.

V. Conclusion

The existence of chemical variation at the population level has to be borne in mind in all chemotaxonomic work. Sampling both within the same population and within different populations is desirable whenever analysing a given plant for its secondary chemistry; at the very least, some 10 to 20 samples should be analysed. Where chemotypes occur, it is apparent that chemical studies can add a new dimension to our understanding of variation, structure and evolution within plant populations. The value of such studies has been illustrated in this chapter with analyses of alkaloids in *Cinchona*, of flavonoids in *Chenopodium*, of depsides in *Ramalina* and *Cetrelia* and of terpenoids in a variety of plants. The range of data obtainable in the case of *Juniperus ashei* terpenoids is especially impressive and bodes well for the future development of chemosystematic studies at the population level.

Whether it is reasonable to consider formal chemonomenclatural status to the various populations discussed above, as might be envisioned by Tétényi (1970) and proponents, is left to the judgement of the reader. But to us formal nomenclature can be reckoned as a convenient structure within which resides a bewildering array of chemical data. In as much as this structure must accommodate the latter, and since ultimately morphology is a reflection of chemical interplay within an organism, form or morphology will most likely reflect phyletic relationships, at least at the specific and infraspecific levels. Besides, one classifies plant populations, not individuals (Flake and Turner, 1968). Below the individual, one is concerned with parts and processes; attempts to classify the latter as something more than the whole seems largely ludicrous.

11

Application of Chemistry at the Specific and Generic Level

I. Plant sampling and quantifying the chemical data 267
II. Use of flavonoids 274
III. Use of volatile terpenes 282
IV. Use of sesquiterpene lactones 286
V. Use of other secondary compounds 288
VI. Conclusion 291

I. Plant sampling and quantifying the chemical data

The previous chapter belaboured the point that considerable biochemical variability existed at the infraspecific level, both qualitative and quantitative. In spite of the plethora of publications documenting this fact, many chemosystematic investigations are undertaken and published as if one individual or, at most, a few individuals of one or but several populations constitute an adequate sample for sound systematic interpretations.

If one bothers to read the 'methods' or 'experimental' sections of most chemosystematic papers that appear in say, Phytochemistry, or yet other reputedly rigorous experimental journals, it will be noted that nearly all such studies are documented with relatively few individuals, not to mention a *dearth* of studies based on population samples (i.e. each population based on 20 individuals from a given site, these analysed separately; for wide-ranging species at least 10 populations should be examined, this amounting to 200 plants collected in the field, all in about the same stage of development and prepared for chromatography with about the same methods).

Good students practising modern monography in plant systematics know that such samples are necessary if they are to make sound statements about the variability of this or that morphological character. Indeed, most good monographs are documented by enormous citations of voucher specimens which were tediously observed and annotated as to identification (often with symbols expressing intergradation or putative hybridization and introgression). Many of these samples are collected in the field by the student himself and serve as the basis for statistical calculations showing relationships or inferred phylogenies. Often the vouchered material is so extensive that journals refuse to catalogue the listings, leaving these to be merely referred to as 'on hand' at this or that herbarium. Contrast all this with the haphazard way many, if not most, of the so-called chemosystematic studies are made: a comment perhaps that the material was obtained from one or a few miscellaneous herbarium sheets, or perhaps from a living population, in which great masses of vegetative material at various stages of development were analysed collectively, the so-called 'bulk-collections'. Or perhaps several such populations were sampled from quite closely situated sites.

Research based on such limited material is not unique to the chemosystematist. Strictly morphological workers, using herbarium sheets, also often possess an inadequate sampling of the variability within a given taxon, especially when field work over a broad region is not undertaken to assess both intra- and interspecific population variability. Nevertheless, morphological workers have an advantage over chemical workers in that most of their characters are easily observed and readily communicated. Still, without a wide spectrum of available material their systematic conclusions and extrapolations must be held suspect; likewise with chemical characters.

One *might* suppose that this or that particular flavonoid, alone or in combination, marks a given taxon, or that its presence in yet another taxon suggests a strong relationship, etc., but, unless a wide survey is made for the character concerned, a convincing case cannot be made as to the utility of that character. This was clearly recognized by the late R. E. Alston (1967; p. 217):

> . . . if variation in secondary products is generally responsive to various factors similar to those factors which govern the morphology of a species, then chemical variation can be described in such a way as to introduce limits to the variation and to discern its pattern and thus, of course, to allow predictability. Since the latter alternative is nearly axiomatic, in my opinion, then the problem is merely to become familiar with chemical variation, its origin and factors affecting it, and its limits and its meaning, if possible, just as any good systematist would attempt to study morphological or cytological variation.

Alston also noted that some plant taxa can be exceedingly uniform in their secondary chemistries, while others might be quite variable. Even within a given genus, such as *Baptisia*, discussed below, certain secondary compounds are spoken of as 'major' (occurring in 80% or more of the *individuals* examined) while others are spoken of as 'minor' (occurring in less than 80% of the individuals examined), and a few are simply labelled 'rare' (sporadic, perhaps artifactual, but in 5% of the individuals or less). No doubt some, if not all, of this variability is due to 'penetrance' phenomena; that is, the 'missing' compounds are present in such small quantities so as not to be detected, as discussed below.

Harborne (1975a) noted, as had several workers before him, that it is difficult to demonstrate the absence of a compound. Most chemosystematists tend to use rather standard procedures for the detection of chromatographic spots. These are subsequently cut out, eluted and identified as to structure. But, as honestly noted by Crawford (1979), 'overloading' of paper chromatograms might yield compounds previously thought absent, simply because they occurred in insufficient amounts for detection by the methods used.

In spite of the methodological problems connected with the detection of trace amounts some workers feel that chemical data are equally diagnostic as morphological characters. Thus, Adesida *et al.* (1971), working with six African species of the genus *Khaya* (African mahogany) which were said to be 'poorly defined, the morphological differences being slight and inconstant', were readily able to compose a chemical key (using limonoids) for the identification of these taxa as follows:

Key for the Chemical Identification of Khaya *Species*

(1)	Timber contains no detectable limonoids in a 100 g sample	2
	Timber contains limonoids	3
(2)	Seeds contain fissinolide and 3-deacetylkhivorin ...	*K. nyasica*
	Seeds contain khayanthone and 3-deacetylkhivorin	*K. anthotheca* (Eastern form)
(3)	Timber contains anthothecol, seeds contain havanensin	*K. anthotheca* (Western form)
	Timber does not contain anthothecol	4
(4)	Timber contains 11β-acetoxykhivorin	5
	Timber does not contain 11β-acetoxykhivorin ...	6
(5)	Timber contains khivorin but not methyl 3β-acetoxy-2-hydroxy-1-oxomeliacate	*K. nyasica*
	Timber contains methyl 3β-acetoxy-2-hydroxy-1-oxomeliacate but no khivorin	*K. madagascariensis*
(6)	Timber contains khayasin	*K. senegalensis*
	Timber contains khivorin	7
	Timber does not contain khivorin or khayasin ...	8
(7)	Seeds contain mainly methyl angolensate	*K. ivorensis*
	Seeds contain mainly khivorin and khayanthone ...	*K. anthotheca* (Eastern form)
	Seeds contain mainly fissinolide	*K. grandifoliola*
(8)	Timber contains 7-ketokhivorin	*K. senegalensis*
	Timber contains methyl angolensate and/or mexicanolide	9
(9)	Seeds contain fissinolide	*K. grandifoliola*
	Seeds do not contain fissinolide	*K. senegalensis*

Table 11.1 Distribution of flavonoids in the leaves of *Anacyclus* (Greger, 1978)

Anacyclus species and provenances[a]	Flavonols						Flavones				
	Q 3-rhamnosylglucoside	Q 3-glucoside	Q 7-glucoside	Isorh 7-glucoside	Q 5-glucoside	Isorh 5-glucoside	Lut 7-rhamnosylglucoside	Lut 7-glucoside	Dios 7-xylosylglucoside	Dios 7-glucoside	Compounds in trace amounts not fully identified
depressus Ball											
cult. (1)			●	●			+		●	+	LG
cult. (2)			●	●			+		●	+	LG
cult. (3)			●	●			+	+	●	+	LG
cult. (4)			●	●			+		●	○	LG
pyrethrum (L.) Cass.											
Spain (5)	○			+ ?	●	+ ?	+	+	●	●	AD,L
cult. (6)	+		+ ?	○	●		+		●		AD.LC
cult. (7)	○			●	●		+		●		AD,L
officinarum Hayne											
Germany (8)					○	●	+		●	+	
Germany (9)					○	●	+		●	○	
maroccanus Ball											
Morocco (10)					○	●	+				K5,AI
radiatus Lois.											
Spain (11)	+	○			●	○	+				K5,AI
Spain (12)					●	○	+				K5,AI
'*purpurescens*'											
cult. (13)					●	+	+				AD
monanthos (L.) C. Christensen											
Algeria (14)	●	●	●	○			+				K7
Libya (15)	●	●	●	○			+				K7
Tunisia (16)	●	●	○	+			+	+			K7
Tunisia (17)	●	●	○				+				K7
Egypt (18)	○	○	●				+	+			
linearilobus Boiss. & Reut.											
Algeria (19)	○	○			○		+				LG
Algeria (20)	○	○			+		+				LG
pigellifolius Boiss.											
Turkey (21)	●	+									K3
atlanticus Lit. & Maire											
Morocco (22)	○	+					○	+			6F,LG
cult. (23)	+						+	○			LG
'*coronatus*'											
cult. (24)					○		+	○			
valentinus L.											
Spain (25)	+	+	○				+	+			K7
Spain (26)			+				+	+			
cult. (27)			○				+	+			
cult. (28)			○				○				
clavatus (Desf.) Pers.											
Italy (29)			+ ?		+ ?		+	+			AD
Spain (30)							+	+			AD
Italy (31)							+	+			
Yugoslavia (32)							+	+			
Algeria (33)							+	+			
cult. (34)							+	+			AD
cult (35)							+	+			AD

Of course, the problem with this key is that it is based upon relatively few samples from a limited range. Also, the authors might, in time, have to modify the key with appropriate 'weasel words', such as this or that compound, or 'usually' or 'more or less'. Exceptions to character states inevitably arise as the number of individuals and populations of those species under scrutiny expand, but we applaud the attempt by these authors to place on record 'unequivocal' chemical leads for taxonomic recognition.

Most workers have been less precise in their attempt to cope with quantitative variation. Variation is usually expressed in tabular form such that broad patterns can be immediately perceived by quick perusal. For example, Greger (1978), working with 35 collections (individuals) of the genus *Anacyclus* (Asteraceae), arranged his flavonoid data in the form of a table (Table 11.1), using symbols to indicate the presence of 'large' (●), 'small' (○), 'trace' (+) and 'uncertain' (+?) amounts. Similar symbols are used by Valant (1978) in her table showing flavonoid variation in approximately 80 taxa of the large genus *Achillea* (Asteraceae). Nowhere in these papers do we have the 'amounts' quantified and we must assume this is a subjective visual process. Even so, the samples are so small relative to the number of individuals and populations that occur in nature that considerable exceptions must be introduced into the chart as sampling is extended.

In one of the more novel studies of this nature, Bierner (1973b) working with *Helenium* (Asteraceae) presented his data as shown in Table 11.2. The flavonoid chemistry of 22 of the 25 North American taxa of this genus were investigated. Twelve 'major' flavonoids were detected. The quantitative expression of these, *within taxa*, were shown by the following symbols: (●), present in more than 75% of the populations examined; (X), present in 25–50% of the populations examined; (○), present in fewer than 25% of the populations examined.

This portrayal is quite different from that presented in the two tables mentioned above in that we are to assume that *entire* populations may lack this or that compound. Unfortunately, Bierner does not give the number of individuals examined in this paper, but consultation of his doctoral work show these to be in excess of several hundred from 160 populations.

The question of what accounts for the presence or absence of a given secondary component has been approached by Fowden (1972) working with non-protein amino acids and particularly with azetidine 2-carboxylic acid. He reckoned that the 200 or more such compounds that are known to occur but sporadically in nature, might 'be synthesized by more types of

mbols: ●, large amounts; ○, small amounts; +, traces; +?, uncertain. Abbreviations: 3,K5,K7, kaempferol 3- 5- 7-glycoside; AD, apigenin 7-diglucoside; LG, additional luteolin glycosides; 6F, presumably 6- or 8-substituted flavonols.
umbers in parentheses refer to the list of provenances in the experimental part.

Table 11.2 Distribution of flavonoids detected in *Helenium* (Bierner, 1973b)

Species / Flavonoids:	Luteolin 6-OMe (I)	Orientin (II)	Isoorientin (III)	Swertiajaponin (IV)	Apigenin 6-OMe (V)	Apigenin 6,7-di-OMe (VI)	Apigenin 7-*O*-glucoside (VII)	Vitexin (VIII)	Isovitexin (IX)	Swertisin (X)	Vicenin-1 (XI)	Vicenin-2 (XII)
Section Helenium												
H. autumnale L.	●	●	●	●	●	●		●	●		●	×
Section Amarum												
H. amarum (Raf.) Rock	●	●	●	●	●	○			●	●	●	●
H. badium (Gray) Greene	●	●	●	●	●				●	×	●	●
Section Leptopoda												
H. vernale Walt.	×	●	●		×			●	●		×	×
H. drummondii Rock		●	●					●	●			
H. pinnatifidum (Nutt.) Rydb.	●	×	●		●				●	×	●	●
H. brevifolium (Nutt.) Wood	●	●	●		●			●	●		●	●
H. campestre Small	●		●		●				●		●	●
H. flexuosum Raf.	●	×	●				●		●		●	●
Section Tetrodus												
H. bolanderi Gray	●	●	●	●	●			×	●	×	●	●
H. bigelovii Gray	●	●	●	×	●			×	●	○	●	●
H. puberulum DC.	●	●	●	●	●				●		×	
H. arizonicum Blake	●	●	●		●			○	●		●	
H. laciniatum Gray	●	●	●	×	●	●		●	●	●	●	●
H. mexicanum HBK	●	●	●	×	●			●	●	○	●	●
H. linifolium Rydb.	●		●	×	●						●	●
H. elegans DC. var. *elegans*	●	×	●		●	×		×	●		○	○
H. e. var. *amphibolum* (Gray) Bierner	●	●	●		●	×		×	●		○	
H. quadridentatum Labill.	●	×	●		●			×	●		●	●
H. thurberi Gray	●	●	●		●			×	●		×	
H. microcephalum DC. var. *microcephalum*	●	●	●		●			●	●			×
H. m. var. *ooclinium* (Gray) Bierner	●	●	●		●			●	●			

plants, or even by all plants, but only in amounts that fall below threshold concentrations that can be recognized in plant extracts without difficulty using routine analytical (normally chromatographic) procedures'.

To test this idea he studied the nitrogenous fraction from commercially processed sugar beets (*Beta vulgaris*), which was made available to him by a company engaged in sugar production. Such extracts are obtained on an extraordinary scale by the sugar-refinery industry where, for example, Fowden (1972) reports that 'one million tons (10^9 kg) of sugar-beets could yield about 500 tons (5×10^5 kg) of a mixture of free amino acids'. Analysis of such extracts led Fowden to conclude that the genetic ability of a given plant to make a wide array of secondary components may be more uniform than has been suspected—'different patterns of product accumulation (from the trace amounts undectable by routine analysis to the massive accumulation associated with some species) may reflect differences in the degree to which particular genes are 'switched on'.' This thought, or possibility, has led several phytochemical workers to suggest that plants might best be referred to as 'accumulators' or 'nonaccumulators' of this or that secondary component, instead of referring to these as either 'present' or 'absent'.

As noted above, perusal of the literature reveals that most workers portray the quantitative variation of flavonoids (and yet other compounds, the notable exception being volatile compounds, which are readily quantified down to the second decimal place using gas chromatography) as occurring in large amounts, small amounts or traces. Sometimes, however, quantitative designations can be confusing; for example, Alston and co-workers (various publications) refer to flavonoid compounds as either 'major' or 'minor', depending upon their likelihood of occurrence in a given taxon. Generally speaking, those that occur in 80% or more of the individuals of a given taxon are said to be major components, those below that are said to be minor components (i.e. too variable for clear-cut systematic purposes). Bierner (1973b), as noted above, used population designations: the presumably major components occurred in more than 75% of the populations examined; presumably minor components in 0–25% of the populations. (Apparently none of his listed compounds occurred with a population frequency of 51–75%!)

Whatever the method used in documenting the existence of quantitative variation within and between populations, it seems reasonable to assume that among closely related species, the 'absence' or 'presence' of a given compound might reflect genetically controlled expression levels, much as the expression of at least a few, if not many secondary components might reflect the complex interplay of multiple alleles (Crawford and Levy, 1978).

Considering much of the above, it might appear somewhat contentious to attempt to place excessive phylogenetic emphasis on this or that secondary compound. More so if, as noted by Crawford (1979) and numerous others before him, a given compound may be synthesized via different metabolic pathways so that the same or similar compounds may arise independently in distantly related plant groups.

Clearly the use of micromolecular components as criteria for assessing relationships among plant groupings is fraught with all kinds of pitfalls (as, also, for morphological characters) and must be approached with an enhanced perspective and a broad data base. With few exceptions, then, (e.g. betalains, chapter 12) secondary compounds are primarily useful at the specific and generic level or among closely related genera.

Having said all of this, it now behoves us to look upon the brighter side of things chemosystematic. Certainly, where morphological and yet other data are equivocal with respect to a given relationship, it is reasonable to use chemical data as an additional criterion for systematic purposes. In the sections that follow, an attempt will be made to call to the fore some of the better documented chemical studies, which have shown the utility of these compounds for systematic purposes at the specific level and higher.

II. Use of flavonoids

A. Flavonoids of *Baptisia* spp.

The late R. E. Alston and students undertook a very ambitious regional study of the flavonoids of the North American *Baptisia* (Leguminosae). Literally thousands of individuals from several hundred populations of the approximately 20 taxa which comprise this genus were examined (Brehm, 1966; Horne, 1965; summarized in Alston, 1967). Results of these studies revealed considerable infra- and interpopulation variation in the flavonoids. For example, Horne (1965) examined 32 populations of *B. nuttalliana* from widely distributed, systematically arranged, sites. Twenty plants were individually examined from each of these localities. In all, he discerned 42 spots using two-dimensional paper chromatography among the 640 plants examined. Twenty of the spots had a frequency of 81.5 to 100% among these individuals; the remaining 22 spots ranged in frequency from 0.31 to 63.5%. Similar studies were made for *B. lanceolata* and results of the frequency of occurrence of 29 spots from four populations of this taxon are shown in Table 11.3. Most of the 'spots' were subsequently identified as noted in Tables 11.4 and 11.5.

Horne could not account for the extraordinary (?) variation found in *Baptisia nuttalliana*, commenting that, 'Whether this variation is the result of random genetic variation, ecological influences, or lack of refined detection

Table 11.3 Frequency of occurrence of flavonoids in four populations of *Baptisia lanceolata*. (From Horne, 1965.)

Spot number	No. of plants in which spot was found[a] Population 1	Population 2	Population 10	Population 20	Total	%
1	20	20	20	20	80	100
2	20	20	20	20	80	100
3	20	20	20	20	80	100
4	20	20	20	20	80	100
5	20	20	20	20	80	100
6	20	20	20	20	80	100
7	7	11	0	0	18	22.5
8	10	16	2	0	28	35.1
9	2	17	20	16	55	68.8
10	20	20	20	20	80	100
12	8	18	20	7	53	66.3
13	20	19	18	19	76	95.0
14	8	9	19	20	56	70.1
15	20	18	12	0	50	62.5
16	20	20	20	20	80	100
17	20	20	20	20	80	100
18	20	17	6	0	43	53.8
19	16	3	0	0	19	23.7
20	6	5	1	17	20	36.2
21	7	7	1	2	17	21.2
22	0	15	5	0	20	25.0
23	20	20	20	20	80	100
24	0	2	0	0	2	2.50
25	0	0	14	0	14	17.5
26	0	0	9	0	9	11.3
27	0	0	1	0	1	1.25
28	0	0	2	0	2	2.50
29	0	0	1	0	1	1.25

[a]Twenty plants were examined from each population. Spot 11 omitted as an artifact.

techniques is not known.' However, he added that most of the spot-variation occurred in *Baptisia nuttalliana* over a region of its range in which hybridization with other species occurred. No doubt this accounts for much of the variation, but it is also likely that 'penetrance' (as discussed above) was a factor, as well as natural genetic variation and problems in technique.

Baptisia was selected for intensive flavonoid analysis because of the utility of these compounds in resolving problems of natural hybridization where, indeed, they proved indispensable (see chapter 13). But in what way have they been of particular use in showing relationships among species? Markham

Table 11.4 Distribution of flavones and flavonols in *Baptisia*. (From Markham *et al.*, 1970.)

Baptisia spp.	Apigenin (Ia)	Apigenin 7-*O*-glucoside (Ib)	Apigenin 7-*O*-rhamnoglucoside (Ic)	Luteolin (IIa)	Luteolin 7-*O*-glucoside (IIb)	Luteolin 7-*O*-rhamnoglucoside (IIc)	Luteolin 7-*O*-galactoside (IId)	7,4′-Dihydroxyflavone (IIIa)	7,4′-Dihydroxyflavone 7-*O*-glucoside (IIIb)	7,4′-Dihydroxyflavone 7-*O*-rhamnoglucoside (IIIc)	7,3′,4′-Trihydroxyflavone (IVa)	7,3′,4′-Trihydroxyflavone 7-*O*-glucoside (IVb)	7,3′,4′-Trihydroxyflavone 7-*O*-rhamnoglucoside (IVc)	Kaempferol (Va)	Kaempferol 3-*O*-glucoside (Vb)	Kaempferol 7-*O*-glucoside (Vc)	Kaempferol 3-*O*-rhamnoglucoside (Vd)	Quercetin (VIa)	Quercetin 7-*O*-glucoside (VIb)	Quercetin 7-*O*-rhamnoglucoside (VIc)	Quercetin 3-*O*-glucoside (VId)	Quercetin 3-*O*-rhamnoglucoside (VIe)	Quercetin 3-*O*-rhamnoglucoside 7-*O*-glucoside (VIf)	Quercetin 3-7-di-*O*-glucoside (VIg)	Quercetin 7-*O*-rhamnoglucoside 3-*O*-glucoside (VIh)	4′,7-Dihydroxyflavonol (VIIa)	4′,7-Dihydroxyflavonol 7-*O*-glucoside (VIIb)	4′,7-Dihydroxyflavonol 3-*O*-glucoside (VIIc)	4′,7-Dihydroxyflavonol 7-*O*-rhamnoglucoside (VIId)	3′,4′,7-Trihydroxyflavonol (VIIIa)	3′,4′,7-Trihydroxyflavonol 7-*O*-glucoside (VIIIb)	3′,4′,7-Trihydroxyflavonol 7-*O*-rhamnoglucoside (VIIIc)	3′,4′,7-Trihydroxyflavonol 3-*O*-glucoside (VIIId)
Group 1																																	
B. perfoliata	●	○	●	●	●	●		●		○	●	●	●							○													
B. sphaerocarpa	○	●	●	○	●	●		●			●																○		○		○	○	
Group 2																																	
B. leucantha														●	●		●	●			●	●					○	●			○	○	●
B. alba														●			●	●			●	●	●	●									
Group 3																																	
B. megocarpa	●	●	○	●	●	○	●	●	●		●	●																					
B. cinerea	●	●		●	●	●	●	●	●		●	●																					
B. bracteata	●	●	●	●	●	●	●	●	●		●	●																					
B. leucophaea	●	●		●	●		●	●	●		●	●			○						○										○		
B. lanceolata	●	●		●	●		●	●	●		●	●															○						
B. nuttalliana	●	●	●	●	●	●	●	●	●		●	●															●				○		
B. australis	●	●		●	●		●	●	●		●	●															○		○		○	○	
Group 4																																	
B. arachnifera	●	●	●	●	●	●	●	●	●		●	●																					
B. simplicifolia	●	●	●	●	●	●	●	●	●		●	●														●	●		○		○	○	○
B. tinctoria	○		○	●	○	●																											
B. lecontei	●	○	●	●	○	●		●	●	●	●	●	●			○			●					○	○			○		●		○	
B. calycosa	●	○	●	●	●	●		●		●	●		●													●							○
B. hirsuta	●		●	●		●																											

Table 11.3 Distribution of isoflavones and coumarins in *Baptisia*. (From Markham *et al.*, 1970.)

Baptisia spp.	Genistein (IXa)	Genistein 7-*O*-glucoside (IXb)	Genistein 7-*O*-rhamnoglucoside (IXc)	Biochanin A (Xa)	Biochanin A 7-*O*-glucoside (Xb)	Biochanin A 7-*O*-rhamnoglucoside (Xc)	Orobol (XIa)	Orobol 7-*O*-glucoside (XIb)	Orobol 7-*O*-rhamnoglucoside (XIc)	6-Hydroxygenistein (XIIa)	6-Hydroxygenistein 7-*O*-rhamnoglucoside (XIIb)	Tectorigenin (XIIIa)	Tectorigenin 7-*O*-glucoside (XIIIb)	Daidzein (XIVa)	Daidzein 7-*O*-glucoside (XIVb)	Daidzein 7-*O*-rhamnoglucoside (XIVc)	Formononetin (XVa)	Formononetin 7-*O*-glucoside (XVb)	Formononetin 7-*O*-rhamnoglucoside (XVc)	3′,7-Dihydroxy-4′-methoxyisoflavone (XVIa)	3′,7-Dihydroxy-4′-methoxyisoflavone 7-*O*-glucoside (XVIb)	3′,7-Dihydroxy-4′-methoxyisoflavone 7-*O*-rhamnoglucoside (XVIc)	Pseudobaptigenin (XVIIa)	Pseudobaptigenin 7-*O*-rhamnoglucoside (XVIIb)	Texasin (XVIIIa)	Texasin 7-*O*-glucoside (XVIIIb)	Afrormosin (XIXa)	Afrormosin 7-*O*-glucoside (XIXb)	Afrormosin 7-*O*-rhamnoglucoside (XIXc)	Scopoletin (XXa)	Scopoletin 7-*O*-glucoside (XXb)
Group 1																															
B. perfoliata	○		○				●		●					●	●		●	●	○	●			○	○			●	○	●	●	●
B. sphaerocarpa			●	●		●	●		●								●	●		○	○	○									
Group 2																															
B. leucantha	○			○	○		●	●	●								○	○					○	○							
B. alba																															
Group 3																															
B. megacarpa	●	●	○				●							●	●	●	●	●	●	●	●	●	○	○			●	●	●		
B. cinerea	●	●					●	●						●	●		●	●		●	●		○	○			●	●		●	●
B. bracteata	○	○												●	●		●	●		●	●		○	○			●	●			
B. leucophaea	●	●					●	●						●	●		○	○		●	●		○	○	○		●	●	○	●	●
B. lanceolata	●	○	○											●	●	○	○	○	○	●	●	○	○	○			●	●		●	●
B. nuttalliana	●	●					●	●				●	●	●	○		●	●		●							●	●		●	●
B. australls														●			●	●							●	●	●	●			
Group 4																															
B. arachnifera														●	●	○	●	●	●	●	●	○	○	○			●	●	○	●	●
B. simplicifolia														●	●	●	●	●	●	●	●	●	○	○			●	●	●	●	●
B. tinctoria	●		●	●		●	●		●			●	○	○			○	○	○	○			●	●			○			●	●
B. lecontei	●	●	●				●		●			●	●	●	●	●	●	●		●	●	●	●	●			●	●		●	●
B. calycosa	●		●	●		●	●		●					●	○	○	●	●		●										●	●
B. hirsuta	●		●	●		●	●		●	●	●		○				○	○		○			●	●						●	●

Symbols: ●, major; ○, trace.

et al. (1970) presented distribution data for 62 of the major flavonoids and two coumarins in all 17 spp. of *Baptisia* (Tables 11.4, 11.5). It is obvious from examination of these Tables that these authors ignored many of the lesser or 'minor' compounds found in this or that individual of *B. nuttalliana*, but as a broad survey of the major 'spots' in flavonoids found in these species, it is an accurate summary. So what does it all mean systematically speaking? Apart from the obvious species-specific profile of this or that species, the data is primarily useful in suggesting that *B. leucantha* and *B. alba* are closely related and that they stand apart from the remaining species. This is surprising since, on fruit characters, such a relationship is not suggested; indeed, emphasis of fruit texture and shape would suggest that *B. alba* is a remote, primitive element of the genus and that *B. leucantha* is related to *B. megacarpa*.

Dement and Mabry (1972, 1975) extended similar-type studies to the North American elements of the closely related genus *Thermopsis*, which is believed to have given rise to *Baptisia* (Turner, 1967). They conclude that the flavonoid chemistry of *Baptisia*, because of 6-hydroxyisoflavones, complex glycosylation patterns and methylenedioxy isoflavones, is the more advanced genus. Further, they conclude that the lack of comparable variation of these flavonoids, such as occurs in *Baptisia*, is further suggestive of its ancestral or relict status, confirming the monophyletic origin of each genus as suggested by Turner (1967).

B. Flavonoids of the Lemnaceae

One of the more interesting studies in which flavonoid data have provided phyletic insights, perhaps unobtainable by purely exomorphic means, is that of McClure and Alston (1966) on the Lemnaceae (Duckweed family). A number of flavonoids occur in the four genera, *Spirodela*, *Lemna*, *Wolffia* and *Wolffiella*, which make up the family, and the distribution of these compounds among the genera is particularly intriguing.

To appreciate the data, however, a brief taxonomic account of the family must be given. It is a wholly aquatic family belonging to the monocotyledonous groups. The included species are highly reduced, both in size and outward complexity. Some of the species (e.g. *Wolffia* spp.) measure less than 2 mm across, being devoid of roots, apparent stems or leaves; indeed, they might be described as minute, green vegetative balls (the species flower but rarely). Most workers agree that the Lemnaceae has been derived from terrestrial ancestors and that within the family there has been a phyletic trend toward increasing simplicity through reduction. Thus *Spirodela*-like species are believed to have given rise to *Lemna*-like species and the latter to the most highly reduced, *Wolffia* and *Wolffiella* (Fig. 11.1).

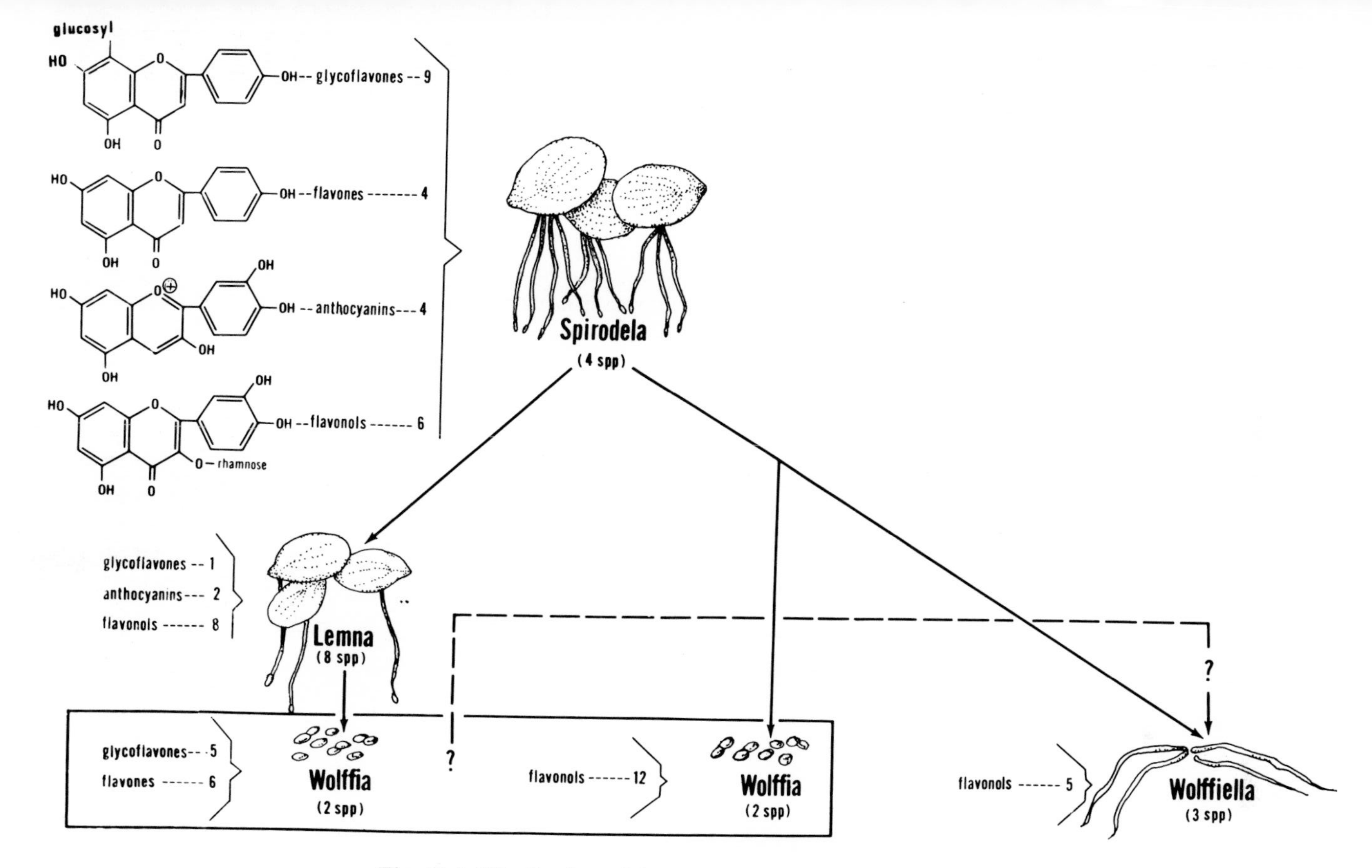

Fig. 11.1 Distribution of flavonoids in the Lemnaceae.

Table 11.6 Distribution of flavonoids in *Wolffia*. (From McClure, 1964.)

	Flavones[a]			Glycoflavones[a]					Flavonols[a,b]											
Wolffia species	1	2	3	1	2	3	4	5	1	2	3	4	5	6	7	8	9	10	11	12
W. punctata									+	+	+	+					+	+	+	+
W. microscopica									+	+	+	+		+	+	+	+			
W. papulifera									+	+	+	+	+	+	+	+	+			+
W. columbiana	+	+		+	+	+	+	+												
W. arrhiza	+	+	+	+		+	+	+												

[a]Numbers refer to different structural types of the compounds concerned; + sign denotes presence of the compound.
[b]*Wolffiella*, sometimes included in *Wolffia*, possesses only flavonols. Flavonols are absent from *Lemma* but present in *Spirodela*, and flavones are found in both genera.

The chemical data are intriguing because these also point to a phyletic reduction series, paralleling that of the morphological. *Spirodela* contains four flavonoid types ((1) anthocyanins, (2) flavones, (3) glycoflavones and (4) flavonols): *Lemna* possesses (1), (2) and (3); *Wolffia* (2) and (3); and *Wolffiella* only (4). The flavonoids of *Wolffia* and *Wolffiella* are particularly interesting. These two genera are distinguished primarily by their vegetative shapes, *Wolffia* being more nearly isodiametric, *Wolffiella* being more nearly linear. *Wolffia*, then, appears to be a taxonomic category that has been erected to include those species of the Lemnaceae that are devoid of exomorphic characters. The flavonoid data (Table 11.6), however, suggest that this grouping is perhaps not a monophyletic one; in fact, it suggests that *Wolffia*, as presently constituted, is biphyletic, containing species derived through a *Lemna*-like line and another stemming from a *Wolffiella*-like element. It will be most interesting to see if a parallel reduction series might not be found in yet other chemical groups within this complex. We have in *Wolffia* an example where the chemical data have suggested a phyletic model which could hardly have been proposed by exomorphic data, but surely an exceptional case among the higher plants generally, for most chemical data are likely to be used to support or deny those phyletic models already proposed by plant systematists.

Mabry (1973b) has argued that flavonoid reduction within a given taxon is the rule in most plant groups. That is, he contends that the more primitive elements within a given taxon will possess a relatively complex array of compounds and that the more recently evolved elements will tend to possess predominantly depauperate or reduced flavonoid profiles. This may be so if one is working with several or more highly diversified plant taxa, and if one takes into account a broad array of secondary compounds such as flavonoids, terpenoids, alkaloids, etc; but 'Mabry's rule' is not likely to hold for a group of closely related taxa where only flavonoids are considered. There is no *a priori* reason why selection pressure might not serve to build up an array of novel flavonoids in this or that highly advanced taxon. After all, novelty in flavonoid production has to *begin* somewhere: diversification and accumulation must follow there upon. Since evolution is a continuing phenomenon in most, if not all, extant taxa, 'invention', accumulation and loss among this or that suite of chemical characters must be rather uniformly occurring throughout the angiosperms generally.

Indeed, several broad studies of flavonoids have suggested that their major utility is at the generic level, or among a group of closely related taxa where morphochemical reductions might be readily detected or at least surmised. Thus Gurni and Kubitski (1981) made a comparative study of the flavonoids found in neotropical genera of the Dilleniaceae concluding that their distribution was reticulate and did 'not permit the recognition of taxa between the level of genus and family'. Nevertheless, it is perhaps reasonable to assume that the possession of a particular compound or group of compounds, which occurs widely in some hypothetical primitive group, is also 'primitive' in the group concerned and that more specialized compounds of more restricted distributions are more advanced. Such is the reasoning of Young and Sterner (1981) for *Degeneria* and *Idiospermum* where they suggest that flavonol glycosides are a 'primitive' feature in Magnoliiflorae and *Degeneria* of the Degeneriaceae but that *C*-glycosylflavones are 'advanced', these occurring in the presumably more advanced Idiospermaceae.

Reduction in chemical diversification as a measure of increased specialization may be reasonable but a more profitable evaluation is perhaps that employed by Gomes *et al.* (1981). They used a biogenetic scheme based upon oxidation/methylation ratios to evaluate flavonoid types in a group of taxa surrounding the genera *Derris* and *Lonchocarpus* which belong to the pantropical subtribe Tephrosieae of the family Leguminosae. Even so, after much complex analysis, they conclude (p. 141) that 'There is no clear-cut correspondence between the chemical clusters and the traditional morphological groups and we do indeed see no reason why chemical and morphological changes should always occur in parallel'. A sound appraisal.

Flavonoid analysis for systematic purposes has been discussed in detail by Harborne (1973a,b, 1975a,b, 1977a,b) and by Swain (1975). A novel or 'improved' method for the evaluation of such compounds has been presented by Crawford and Levy (1978) in which considerations are given to minimum biosynthetic-step distance and biosynthetic-step identity. This allows for the computation of 'a weighted evaluation of similarities among members of a family of biochemicals in contrast to compound-identity indices which treat all flavonoid end products as if they were uniformly complex and unrelated biosynthetically'. This is not an unreasonable way to approach the systematic evaluation of flavonoids if one is comparing closely related taxa (i.e. infrageneric groupings). As noted by Crawford and Levy, there are pitfalls to such 'weighted' evaluations but it does appear to offer a more refined approach in that the procedure allows consideration of the relative complexities of both disparate and shared components of any given flavonoid profile.

III. Use of volatile terpenes

As noted above, one of the advantages of volatile components for systematic purposes is that instrumentation has permitted ready quantitative analysis. Although recent developments in liquid chromatographic techniques will certainly permit easy quantitative work on flavonoids in the future, most published studies have not focused on quantitative variables, at least as expressed as a percentage of the flavonoids detected.

In addition to their relative ease of quantification, volatiles occur across a wide spectrum of plant groups from algae (Crews *et al.*, 1976) to angiosperms, and when found, they are often abundant as to kind. For example, the volatile constituents of *Passiflora edulis* (purple-skinned passion fruit) was examined by GLC and found to contain in excess of 250 components: 71 of these were identified as esters and terpenes; the remainder were unresolved. In fact, GLC examination of most extracted plant parts, even relatively non-aromatic foliage such as occurs in *Baptisia*, will yield an array of volatile compounds which are highly species-specific, both qualitatively and quantitatively. Such a wealth of characters (e.g. 250 'blips' on a gas chromatographic trace), all identified and measured to the second decimal place within 30 minutes by the GLC, is a real boon to the numerical systematist attempting purely phenetic classifications.

In spite of the ease of quantification, terpenoid studies are often reported upon as if they varied in the manner of flavonoids. Thus, Stangl and Greger (1980) studying the monoterpenes of 71 species of *Artemisia* from 166 provenances of Eurasia and North America, focused on about 16 compounds, listing these in chart form as occurring in either small amounts (0.02 ml for each 100 g fresh weight or less); moderate amounts (0.02–0.04 ml per 100 g); or large amounts (0.4 ml per 100 g or more). Their tabulation revealed relatively weak subgeneric and sectional correlations. Clearly a study of this magnitude would be difficult to complete using the more refined population approaches employed, for example, by Zavarin and his colleagues, von Rudloff, and those of Robert Adams, discussed below. However, if chemosystematists wish to compete with the morphological worker at the specific level, they must learn to think in terms of large population samples which permit the acquisition of reliable quantitative (statistical) data, most species being largely grouped by suites of quantitative characters; in the case of terpenoids, 'absence' of a compound may reflect but 'trace' amounts below the level of detectability.

Because of their economic importance and aromatic properties, cone-bearing plants, or conifers, have long been singled out for terpenoid investigation. Most early work was primarily of the survey type and led to a number of reviews, most of these more notable for the array of compounds detected than for their systematic import (e.g. Erdtman and Norin, 1966; Norin, 1972). What were thought to be valid chemical differences between this or that suprageneric grouping, with subsequent sampling revealed exceptions, unexpected terpenoids occurring almost capriciously among closely related and even distant genera, so much so that von Rudloff (1975b) was compelled to note that the volatiles of most conifers are 'best applied in chemosystematic studies at the species and subspecies level'.

Examples of interspecific studies within a given genus using volatiles are fairly common in the literature but, as noted above, there are relatively few broad *population* studies which match data accessible to the morphological worker. That need not be the case, however, as shown by Zavarin *et al.* (1971b) in their study of five species of the conifer genus *Cupressus*. Working with about 20 terpenes collected from 250 trees from 20 localities in the western U.S.A., they compiled statistical data to show that these taxa, which are largely relicts, are classified with difficulty by the morphological worker and could be grouped chemically in three distinct groups. They related their data to post-Pleistocene events in the region concerned and added a dimension to the phyletic history of these taxa that would be difficult to construct with purely morphological data.

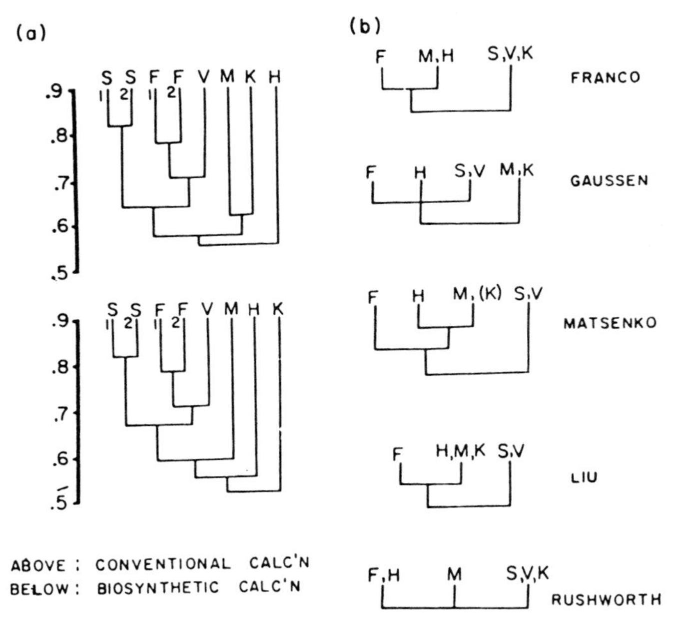

Fig. 11.2 Dendrograms for taxa of *Abies* constructed from chemical data (a) and those proposed by five different workers using purely morphological data (b). For additional discussion see the text. (From Zavarin *et al.*, 1978.)

A similar but mathematically more sophisticated treatment was carried out by Zavarin *et al.* (1978) on 73 individuals of six species of the conifer genus *Abies* from Japan and Taiwan. Using cortical essential oils which yielded about 60 identifiable volatiles (about 50% monoterpenes and 50% sesquiterpenes), these workers were able to circumscribe chemically nearly all of the taxa, a feat not accomplished by most morphologically oriented systematists, or if so, not agreed upon by yet other morphological workers.

A novel comparison of the work of Zavarin *et al.* (1978) with that of previous workers has been provided in Fig. 11.2. It will be noted that Zavarin conconstructed two dendrograms based on chemical data: one derived from 'conventional' calculations based on terpenoid percentages, an arbitrary procedure which depends in large measure on the methodology used; and the other derived from biogenetic knowledge of the pathways involved in the synthesis of terpenoids (Fig. 11.3). As noted by these authors:

> Here chemosystematics occupies a particularly advantageous position (as compared with morphology) with a large amount of information available on biogenesis of natural compounds. This allows us to express the chemical data in concordance with the way the chemical compounds are formed in nature.

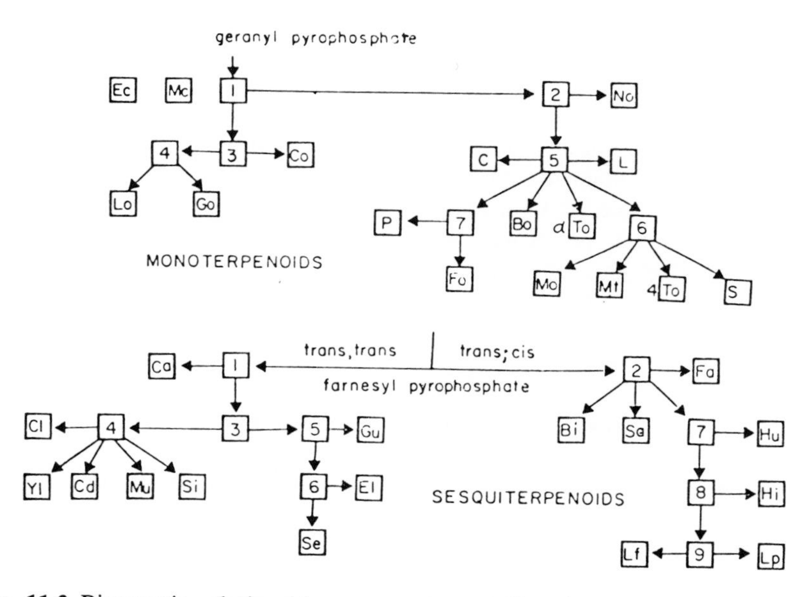

Fig. 11.3 Biogenetic relationships among terpenoids adopted for calculation of similarity coefficients in Fig. 11.2. Abbreviations: Ec, ethylcaprylate; Mc, methylchavicol. **Monoterpenoids:** Lo, linalol, linalyl acetate; Go, geraniol, geranyl acetate, myrcene; Co, citronellol, citronellyl acetate, citronellal; No, nerol, neryl acetate; L, limonene; C, 3-carene, α-phellandrene; P, α-pinene, β-pinene, β-phellandrene; Fo, α-fenchyl alcohol; Bo, borneol, bornyl acetate, camphene, tricyclene; aTo, α-terpineol; Mo, menthyl acetate; Mt, methyl thymol; 4To, 4-terpineol; S, sabinene, terpinolene, γ-terpinene. **Sesquiterpenoids:** Ca, caryophyllene, α-humulene; Cl, calamenene; YI, β-ylangene; Cd, β_1-cadinene, γ-cadinene, ε-cadinene; Mu, γ-maurolene, δ-cadinene; Si, sibirene; Gu, α-guaiene, δ-guaiene; El, β-elemene, δ-elemene; Se, α-selinene, β-selinene, selina-4(14)7(11)-diene, selina-37(11)-diene; Fa, β-farnesene; Bi, β-bisabolene, *cis*-α-bisabolene; Sa, sativene, cyclosativene; Hu, γ-humulene; Hi, α-himachalene, β-himachalene; Lf, longifolene, longicyclene; Lp, longipinene. (From Zavarin *et al.*, 1978.)

Bailey *et al.* (1982), using similar approaches, studied a more restricted species-problem in the *Pinus cembroides* complex of north-western Mexico. Various workers regarded the complex as being composed of two species; two allopatric varieties; or possibly two taxa with hybrid derivatives. Studying the terpenoids of 60 trees from two mixed stands and 80 trees from eight pure stands Bailey *et al.* (1982) were able to conclude that two species existed and, when growing together, showed no tendency to hybridize.

Adams and co-workers, using both terpenoids and morphological features, have persisted in a thorough population study of the moderately large conifer genus *Juniperus*. Building up an intricate data base at the infraspecific level (cf. chapter 10), Adams expanded his terpenoid investigations to include most of the 30 or more taxa of *Juniperus* in North America (Zanoni and Adams, 1976; Adams *et al.*, 1980, 1981; and unpublished work). Using 80–100 terpenoid characters taken from population samples and treating the derived data statistically, they were able to construct an hypothetical phylogeny for the more southern species of *Juniperus* on that continent. They compared this with a morphologically based classification concluding that, in general, the two systems agreed in their major groupings but that 'differences between more closely related species were more apparent with the chemical data'. In other words, like the infraspecific studies of *Juniperus*, computer analysis of volatile oils proved to be a more sensitive approach when working with closely related species.

IV. Use of sesquiterpene lactones

Because of their structural diversity, relative ease of identification, and widespread occurrence among genera of the Compositae, sesquiterpene lactones have been studied extensively, the most recent review of their systematic import being that of Seaman (1982). He comments on the systematic obstacles to their use as follows: (1) inordinate infraspecific variation; (2) lack of quantitative variation, i.e. their tabulation as present or absent; and (3) the ease with which alternative biosynthetic pathways are activated.

Like Zavarin and his colleagues in their work with terpenes generally (discussed above), Seaman would emphasize in systematic evaluations the biogenetic pathways leading to the production of sesquiterpene lactones. Unfortunately, experimental proof of biosynthetic routes is largely lacking and one must rely on the distribution of skeletal types among related taxa to infer likely biogenetic pathways. This is a rational process; certainly, for phylogenetic classifications, better than the mere phenetic evaluation of tabulated differences between this or that taxon. Seaman rightly notes that 'Biogenetic cladistic analysis seems to be the best solution to problems

Table 11.7 Species groupings in *Parthenium* and their relationships to habit, ecogeography and sesquiterpene lactones. (From Rodriguez, 1977.)

Taxon	Habit	Habitat	Sesquiterpene lactones
Section Parthenichaeta			
P. tomentosum	Small, deciduous trees	Tropical deciduous and/or thorn	Parthenolides
P. fruticosum	(2–7 in high) entire leaves	forest, elev. 500–1500 m, latitude	Xanthanolides
P. schottii	corymbose inflorescence	20°, on limestone soil	Ambrosanolides
P. cineraceum (Bolivia)			
P. lozanianum			
P. argentatum	Shrubs (3–8 dm), intricately	Chihuahuan desert, elev.	Ambrosanolides
P. incanum	branched, densely pubescent	1000–2300 m, latitude 25–30°,	None
P. rollinsianum	leaves	primarily on limestone soil	
Section Bolophytum			
P. alpinum	Plants acaulescent, caespitose,	Montane, Rocky Mts. of Colorado,	Parthenolides
P. alpinum var. *tetraneurans*	perennials less than 4 cm high,	Wyoming and Utah, elev. 2500–3000	
P. ligulatum	solitary heads	m, latitude 39–42°, endemic to	
		gypsum outcrops	
Section Partheniastrum			
P. integrifolium	Tuberous, perennials herbs	Grassland prairies and Temperate	Parthenolides
P. hispidum	3–10 dm, stems sulffrutescent,	deciduous forests, elev. 500 m,	
P. hispidum var. *auriculatum*	heads in flat-topped corymb	latitude 36–40° on sandy and limestone soil	
Section Argyrochaeta			
P. hysterophorus	Annual and perennial herbs,	Temperate lowlands, on limestone	Ambrosanolides
P. confertum varieties	2–7 dm, heads loosely paniculate	and sandy soil, elev. 300 m, latitude	Parthenolides
P. densipilum		25–30°	
P. bipinnatifidum	Annuals and perennials, 1–5 dm,	Highlands of Mexico and Moist	Ambrosanolides
P. glomeratum (Argentina)	taproot in alpine species	'Puna' of Argentina elev.	Parthenolides
		2300–3500 m, latitude 27°	

associated with (1) infraspecific variation and (2) the retrieval of taxonomically useful information from complex molecules'. He is aware that the genetic loss of sesquiterpene lactones is a common occurrence and that cladistic analysis without such an awareness might lead to erroneous groupings. Nevertheless he believes that this problem can be alleviated by limiting one's cladistic analysis to sesquiterpene lactone-rich groups such as occur in *Ambrosia*, *Iva* and related genera of the tribe Ambrosiinae (cf. Seaman and Funk, 1983).

No doubt ecogeographic and coevolutionary forces are powerful determinants of the secondary chemistry of a given species group, as pointed out by Harborne (1977a) and yet other workers. This being so, it is not surprising to find compounds and their metabolic pathways largely confined to related species occupying similar ecogeographic regions. Thus Rodriguez (1977) examined most of the known taxa of the American genus *Parthenium* for sesquiterpene lactones, flavonoids and alkaloids. The tropical arborescent species were found to have three different sesquiterpene lactone types and flavonol derivatives; the desert shrubs were noteworthy for their methylated quercetagetin derivatives and alkaloids and general absence of sesquiterpene lactones; the perennial temperate and more mountain elements contained mainly lactones and flavonoids in extremely low concentration; whereas the weedy annuals contained pseudoguaianolidic lactones only, together with the flavonols, quercetin and kaempferol *O*-glycosides. The more tropical perennial herbs were similar to the annuals except that they were more abundantly supplied with the compounds concerned. The distribution of sesquiterpene lactones among the approximately 20 species of *Parthenium* is summarized in Table 11.7; it will be noted that the above mentioned habital and ecographical patterns fall into previously recognized morphological or sectional groupings within the genus.

V. Use of other secondary compounds

A. Non-protein amino acids

Numerous chemosystematic studies at the specific level have been made using yet other compounds, but most of these are limited in scope, sample size, techniques, phenetic analysis or inferred significance. Some of the more thoroughly investigated compounds include those of non-protein amino acids of legumes generally (Bell, 1971; and numerous other references), Krauss and Reinbothe (1973) for species of Mimosoideae, Evans and Bell (1978) for *Acacia*, and Watson and Fowden (1973) and Evans and Bell (1978) for *Caesalpinea* and related groups. In fact, the legumes are notoriously rich in such compounds, a few of which have been intensively studied. Canavanine, a nitrogen-rich amino acid which is important in food storage and defence in the plants that possess it, is probably the best surveyed (Bell *et al.*, 1978); the systematic significance of its distribution is speculative at best, since it may be present or absent among species of the same genus, but Lackey (1977) shows that it is most often present in two subtribes of the Phaseoleae (Diocleinae and Kennediinae) and is only sporadic in its occurrence elsewhere (see also chapter 6).

Except for the legumes, relatively few genera have been intensively studied for non-protein amino acids. One of the more notable studies is that of Fowden and Pratt (1973), who surveyed a wide range of species belonging to the genus *Acer* (maples). They found that the distribution patterns of these compounds correlated well with groupings based on morphological data. This contrasts with a little cited paper by Reddi and Phipps (1972) who studied amino acids in the grass tribe Arundinelleae and conclude that the patterns of variation were unrelated to taxonomic groupings, and that such compounds 'should be considered taxonomic noise' and further that they 'may not be of general value in higher plant systematics'. Strong words for such a poorly conceived study but perhaps not surprising in that most of the compounds they report are *free* protein amino acids, which can be expected to vary for a number of environmental and metabolic reasons. Nevertheless, Watson and Creaser (1975) found 'taxonomically-intelligible patterns' in their brief review of non-free protein amino acids among angiosperms generally, including grasses. But they note that the intelligible pattern obtained might actually reflect the presence of free protein amino acids!

B. Leaf waxes (alkanes and other hydrocarbons)

Beginning in the early 1960s with the development of GLC techniques, a number of interesting studies were made to ascertain the utility of leaf wax volatiles as systematic characters. Eglinton *et al.* (1962; Eglinton and Hamilton, 1963), in particular, surveyed a broad range of species belonging to the succulent genus *Aeonium* (Crassulaceae). Although the species were not sufficiently sampled, they noted a fairly constant chemical profile for each and there seemed to be a rough parallelism of their data with sectional breakdowns within the genus.
example, Scora *et al.* (1975) note that among 22 species of *Persea* (Lauraceae) the alkane profiles give 'good agreement with established taxonomy', whereas Nordby and Nagg (1977), working with fruit peels of 48 cultivars of *Citrus* (Rutaceae) found that each exhibited significant terpenoid differences and that chemical profiles obtained from the fruits matched those obtained from their leaf waxes.

Less enthusiastic results have been obtained from grasses (Russell *et al.*, 1976, on *Chionochloa*; Tulloch and Hoffman, 1979, on *Andropogon*); indeed, at least one recent report asserts that among taxa of the grass genus *Saccharum* (using a number of species and a large number of cloned cultivars) 'No chemotaxonomic relationships could be derived from the compositions (mainly alkanes), as the intraspecific variation was greater than the interspecific variations'. The same kind of conclusion was made by Corrigan *et al.* (1978) in their study of the leaf waxes of *Picea* (Pinaceae) in which they conclude that 'It was not possible to ensure that the compositions of the wax [mainly alkanes] from each of the twenty-eight species did not represent an extreme of intraspecific variation'. This being so, it is interesting that they should contend that their results 'support the view that the genus should not be divided into sections'.

Clearly the efficacy of leaf waxes as systematic markers must await the work of an investigator who is willing to sample natural populations with sufficient care so that quantitative variations within and among populations can be treated as a statistic instead of an individual, much as indicated in the above accounts for flavonoids and terpenoids.

C. Glucosinolates (mustard oil glucosides)

This group of compounds, mostly occurring in the Cruciferae and related families, has been studied by a number of workers, the most recent review being that of Rodman (1981). The latter worker notes that glucosinolates do not occur randomly among these groups and may be taxon-specific at any taxonomic level from the family to the species. Indeed, in a survey of the closely related genera *Caulanthus* and *Streptanthus* (Cruciferae), Rodman *et al.* (1981) found that most of the approximately 40 species possessed 'a characteristic chemical profile distinguishable from that of morphologically similar taxa'. Nevertheless, Rodman (1981) feels obliged to note that while 'the population biology of glucosinolates is probably the best studied of that of any comparable diverse class of secondary plant metabolites' it is, nevertheless, 'sobering to realize . . . that we remain largely ignorant of the patterns of variability in glucosinolates within natural plant populations . . .'. This is a reiterative theme for most metabolites but, considering the published information since 1963, there has been a near exponential increase in such data for plant species generally.

In addition to the selected papers for the various compounds mentioned above, numerous other secondary metabolites have been applied to systematic problems at the specific level. These include alkaloids (e.g. Mabry and Mears, 1970, Stermitz, 1968, Stermitz *et al.*, 1975 and Bandoni *et al.*, 1975, on *Argemone*; Bui *et al.*, 1977, on *Hazunta*), steroids (Nielsen *et al.*, 1979, on

Euphorbia), sapogenins (Bohannon *et al.*, 1974, on *Trigonella* and related genera), iridoids (Jensen *et al.*, 1975b, on *Cornus*), cyanogenic glucosides (Secor *et al.*, 1976 and Siegler *et al.*, 1978, on *Acacia*), fatty acids (Rogers, 1972, on *Linum*; Morice, 1975a,b, on *Astelia*; Seigler *et al.*, 1978, on *Krameria*), sugars (Fischer and Kandler, 1975, on *Selaginella*; Rix and Rast, 1975, on *Fritillaria*; Sellmair *et al.*, 1977, on *Primula*) and yet other compounds. Most of these studies have been of the survey-type and the data are mostly treated phenetically. No doubt a more inclusive study, along with biosynthetic information relating to their formation and more sophisticated phyletic analyses, will enhance their systematic contributions.

VI. Conclusion

At the present time at least, it is clear that secondary compounds are primarily useful in taxonomic studies at the species and generic level. The data, however, can only be systematically useful if care is taken to sample adequately the plant group under investigation to allow for variation that undoubtedly will occur below the species level. In favourable cases (e.g. in the genera *Baptisia* and *Wolffia*), new information may accrue from chemical studies which will allow the revision of a group where morphological and other biological data may be at variance.

There is no limit to the type of secondary compound that might be usefully employed in surveys at the specific and generic level. Flavonoids have received most attention, probably because they are universal in occurrence (at least, within vascular plants) and they are relatively easily analysed. The volatile terpenoids are mainly of interest at the population level (chapter 10) but, as mentioned above, they can be of interest when arranging species within a genus (e.g. in *Abies* and *Juniperus*).

Distribution data on more than one class of metabolite are rarely achieved but such a multi-dimensional approach, as has been attempted in the genus *Parthenium* (where alkaloids, flavonoids *and* sesquiterpene lactones were documented), has much to commend it.

12

Application of Chemistry at the Familial Level

I.	Introduction	292
II.	Flavonoid surveys	297
III.	Alkaloid surveys	298
IV.	Terpenoid surveys	303
V.	Betalain surveys	306
VI.	Miscellaneous surveys	308
VII.	Secondary compounds ancestral to the angiosperms	309
VIII.	Conclusion	312

I. Introduction

Early workers using micromolecular substances were enthusiastic in their belief that knowledge of the distributions of this or that group of compounds might yield considerable insight into familial relationships, especially as related to phyletic groupings. With the sole exception of the betalains, reviewed briefly below, the utility of such compounds for these purposes has not proven particularly significant. In fact, as noted in the two previous chapters, secondary compounds have proven most useful for systematic purposes at the generic level and lower. The reasons for this appear obvious: like the distribution of 'superficial' (or secondary) morphological structures such as vestuture (hairs), leaf-lobing, branching habits etc., any attempt to cluster families on the mere appearance of this or that character is bound to produce artificial clusterings. If one were to know, however, the metabolic pathways and associated enzymes which gave rise to these superficial structures or characters, one would be able to discern the occasional artifact or convergent character, thus permitting a more informed classification. Unfortunately, like 'secondary' morphological characters, the metabolic pathways leading to the development of most secondary compounds in

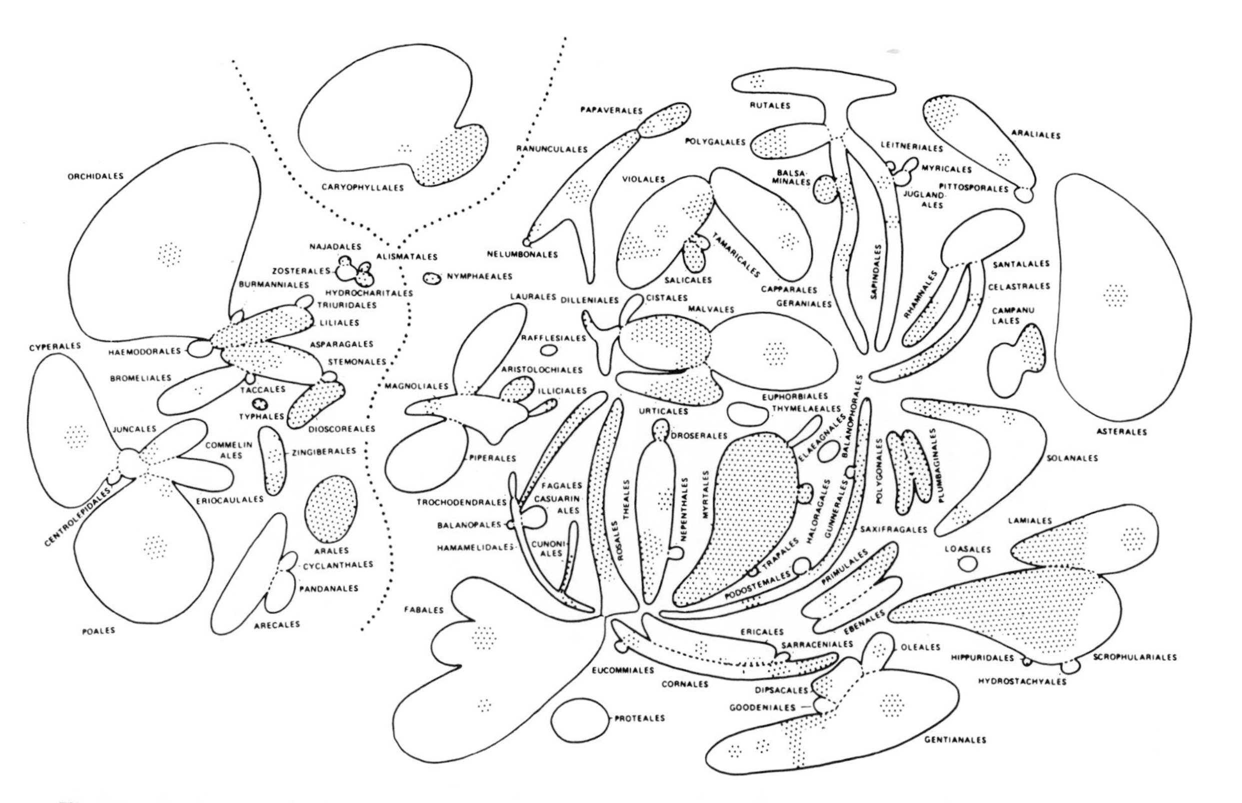

Fig. 12.1 Distribution of cyanidin and/or pelargonidin among Angiospermae. (From Gornall *et al.*, 1979.)

distantly related or even closely related plant groups are not usually known. Consequently it is difficult, if not impossible, to infer that identical secondary compounds in distantly related groups are necessarily homologous (i.e. have been derived via metabolic pathways mediated by homologous enzymes).

Fowden (1965) clearly brought this point to the fore in noting that very little detailed information is available concerning the enzymes involved in the biosynthesis of 'secondary plant products' but suggested, however, that 'synthesis may sometimes be the result of non-specific catalysis by enzymes whose normal function is in association with the basic metabolic processes of plant cells'. A few examples of what might be considered independent or convergent evolution of identical micromolecules are now known, especially among alkaloids and quinones (see chapters 6 and 7).

In connection with the above introduction, it is interesting to consider the distribution of flavonoids among angiosperms generally, perhaps the most thoroughly surveyed group of secondary compounds (Harborne, 1977a,b). Gornall *et al.* (1979) have provided a series of figures showing the distribution among flowering plants of delphinidin, cyanidin-pelargonidin, *O*-methylated anthocyanidins, acylated anthocyanidin glycosides, 3-desoxyanthocyanidins, myricetin, luteolin-apigenin, tricin, and yet other compounds (see Figs 12.1 to 12.3, for selected examples). Viewing these (and even ignoring the uneven and limited sampling among plant families) one must be more impressed with the sporadic appearance of this or that flavonoid among angiosperms generally, than to their restriction to any given familial grouping. In attempting to summarize the systematic import of these data for the higher systematic categories, Gornall *et al.* (1979) conclude 'that many of our flavonoid structural classes are polyphyletic and therefore their systematic importance at the suprafamilial level is lessened'.

Crawford (1979), in a clearly stated review of the subject (flavonoid chemistry and angiosperm evolution), finds the presently available data to be of little potential for systematic purposes at the familial level or higher, quoting Alston (1967) to the effect that 'there are hundreds of examples of interesting chemical and taxonomic correlations, yet the vast majority of these have little or no systematic value beyond adding to the preexisting pile, and it is extremely doubtful that the pile will ignite spontaneously and yield some Promethean taxonomic overview'. Crawford (1979) adds that 'in the more than a decade since this was written, despite numerous surveys and the isolation and identification of hundreds of flavonoids from hundreds of species, the pile is in reality no closer to igniting now than it was then'. So it is in 1984.

Lacking reasonable information on the metabolic pathways and mediating enzymes that give rise to this or that secondary component among disparate groups, perhaps the next best way to evaluate the import of such data is to do

Fig. 12.2 Distribution of myricetin among Angiospermae. (From Gornall *et al.*, 1979.)

Fig. 12.3 Distribution of luteolin and/or apigenin among Angiospermae. (From Gornall *et al.*, 1979.)

exactly what the morphological systematist does: look at general trends and suites of chemical characters, especially as these might be correlated with yet other substances, etc. which, of course, is what Kubitzki (1969), Gottlieb (1972), Gomes and Gottlieb (1980), Gershenzon and Mabry (1983) and a few other workers have attempted to do, as noted below.

In the sections that follow, an attempt is made to present a cross section of some of the more recent familial and interfamilial review papers for selected groups of compounds. Some of these are based on relatively detailed surveys within or among closely related families or on broad surveys among families generally for a given group of compounds. Examples of the former are more common and these yield more meaningful systematic data.

II. Flavonoid surveys

Largely because of the ease with which herbarium material can be utilized, this group of compounds has been extensively surveyed at the tribal level or higher. Many such studies might be thought of as 'so what?' papers in that relatively little new insight is shed upon the systematic problems at hand. Still, a number of these have yielded interesting suggestions as to relationships, especially among problem groups where morphological criteria are difficult to assess.

Some of the better known surveys with their relevant 'contribution' include those of Moore *et al.* (1970) on the Empetraceae (it's related to the Ericaceae!), Williams *et al.* (1971) on the Gramineae and related families (grasses and palms are especially close!), Harborne and Williams (1973a) on the Ericaceae (supports the transfer of *Diplache* from the family Diapensiaceae to the Ericaceae; also the separation of *Calluna* and *Cassiope*; and the retention of the subfamilies Pyroloideae and Monotropoideae), Williams *et al.* (1973) on the Palmae (numerical analyses of chemical data—along with morphological—produces a 'new classification'), Harborne (1975b) on the Bixaceae and related families (*Cochlospermum* and *Bixa* are not closely related as suggested by Cronquist), Williams and Harborne (1977b) on the Zingiberales (the Marantaceae is chemically diverse and flavonoid patterns do not support the existing classifacatory scheme), Williams (1978) on the Bromeliaceae (flavonoids, overall, suggest an isolated position of the family among monocotyledons), Harborne and Green (1980) on the Oleaceae (flavonoids correlate with chromosomal data suggesting ancestral amphidiploidy for $x = 23$ lines), Kubitzki (1968) on the Dilleniaceae (*Dillenia* and *Tetracera* are chemically advanced but belong to different phyletic lines), Jay (1969) on the Pittosporaceae (homogeneous chemically and perhaps best related to the Umbelliflorae), Grieve and Scora (1980) on the Aurantioideae of the Rutaceae (flavonoids support Swingle's taxonomic concept of subfamily Aurantioideae), Gray and Waterman (1978) on the coumarins of the Rutaceae (data negate familial status for the subfamily Flindersioideae and suggests that the family is phyletically related to the Umbelliferae; 'if a direct phylogenetic link were to prove untenable [this] would represent a major case of chemical convergence'), Challice (1973, 1974) on the Rosaceae, subfamily Pomoideae (chemistry suggests that allopolyploid origin of group is correct), Venkataraman (1972) on the Moraceae ('phenolic constituents can be useful in the taxonomy of the Moraceae if the investigation [were] extended to other genera and species'; how true!), Bate-Smith *et al.* (1975) on the Cornaceae (order Cornales shows affinities to the Saxifragales and Dipsacales and should exclude Araliaceae, Umbelliferae and Rhizophoraceae), Lowry (1976) on the Melastomaceae, Myrtaceae and related families (separation of family Lecythidaceae from the order Myrtales supported), Sterner and Young (1980) on the Calycanthaceae (*Idiospermum* 'is sufficiently distinct' in its flavonoid chemistry to be placed in a separate family).

III. Alkaloid surveys

As for the flavonoids, a number of surveys for specific alkaloids may be found in the literature. In the early years, these were mostly long lists with notation of mere presence (e.g. Lawler and Slaytor, 1969) or else presence of a general type of alkaloids (e.g. lupine-type alkaloids etc.), but several recent survey papers are quite specific and thorough and suggest strongly that alkaloids, properly surveyed for and considered within the context of their biosynthetic pathways, might yield considerable phyletic insight. One of the better such contributions is that of Waterman (1975), working with alkaloids of the Rutaceae. He notes that the alkaloid data is largely inconsistent with the classification of the family provided by classical workers and consequently proposes an alternative treatment, based on alkaloids and other secondary metabolites, that he feels is superior (Fig. 12.4). It is to his credit that he considers this to be a 'working hypothesis' with the 'hope that it will stimulate further research and discussion'. Other alkaloid surveys are less integrated and more difficult to interpret, for example, Thornber's (1970) account of alkaloids in the Menispermaceae. Even in the latter, however, an informed systematist, familiar with metabolic pathways, might well appreciate the implication of his charts 1 and 2 (Fig. 12.5) which purport to show the sequential transformation of ancestral skeletal types into their various derived states. If correct, assignment of phyletic weights to the sequence would seem perfectly sound, and perhaps better justified than assignment of intuitive weights to hypothetical sequential 'pathways' for this or that morphological state.

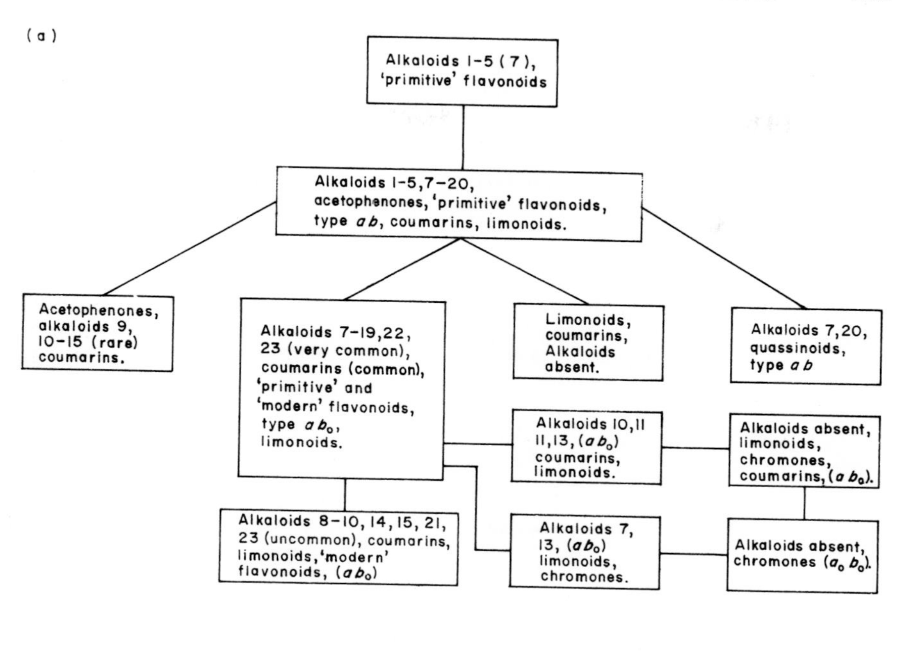

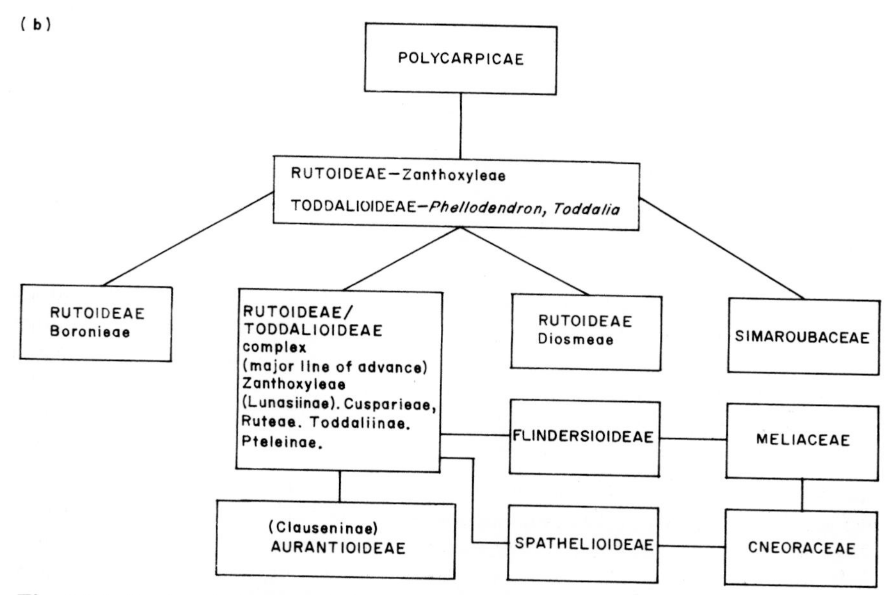

Fig. 12.4 Biochemical interrelationships between the subfamilies of the Rutaceae and their close allies in the Rutales. (From Waterman, 1975.)

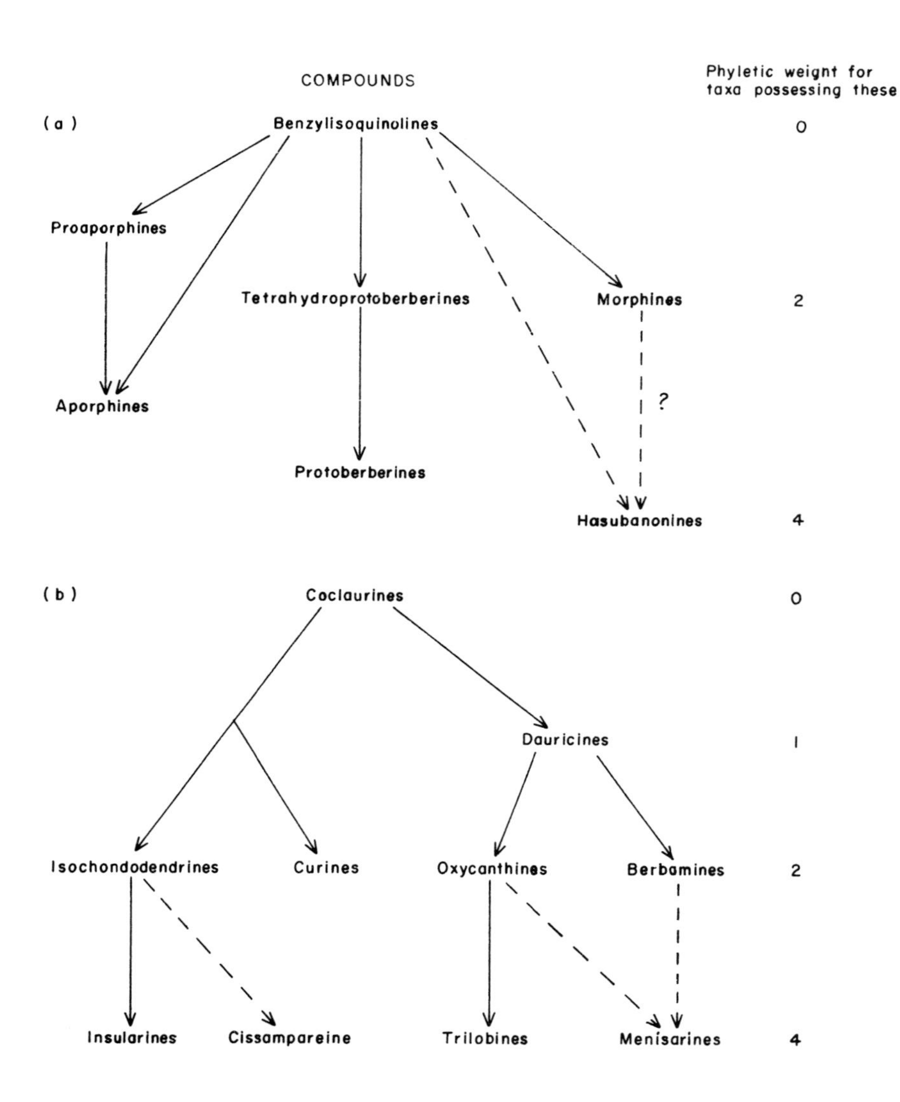

Fig. 12.5 Relative metabolic position of alkaloids in the family Menispermaceae. (a) Assuming the benzylisoquinolides are primitive, one might arbitrarily assign phyletic weights of 0, 2, 4 etc. to their successive derivatives to ascertain, on alkaloid data, hypothetical primitive taxa. (b) Similar chart for the coclavrines of the Menispermaceae. (From Thornber, 1970.)

Fig. 12.6 Distribution of tropane alkaloids in the angiosperms. (From Romeike, 1978.)

Hegnauer (various references) has summarized much of the broad systematic implications of alkaloid distribution among angiosperms. Viewed at the suprafamilial level it would appear that the general alkaloid types follow a 'pattern' similar to that of flavonoids: they can appear sporadically in various remotely related groups, presumably as a result of convergence (i.e. via independent metabolic pathways). Thus tropane alkaloids (Romeike, 1978) which occur chiefly in the Solanaceae, occur in a restricted but wide smattering of angiosperms, both di- and monocotyledons (Fig. 12.6).

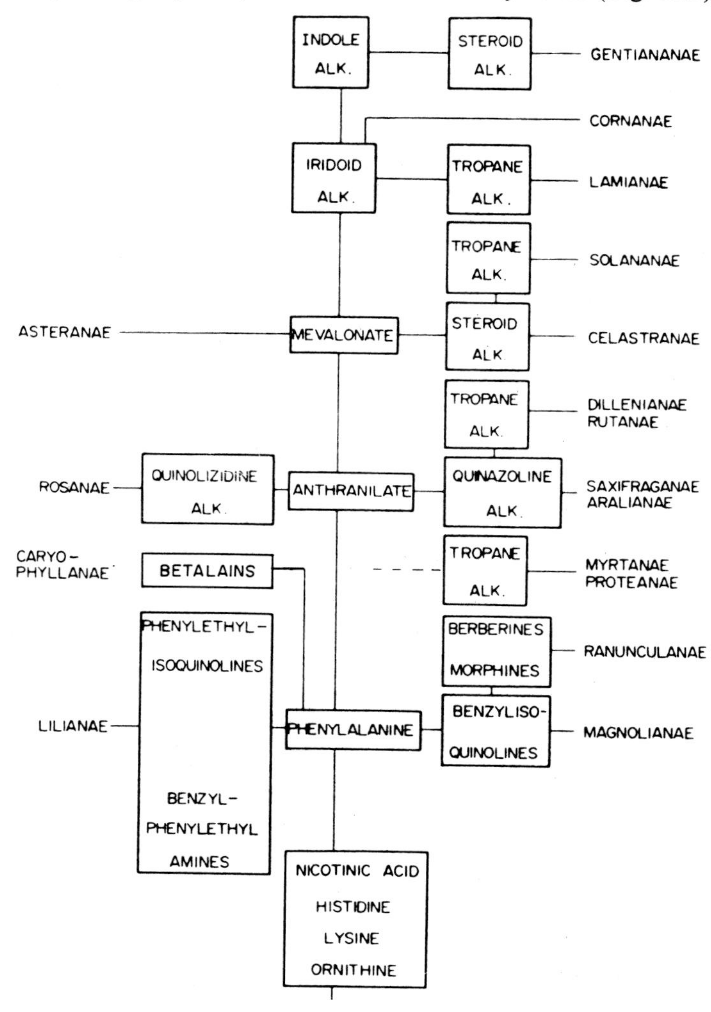

Fig. 12.7 Phylogenetic tree of angiosperm super orders (*sensu* Dahlgren) based on the evolution of metabolic pathways and on the distribution of alkaloids. (From Gomes and Gottlieb, 1980.)

Finally, it should be noted that Gomes and Gottlieb (1980) provided a recent summary of the evolution of alkaloids and their biosynthetic pathways among angiosperms generally. They conclude that such data suggest a partitioning of the Angiospermae into three major subclasses: (1) Magnolidae; (2) Rosidae; and (3) Asteridae as shown in Fig. 12.7. This is in contrast with the views of Mabry and Mears (1970) who considered the systematic value of alkaloids to be largely applicable at the familial level and lower.

IV. Terpenoid surveys

As noted in chapter 10, terpenoids are widespread in plant families generally and, in chemical complexity and structural variability, they rival the flavonoids. For example, the essential oil of the fruits of *Passiflora edulis* reportedly contains over 250 volatile components (K. E. Murray *et al.*, 1972). Fortunately, GLC provides an excellent method for isolating and quantifying such volatile oils. Because of this, numerous surveys and some in-depth studies have been made for this group of compounds.

Among closely related families, plants belonging to the conifer groups (gymnosperms) have been the most extensively surveyed for terpenoids. Von Rudloff (1975b) has provided an excellent over-view of the suprafamilial value of leaf-oil terpenoids as systematic markers, concluding that 'whereas differentiation by means of leaf-oil terpene patterns is possible at the species and subspecies levels, no clear-cut chemical differences are discernible at higher taxonomic levels'. For example, Von Rudloff notes (1975b, p. 167) that close examination of the oils of Pinaceae reveals that two biosynthetically related groups of terpenes stand out: the fenchone–fenchal group (cf. Fig. 12.8) in *Abies* and the correlation of sabinene with terpinolene and terpinen-4-ol (cf. Fig. 12.9) in *Pseudotsuga*. This correlation is also found in some of the leaf volatiles of *Juniperus* (Cupressaceae) and von Rudloff concludes that 'the leaf oils provide only limited chemical data to differentiate genera'.

Terpenoids, at least the low molecular-weight mono-, bi- and tricyclic types, are widely distributed among plant families from algae to the Compositae. This being so, the biosynthetic potential for their *accumulation*, if not production, is perhaps latent in most plant taxa. It is not surprising, then, that terpenoids, as such, are not especially good markers for relating taxa at the generic level or higher. As noted in chapter 11, they are robust characters for population and specific studies but as familial markers they have some of the same drawbacks possessed by flavonoids: the metabolic pathways leading to their production must be widespread and often capricious in their activity.

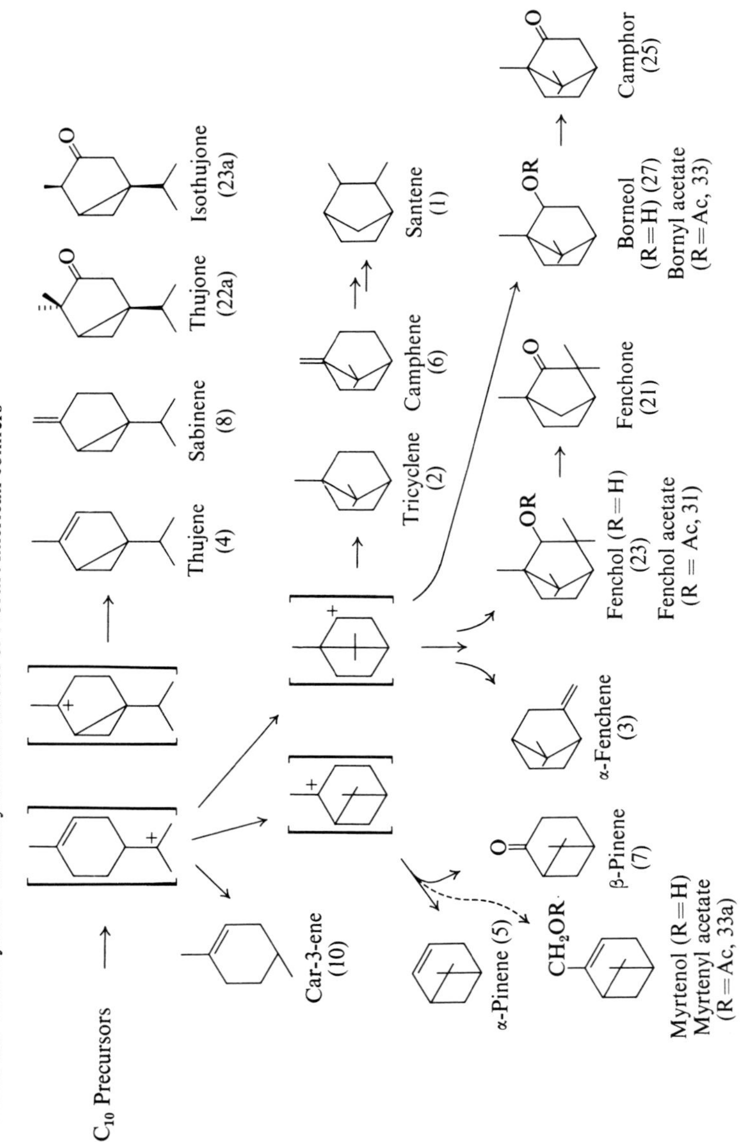

Fig. 12.8 Some chemical and biochemical interrelationships of bi- and tricyclic monoterpenes in the volatile oils of North American conifers. (From von Rudloff, 1975.)

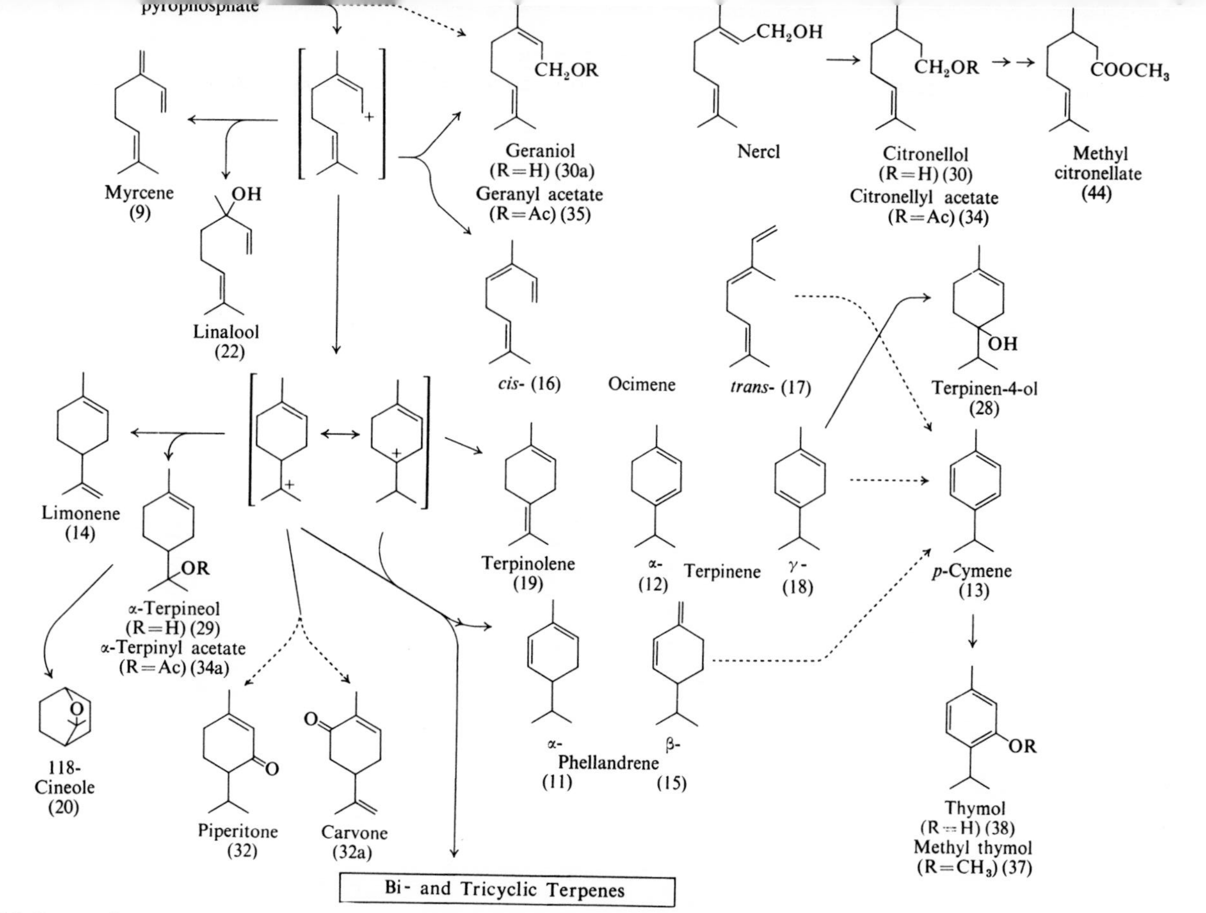

Fig. 12.9 Some chemical and biochemical interrelationships of acyclic and monocyclic monoterpenes found in the volatile oils of conifers. (From von Rudloff, 1975.)

V. Betalain surveys

Probably no group of secondary compounds has provided as much taxonomic impact at the familial level (and phyletic controversy) as have the betalains. Unlike the several classes of compounds discussed above, the betalains are restricted to a given group of closely related families in the order Centrospermae. In fact, this group of yellow and purple-red nitrogenous pigments are mutually exclusive with the more common non-nitrogenous flavonoid pigments, the anthocyanins.

Mabry (1977) has discussed the chemistry and systematics of the Centrospermae in some detail. Consequently, it need only be noted here that he perceives the Order as consisting of two phyletic suborders, the Chenopodiinae (betalain families) and the Caryophyllineae (non-betalain or anthocyanin families). This is contrary to much of the phyletic thinking of at least a few current phyleticists working at the familial level and higher who would like to relate these two groups as somewhat closer than the rank 'suborder implies'. Indeed, many would like to evolve the betalain groups out of the non-betalain or anthocyanin groups, or at least treat them as much more reticulately related than is suggested by Mabry *et al.* (1963).

It is a moot question whether or not the 10 or more families which make up the suborder Chenopodiinae ever possessed anthocyanins. By all accounts the Centrospermae is an old and isolated phyletic line of the angiosperms (Fairbrother *et al.*, 1975). This being so, it would appear reasonable that the chenopoid line separated from the caryophylloid line at a very early time, the latter being somewhat closer to the phyletic stock(s) leading to the dicotyledons generally. Subsequently the two orders presumably remained closely parallel. In short, Mabry and co-workers would argue that genes for the production of enzymes that produce betalains were not shuffled into the mainstream of the dicotyledonous families; and conversely, enzymes for the production of anthocyanins never developed or became incorporated into the betalain lines. If, over time, betalains *replaced* anthocyanins in the chenopoid lines, as suggested by Ehrendorfer (Mabry, 1976, p. 217) and Cronquist (1977a,b), then selection-pressure must have been severe at an early time, perhaps millenia ago when the dicotyledonous group were in their infancy. Or else why would recurrent mutation not subject the 'muted genes', or enzymic pathways leading to anthocyanins, to repeated reselection as this or that anthocyanic mutant (or gene recombination) arose in the thousands of species and millions of individuals which make up the extant chenopoid line? The simplest explanation would appear to be that the taxa which make up the chenopoid order were essentially without pigments until certain members of its ancestral populations developed betalains. These latter compounds, which act remarkably like anthocyanins, were possibly siezed upon, filling

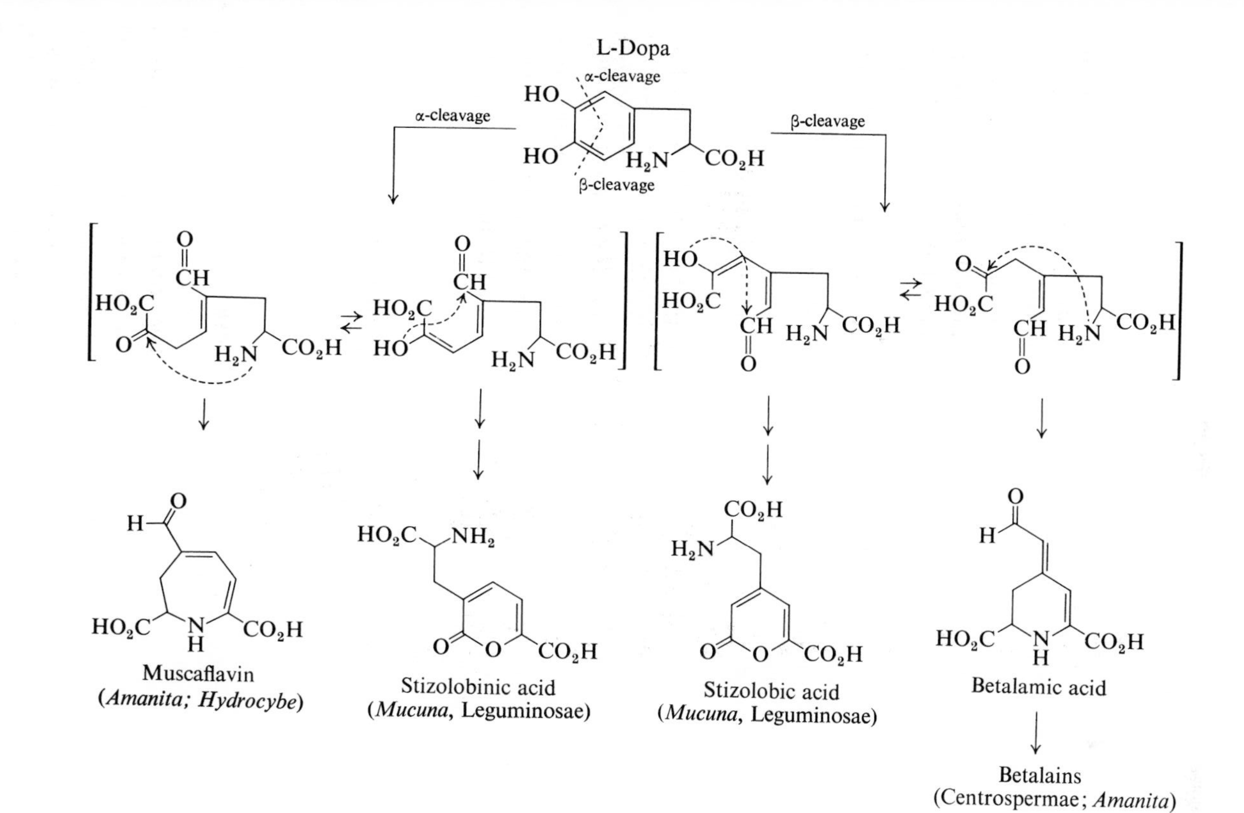

Fig. 12.10 L-Dopa is known to undergo 2,3-1 (α-cleavage) or 4,5- (β-cleavage) extra diol cleavage in different organisms. Only in the centrospermae and in mushrooms does L-dopa lead, via the β-cleavage pathway, to betalamic acid, the precursor of all other betalains. (From Mabry, 1977.)

the chemical 'niche' which might have befitted an earlier and more enterprising genetic mechanism for the production of anthocyanins, which event or events must have occurred in the caryophylloid lines.

The hypothetical metabolic schemata leading to the production of betalains is shown in Fig. 12.10. Referring to this figure, it should be noted that molecules such as stizolobinic or stizolobic acids, which are superficially similar to betalamic acid, have been found in the Leguminosae. This might suggest that only an enzyme or two might yet permit the conversion of the precursors of such acids into betalains. But perhaps the critical enzymes for such conversions do not exist; indeed, occurrence of the supposed intermediary compounds leading to these compounds in *Mucuna*, might reflect the activity of enzymes non-homologous with those leading to the production of betalamic acid. A fruitful research programme might involve the isolation of such enzymes with an eye toward elucidation of their amino acid sequences to ascertain homology.

Cronquist (1977b), in his review of the taxonomic significance of secondary metabolites in angiosperms generally, state that the Centrospermae 'are chemically noteworthy for the [complete?] substitution of betalains for anthocyanins in most families' and . . . 'that a search for the biological significance of the betalains should concentrate on their repellent (and fungicidal) properties, rather than on their function as flower pigments'. However, unlike terpenoids, alkaloids and flavonoids, the betalains do not reappear willy nilly among this or that phyletic line of the angiosperms. The mutual exclusion of anthocyanins and betalains in the Centrospermae and their obvious isolated and old phyletic status (Fairbrother *et al.*, 1975) strongly suggest that betalains have had a phyletic history quite different from that of most other secondary compounds.

VI. Miscellaneous surveys

A wide assortment of surveys for numerous other compounds for specific groups may be found in the literature but, because of space limitations, no attempt has been made in this chapter to treat these comprehensively (but see also chapters 5–8). Some of the more noteworthy contributions are those of Bell *et al.* (1978) on the amino acid canavanine (unique to the Leguminosae, subfamily Papilionoideae), fatty acids in the Proteaceae (Vickery, 1971), limonoids in the Rutaceae (Dreyer *et al.*, 1972), triterpenoids in the Gramineae (Ohmoto *et al.*, 1970), cyanogenic glycosides in angiosperms (Hegnauer, 1973), sterols among angiosperms (Tétényi, 1976) and among higher fungi (Yokoyama *et al.*, 1975), hydrocarbons of cuticular wax of leaves (Eglinton *et al.*, 1962), the sugars mannoheptulose and sedoptulose among vascular plants, and lunularic acid and related compounds among plants generally (Pryce, 1972; Gorham, 1977).

In addition to the systematically oriented studies of specific compounds mentioned above, review articles of this or that group of compounds are periodically presented in chemical journals, the emphasis being on the compounds themselves with little emphasis on their phyletic significance. Much of the tabulated data in these articles, while poorly documented as to taxonomic identification, are nevertheless quite useful to the systematist since they pull together a diverse and largely foreign literature (to plant systematists) and provide insight into the likely value of such compounds for systematic purposes. Some of the more noteworthy of such reviews are those for triterpenoids (Pant and Rastogi, 1979), for friedelin and associated triterpenoids (Chandler and Hooper, 1979), and for sesquiterpene lactones (Yoshioka *et al.*, 1973; Fischer *et al.*, 1979).

VII. Secondary compounds ancestral to the angiosperms

As may be ascertained from the above, the capacity for plants to develop flavonoids, if not most secondary compounds, undoubtedly developed quite early among vascular plants. Indeed, a wide array of flavonoids, mostly flavones, has been reported for the primitive liverwort *Takakia* (Markham and Porter, 1979), so one might infer that metabolic pathways leading to the synthesis of flavonoids appeared quite early in the evolutionary record of plants, perhaps even before the development of vascular tissue. Lowry *et al.* (1980) have discussed this problem in a somewhat speculative scheme (Fig. 12.11), suggesting that, in addition to predatory defence, phenolics, and ultimately flavonoids, were also selected for their protection against ultraviolet damage and autotoxicity. They further note that 'there is a correlation between the incidence of leaves with high concentrations of flavonoids (particularly at early developmental stages) and the 'primitiveness' of taxa in tropical rain forests . . .'. An intriguing but, as yet, undocumented observation.

According to many workers the most 'primitive' living vascular plant family is the Psilotaceae, a non-flowering taxon which first appears in the fossil record during the early Palaeozoic. Its phyletic position has been variously interpreted. Conventional thought reckons that elements of the Psilotaceae gave rise to the ancestral populations which in turn led to the present-day vascular plants. At least one respected worker, however, has suggested that the family belongs with the filicalean ferns and as such could not have been ancestral to the present day vascular plants. Wallace and Markham (1978) have shown that the only two extant genera of the Psilotaceae, *Psilotum* and *Tmesipteris*, both contain amentoflavone as the predominant flavonoid in addition to lesser amounts of yet other flavonoids, including *C*-glycosides which they interpret as providing some link with extant ferns.

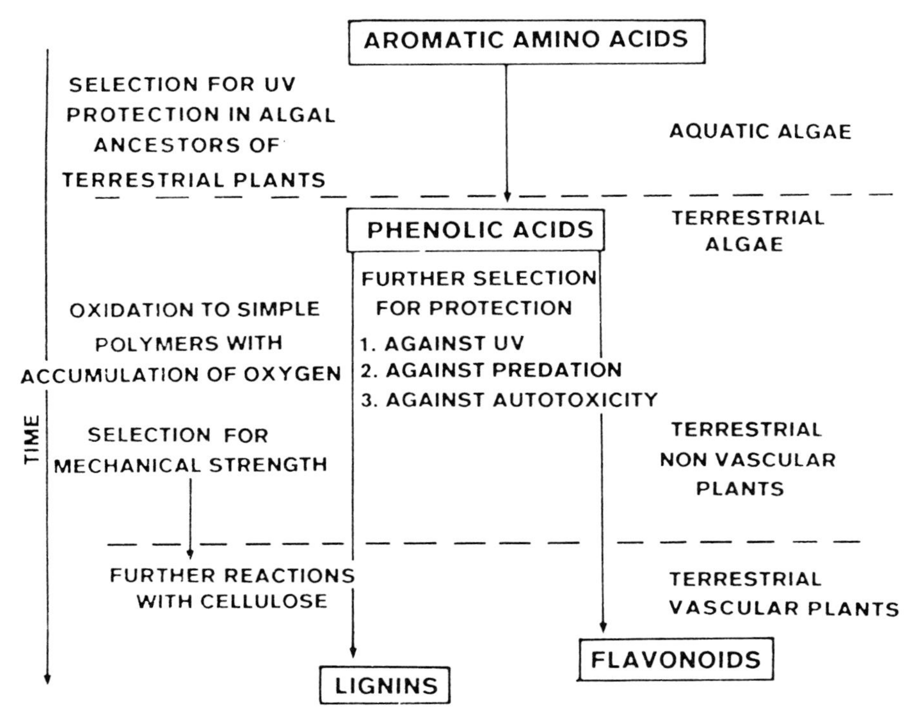

Fig. 12.11 A general scheme showing the development of flavonoids, lignins and cellulose in land plants. (From Lowry *et al.*, 1980.)

Regardless of the interpretations accorded the above, one must be impressed that amentoflavones are so widely distributed among other supposedly primitive vascular plants such as the Cycadales (Dossaji *et al.*, 1975) and gymnosperms generally. Flavones are also known to occur in the Lycopodiales (Voirin and Jay, 1978a) and yet other morphologically less advanced groups, including the Isoetales (Voirin and Jay, 1978b).

Among the several compounds that have been reasonably surveyed for, only a few appear to show significant phyletic localization. Dahlgren (1980), for example, has singled out the distribution patterns of iridoids, polyacetylenes, glucosinolates and benzylisoquinoline alkaloids among the angiosperms as being especially noteworthy (Fig. 12.12). Clearly their restricted occurrence to given hypothetical familial assemblages seems significant, but much of the apparent 'order' may reflect sampling lacunae, or merely bias as to the arrangement of family groupings.

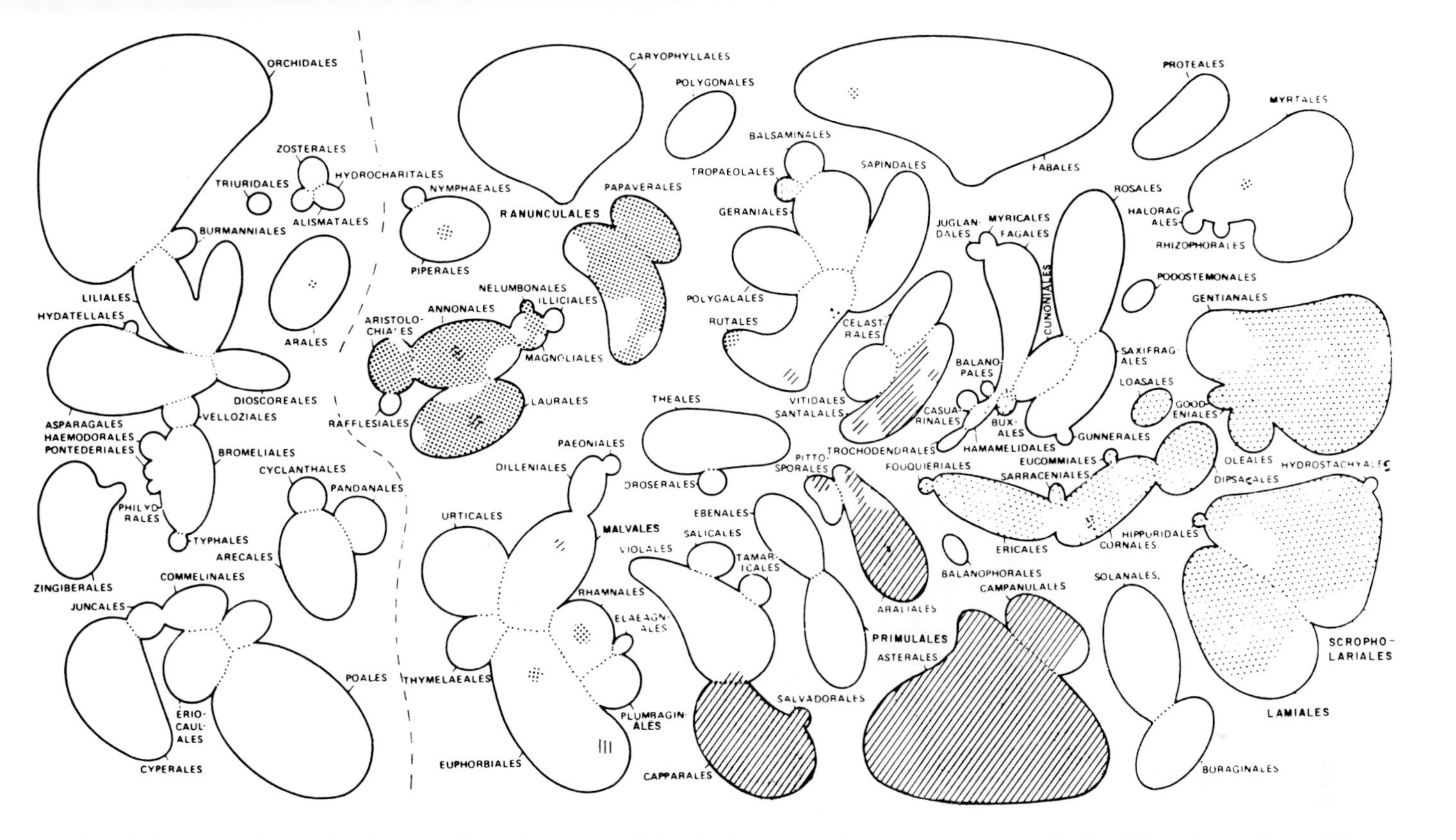

Fig. 12.12 Approximate distribution of certain groups of chemical compounds in angiosperms: iridoids (⁙), polyacetylenes (▨), glucosinolates (▥) and benzylisoquinoline alkaloids (⁘). (From Gornall *et al.*, 1979.)

VIII. Conclusion

Familial and orderial circumscription and their phyletic positions one to the other has been and will continue to be the bane of the angiosperm systematist, as aptly noted by Turner (1977). Thus Dahlgren's (1980) most recent revised system of classification has been assailed by Corner (1981) with the introductory sentence 'Despite many contributions to this subject [angiosperm classification], the fact remains, and needs continual emphasis, that the whole history of the angiosperms is still a mystery'. But the abominable mystery is not likely to remain so for powerful new macromolecular techniques have been developed to sort out the various phyletic branches and the relative times of their divergence from yet other branches (Turner, 1977; Holmquist and Pearl, 1980; cf. chapter 17).

If one considers the relatively few years that have expired since the inception of the Chemosystematic Period of plant systematics (see chapter 3) and the extraordinary developments that have occurred in methodology, instrumentation and concept in chemosystematics since that time, it appears obvious that chemical approaches, both micromolecular and macromolecular, will soon yield a classificatory bulwark which is likely to withstand the whim and whine of any number of newly inspired intuitional workers using the long-overworked, now classical, megamorphic approaches.

13
Chemical Documentation of Hybridization

I.	Introduction	313
II.	Historical	314
III.	Micromolecules versus macromolecules	318
IV.	Micromolecular documentation	319
V.	Documentation of sympatric hybridization	320
VI.	Documentation of allopatric introgression	327
VII.	Polyploidy—its effect on chemical constituents	339
VIII.	Conclusion	341

I. Introduction

Natural hybridization among plants is reportedly a relatively common event. At least several thousand references exist which purport to document F_1, F_2 and backcross individuals between this or that species (Knobloch, 1972; Stace, 1975). Most such studies are supported by morphological data which, as will be shown below, are often equivocal. If gathered with sufficient care and corroborated with experimental crosses, morphological data can be very effective in the resolution of problems involving hybridization. But it can also be very misleading, especially when attempting to assess the extent of hybridization and the influx of genes from one taxon into another.

In spite of the published literature on interspecific hybrids and, by inference, the degree of hybridization among species to be expected under field conditions, natural hybridization is a relatively rare phenomenon. In fact, the large number of articles documenting its occurrence can be attributed to its novelty. That is, it is sufficiently uncommon at any one site to bring out the enthusiasm of its discoverer: the exception attracts and excites.

II. Historical

Terpenes were among the first compounds to be used in the chemical documentation of hybrids, the work of Zobel (1951) and Mirov (1956) on *Pinus* being especially noteworthy. The first significant use of flavonoids was that by Turner and Alston (1959). They used chromatographic techniques in their analysis of a complex series of hybrids in the genus *Baptisia*. This hallmark paper ushered in much of the subsequent chemotaxonomic interest in plants, at least in North America (e.g. Smith and Levin, 1963, on *Asplenium*; Stebbins *et al.*, 1963, on *Viola*). In fact, a 100 or more papers have appeared since 1959 variously documenting hybridization, mostly using flavonoids (Table 13.1).

Table 13.1 Selected chemical studies documenting hybridization in plants

Year	Workers (Reference)	Compound used	Taxa studied
1951	Zobel	Terpenes	*Pinus* (Pinaceae)
1955	Williams	Phenolics	*Malus* (Rosaceae)
1956	Mirov	Terpenes	*Pinus*
1958	Pryor and Bryant	Terpenes	*Eucalyptus* (Myrtaceae)
1959	Bannister *et al.*	Terpenes	*Pinus*
	Schwarze	Flavonoids	*Phaseolus* (Leguminosae)
	Turner and Alston	Flavonoids	*Baptisia* (Leguminosae)
1962	Alston *et al.*	Flavonoids	*Baptisia*
	Alston and Simmons	Flavonoids	*Baptisia*
	Alston and Turner	Flavonoids	*Baptisia*
1963	Alston and Turner	Flavonoids	*Baptisia*
	Harney and Grant	Flavonoids	*Lotus* (Leguminosae)
	Smith and Abashian	Alkaloids	*Nicotiana* (Solanaceae)
	Smith and Levin	Flavonoids	*Asplenium* (Polypodiaceae)
	Stromnaes and Garber	Flavonoids	*Collinsia* (Scrophulariaceae)
	Stebbins *et al.*	Flavonoids	*Viola* (Violaceae)
1964	Alston and Hempel	Flavonoids	*Baptisia*
	Forde	Terpenes	*Pinus*
	Harney and Grant	Flavonoids	*Lotus*
	McHale and Alston	Flavonoids	*Baptisia*
	Scora and Wagner	Flavonoids	*Dryopteris* (Polypodiaceae)
	Torres and Levin	Flavonoids	*Zinnia* (Compositae)
1965	Alston	Flavonoids	*Baptisia*
	Alston *et al.*	Flavonoids	*Baptisia*
	Brehm and Ownbey	Flavonoids	*Tragopogon* (Compositae)
	Marks *et al.*	Alkaloids, flavonoids	*Solanum* (Solanaceae)
	Stone *et al.*	Fatty acids	*Carya* (Juglandaceae)

Year	Workers (Reference)	Compound used	Taxa studied
1966	Brehm	Flavonoids	*Baptisia*
	Frost and Bose	Phenolics	*Corchorus* (Tiliaceae)
	Harney	Flavonoids	*Pelargonium* (Geraniaceae)
	Jaworska and Nybom	Flavonoids	*Saxifraga* (Saxifragaceae)
	Levin	Flavonoids	*Phlox* (Polemoniaceae)
	Schantz and Juvonen	Terpenes	*Picea* (Pinaceae)
	Widen and Sorsa	Phloroglucinol	*Dryopteris*
1967	Critchfield	Terpenes	*Pinus*
	Dass and Nybom	Phenolics	*Brassica* (Cruciferae)
	Emboden and Lewis	Terpenes	*Salvia* (Labiatae)
	Hunter	Flavonoids	*Vernonia* (Compositae)
	Levin	Flavonoids	*Liatris* (Compositae)
	Levin	Flavonoids	*Phlox*
1968	Baetcke and Alston	Flavonoids	*Baptisia*
	Bate-Smith *et al.*	Flavonoids	*Hieracium* (Compositae)
	Fahselt and Ownbey	Flavonoids	*Dicentra* (Fumariaceae)
	La Roi and Dugle	Flavonoids	*Picea*
	Levin	Flavonoids	*Liatris*
	Levin	Flavonoids	*Phlox*
	Novotny *et al.*	Terpenes	*Petasites* (Compositae)
	Ogilvie and Von Rudloff	Terpenes	*Picea*
	Von Rudloff and Holst	Terpenes	*Picea*
1969	Albach and Redman	Flavonoids	*Citrus* (Rutaceae)
	Binns and Blunden	Flavonoids	*Salix* (Salicaceae)
	Carter and Brehm	Flavonoids	*Iris* (Iridaceae)
	Flake *et al.*	Terpenes	*Juniperus* (Cupressaceae)
	Habeck and Weaver	Terpenes	*Picea*
	Hanover and Wilkinson	Terpenes	*Picea*
1970	Adams and Turner	Terpenes	*Juniperus*
	Ballard *et al.*	Flavonoids	*Capsicum* (Solanaceae)
	Chalice and Williams	Flavonoids	*Pyrus* (Rosaceae)
	Crawford	Flavonoids	*Coreopsis* (Compositae)
	Gerhold and Plank	Terpenes	*Pinus*
	Hanover and Wilkinson	Terpenes	*Picea*
	Hinton	Flavonoids	*Physalis* (Solanaceae)
	Kaltsikes and Dedio	Phenolics	*Aegilops* (Gramineae)
	Widen *et al.*	Phenolics	*Dryopteris*
1971	Belzer and Ownbey	Flavonoids	*Tragopogon*
	Bjeldanes and Geissman	Sesquiterpenes	*Encelia* (Compositae)
	Hess	Flavonoids/carotenoids	*Torenia* (Scrophulariaceae)

Table 13.1 (*continued*)

Year	Worker (Reference)	Compound used	Taxa studied
	Hunziker	Flavonoids	*Larrea* (Zygophyllaceae)
	Levy and Levin	Flavonoids	*Phlox*
	Taylor	Phenolics	*Tiarella* (Saxifragaceae)
	Widen and Britton	Phloroglucinol	*Dryopteris*
1972	Challice	Flavonoids	*Pyrus*
	Crawford	Flavonoids	*Coreopsis*
	Dass	Phenolics	*Triticum*
	Jones, S. B.	Flavonoids	*Vernonia*
	Murray and Hefendehl	Terpenes	*Mentha* (Labiatae)
	Murray	Terpenes	*Mentha*
	Santamour	Flavonoids	*Ulmus* (Ulmaceae)
	Taylor	Phenolics	*Tsuga*
	Von Rudloff	Terpenes	*Pseudotsuga* (Pinaceae)
1973	Culberson and Hale	Lichenic acids	*Parmelia* (lichen)
	Flake and Turner	Terpenes	*Juniperus*
	Flake *et al.*	Terpenes	*Juniperus*
	Halim and Collins	Terpenes	*Myrica* (Myricaceae)
	Harborne and Williams	Flavonoids	*Asplenium*
	Hsiao and Li	Flavonoids	*Aesculus* (Hippocastanaceae)
	Irving and Adams	Terpenes	*Hedeoma* (Labiatae)
	Mills	Diterpenes	*Larix* (Pinaceae)
	Ornduff *et al.*	Flavonoids	*Lasthenia* (Compositae)
	Steussy *et al.*	Flavonoids	*Picradeniopsis* (Compositae)
	Whiffin	Flavonoids	*Heterocentron* (Melastomataceae)
	Zavarin and Snajberk	Terpenes	*Pseudotsuga*
1974	Britton and Widen	Phloroglucinol	*Dryopteris*
	Crawford	Flavonoids	*Populus* (Betulaceae)
	Gardner	Flavonoids	*Cirsium* (Compositae)
	Hunt and Von Rudloff	Terpenes	*Abies* (Pinaceae)
	Levy and Levin	Flavonoids	*Phlox*
	Natarella and Sink	Phenolics	*Petunia* (Solanaceae)
	Parks and Kondo	Flavonoids	*Camellia* (Theaceae)
	Williams and Harborne	Flavonoids	*Saccharum* (Gramineae)
1975	Adams	Terpenes	*Juniperus*
	Dass *et al.*	Flavonoids	*Annona* (Annonaceae)
	Goodwin	Flavonoids	*Dodecatheon* (Primulaceae)
	Hinton	Flavonoids	*Physalis* (Solanaceae)
	Jones and Siegler	Flavonoids	*Populus*
	Kelsey *et al.*	Sesquiterpenes	*Artemisia* (Compositae)

Year	Workers (Reference)	Compound used	Taxa studied
	King *et al.*	Flavonoids	*Rhododendron* (Ericaceae)
	Lawrence *et al.*	Terpenes	*Cupressus* (Cupressaceae)
	Levy and Levin	Flavonoids	*Phlox*
	McMillan *et al.*	Sesquiterpenes	*Xanthium* (Compositae)
	Ornduff and Bohm	Flavonoids	*Tracyina*, *Rigiopappus* (Compositae)
	Scora *et al.*	Alkanes	*Persea* (Lauraceae)
	Star *et al.*	Flavonoids	*Pityrogramma* (fern)
	Von Rudloff	Terpenes	*Juniperus*
	Wollenweber	Flavonoids	*Populus*
1976	Baker and Baker	Amino acids	*Aloe* (Agavaceae)
	Baker and Baker	Amino acids	*Armeria* (Plumbaginaceae)
	Baker and Baker	Amino acids	*Cercidium* (Leguminosae)
	Baker and Baker	Amino acids	*Oxalis* (Oxalidaceae)
	Baker and Baker	Amino acids	*Silene* (Caryophyllaceae)
	Conner and Purdie	Triterpenes	*Cortaderia* (Gramineae)
	Guppy and Bohm	Flavonoids	*Hieracium* (Compositae)
	Hardman and Benjamin	Ecdysones	*Helleborus* (Ranunculaceae)
	Kallunki	Flavonoids	*Goodyera* (Orchidaceae)
	McMillan *et al.*	Sesquiterpenes	*Xanthium*
	Simmons and Parsons	Terpenes	*Eucalyptus*
	Zavarin *et al.*	Terpenes	*Pinus*
1977	King	Flavonoids	*Rhododendron*
	Martinez and Swain	Flavonoids	*Gibasis* (Commelinaceae)
	Tateoka *et al.*	Flavonoids	*Calamagrostis* (Gramineae)
	Von Rudloff	Terpenes	*Picea*
	Whiffin	Terpenes	*Correa* (Rutaceae)
1977	Widen *et al.*	Terpenes	*Thymus* (Labiatae)
	Zavarin *et al.*	Terpenes	*Abies*
1978	Baker *et al.*	Amino acids	*Campsis* (Bignoniaceae)
	Bragg *et al.*	Flavonoids	*Prosopis* (Leguminosae)
	Flake *et al.*	Terpenes	*Juniperus*
	Gibby *et al.*	Phloroglucinols	*Dryopteris*
	Hiraoka	Flavonoids	*Dryopteris*

Table 13.1 (*continued*)

Year	Worker (Reference)	Compound used	Taxa studied
	Kjaer *et al.*	Isothiocyanates	*Tropaeolum* (Tropaeolaceae)
	Leach and Whiffin	Flavonoids	*Acacia* (Leguminosae)
	Nachit and Feucht	Phenolics	*Prunus*
	Semple and Semple	Flavonoids	*Borrichia* (Compositae)
	Widen *et al.*	Phloroglucinols	*Dryopteris*
1979	Herz *et al.*	Sesquiterpenes	*Eupatorium* (Compositae)
	Hunt and Von Rudloff	Terpenes	*Abies*
	Ikenaga *et al.*	Alkaloids	*Duboisia* (Solanaceae)
	Saleh and El-Lakany	Flavonoids	*Casuarina* (Casuarinaceae)
	Von Rudloff and Nyland	Terpenes	*Pinus*
	Weimarck *et al.*	Phenolics	*Heracleum* (Umbelliferae)
1980	Ali an Qaiser	Flavonoids	*Acacia* (Leguminosae)
	Crawford and Smith	Flavonoids	*Coreopsis*
	Crawford *et al.*	Flavonoids	*Coreopsis*
	Heckard *et al.*	Amino acids	*Castilleja* (Scrophulariaceae)
	Irving	Terpenes	*Hedeoma*
	Luteyn *et al.*	Flavonoids	*Cavendishia* (Ericaceae)
	Parfiti	Flavonoids	Cactaceae
	Rodman	Glucosinolates	*Cakile* (Cruciferae)
1980	Scheffer *et al.*	Terpenes	*Cupressocyparis* (Cupressaceae)
	Wollenweber and Dietz	Flavonoids	*Pityrogramma* (fern)
	Zavarin *et al.*	Terpenoids	*Pinus*
1981	Rodriguez *et al.*	Sesquiterpenes	*Parthenium* (Compositae)
	Whiffen	Terpenoids, flavonoids	*Eucalyptus*
1982	Adams	Terpenoids	*Juniperus*
	Eckenwalder	Flavonoids	*Populus*
	Huizing *et al.*	Alkaloids	*Symphytum* (Boraginaceae)
	Snajberk *et al.*	Terpenoids	*Pinus*

III. Micromolecules versus macromolecules

Both micromolecules and macromolecules have been employed in the detection, or documentation, of hybridization in plants. The former have been more widely used, largely because of their simplicity and ease of isolation and detection. Macromolecules, especially isoenzymes, have also been much used in the detection of hybrids or gene flow among animal groups, but less so in plants. This is perhaps due to the large, highly uniform tissues or cell types found in the organs of animals (e.g. liver tissue and blood) and the fact that less environmental variability has been encountered in the analysis of such tissue systems.

In any case, plant systematists have not taken to macromolecular techniques with as much enthusiasm as animal workers, presumably because their data have not been as readily obtained with as much repeatability. Nevertheless, as the techniques of slab-gel electrophoresis catch on among systematists generally, and as the variables become more readily understood, a plethora of papers may be expected, both for the documentation of hybridization and the detection of gene flow between populations following hybridization.

In the section that follows micromolecular data or secondary compounds will be discussed. Macromolecular data bearing upon hybridization will be discussed in chapters 17 and 18. More emphasis is placed on micromolecular data, simply because most of the published work has involved such compounds.

IV. Micromolecular documentation

In principle, any of the numerous micromolecules discussed in this text can be used to detect hybridization; flavonoids, however, have been used more extensively than any other class of compounds (Table 13.1). No doubt this is due, in large measure, to their wide distribution among plant groups generally and because of their relative ease of structural elucidations. This contrasts with alkaloids and terpenes, for example, which are more restricted in their phyletic distributions and are somewhat more difficult to work with.

Anderson (1949) pointed out that most natural hybridization is of the sympatric (or parapatric) type in which the species involved occupy the same general regional area but are themselves each adapted to different ecosystems or habitats within that area. Their immediate proximity may permit hybridization but, in that the parental taxa are generally better adapted to their particular habitats, F_1 and/or backcrosses are hard pressed to obtain footholds, except in narrow zones of an ephemeral nature between the habitats concerned which are maintained by periodic disturbance. Because of these experimental (and theoretical) considerations Anderson thought that sympatric interspecific hybridization was not an important factor in evolutionary processes. Instead, he believed that allopatric hybridization was more important in evolutionary processes largely because the absence of strong reproductive barriers and relatively broad zones of regional contact or peripheral overlap would permit the ready influx of genes from one taxon into another.

In the section that follows both sympatric and allopatric introgression will be discussed, the former using *Baptisia* (Leguminosae) as an example and the latter using *Juniperus* (Cupressaceae).

V. Documentation of sympatric hybridization

As noted above, Turner and Alston (1959), using one-dimensional paper chromatography, first demonstrated the utility of flavonoids in the resolution of complex hybridization in the genus *Baptisia*. In this work they demonstrated the occurrence of three species-specific flavonoids in each of the two species, *B. leucophaea* and *B. sphaerocarpa*. All six of the compounds occurred together in the putative hybrids. In short, the hybrids tended to produce, in a single plant, the compounds characteristic of the parental taxa. This so-called 'complementation' effect in the hybrids permitted relatively easy detection of F_1 hybrids, at least as contrasted with quantitative morphological features, for the latter are most often controlled by multiple gene systems with equal and additive effects on the character expression which tend to 'average-out' in the hybrids. Stated simply, if one is dealing with a distinguishing morphological feature, say petiole length, which varies from 3 to 5 mm in one species and from 6 to 8 mm in another, the petiole is likely to vary from 4 to 7 mm in the hybrid. Detection of such F_1 hybrids on this character alone would prove difficult. If, however, the two species differ by three (3) flavonoids each, the putative hybrid should yield an additive profile $3 + 3 = 6$. Thus F_1 hybrids can be readily detected, providing that the hybridizing species possess chemical differences and that the absence or presence of the latter is a stabilized feature of the taxa under consideration.

Subsequently, Alston and Turner (1962, 1963) used two-dimensional chromatography to document complex hybridization among both two-way and three-way species-hybrids in *Baptisia*.

When the species concerned are quite different in their chromatographic profiles, F_1 hybrids can be readily detected; indeed, with more assurance than through the use of morphological features. Perhaps the best published example (Alston and Hempel, 1964) is that involving hybridization between *Baptisia leucantha* and *B. sphaerocarpa*. Chromatographic documentation (Fig. 13.1) is readily accomplished in these two species since they each differ by a large array of flavonoids which are mostly characteristic of the species throughout most of their ranges. As noted by Alston (1965) the applicability of flavonoids for this purpose rests largely on the fact that these compounds, qualitatively speaking, are genetically controlled, Mendelian dominance for the appearance of a specific compound being the rule. This being so, *in vitro* 'hybrids' may be produced in the laboratory simply by extracting flavonoids from both of the species concerned (either separately and mixing the aliquots, or by mixing crushed leaves prior to extraction). However, the chromatographic spots revealed within an *in vitro* hybrid need not match the actual spots found in the *in vivo* or natural hybrid (or even a synthetic hybrid), since anomalous compounds may appear in the leaves of living hybrids that do not normally appear in the leaves of the species themselves (although these compounds might appear elsewhere; for example, in cotyledons or floral parts).

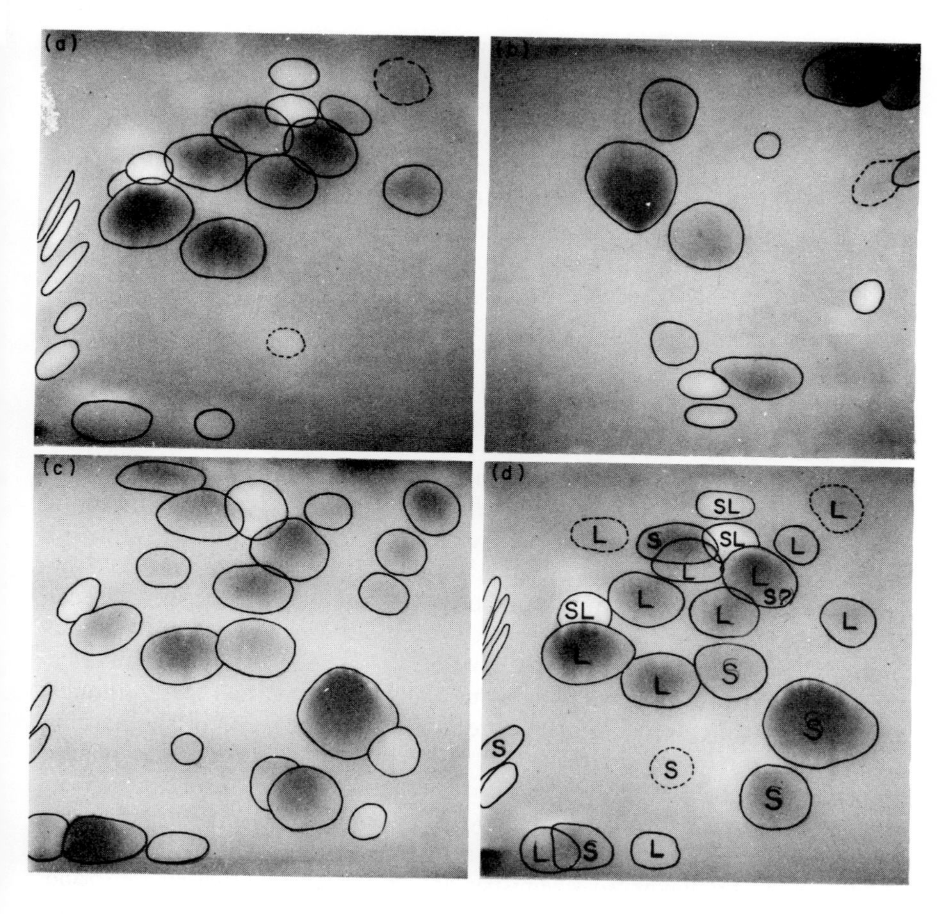

Fig. 13.1 Chromatographic patterns of leaf extracts from *Baptisia leucantha* (a) and *B. sphaerocarpa* (b). (c) shows an *in vitro* 'hybrid profile' resulting from a mixture of extracts from the two species. (d) shows the chromatographic profile obtained from a natural hybrid of the two species. Note the 'novel' compounds occur in the natural or *in vivo* hybrid profile. (From Alston and Hempel, 1964).

The appearance of these so-called 'new' compounds has been neatly documented by Alston and Simmons (1962) in their study of natural hybrids in *Baptisia sphaerocarpa* (=*B. viridis*) × *B. leucantha.* In 22 putative F_1 hybrids between these species, they invariably found four distinctive compounds which were not found in the leaves of the parental types at the sites concerned. This is shown in Fig. 13.1, which contrasts the so-called *in vitro* 'hybrids' with the *in vivo* or actual living hybrids. Other workers have also reported novel compounds in interspecific hybrids of yet other plants; for example, in *Dicentra* (Fahselt and Ownbey, 1968), in *Zinnia* (Torres and Levin, 1964), in *Encelia* (Bjeldanes and Geissman, 1969) and in *Vernonia* (Jones, 1972).

Species-specific chemical markers, assuming their reliability, are especially useful in resolving more complex hybridization. For example, in Texas, three, and rarely four, species of *Baptisia* may occur together, or within close proximity, at a given site. Morphologically, the species are quite different and it might be assumed that F_1 hybrids and their various backcrosses could be easily recognized, but because of the tendency of the distinguishing characters to 'blend' among the hybrids (as a result of polygenic inheritance and variability among the individuals), putative F_1, F_2 and/or backcross plants are difficult to detect. Thus, in a situation involving three species of *Baptisia*, Alston and Turner (1962) reckoned that the morphological data suggested the occurrence of three kinds of F_1 hybrids: *B. leucophaea* × *B. sphaerocarpa*, *B. leucophaea* × *B. leucantha* and *B. leucantha* × *B. sphaerocarpa*, including various F_2 individuals and backcrosses. However, when analysed chemically, using the same individuals, as exemplified in Fig. 13.2, what appeared to be an exceedingly complex hybrid swarm on morphological grounds proved to be relatively simple when analysed by flavonoid comparisons. Similar methods were used by Alston and Turner (1963a,b) to document, at a single site where four species grew within close proximity, the existence of all six possible F_1 hybrids. It is doubtful that morphological comparisons alone would have permitted this feat, although such comparisons were necessary to establish at least some of the hybrids since two of the species concerned (*B. leucophaea* and *B. nuttalliana*) possess essentially identical flavonoid profiles so that at least one of the hybrid combinations must be detected solely by morphological considerations.

Other natural hybrid combinations in *Baptisia* which have proved difficult to document by flavonoid chemistry are *B. tinctoria* × *B. perfoliata*, and *B. alba* × *B. leucantha. In theory*, intensive population studies of the chemistry of these species-pairs might make detection of such hybrids possible. However, under natural conditions, all individual hybrids are not alike chemically and, when one considers the fine-line criteria by which species must be implicated in the origin of a particular hybrid (for example, one or a few

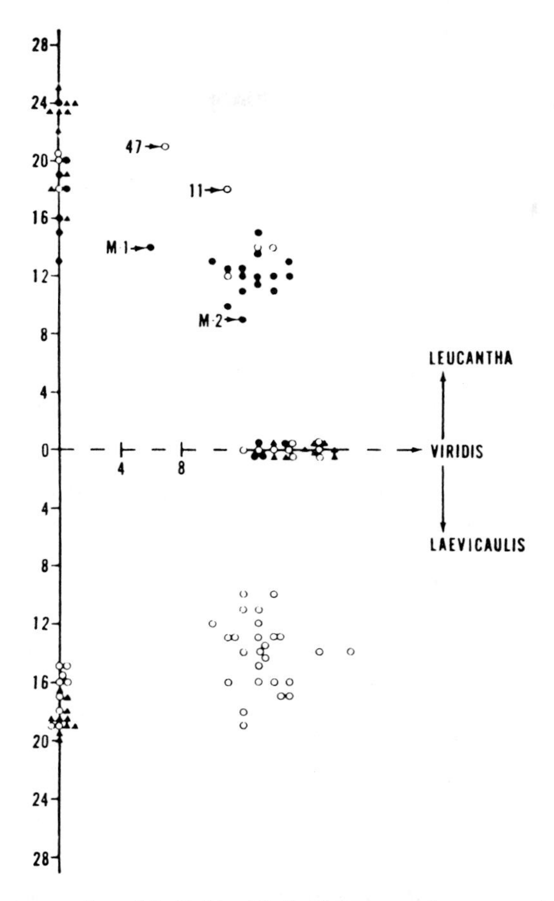

Fig. 13.2 A three-way plot of individual hybrid types and pure species. ○, indicate plants from tri-hybrid population; ▲, miscellaneous supplementary plants from pure populations; ●, additional (supplementary) *B. leucantha* × *B. viridis* hybrids. Points along the X-axis represent the number of compounds recognized of *B. viridis*; points along the Y-axis represent (above) the number of compounds recognized of *B. leucantha* and (below) the number of compounds recognized of *B. laevicaulis*. Hybrids fall at some angle between X and Y axes. (From Alston and Turner, 1962.)

minor components, or quantitative variation of a particular component), it would be rash to guarantee one's ability to do so. In short, hybrids involving any two species with very different chromatographic patterns are relatively easy to identify; but among two species with quite similar patterns, it may or may not be easy to identify a particular hybrid. As noted by Alston (1965), chemical data are neither inferior nor superior to morphology. Rather each case must be judged on its particular set of attributes, the important point being that in combination the total effectiveness is greatly increased. This conclusion has also been voiced by Levin (1968a) in his work with *Liatris*, discussed below.

It seems important to note here that flavonoid patterns need not be obtained from living plants in the field. Careful collection and drying of such material will make the detection and identification of flavonoids relatively simple. Also, leaf material from herbarium sheets several hundred years old may be readily used to document hybrids and this has been routinely called upon to detect putative hybrids of *Baptisia* spp. in the herbarium itself. One of the better examples is that of *B. serenae* a very old specific name applied to the relatively rare hybrid *B. alba* × *B. tinctoria*. This 'species' was first described a hundred years ago or more, but combined chromatographic–morphological studies, both in the field and from sheets in the herbarium, reveal the plant to be an F_1 hybrid.

A more provocative question is that of whether flavonoid data are superior to other data in establishing introgression (the incorporation of genes from one taxon into another via hybridization and backcrossing). The following quote sums up one worker's feelings on the topic (Alston, 1965):

> In *Baptisia* when we have been specifically interested in the question of introgression, we have not obtained strong evidence supporting introgression. However, individual examples of what I judge to be introgression occur. For example, Mr. Karl Baetcke and I once collected a typical *Baptisia leucantha*, growing alone in a field in McCurtain County, Oklahoma. Four other species of *Baptisia* were collected in that county on that trip. This species is quite distinctive, with white flowers, and its hybrids always have pale-yellow flowers. The plant in question, having pure white flowers, was chromatographically a typical *B. leucantha* plus the addition of apigenin-7-monoglycoside, a compound found in at least three of the sympatric species, but never found in *B. leucantha*. I have given much thought to this situation and can only conclude that the apigenin-7-monoglycoside represents introgression from an unknown species.

To ascertain the efficacy of flavonoid chemistry in resolving sympatric introgression in *Baptisia*, Baetcke and Alston (1968) subsequently made a detailed study of an hybrid swarm involving *B. sphaerocarpa* and *B. leucophaea*. In an area measuring 240 ft × 300 ft, which contained about 1100 plants, they studied each plant both chemically and morphologically. The population was found to consist of 470 *B. leucophaea*, 561 *B. sphaerocarpa*, 83 F_1 hybrids (including possible F_2 hybrids), 37 backcross types to *B. leucophaea* and 19 backcross types to *B. sphaerocarpa*. The percentage of hybrid types in the population was calculated to be about 12%. They concluded that the results of detailed knowledge of the chemical markers indicated that there were certain situations in which, on theoretical grounds, one enzyme might introduce more than one chemical difference. However, segregation of such characters was observed among the backcross types in such a way as to allow the conclusion that it was legitimate to count each species-specific compound as a single character. Their data strongly suggested that introgression of the sympatric type was occurring at the locality concerned. Whether or not gene flow might be occurring beyond the immediate limits of the hybrid swarm was not answered.

Levin (1968a) studied complex hybridization at a single site where three quite different species of *Liatris* (Compositae) grew in close proximity. Using both chemical and morphological attributes, he also noted difficulty in correlating data from these two character-sets for occasional putative hybrids between *L. aspera* × *L. spicata*, although approximately 70% of the putative hybrids exhibited complementary chromatographic patterns. He notes, however, that some hybrids 'although manifestly intermediate in morphology, possessed only the compounds of a single species'. He also noted that most of the *putative* backcrosses contained only compounds of a single species, that of the recurrent parent. As in the Baetcke and Alston (1968) study, a few plants reportedly displayed complementary chromatographic patterns, yet did not show morphological evidence of hybridity. Levin noted similar findings among the natural hybrids of *L. spicata* × *L. cylindracea*, but noted that in advanced backcross generations, there was no evidence of recombination of the species-specific compounds: rather the compounds of the two species appeared 'to be inherited as discrete blocks, a block either being present or absent'. He concludes that 'most of the chromatographic characters are linked by a strong coherence mechanism'. This seems reasonable, since the same phenomenon holds for morphological characters and there are no *a priori* reasons that coherence might not also apply to suites of chemical characters. Indeed, it is just such coherence that permits the detection of hybridity in natural populations, as noted by Anderson (1949) and as shown by studies on allopatric introgression in *Juniperus*, discussed below.

Hess (1971), in controlled crosses between the scrophulariod species, *Torenia fournieri*, which contained several carotenoids but no flavonoids, and *T. baillonii*, which contained several flavonoids but no carotenoids, demonstrated that the F_1 hybrid showed only the flavonoid pattern. This was explained by the assumption that the carotenoid precursors in the hybrid (e.g. malonyl-CoA) were drawn off from carotenoid to flavonoid biosynthetic pathways. Similar switches in pathway synthesis might account for 'block' expression of certain suites of compounds in yet other species and need not reflect character-coherence *per se*.

It is likely that much of the difficulty in finding absolute correlation of morphological and chemical characters in natural hybrids is due to the disproportionate genetic variability found among individuals in natural populations of a given species. Thus Murray *et al.* (1972) noted that, in synthetic crosses between *Mentha spicata* (an outcrosser) and *M. aquatica* (an inbreeder), the considerable variability in morphologic appearance of the F_1 hybrids was largely due to the heterozygosity of the *M. spicata* parent, since the self-pollinated progeny of *M. aquatica* were essentially alike. That is, among the several synthetically produced F_1 hybrids between these two species, some were readily identified as morphological hybrids, 'whereas others resemble *M. aquatica* and might be confused with it'. Similar statements for similar reasons could be made for the quantitative variation of the 14 or so terpenes which characterize the two species. Undoubtedly, similar reports of equivocal complementation among natural interspecific hybrids (e.g. King *et al.*, 1975, in *Rhododendron*; Wollenweber, 1975, in *Populus*) can be explained by genetic variability controlling the production of chemical constituents of the parent populations contributing to the individual hybrids.

Most chemical studies of interspecific hybrids have involved flavonoids (Table 13.1), but F_1 complementation, also holds for most other compounds (e.g. Baker and Baker, 1976, using amino acids), with occasional exceptions. Conner and Purdie (1976), in synthetic crosses among species of *Cortaderia* (Gramineae), note that triterpene methyl ethers are mostly expressed as dominants in the F_1. However, the hybrid *C. richardii* $\times$ *C. toetoe* proved an exception in that synthesis of α- and β-amyrin methyl ethers was suppressed in F_1 and F_2 individuals, but was restored in the backcross $F_1 \times$ *C. toetoe*. This backcross generation proved heterozygous for genes producing amyrin methyl ethers and, on selfing, segregated in simple Mendelian ratio.

Complementation of chemical compounds in F_1 hybrids is generally the rule, as noted above, but it also appears likely that certain classes of secondary compounds might be more prone to plasticity than others. For example, Ornduff *et al.* (1973), working with synthetic interspecific hybrids in *Lasthenia* noted that inheritance of anthochlors, kaempferol and some patuletin glycosides were additive, whereas luteolin and quercetin glycosides were not always additive; in addition, in some progenies, quercetin glycosides were produced that did not occur in the plants.

Most secondary compounds are probably under relative severe genetic control (cf. chapter 9). That this genetic expression need not be expressed in hybrids between disparate species should not be surprising. As noted by Lokki *et al.* (1973), studying the genetic control of monoterpenes within *Chrysanthemum vulgare*, 'Most probably [interspecific crosses] may disturb the inbuilt balance of gene interactions and bring about unnecessary non-additive variation into the control of monoterpene biosynthesis'. These authors showed, starting with eight parent plants of this species and producing 31 individual crosses, that the progeny, containing 513 individuals contained at least three chemotypes among the offspring. By creation of a model to account for terpenoid variations among the crossings and selfings produced, they concluded that at least seven of the loci controlling the biosynthesis of camphor and thujone must be assumed to be polymorphic. Such polymorphism, combined with the imbalance imposed by the interplay of disparate genomes, undoubtedly account for the occasional non-complementation of chromatographic profiles among interspecific hybrids generally, as suggested by Fahselt and Ownbey (1968) and yet other workers.

That changes in secondary components following hybridization need not involve genetic masking on complementation is implicit in the results obtained from hybrids in the fruit fly *Drosophila* (Wright and Shaw, 1970). An examination of the patterns of enzyme expression in the hybrids showed that, in three hybrids, absence of an enzyme from a specific tissue was dominant to its presence, suggesting that pattern differences in the parental species were due to diffusable factors that 'affected expression of the relevant structural genes rather than to differences in the genes themselves or in *cis*-acting or regulatory sites'.

VI. Documentation of allopatric introgression

Nearly all of the chemical studies to date, which purport to document introgression for this or that hybrid situation among plants, have been with sympatric or parapatric species (see Heiser, 1973, for review; and e.g. Lee, 1975, in *Typha*; Olivieri and Jain, 1977, in *Helianthus*) but, as noted by Anderson (1949), because of the integrated, highly coherent, genomes of sympatric species, such introgression is not likely to cause any considerable influx of genes from one species into another. In contrast, allopatric hybridizing species because of their more weakly cohering genomes and their more extensive regional contact and, presumably, more poorly developed reproductive barriers, provide an ideal setting for extensive regional introgression and consequent speciational events in the taxa concerned.

Anderson (1949) in an epilogue to his brilliantly conceived text, *Introgressive Hybridization*, made the following statement:

> How important is introgressive hybridization? I do not know. One point seems fairly certain: its importance is paradoxical. The more imperceptible introgression becomes, the greater is its biological significance. It may be of the greatest fundamental importance when by our present crude methods we can do no more than to demonstrate its existence. When, on the other hand, it leads to bizarre hybrid swarms, apparent even to the casual passerby, it may be of little general significance .... Only by the exact comparisons of populations can we demonstrate the phenomenon .... The wider spread of a few genes (if it exists) might well be imperceptible even from a study of population averages, but it would be of tremendous biological import .... Hence our paradox. Introgression is of the greater biological significance, the less is the impact apparent to casual inspection.

In spite of these reflections from the foremost proponent of introgressive hybridization, few well-documented studies have been forthcoming on allopatric introgression. In fact, until quite recently the best documented case in the literature for allopatric introgression is reportedly that involving *Juniperus virginiana* and *J. ashei* (Anderson, 1953; Davis and Heywood, 1963). However, in a number of detailed studies, the existence of F_1 hybrids or their immediate derivatives could not be detected, even at sites where large populations of both species grew intermixed, and in *no instance* could the existence of introgression be inferred from the data accumulated (von Rudloff *et al.*, 1967; Flake *et al.*, 1969, 1973; Adams and Turner, 1970; Flake and Turner, 1973; Adams, 1975). In short, what was taken to be a well-documented case study of allopatric introgression turned out to be a situation in which clinal intergradation in habital features over a broad region occurred such that, superficially, hybridization and introgression might be inferred.

In hindsight, it now seems rather reasonable to have viewed the case of introgression between *J. virginiana* and *J. ashei* with considerable doubt, for the two species are readily distinguished by a number of morphological features and are placed in different species groups (sections) of the genus, and the character used for such taxonomic segregation (cilia along the leaf margins) does not segregate in putative hybrid swarms. In fact, Barber and Jackson as early as 1957 suggested that the morphological variation found in *J. virginiana* was clinal, i.e. the species had formed or is in the process of forming regional races as a result of adaptation mechanism arising out of its own gene pool, this being unrelated to the possible influx of genes from the largely allopatric *J. ashei*. Subsequent studies have substantiated fully these suppositions (Flake and Turner, 1973).

Fortunately, however, there has been a recent, carefully conceived, population study of *Juniperus virginiana* and *J. scopulorum* in the Missouri River Basin of the north central United States by van Haverbeke (1968a) which appears to be a situation involving allopatric introgression of the type Anderson felt to be so important in evolutionary processes. The study seems to be unusually well documented. van Haverbeke made very accurate records of the population sites, including precise data on ten individually marked trees which were selected for study at each site. These included photographs and detailed field notes. In short, the *J. virginiana–J. scopulorum* complex provided an ideal case study of allopatric introgression using the chemo-numerical methods that proved so effective in disproving the occurrence of this phenomenon in the *J. ashei–J. virginiana* 'complex'.

Initial investigation of the terpenes of *Juniperus scopulorum*, unlike that of *J. ashei*, showed that its volatile components were essentially those of *J. virginiana*, differing only in their quantitative expression (von Rudloff,

1975a). Subsequent population analysis showed that regional intergradation of these chemical characters occurred across the Missouri River Basin, much as found by Fassett (1944) and van Haverbeke (1968a) for morphological features.

Three models might be proposed to account for the variation found in this region:

1. *Ancestral gene pool. Juniperus scopulorum* and *J. virginiana* may have arisen from ancestral populations largely endemic to the Missouri River Basin. Subsequent evolutionary divergence to the west and east respectively might have occurred, leaving a residuum of genes common to each in the area concerned.
2. *Allopatric introgression.* The variability is due to extensive gene flow from *J. scopulorum* into *J. virginiana* as a result of hybridization and backcrossing in peripheral regions of contract and areas of sympatry.
3. *Migratory tailings.* The River Basin was an ancestral migratory route through which *J. scopulorum*-like populations passed on their way to becoming what's now known in the eastern United States as *J. virginiana.* In van Haverbeke's words (1968b) 'Thus, rather than being considered as an introgressive series, this juniper population [those of the Missouri River Basin] can alternatively be interpreted as a divergent evolutionary series which has not yet completely separated'.

It should be emphasized that in the investigation by van Haverbeke about 50 morphological characters were selected for measurement and numerical analysis. They were obtained from some 700 trees from 72 sites scattered throughout the River Basin area. In spite of this excellently conceived, carefully documented, laborious study, the investigator was unable to decide, unequivocally, between models 2 and 3; in fact, he believed that his data best fit the migratory tailings model. (Model 1 was not tested, presumably because of its implausibility, considering the recent biogeographic history of the Basin region; i.e. Pleistocene glaciation.)

The studies by Flake *et al.* (1978), in which terpenoid components were detected using combined gas chromatography (GLC) and mass spectroscopy, strongly suggested that allopatric introgression over a large region of the Missouri River Basin was occurring. They submitted their data to sophisticated, rather laborious, cluster analyses (Flake and Turner, 1968) and their results, in part, are illustrated in Fig. 13.3. The populations were characterized by 43 volatile GLC peaks averaged for the individuals in each stand sampled. A similarity measure derived from a Euclidean metric of these 43 characters formed the basis of the cluster analysis.

The general pattern of Fig. 13.4 indicates that the transect of populations consists largely of two main groups with some smaller subdivisions within each. One group is predominantly within the recognized distribution range of *J. scopulorum* and the other within the *J. virginiana* range. In addition, there are two significant clusters of populations at the highest level of similarity. One of these clusters (B,C,D) is well within the *J. virginiana* range of the transect and the other (I,J,K) within the *J. scopulorum* range. The pooled data from within each of these two latter groups were selected as representative of the *J. virginiana* and *J. scopulorum* parental types for subsequent discriminant analysis of the data.

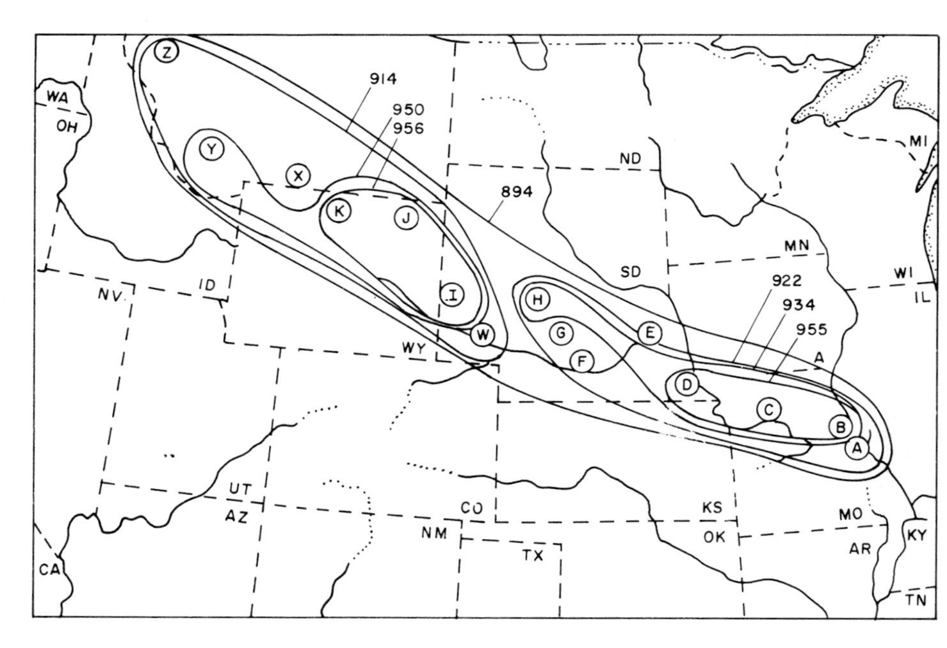

Fig. 13.3 Clustering results using terpenoid characters from 15 populations of *Juniperus scopulorum–J. virginiana* taken across approximately 1500 miles from St. Louis, Missouri (A) to Kalispel, Montana (Z) through a region of allopatric contact (population I, W). Introgression and putative gene flow occurs predominantly across the Missouri River Basin (population H through A) as shown in Fig. 13.4. Additional discussion in text. (From Flake *et al.*, 1978.)

Selected histograms of the number of individuals versus a hybrid scale derived from the data are shown in Fig. 13.4. The arrow below the hybrid scale axis in these figures indicate the values of the hybrid scale means for the histograms of the two parental groups of populations (B,C,D) and (I,J,K). The dashed lines extending above each histogram indicate the mean hybrid scale value for the individuals sampled in the corresponding population. A

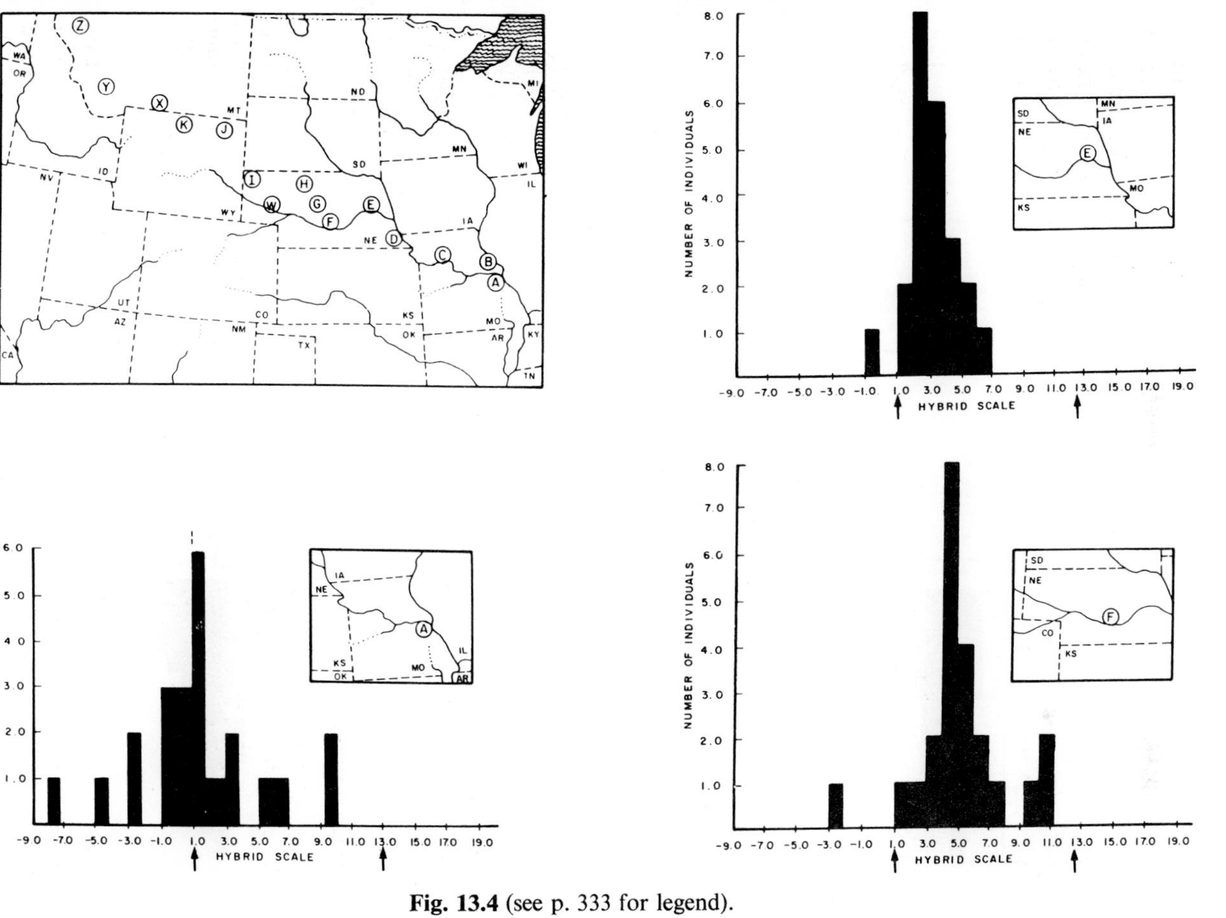

Fig. 13.4 (see p. 333 for legend).

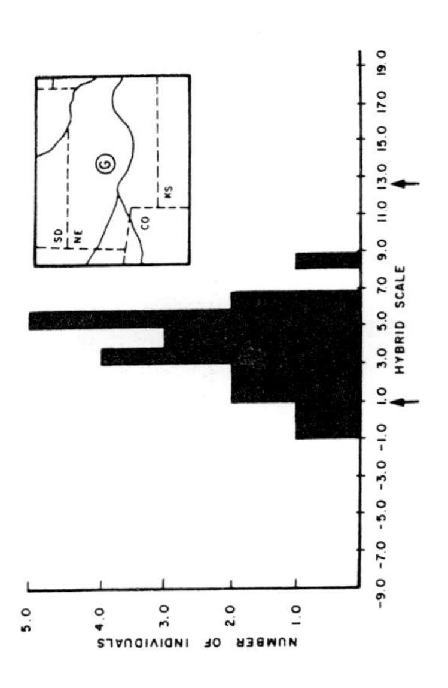
NUMBER OF INDIVIDUALS
HYBRID SCALE
SD
NE
CO
KS
G

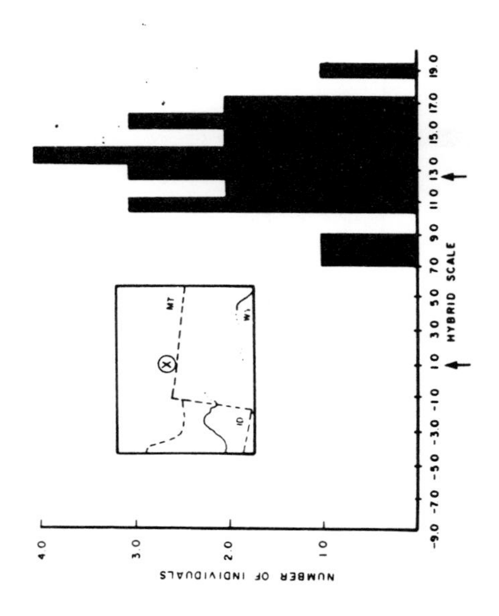
NUMBER OF INDIVIDUALS
HYBRID SCALE
MT
ID
X

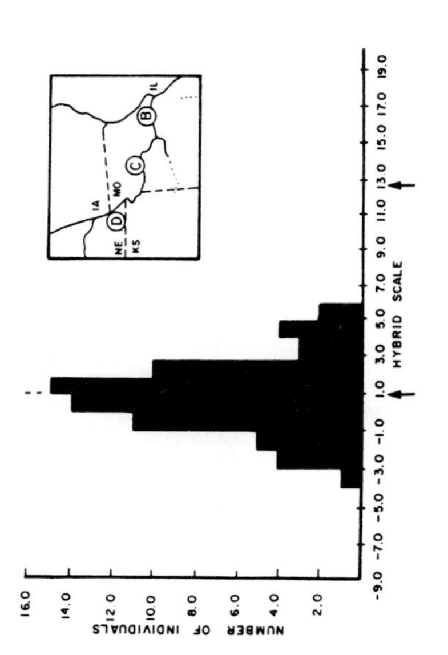
NUMBER OF INDIVIDUALS
HYBRID SCALE
IA
MO
NE
KS
IL
B
C
D

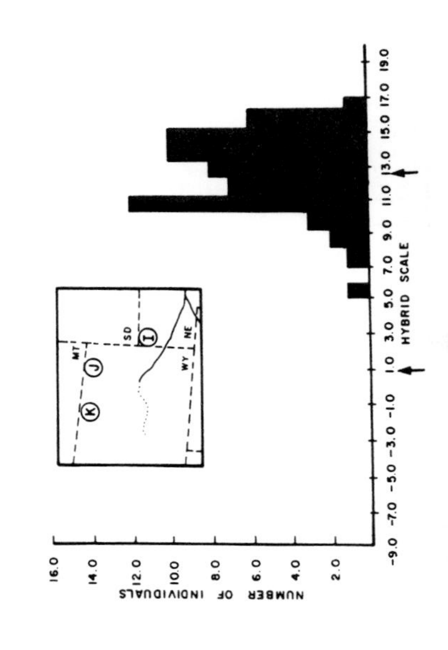
NUMBER OF INDIVIDUALS
HYBRID SCALE
MT
SD
WY
NE
I
J
K

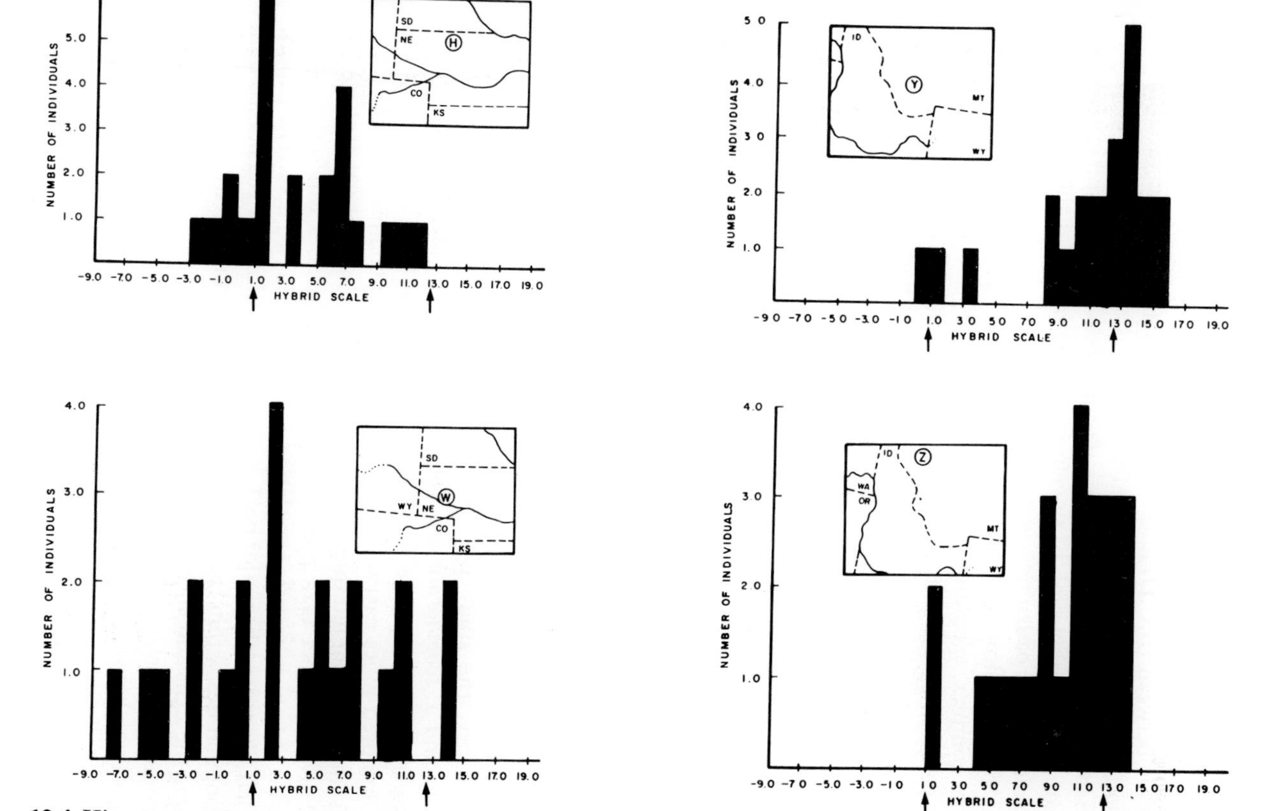

Fig. 13.4 Histograms showing population variation among terpenoids as an indicator of introgression between *Juniperus scopulorum* and *J. virginiana*. Sites sampled are those shown in Fig. 13.3. Twenty trees were sampled at each site. In the calculation of a 'representative chemical profile' for *J. scopulorum*, populations I,J,K were pooled (i.e. statistical data for 60 plants from three sites were used); for *J. virginiana* this was accomplished with populations B,C,D. Additional explanation in text. (From Flake *et al.*, 1978.)

shift in the population hybrid scale means is visually apparent when comparing histograms for the individual populations along the transect. Significant broadening of the histograms can also be noted for some of the populations between (B,C,D) and (I,J,K), including a dramatic variation in population W which is in the zone of contact of the two species. This latter variation in pattern is clearly indicative of transgressive segregation in a hybrid population. The combination of the change in the means of the histograms, accompanied by broadening, is indicative of a hybrid population (W) followed by gene flow via backcrossing, from *J. scopulorum* into the *J. virginiana* complex extending at least to population F, with some effect possibly detected by the shift in the mean histogram of E.

The complications in the histogram and clustering patterns of the populations at the northwest end of the transect are possibly caused by the influence of hybridization of *J. scopulorum* with a third species (*J. horizontalis*) in that region. Hybrid populations of these species near the locations of populations X and Y have been reported by von Rudloff (1975b). The study of Flake *et al.* (1978) did not include the extensive sampling of *J. horizontalis* populations which would be required to verify hybridization or a pattern of gene flow in this instance. In contrast, there is no obvious explanation for the apparent variation in the histogram of population A at the southern extreme of the transect and the relative dissimilarity in its clustering pattern with neighbouring populations B,C and D. Flake and Turner (1973) suggest the existence of an Ozark race in *J. virginiana* extending from northern Arkansas up near the southern tip of the present transect, which could possibly account for the somewhat aberrant nature of population A.

In summary, the study by Flake *et al.* (1978) indicates that *Juniperus virginiana* and *J. scopulorum* are not completely isolated. That is, there is evidence of significant gene flow from *J. scopulorum* into *J. virginiana* for several hundred miles eastward into the *J. virginiana* complex. Under these circumstances one should question the opposite view: namely, that there are two relatively distinct taxa present. Figure 13.5 addresses this question by comparing the characteristic correlation patterns of the two taxa. The pattern on the left illustrates the major peak-to-peak correlations for the pooled data of the populations within the recognized *J. virginiana* range (A,B,C,D,E,F,G) indicating the signs and magnitudes of the correlation coefficients. The heavier lines represent correlation values exceeding 95% and the lighter weight lines are for values between 80 and 95%. Lines extending to the right are for positive correlations and those to the left for negative. The pattern in the centre of the figure exhibits the same correlation information for the populations in the *J. scopulorum* range (I,J,K,Y,Z). Similarly, the correlations pattern for the pooled data of all 15 populations is shown on the right of the figure. It is clear from these diagrams that the peak-to-peak correlation pattern for *J. virginiana* is quite distinct from that of *J. scopulorum*. Furthermore, when all the data are pooled, the strong correlations that appeared separately within the two groups of populations almost completely disappear. These workers present this as clear evidence that, although the two groups of populations are not completely isolated, they are still sufficiently distinct, in spite of the apparent gene flow, to be two separate taxa, and not simply geographic races of the same species, as otherwise might be suggested.

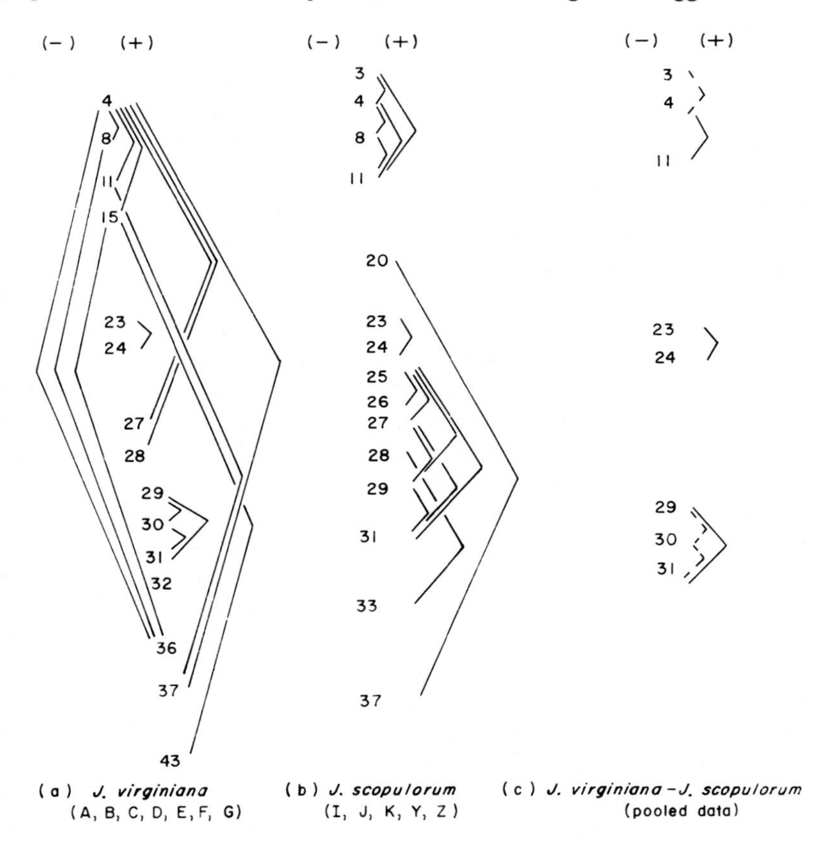

Fig. 13.5 Comparison of characteristic correlation patterns of *Juniperus*. The pattern in (a) illustrates the major peak-to-peak correlations for the pooled data of the *J. virginiana* populations (A,B,C,D,E,F,G) indicating the signs and magnitudes of the correlation coefficients. The heavier lines represent correlation values exceeding 95% and lighter weight lines for values between 80 and 95%. Lines extending to the right are for positive correlations and those to the left for negative. (b) exhibits the same correlation information for the *J. scopulorum* populations (I,J,K,Y,Z). Similarly, correlation pattern for the pooled data of 15 populations is shown in (c). Additional discussion is in the text. (From Flake *et al.*, 1978).

Habeck and Weaver (1969) also used gas chromatography in the analysis of putative hybrid populations of *Picea engelmannii* and *P. glauca* in western Montana. Their study differs from that of Ogilive and von Rudloff (1968) in that they compared their regional hybrid sites with relatively remote populations of the putative parents (*P. glauca* from Manitoba, Canada and *P. engelmannii* from Wyoming). Unfortunately, the individuals sampled at any one site (two to eight for the parental taxa) and the number of characters (seven of ten monoterpenes obtained from resin blisters on the bark) were too few to make their study especially convincing as regards introgression. Nevertheless, they were able to conclude that 'throughout most of western Montana, it is difficult to designate any spruce as typical *P. engelmannii*', as far as their terpene chemistry is concerned. This suggests regional introgression but is not convincing since character coherence over a broad region for the taxa concerned, and statistical treatment of the population data, were not sufficient to show *in situ* perturbation of such characters near the region of contact, followed by perceptible gene-flow away from the highest region. For *Picea engelmannii* this might be difficult to accomplish since the species reportedly hybridizes with yet other taxa in the more northwestern portions of its range (e.g. with *P. sitchensis*, as noted by Hanover and Wilkinson, 1970, using flavonoid data). However, gene flow in the direction of *P. glauca* ought to be readily documented. Unfortunately, the only detailed chemical study of populations in the more eastern regions (La Roi and Dugle, 1968) used flavonoid profile patterns obtained from leaves. They examined 73 specimens of *P. engelmannii* and *P. glauca*, 16 of these from herbarium sheets. Along with morphological features, they applied various statistical analyses to their data concluding 'that most of the Engelmann spruce populations in the subalpine zones of Alberta, Canada, have demonstrable evidence of introgressive gene flow from white spruce'. Similarly, 'most white spruce populations of the Alberta Rockies and foothills show chemical and morphological evidence of introgressive gene flow from Engelmann spruce'. Which is not to say that regional introgression over a broad region such as found in *J. virginiana* occurs, rather the workers are clearly aware that *P. glauca* may have a very different population make-up in the more eastern regions and that these may not be affected by introgression from *P. engelmannii* to any considerable extent. Obviously, much additional work on populations will be needed before allopatric introgression can be said to have been established.

Regional intergradation due to introgression between yet other coniferous taxa has been reported by Hunt and von Rudloff (1979), the best documented being that involving *Abies lasiocarpa* (coastal Alpine Fir) and *A. bifolia* (Rocky Mountain Alpine Fir). The latter taxon has long been considered a synonym of the former, largely because these were known to intergrade in the region of allopatry. Their data, using GLC acquisition (18 identified and fully quantified terpenes), obtained from five or ten trees at each population site (summarized in Fig. 13.6), strongly suggested that this intergradation was due to introgression in the region concerned and that the species pair, as far as their terpenoid compounds were concerned, were as worthy of recognition as, for example, *Abies lasiocarpa* and *A. balsamea* which, according to these same authors (Hunt and von Rudloff, 1974), introgress in yet other regions of the Pacific Northwest.

As will be noted by the examples discussed above and as emphasized by Endler (1977) introgressive hybridization is exceedingly difficult to document. In fact, the latter author contends that it is essentially impossible to distinguish between primary intergradation (population divergence via gradual gene partitioning at regions of allopatry) and secondary intergradation (introgressive hybridization) because they result in the same types of geographic intergradation and may evolve from gradient clines in a similar fashion. Using detailed population data, similar to those of Flake *et al.* (1978), and adapting these to sophisticated computer analysis to detect population perturbations as one approaches regions of intergradation or contact, it should be possible to distinguish between these two phenomena.

All such studies, however, need to be oriented towards population samples. One cannot cluster individual trees with any meaning (Flake and Turner, 1968, 1973), nor do systematists. What the systematist attempts to do is work with individuals as if they were parts of a population, intuitively working from subjectively arrived at population statistics such as ranges, means and standard deviations and interpopulation correlations among and between these estimates (Flake and Turner, 1968). Of course, the systematist using but one or a few variable characters is not likely to be very sophisticated in his analyses, but using terpenoid data or yet other variable characters which may be readily quantified, and treatment of these via sophisticated algorithms and computer analyses, one can make very incisive explorations.

How many individuals must be examined at a given site and over what portion of the region, depends on the organism concerned, its intrinsic variation as determined by preliminary sampling, and the extent of geographical intergradation. It would appear from the relatively few studies made to date, that 20 individuals is perhaps sufficient to characterize a particular population, but intrapopulational variability precludes such small samples as nine or fewer individuals, at least if one hopes to obtain levels of confidence sufficient to impress most biometricians and scientists. von Rudloff (1975b) found that leaf samples from ten trees at a given site were sufficient to characterize populations of *Tsuga heterophylla*, but Flake and Turner (1973) note that ten individuals are perhaps inadequate, especially in widespread, variable, successional species such as *Juniperus virginiana*.

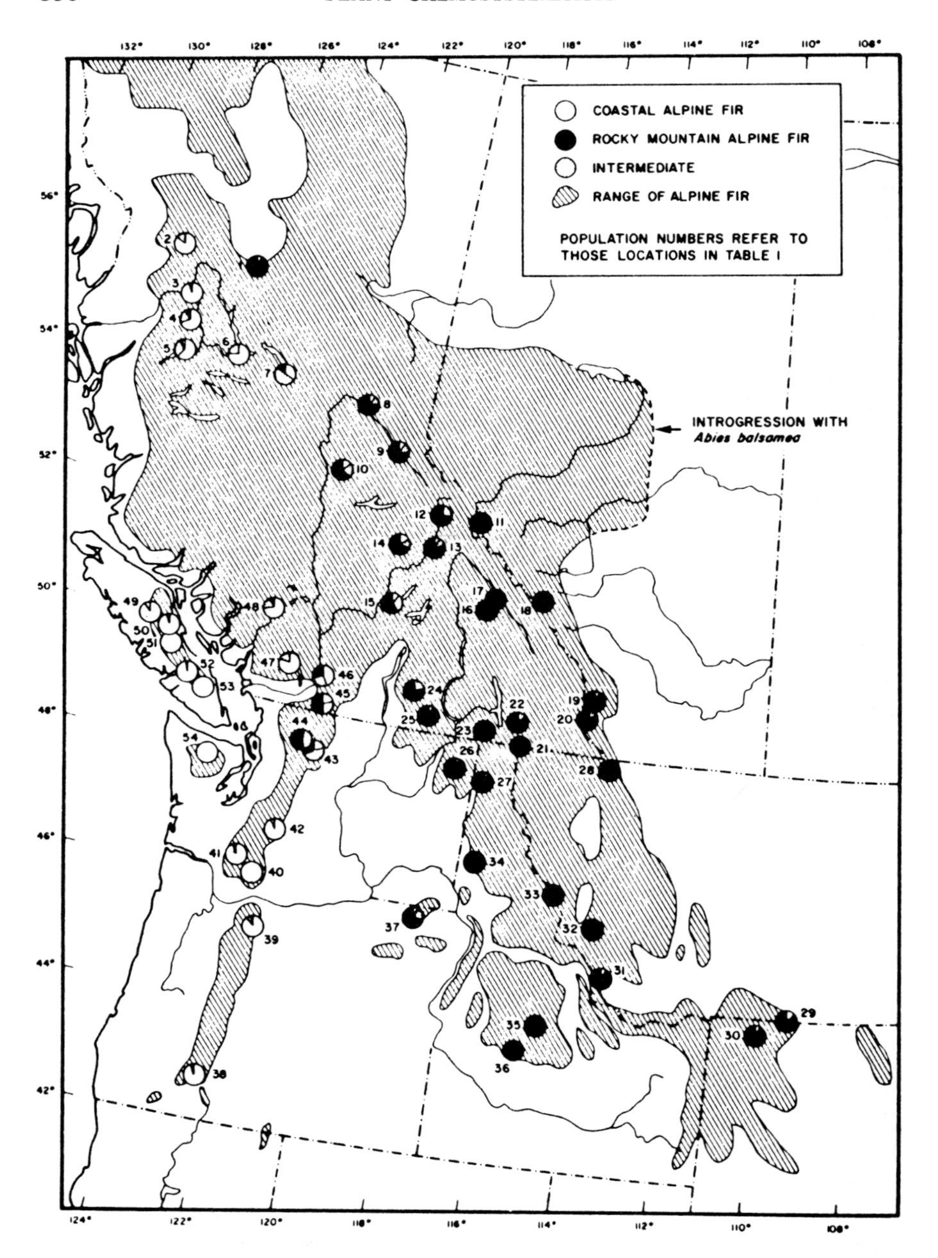

Fig. 13.6 Distribution of alpine fir showing Coastal, Rocky Mountain and Intermediate forms based on terpene hybridization index. (From Hunt and von Rudloff, 1979.)

VII. Polyploidy—its effect on chemical constituents

Some of the earliest studies of polyploidy and its effect on chemical constituents have been made by agronomists and pharmacists interested in the mere yield of a given compound or group of compounds. In general, it can be assumed that polyploidy will increase somewhat chemical yield, at least this has been found to be so for most agronomic or medicinal plants (Löve and Löve, 1957). But whether these yields reflect the usually more robust character of such polyploids, or is a reflection of increased biosynthetic production of these compounds per unit cell due to increased 'gene dosage' (i.e. equal and additive effects), is problematical.

There are a few studies which show that colchicine-induced tetraploids have an increased production of specific compounds. Thus, Kowatani *et al.* (1954) found increased pyrethrin production in synthetic autotetraploids of *Chrysanthemum cinerariaefolium* and similar results were also reported for other constituents in experimentally produced autoploids of *Lobelia inflata*. Nevertheless, this need not be so as shown by Grant and Sidhu (1967) who report that colchicine-induced tetraploids of *Lotus* gave a lower HCN reaction than their diploid counterparts, confirming the work of Roo (1963) who produced 34 colchicine-induced autoploids of *Trifolium repens* and noted that the plants varied considerably in their reaction to the HCN test, from non-reacting to strongly positive.

In reality, however, relatively little is known about the effects of infraspecific polyploidy on chemical constituents under natural or field conditions. The chemistry of amphiploids, or polyploidy through intergenomic hybridization, is relatively better known, such polyploids usually exhibiting a complementary chemical profile (e.g. Smith and Levin, 1963). Complementation would also be expected in the hybrids of chemically different, infraspecific polyploids (or segmental allopolyploids in the sense of Stebbins *et al.*, 1953). However, from mere field observations, it is difficult, if not impossible, to distinguish between this kind of ploidy and autoploidy. Technically one might define autopolyploidy as plants that self-pollinate with unreduced gametes, or at most as polyploids derived from closely adjacent members of a population at any one site. In natural randomly breeding plant populations, however, most autopolyploidy probably arise as a result of hybridization between quite heterogeneous individuals, and perhaps many from interpopulational hybridization.

Sporadic polyploids within an otherwise diploid population will probably have pretty much the same chemistry as the diploids, providing intrapopulational variability is not extreme. At least a few workers (Levin, 1968b, in autotraploid *Phlox pilosa*) have been unable to detect such polyploids (or indeed aneuploids) using conventional analytical procedures.

Williams and Murray (1972), however, report that eight natural autotetraploids of the grass *Briza media* accumulate orientin and iso-orientin. The two chromosome forms or 'races' of this species are reportedly quite similar morphologically, but are readily detected using flavonoids. It is interesting that chromatographic detection reportedly depends on *accumulation* of orientin and iso-orientin, for the autotetraploids produce all seven of the compounds of its putative diploid parent, plus three compounds (iso-orientin, orientin 4′-glucoside and iso-orientin 4′-glucoside) *not* detected (but presumably present in very low concentrations) in the diploids. The authors attribute this to disturbances in gene-controlled physiological processes resulting from the doubled genome. In fact, Murray and Williams (1973) subsequently produced synthetic triploids of *B. media* (diploid × tetraploid) as well as colchicine-induced polyploids. All these were found to have identical chromatographic patterns showing that the addition of a single genome was sufficient to produce the change in flavonoid synthesis. Addition of yet more duplicate genomes had no further effect. In a somewhat more sophisticated study of aneuploids within *B. media*, Murray and Williams (1976) thought it probable that the gene or genes responsible for the switch in flavonoid synthesis occurred on a particular acrocentric chromosome.

That intrapopulational enzymic variability for a given species is sufficient to account for such variability is implicit in the work of Brehm and Ownbey (1965) on *Tragopogon mirus* in which the proven allotetraploid (*T. mirus*) had two chemical colonies. In addition to additive chromatographic patterns, one race lacked luteolin 7-*O*-monoglucoside, whereas the other race synthesized the compound. The colonies were within 30 miles of each other. The authors interpreted these results (Brehm, 1969) as evidence that *T. mirus* in the two colonies were derived from distinct hybrids (i.e. differing genotypes contributing to each of the hybrid combinations). In short, that spontaneous interspecific alloploidy leading to the formation of *T. mirus* occurred twice within the area concerned, both within relatively recent time.

The suggestions that autoploidy might cause physiological imbalance of regulatory mechanisms is interesting in that at least a few workers have interpreted distinct flavonoid profiles in polyploids among closely related or seemingly near-identical diploid–polyploid races as due to complementation of 'ancestral' genomes. Thus, Melchert (1966) found that tetraploids of *Thelesperma simplicifolium* had the ability to make aurones and chalkones whereas the diploids could not. He therefore assumed that the tetraploid races did not develop from *extant* diploids but rather that both the diploids and tetraploids had a common origin in the past. That is, the phenolic chemistry of each evolved independently, in marked contrast to what the morphology suggested.

Similar suggestions on essentially the same grounds have been put forward for the origin of natural intraspecific polyploidy in the solanaceous genus *Chamasaracha* by Averett (1970).

Murray and Williams (1976), also induced polyploidy in seven additional species of *Briza* and the leaf flavonoids of the polyploids and their progenitors were compared. In six of these no significant differences were noted among the various paired cytotypes. In one, however, tricin 5-glucoside, a common grass constituent, was present only in the diploids.

Unfortunately, relatively few chemical studies, other than those involving flavonoids, have been made on *synthesized* autoploids to ascertain what might happen following genome doubling. Hefendehl (1977) reported that the oil composition of a colchicine-induced tetraploid strain of the diploid *Mentha longifolia* ($2n = 24$) did not differ greatly from the latter, ascribing any minor changes as being due to normal ontogenetic anomalies. On *a priori* grounds there is little reason to believe that synthetically induced autoploidy *per se* is likely to lead to the production of novel compounds, but quantitative shifts might be expected.

Autoploidy in natural plant populations, however, is a different matter, for it has been shown to occur quite widely, both within and among a large number of species, largely through the serendipitous combination of unreduced gametes. Considering the polymorphic nature of the enzyme systems of most outcrossing plant species, one could hardly expect natural autoploids to be especially uniform.

The work by Murray and Williams (1976), mentioned above, is especially useful in that it provides some insight into what might be anticipated in the way of chromatographic patterns among infraspecific polyploids. Clearly the work on *Briza media* shows that autoploidy, or perhaps, more succinctly, that gene dosage can, on occasion, significantly affect chromatographic profiles.

VIII. Conclusion

The use of flavonoids, terpenoids or nitrogenous constituents for documenting hybridization among natural plant populations is still in its infancy, but the results to date suggest that these compounds can be employed, particularly when the morphological data are equivocal, for identifying hybrids and in determining whether sympatric or allopatric introgression has occurred. The now classic chemical studies of hybridization in *Baptisia* (Leguminosae) and in *Juniperus* (Cupressaceae) have been considered here in detail. A range of other examples are also mentioned or tabulated. The effect of polyploidy on chemical constituents has also been discussed and a few cases (e.g. in

Briza) where changes have taken place on increasing ploidy levels have been described. In general, however, the effects of varying ploidy levels on secondary chemistry have been rather poorly documented and further investigations are highly desirable.

14

Comparative Biosynthetic and Metabolic Pathways

I. Introduction 343
II. Variation in biosynthetic pathways 345
III. Variation in conjugation and degradation 355
IV. Conclusion 360

I. Introduction

The systematic study of low molecular-weight metabolites in plants depends essentially on the screening of groups of related taxa for their presence or absence. Most compounds of interest are products of secondary metabolism, although some are closely related to primary metabolism and function (see chapters 5–8). In making such chemical comparisons between plants, it has to be borne in mind that a given compound is produced by the plant along a given pathway, being formed ultimately from carbon dioxide taken in from the atmosphere. Thus, comparisons that are being drawn may concern the presence or absence of a particular biosynthetic pathway, involving a range of intermediates and of enzymes catalysing the various steps along that pathway.

In practical terms, there are usually sufficient difficulties in determining whether a given compound, such as the alkaloid morphine, is present or absent from a given group of plants without worrying about whether it is actually formed along the same biosynthetic pathway in all the plants under study. Fortunately, there is sufficient evidence from comparative feeding experiments using labelled precursors to know that within flowering plants most secondary compounds are formed by the same route, whatever family the plant being examined belongs to. Significant exceptions do occur (e.g. with certain naphthoquinones) and in such cases it is the detection of the pathway by which the substance is elaborated that may be more rewarding or significant systematically than the simple determination of its presence or absence. Indeed, there may be variation in biosynthetic pathways leading to both secondary and primary metabolites. Differences in the latter pathways may not be obviously apparent, since primary products being essential to the life of the cell are produced universally in similar amounts in all organisms. An indirect, probably complex experimental approach will be needed to detect biosynthetic diversity among primary metabolites.

In the case of essential plant metabolites, the best known example of biosynthetic diversity is that of the protein amino acid lysine. This is formed *in vivo* by one of two mutually exclusive routes, one starting from α-amino-adipic acid (the so-called AAA pathway) and the other from diaminopimelic acid (the DAP pathway). From recent experiments, it is becoming clear that a second protein amino acid, tyrosine, may be formed by two slightly different routes; in this case, it is not an either/or situation since at least one plant *Vigna radiata* has been found to have the enzymes of both routes.

An even more fundamental division in primary metabolite synthesis than that of lysine or tyrosine synthesis that can be discerned among angiosperms is in the path of carbon in photosynthesis. In most plants, there is a direct route from carbon dioxide to sugar via ribulose bisphosphate carboxylase and the Calvin cycle. A significant number of angiosperms, especially those of subtropical or desert habitats, have an additional biochemical pathway by which the inspired carbon dioxide is first converted into organic acid via phospho*enol*pyruvate carboxylase and the Hatch–Slack pathway. The carbon trapped in oxaloacetic and malic acids is then passed onto the Calvin cycle, with sugar synthesis occurring in the usual way. These latter plants are known as C_4 plants as distinct from the more usual C_3 plants and this dichotomy in the path of sugar biosynthesis is of taxonomic interest, as will be discussed below.

Several variations in pathways to secondary metabolites are now well documented. Some of these different pathways have already been mentioned in appropriate places in earlier chapters. The most striking case is that of naphthoquinone synthesis; these pigments may be formed by any one of four different routes (see chapter 7, p. 167). Piperidine-type alkaloids are also known to be formed from different precursors in the Punicaceae and in the Umbelliferae (see chapter 6, p. 81).

Besides obviously different pathways involving different precursors and enzymes, there may be more subtle variations in a given biosynthetic route when one compares one organism with another. The order in which two related steps in synthesis occur may be switched, for example, in tyrosine biosynthesis, as already mentioned above. Also, there may be more than one form (isozyme) of the enzyme catalysing a particular step in the pathway and plants may vary in the number and nature of such isozymes. Examples of these variations will be mentioned here taken from the shikimate pathway, the major route in all plants for the production of aromatic compounds.

One other comparative metabolic study of some taxonomic interest in plants is concerned with the different ways that organisms can metabolize both natural and foreign compounds when they are applied exogenously to the tissues. The substances may be absorbed unchanged or more likely they will undergo metabolism, and suffer conjugation (to reduce toxicity) or be degraded. Such studies may be carried out in lower organisms by direct addition of the foreign compound to the growth medium. It is possible, for example, to separate *Pseudomonas* species according to the way they degrade a given aromatic compound supplied to them in the medium. With higher plants, foreign substances can be fed through the petiole, but there are practical difficulties in following the metabolic fate of such compounds. More recently, it has been found that tissue cultures of higher plants are ideal systems for observing the turnover and metabolism of exogenously applied organic constituents and some interesting results of taxonomic value have emerged.

The various themes mentioned above form the subject matter of the present chapter. These will be discussed briefly in some sort of order and a discussion of the systematic input of such information will conclude this presentation.

II. Variation in biosynthetic pathways

A. Lysine biosynthesis

The fact that living organisms may synthesize lysine, one of the 20 protein amino acids, by two different routes, starting from either α-aminoadipic acid (AAA) or from diaminopimelic acid (DAP) (Fig. 14.1) has been known for some time. It was Vogel (1963, 1964) who realized that the two routes are generally mutually exclusive so that organisms can be grouped according to whether they use the AAA or DAP pathway (Table 14.1). Why there should be two pathways to this one particular amino acid is not entirely clear and it is also not apparent whether one pathway can be considered more 'primitive' than the other or less efficient. The fact that procaryotes (Table 14.1) use the DAP pathway might suggest that this one evolved first. On the other hand, its efficiency vis-a-vis the AAA pathway must be equal, since the DAP pathway seems to be the major pathway in the more highly evolved angiosperms (Møller, 1974).

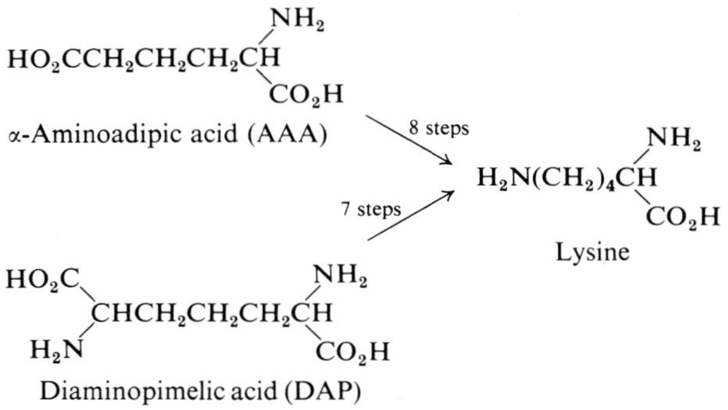

$HO_2CCH_2CH_2CH_2CH(NH_2)CO_2H$
α-Aminoadipic acid (AAA) → 8 steps → $H_2N(CH_2)_4CH(NH_2)CO_2H$ Lysine

$HO_2C(H_2N)CHCH_2CH_2CH_2CH(NH_2)CO_2H$
Diaminopimelic acid (DAP) → 7 steps → Lysine

Fig. 14.1 Alternate pathways of lysine biosynthesis in plants.

Within the fungi, both pathways are present and here there seems some sense in the replacement of one pathway by the other. Thus, the occurrence of the AAA pathway is correlated with the replacement of the more usual cellulose cell wall by the amino sugar-based polymer, chitin. Lé John (1971) suggests that the reason why DAP-lysine synthesis does not occur in these fungi is because it interferes in some unspecified way with chitin biosynthesis. Thus, the replacement of DAP pathway by the AAA pathway is an indirect effect of natural selection, the synthesis of chitin cell wall being advantageous to the organism because of its greater resistance to bacterial attack.

Table 14.1 Natural distribution of lysine biosynthesis pathways

DAP pathway	AAA pathway
Bacteria	
e.g. *Bacillus subtilis*	
Algae	Algae
Chlorella vulgaris	*Euglena gracilis*
Two fungal groups	Most fungal groups
Oomycetes	e.g. Ascomycetes
Hypochytridiomycetes	Basidiomycetes
Ferns	
Azolla carolina	
Angiosperms	
Hordeum vulgare	

From a chemotaxonomic viewpoint, lysine biosynthesis is most interesting in relation to the evolution of fungi, since it clearly separates the Oomycetes and Hypochytridiomycetes (both with the DAP pathway) from other fungal groups. These two fungal classes are distinctive in many other biochemical features besides the cell wall type already mentioned. Thus they vary in the nature of the glutamic dehydrogenases, lactic dehydrogenases and tryptophan synthesizing enzymes present (Table 14.2). It may also be noted that the Oomycetes and Hypochytridiomycetes are sometimes separated on morphological grounds from other fungi and are referred to as achlorophyllous algae, since they have distinctive 'tinsel' flagellae.

Table 14.2 Correlations of lysine pathway with other biochemical markers in the fungi

Fungal class	Lysine pathway	Cell wall polysaccharide	GDH[a] type	LDH[a] type	TRY[b] type
Myxomycetes	?	Cellulose	III	I	n.d.
Hypochytridiomycetes	DAP	Cellulose	III	I	n.d.
Oomycetes	DAP	Cellulose	III	I	IV
Chytridiomycetes	AAA	Chitin	I, II	II	I
Zygomycetes	AAA	Chitin	I, II	II	III
Ascomycetes	AAA	Chitin	n.d.	II	I, II
Basidiomycetes	AAA	Chitin	n.d.	n.d.	II, III

[a]GDH (glutamic dehydrogenase) type I is NAD+-linked, unregulated, type II is NAD+-linked, regulated and type III is NADP+-linked; LDH (lactic dehydrogenase) type I is allosterically inhibited by GTP while type II is not inhibited (Le John, 1971).
[b]TRY types (tryptophan synthesizing enzymes) vary according to their sedimentation patterns (Bartnicki-Garcia, 1970).

The possibility exists that both lysine pathways may operate within the angiosperms. There are experimental problems in establishing which pathway operates in a higher plant and most of the earlier experiments were insufficiently critical. It is now clear, for example, that the demonstration of the presence or absence of one or other of the two key substrates, AAA and DAP, is not a satisfactory criterion. The only sure evidence available seems to be for barley, *Hordeum vulgare*, where Møller (1974) has shown unambiguously by labelling experiments that the DAP pathway operates. Even in barley, it is possible to show some incorporation of label from exogenous AAA by what is essentially a detoxification mechanism. It is necessary to establish by chemical degradation of the lysine, after suitable feeding experiments, that radioactivity is incorporated in the expected pattern from the precursors of the one pathway and not the other. Such detailed studies in other species besides barley would be welcome in order to ascertain whether or not all green plants use the DAP route.

B. The path of carbon in photosynthesis

The idea that characters derived from such a universal process as photosynthesis should be of assistance in systematics is, on the face of it, a very unexpected one. And yet, recent studies of plants with C_4 photosynthesis (as distinct from the more common C_3 plants) indicate that such plants may be taxonomically related to each other. As is now well known, plants with C_4 photosynthesis have a distinctively different carbon pathway, because of the presence of an additional biochemical cycle, the so-called Hatch–Slack pathway. In such plants, carbon dioxide from the atmosphere is first incorporated into the synthesis of two organic acids, oxaloacetate and malate, before being 'passed on' into sugar and starch, via the Calvin (C_3) cycle. C_4 photosynthesis represents an adaptation to tropical climates, is a means of retaining photosynthetic efficiency at high ambient temperatures and is restricted to a relatively small number of angiosperm species.

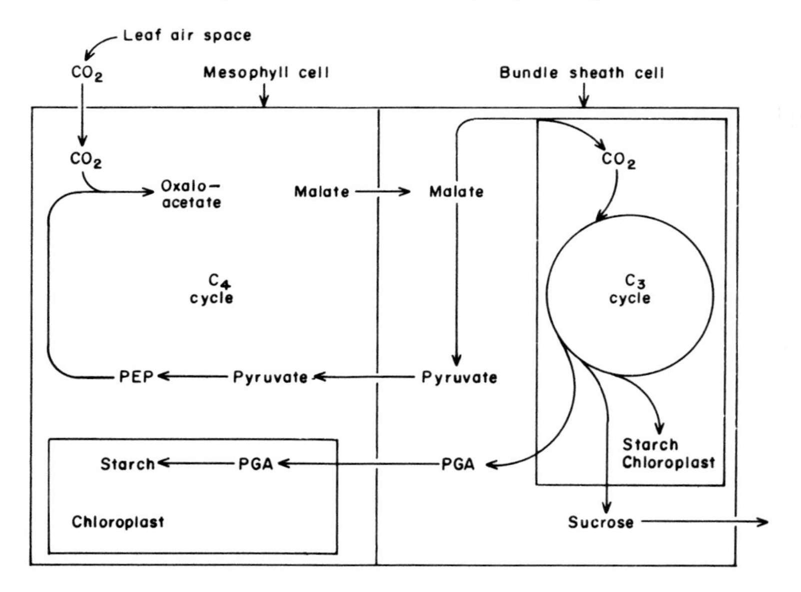

Fig. 14.2 C_4 pathway of photosynthesis, showing the compartmentation of the C_4 and C_3 cycles. *Key*: PEP, phospho*enol*pyruvate; PGA, phosphoglyceric acid

C_4 plants are also anatomically distinct from C_3 plants, the various anatomical features distinguishing such plants being referred to as the Kranz syndrome. Such plants are characterized by the presence of specialized sheaths around each vascular bundle. Although the typical Calvin cycle probably operates within the parenchyma sheath, the additional C_4 pathway is restricted to the mesophyll cells (see Fig. 14.2).

Depending on one's viewpoint, the two classes of photosynthetic species are termed C_4 or Kranz and C_3 or non-Kranz. Both biochemical measurements and microscopic techniques can be used to determine the condition.

Furthermore, both methods can be applied successfully to small leaf fragments from herbarium sheets, so that wide surveys can be readily achieved. Biochemical measurement involves determining the proportion of ^{13}C in the carbon laid down in the plant. While C_4 plants have $\delta\ ^{13}C$ values between —10 and —18 parts 10^{-3}, C_3 plants have values between —23 and —34 parts 10^{-3}. The distinction is usually clear-cut. There are actually species which are biochemically intermediate between C_3 and C_4, for example, *Panicum milioides*, but these are fortunately quite rare.

As a result of a number of surveys, it is now known that C_4 plants are found with some regularity in about 11 angiosperm families (Table 14.3). There are also usually single records in another seven families. In all, about 2000 species from 200 genera have been found to be C_4 plants (Raghavendra and Das, 1978). Of the 11 major families, no less than six belong to the same natural order, the Centrospermae, a group of plants already distinguished by unusual pigmentation based on betalains and by the presence of distinctive sieve-tube plastids. In addition, two of the monocotyledonous families in the list – namely the Gramineae and Cyperaceae – are recognized as being closely related in many morphological features. All families with C_4 plants are relatively advanced and tend to be herbaceous, so that this photosynthetic adaptation is apparently a specialized feature within the angiosperms as a whole.

Table 14.3 Plant families with significant numbers of species which possess the C_4 photosynthetic cycle[a]

Centrospermae	Euphorbiales
Aizoaceae	Euphorbiaceae
Amaranthaceae	Geraniales
Chenopodiaceae	Zygophyllaceae
Molluginaceae	Asterales
Nyctaginaceae	Compositae
Portulaccaceae	Poales
	Gramineae
	Cyperales
	Cyperaceae

[a]There are also occasional (usually single) records of C_4 plants in the following families: Acanthaceae, Asclepiadaceae, Boraginaceae, Capparaceae, Polygalaceae, Scrophulariaceae (dicotyledons) and Liliaceae (monocotyledons)

Although the overall distribution of C_4 photosynthesis in plants, as presently understood, is clearly not of major taxonomic significance, it is apparent that at the narrower confines of subfamily, tribe and genus the character becomes of interest. Thus, in the family Gramineae, it occurs in the whole of the subfamily Eragrostoideae without exception (over 60 taxa surveyed). It also occurs regularly in the subfamily Panicoideae, being represented in all tribes except for the small group Isachneae and some genera of the Paniceae. Within Paniceae, the genera considered to be the most primitive on other grounds are nearly all non-Kranz (Smith and Brown, 1973). At the generic level in *Panicum*, the character is of considerable interest in relation to recent taxonomic revision. Because of heterogeneity within this genus, as defined by Linnaeus, many taxa have been separated and described under new generic names, such as *Dichanthelium* and *Hymenache*. The correctness of these decisions is nicely reflected in the fact that all recently removed species are non-Kranz, while all remaining true *Panicums* are Kranz (Brown and Smith, 1975).

Surveys for the C_4 condition in the Cyperaceae (Raynal, 1973), Compositae (Smith and Turner, 1975) and the Euphorbiaceae (Webster *et al.*, 1975) have all yielded results of taxonomic value. Within the Compositae, the only known Kranz species are in the genera *Chrysanthellum*, *Eryngiophyllum*, *Isostigma*, *Glossocardia* and *Glossogyne* all of the tribe Heliantheae, in *Flaveria* (tribe Helenieae) and in *Pectis* (Tageteae). Although the presence of C_4 in *Pectis* is clearly unrelated to the other occurrences, it is interesting that this is the only genus in the Tageteae to have the syndrome. As Smith and Turner (1975) point out, Bentham in 1873 regarded *Pectis* as occupying a remote position in the tribe so it is interesting that the new biochemical evidence confirms such a positioning.

As a result of surveys in *Euphorbia* (Webster *et al.*, 1975), it is clear that the presence or absence of the Kranz syndrome is useful for deciding the taxonomic boundaries of the subgenus *Chamaesyce*. It is interesting (see below) that Crassulacean acid metabolism (CAM) is also present in the genus *Euphorbia* but in an unrelated subgenus, namely *Euphorbia*. Apparently, the two photosynthetic specializations arose independently in different geographic zones: C_4 in North American subtropics, CAM in African tropics. Finally in the Cyperaceae, it has been found that whole groups of genera of the tribes Cypereae and Fimbristylideae are Kranz (Table 14.4). There are significant correlations here between the distribution of the C_4 character and other morphological features in these plants (Raynal, 1973).

From these examples, it is apparent that the C_4 character, which is basically an adaptive feature in photosynthesis to subtropical environments, is nevertheless a useful taxonomic marker in plants. Further surveys are clearly called for, because of the systematic potential.

Another photosynthetic adaptation in Xeric plants, which is associated with the need to conserve moisture in desert habitats, is the so-called CAM. Such CAM plants accumulate organic acids, especially malic acid, during the night and then convert them into sugars via the Calvin cycle during the daytime. As a result, there is a large diurnal variation in organic acid in the plant leaves and the condition can be recognized by recording an extremely acid pH in the cell sap at dawn (Kluge and Ting, 1978).

Table 14.4 Distribution of C_4 pathway in the subfamily Cyperoideae of the Cyperaceae

Tribe	Pathway			
	C_3	C_4		
Cypereae	*Androtrichum* *Courtoisia*	*Ascolepis* *Cyperus* in part	*Lipocarpha* *Mariscus*	*Remirea* *Scirpus* section squarrosi
	Cyperus in part	*Hemicarpha*	*Pycreus*	*Torulinium*
	Dulichium	*Kyllinga*	*Queenslandiella* (600 spp.)	*Volkiella*
Fimbristylideae	*Eleocharis* *Websteria*	*Abilgaardia* *Bulbostylis*	*Crosslandia* *Fimbristylis* (350 spp.)	*Nelmesia* *Nemum*
Scirpeae	All genera			

First described in *Bryophyllum* and *Kalanchoe* (Crassulaceae), this photosynthetic modification is widespread in succulent plants. According to a recent survey (Szarek and Ting, 1977), CAM has been reported in 18 families of angiosperms (109 genera). It is also present in one gymnosperm, *Welwitschia mirabilis*, and in two epiphytic Filicophyta. Not surprisingly, in view of its close biochemical similarity to C_4 photosynthesis, the presence of CAM is correlated to some extent with the presence of C_4 photosynthesis at the ordinal level (Table 14.5).

Table 14.5 Taxonomic distribution of CAM plants

Centrospermae	Celastrales
Aizoaceae	Geraniaceae
Cactaceae	Oxalidaceae
Didiereaceae	Piperales
Portulaccaceae	Piperaceae
Euphorbiales	Lamiales
Euphorbiaceae	Labiatae
Rhamnales	Asterales
Vitaceae	Asteraceae
Gentianales	Liliales
Asclepiadaceae	Agavaceae
Cucurbitales	Liliaceae
Cucurbitaceae	Orchidales
Saxifragales	Orchidaceae
Crassulaceae	Bromeliales
	Bromeliaceae

The fact that CAM has been found in several non-succulents is perhaps a most significant recent finding for plant systematists. It suggests that the character may turn out to be of chemotaxonomic value. No deliberate surveys, outside succulent groups, have yet been attempted; these new findings hint that such surveys might well yield data of systematic interest.

C. The shikimic acid pathway

The shikimic acid pathway is restricted to the plant kingdom and provides the main biosynthetic route by which aromatic compounds are produced from carbohydrates. Some ten steps are involved in the production of phenylalanine and tyrosine from phospho*enol*pyruvate and erythrose 4-phosphate, which are derived from the Calvin cycle through photosynthesis. The pathway owes its name to shikimic acid, one of the pre-aromatic intermediates along the route. The pathway is important not only because it provides the essential aromatic amino acids for protein synthesis, but also because these same acids are further employed for the production of almost all the secondary aromatic substances—polyphenols, flavonoids, coumarins—of the plant kingdom. Although the pathway occurs in all plants, a number of variations in the enzymology of the main pathway have been encountered and many of these variations appear to be of undoubted systematic interest.

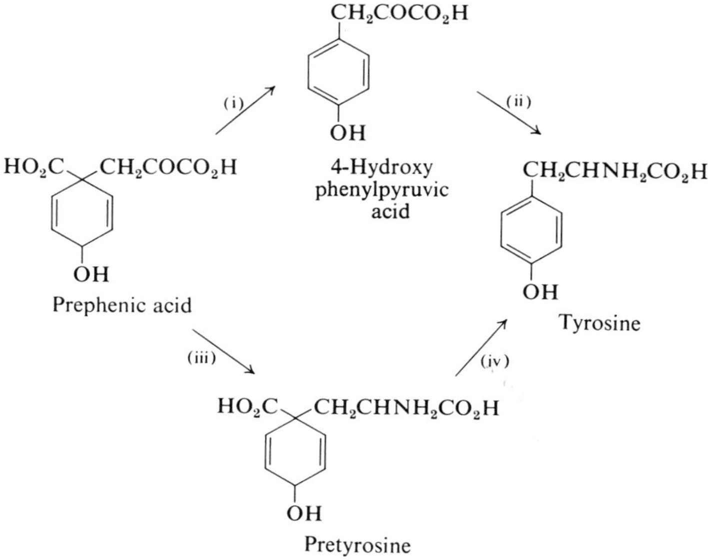

CH_2COCO_2H
(i) (ii)
OH
HO_2C CH_2COCO_2H
4-Hydroxy phenylpyruvic acid
$CH_2CHNH_2CO_2H$
OH
Prephenic acid
OH
Tyrosine
(iii) (iv)
HO_2C $CH_2CHNH_2CO_2H$
OH
Pretyrosine

Fig. 14.3 Alternative pathways to tyrosine biosynthesis in plants. Enzymes: **i,** prephenate dehydrogenase; NAD^+; **ii,** 4-hydroxyphenylpyruvate transaminase, pyridoxal 5′-phosphate; **iii,** prephenate transaminase, pyridoxal 5′-phosphate; **iv,** pretyrosine dehydrogenase, NAD^+.

From the phyletic viewpoint, one of the most intriguing variations that has been found concerns the last two steps in the synthesis of tyrosine from prephenic acid (Fig. 14.3), where a switch in the order of these two steps has occurred during evolution. It is apparent that in primitive plants, notably in blue-green algae, prephenic acid is first transaminated at the side chain to give pretyrosine and this is then aromatized (or dehydrogenated) to yield tyrosine. An alternative system is for aromatization to precede transamination with 4-hydroxyphenylpyruvic acid as intermediate (Fig. 14.3) and this is the enzymology present in more advanced organisms.

It is interesting that intermediate stages between the two extremes have been observed (Jensen and Pierson, 1975) so that it is possible to draw a scheme by which one system has evolved from the other (Fig. 14.4). Thus, in *Pseudomonas* species, it is apparent that both pathways operate between prephenic acid and tyrosine. On the other hand, in *Neurospora crassa* it appears that the blue-green algal route is only partly functional and although pretyrosine is synthesized, the enzyme for dehydrogenating this to give tyrosine has been lost.

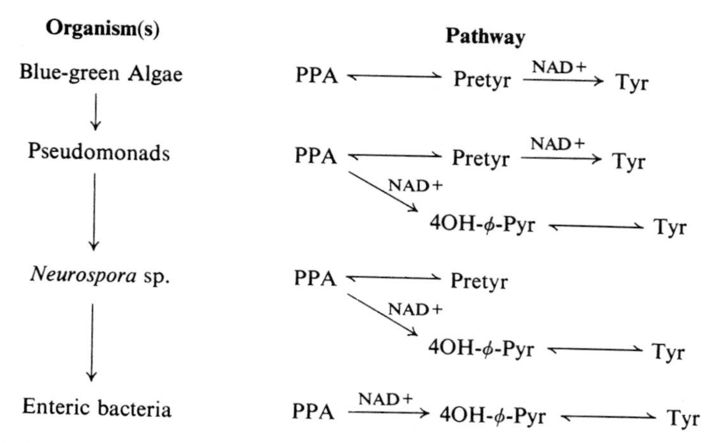

Fig. 14.4 Evolution of 4-hydroxyphenylpyruvic acid pathway of tyrosine synthesis. *Key*: PPA, prephenic acid; pretyr, pretyrosine; 4OH-φ-Pyr, 4-hydroxyphenylpyruvic acid; tyr, tyrosine.

From the scheme illustrated in Fig. 14.4, one might assume that, in higher plants, the only route to tyrosine would be via 4-hydroxyphenylpyruvic acid. Surprisingly, however, recent experiments (Rubin and Jensen, 1979; Byng *et al.*, 1981) have shown that in both higher plants studied enzymes of the pretyrosine route are present. Also, in one of them, an enzyme of the 4-hydroxyphenylpyruvic acid route is present so that it is similar to pseudomonads (Fig. 14.4) in having both pathways. A further variation (Table 14.6) in the higher plants is in the cofactor dependence of the dehydrogenating enzyme acting on prephenic acid. These new results indicate that the enzymology of tyrosine synthesis in higher plants is even more complex than in bacteria, algae and fungi and that further exploration of the pathway from a comparative viewpoint should be most rewarding.

Table 14.6 Enzymology of tyrosine biosynthesis in higher plants

Plant	Prephenate dehydrogenase	Pretyrosine dehydrogenase
Zea mays	—	NAD^+-dependent[a]
Vigna radiata	$NADP^+$-dependent	$NADP^+$-dependent

[a]Sensitive to feedback inhibition by L-tyrosine; the *Vigna* enzyme by contrast is not effected by this inhibition. Studies in suspension-cultured cells of *Nicotiana sylvestris* indicate that this plant, like *Zea mays*, used the pretyrosine dehydrogenase route exclusively (Gaines *et al.*, 1982)

In studies of the enzymology of the shikimate pathway, other types of variation apart from that described above have been noted. For example, there is variation in the way that the enzymic activities are linked together; in lower plants such as in the bread mould *Neurospora crassa* up to five enzymic activities of the pathway may be linked together and form a single polypeptide chain. By contrast, in higher plants nearly all the enzymes of the pathway can be isolated as individual proteins (Harborne, 1980). There is also variation in the number of different forms or isozymes of certain enzymes, particularly in higher plants. Such isozymes are electrophoretically different and yet they catalyse the same reaction along the pathway. Such isozymes may be further distinguished in that their activities may be differentially repressed or stimulated by the products of the pathway, i.e. phenylalanine, tyrosine or cinnamic acid, and this duplication in enzymic activity may be important in the regulation of the pathway.

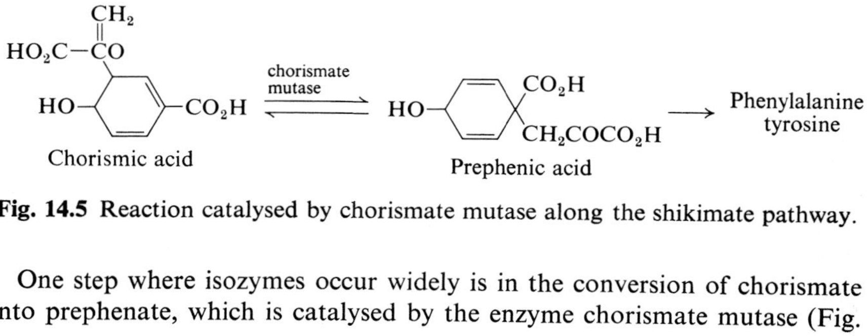

CH_2
HO_2C-CO
HO— —CO_2H
Chorismic acid
chorismate mutase
HO— CO_2H
CH_2COCO_2H
Prephenic acid
→ Phenylalanine tyrosine

Fig. 14.5 Reaction catalysed by chorismate mutase along the shikimate pathway.

One step where isozymes occur widely is in the conversion of chorismate into prephenate, which is catalysed by the enzyme chorismate mutase (Fig.

14.5). Two isozymes are present in most lower plants examined and also in *Selaginella*, in a fern and in a pine. By contrast, those angiosperms that have been tested have a third isozyme as well, except for three members of the Leguminosae which only have the original two forms (Woodin *et al.*, 1978). There is clearly evidence here that evolutionary advancement has led to an increase in isozyme complement and hence greater flexibility in the regulation of the pathway, but wider sampling is still needed to confirm this hypothesis.

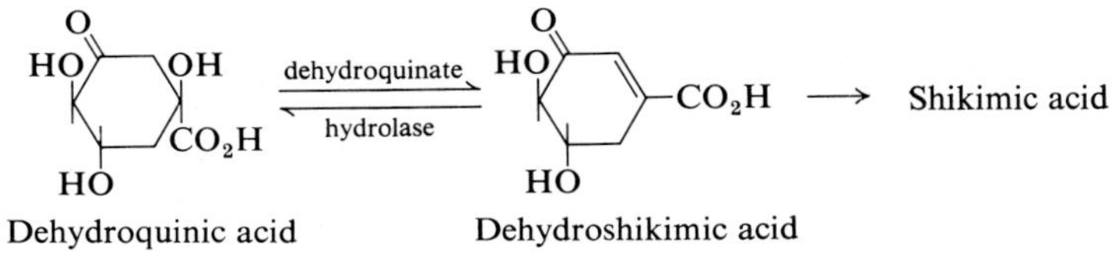

Dehydroquinic acid ⇌ (dehydroquinate hydrolase) Dehydroshikimic acid $\longrightarrow$ Shikimic acid

Fig. 14.6 Reaction catalysed by dehydroquinate hydrolase along the shikimate pathway.

A second example of isozyme variation is in the conversion of dehydroquinate into dehydroshikimate, which is the third reaction along the pathway (Fig. 14.6). Here there is normally only one form of the enzyme, dehydroquinate hydrolase, present. Boudet and his co-workers (1977) have recently found a second form to occur specifically in monocotyledons and not in dicotyledons. It occurs widely in three related families, the Juncaceae, Gramineae and Cyperaceae; there are also occasional records in the Liliaceae and Iridaceae. It differs from the common form of the enzyme in that it is activated by shikimic acid. It may thus, in some way, be of physiological value to the plant in which it occurs. Its restriction to the monocotyledons and mainly to three closely related families suggests that it could be a biochemical character of some taxonomic value.

III. Variation in conjugation and degradation

It has been appreciated for many years (see e.g. Towers, 1964) that plants vary in the way they may detoxify or conjugate substances fed to them through petiole or stem or by spraying onto the leaf. The growth hormone indole 3-acetic acid under these circumstances may be conjugated as the glucose ester or as the aspartic acid derivative. Simple phenols are converted into the corresponding monoglucosides within a few hours of being introduced into plant tissues (Pridham, 1964). The glucosylation reaction may be more complex if the phenol contains two hydroxyl groups of similar reactivity.

For example, 6,7-dihydroxycoumarin (aesculetin) is converted into the 6-glucoside, to a mixture of 6- and 7-glucoside, or to the 6-diglucoside, depending on the plant species examined (Harborne, 1963b).

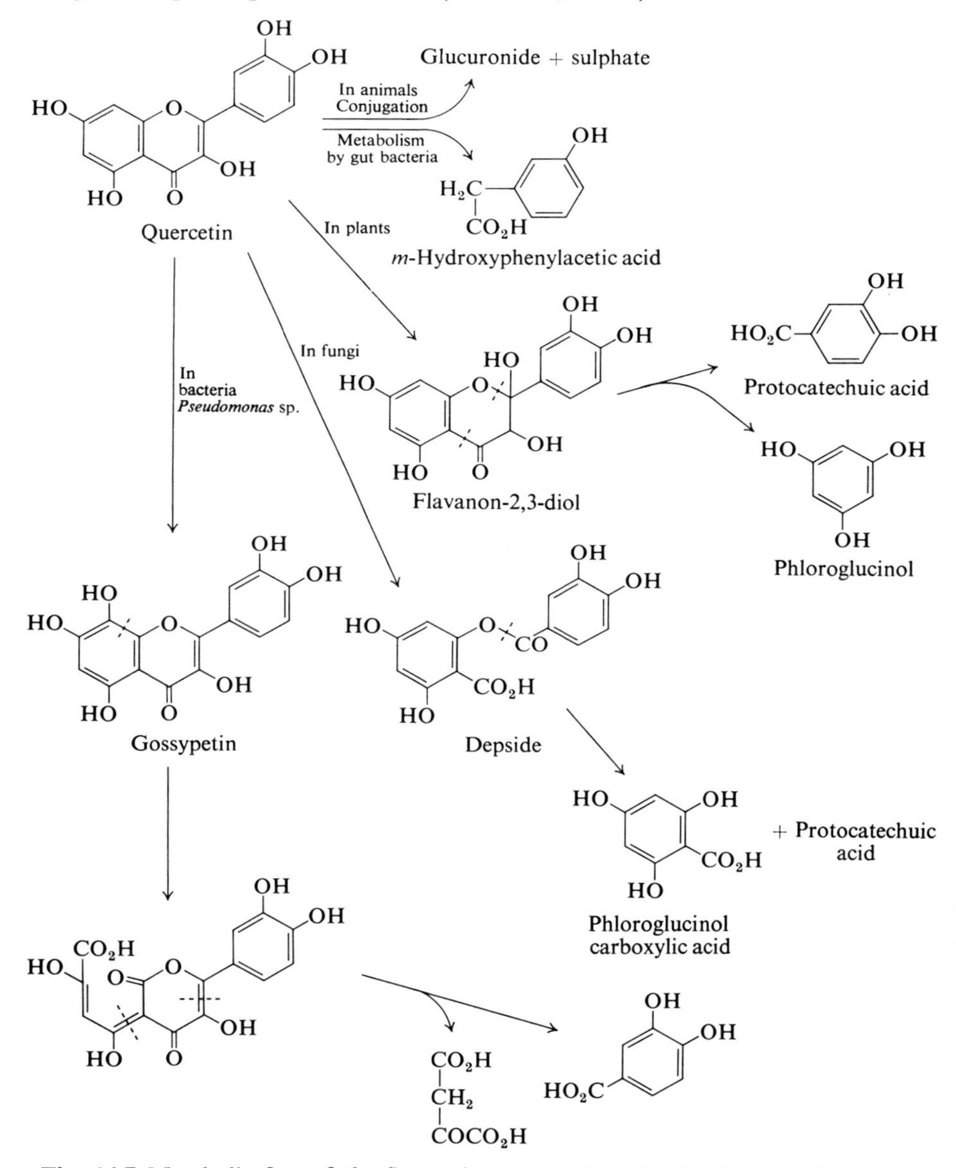

Fig. 14.7 Metabolic fate of the flavonol quercetin in animals, plants and micro-organisms.

Herbicides and fungicides are also well known to be variously metabolized *in vivo* by plants (Baldwin, 1976; Naylor, 1976). However, it is usually not clear from the experiments that have been carried out whether real differences exist between plants in the metabolites or conjugates formed. Foreign compounds so fed to plants may also be degraded and ring cleavage is a common fate of aromatic substances. Ring cleavage reactions were first well studied in microbial systems but more recently they have been extensively investigated in higher plants (Ellis, 1974).

Variation in metabolism may be of systematic interest at both the higher and lower levels of classification. For example, the fate of the flavonol quercetin differs in bacteria, fungi, plants and animals (Fig. 14.7), although the end products of metabolism are generally similar. At the other extreme, the route of metabolism of a phenolic substrate may be employed to separate two species within the same genus. Thus *Pseudomonas aeruginosa* metabolizes *p*-hydroxybenzoic acid by *ortho*-ring cleavage to succinate and acetyl-CoA, while *Ps. acidovorans* converts the same substrate by *meta*-ring cleavage into two molecules of pyruvate (Fig. 14.8). Such a diversity in metabolism can be used to separate *Pseudomonas* species into two groups (Stainer, 1968). The reaction is sufficiently reliable for it to be used in species identification, although in practice it has not yet been so applied (Hendrie and Shewan, 1979).

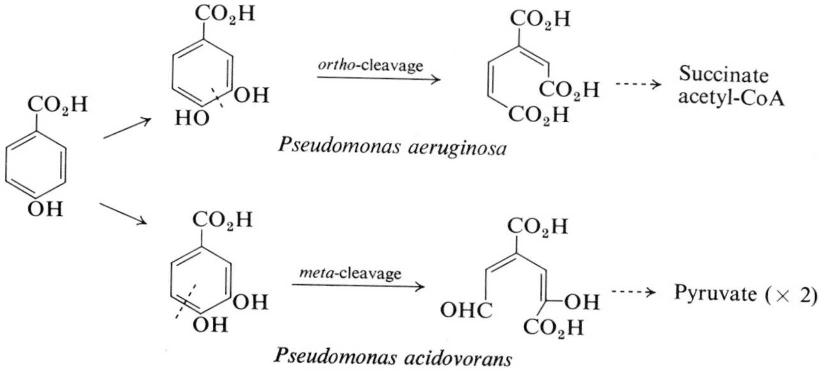

ortho-cleavage ⟶ Succinate acetyl-CoA

Pseudomonas aeruginosa

meta-cleavage ⟶ Pyruvate (× 2)

Pseudomonas acidovorans

Fig. 14.8 Alternative pathways of ring cleavage in the bacterial *Pseudomonas* genus.

In spite of the information available on comparative metabolism in plants generally, few attempts have been made to deliberately challenge plants with foreign compounds in order to elicit chemotaxonomic data. Until recently, the only taxonomic approach by feeding experiments seems to have been with unnatural D isomers of the standard protein amino acids. Results have been reported with D-serine and D-tryptophan (Zenk and Scherf, 1964;

Ladesic *et al.*, 1971) but the most interesting data have emerged from experiments with D-methionine. This sulphur amino acid is acylated in vascular plants to the *N*-malonyl derivative, while in fungi and lichens it is converted into the *N*-acetate. By contrast in bacteria and algae, it is not conjugated but instead is metabolized immediately by deamination.

All the above phyla are uniform in the way they dispose of D-methionine, but the bryophytes are exceptional in having all three routes (Table 14.7). Thus some bryophytes deaminate it, some conjugate it with acetic acid and others with malonic acid. The results of a feeding survey in the Bryophyta have yielded valuable data of chemotaxonomic significance, particularly for subclass classification (Pokorny, 1974). Some of the inferences that can be drawn about phyletic relations from these feeding experiments are as follows. First, the bryophyta that metabolize the L and D anomers identically (e.g. the Marchantiales) may be considered the more primitive, since this metabolism occurs in fungi and algae. Second, the fact that the class Hepaticae (including the Mniaceae) metabolize D-methionine by deamination links them with the algae and supports the views of those who regard the Hepaticae as having evolved from an algal-like ancestor. Third, the malonyl reaction observed in the Anthoceratales separates this group from the Hepaticae with which they have often been associated and links them more nearly with the vascular plants.

Table 14.7 Variations in metabolic pathways of D-methionine in the Bryophtya

Metabolism without conjugation[a]	Conjugation as *N*-acetate	Conjugation as *N*-malonate
	Hepaticae	
Sphaerocarpales		
Jungermanniales		
Marchantiales		
	Musci	
Mniaceae	Sphagnales	Polytrichaceae[b]
	Andreales	
	Buxbaumia[b]	
	Amblystegium[b]	
	Acrocladium[b]	
	Hylocomium[b]	
	Cinclidotus[b]	
	Brachythecium[b]	
	Anthoceratae	
		Anthocerotales

[a]All Bryophyta metabolize L-methionine without conjugation
[b]In these taxa, conjugation is incomplete with varying amounts of deamination occurring

In the experiments with unnatural amino acids in higher plants, feeding was carried out through cut surfaces and there are always problems in following the fate of the compounds due to incomplete translocation and so on. One approach which avoids these difficulties in higher plants is to use tissue culture and such a technique may well be of chemotaxonomic importance. Thus Willeke *et al.* (1979) have found that when nicotinic acid is added to suspension cultures of higher plants, one of two alternative pathways come into operation. It is *either* converted by methylation into trigonelline *or* it is conjugated with sugar and bound as the *N*-arabinoside (Fig. 14.9). Only when conjugated as the arabinoside does the molecule then undergo further metabolism and degradation.

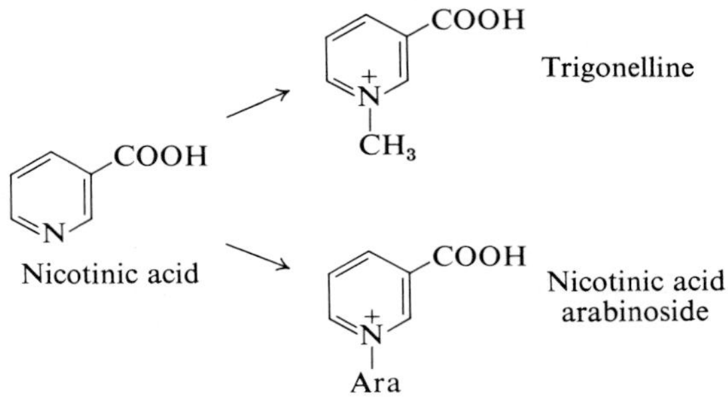

Fig. 14.9 Alternative pathways of nicotinic acid conjugation in suspension cultures of higher plants.

Some fifty species of higher plant that were available in cell culture were surveyed by these workers for their ability to metabolize nicotinic acid and it was found that the arabinoside is, with few exceptions (Table 14.8), formed exclusively in members of the subclass Asterideae—all other plant groups generally follow the trigonelline route. This metabolic character thus appears to be a marker at the level of order or subclass. The production of the arabinoside in a few members of the Rosidae and Dilleniidae (see Table 14.8) supports the view of some taxonomists of the presence of a phylogenetic line linking these orders with the Asteridae. The fact that there are these exceptional taxa makes the character a more interesting one, since its distribution pattern questions the validity of the existing system of plant classification above the family level.

It is important to realize that this character is strictly a property of suspension cultures. Although the arabinoside has not yet been demonstrated unequivocally as a product of intact plants, the methyl derivative trigonelline is a well known natural product. In fact, it occurs in whole plants of several members of the Asteridae, e.g. in *Coffea arabica* (Rubiaceae), and *Solanum*

Table 14.8 Distribution of different metabolic pathways of nicotinic acid in the dicotyledons

Super order	Trigonelline production	Arabinoside production
Magnoliidae	*Papaver*	—
Hamamelididae	*Cannabis*	—
Rosidae	*Sedum, Rosa, Mucuna, Phaseolus, Cicer, Glycine, Arachis, Drosophyllum, Ruta, Aesculus, Parthenocissus, Euphorbia.*	*Cornus, Daucus Petroselinum*
Dilleniidae	*Sinapis, Cucumis, Bryonia*	*Anagallis*
Caryophyllidae	*Chenopodium*	—
Asteridae	—	*Galium, Catharanthus, Nicotiana, Duboisia, Tectona, Mentha, Ocimum, Tagetes, Haplopappus, Tanacetum*

Data modified from Willeke *et al.* (1978). In most cases, only single species were studied in any given genus. The three gymnosperms and seven monocotyledons studied all gave trigonelline.

tuberosum (Solanaceae), the tissue cultures of which synthesize the arabinoside. The difference in metabolism is thus something that can only be observed in cell culture. This limits its widespread use as a taxonomic tool, since there are still only relatively few plants which have been grown successfully in suspension culture. However, in the future, more and more species are likely to be cultivated in this way for other purposes. A more representative sample of angiosperm species should thus become available for exposure to foreign compounds in order to test their metabolic abilities and hence yield results of possible taxonomic value.

IV. Conclusion

In this chapter, it is shown that there are significant variations in the way that low molecular-weight substances—both primary and secondary metabolites—are formed in plants and also in the way that they are further metabolized. There are also variations in the way that plants conjugate or degrade unnatural metabolites when they are absorbed into the living tissues. Most of this variation is potentially of taxonomic interest. Some differences are apparent at the level of phyla, e.g. the two pathways of lysine biosynthesis; others are apparent at the genus and species level, e.g. the presence or absence of C_4 photosynthesis.

Most of the information on comparative synthetic and metabolic pathways has accrued quite incidentally to chemosystematic efforts. Indeed, the exploration of biosynthetic pathways requires expensive radioactive precursors and is highly time consuming since complex experimentation is essential. It is highly unlikely that such operations could be absorbed readily into chemosystematic programmes. On the other hand, variations in the enzymology of biosynthesis, where they exist, are relatively easy to detect. It is clearly possible to screen plants fairly rapidly for the presence or absence of isozymes concerned at particular stages along the shikimate pathway. Of especial interest are variations at the later stages prior to tyrosine formation and here comparative studies, which are just beginning, to see which of the two routes to tyrosine dominate in higher plants will be awaited with considerable anticipation.

Finally, one may note the possibility of obtaining significant comparative information on plant biochemistry by following the metabolic fate of secondary substances applied artificially to cell suspension cultures of higher plants. This approach has already been successful with nicotinic acid (see Table 14.8) and many other molecules could readily be studied in the same context. There is the fascination here of obtaining completely novel comparative information on plant species which cannot be obtained by any other means—information hidden in the genome which is only expressed when cells are cultured in the test tube.

15

Phytoalexin Induction

I.	Introduction	362
II.	The experimental approach	364
III.	Phytoalexin variation at the family level	366
IV.	Phytoalexin variation below the family level	367
V.	Conclusion	372

I. Introduction

Phytoalexins are fungitoxic organic compounds which by definition are formed *de novo* in plant tissues in response to microbial attack. Evidence that fungitoxins are produced in response to fungal disease was obtained early this century by Bernard (1911), working with orchid bulbs. The phytoalexin concept of disease resistance was formally propounded 30 years later by Müller and Börger (1941) following their experiments on resistant factors formed in potato tissue infected with the blight pathogen *Phytophthora infestans*. The first phytoalexin to be fully characterized chemically, however, was pisatin, an isoflavonoid formed in pea plants in response to infection with a range of fungal organisms, both pathogenic and non-pathogenic on pea (Cruickshank and Perrin, 1960).

The idea that phytoalexins are of importance in the natural resistance of higher plants to fungal and bacterial diseases was the spur for many subsequent investigations of phytoalexins in other plant groups. As a result, it is now well established that many, if not most, higher plants respond to microbial invasion by the synthesis, at the site of infection, of organic toxins. These phytoalexins are typically secondary metabolites in terms of their biosynthesis but are normally absent from healthy plants. They cannot thus be isolated as part of the usual secondary metabolism occurring in a given plant.

Although occasionally produced in plant tissues by stress situations of a non-microbial nature, phytoalexins are only produced consistently and in high concentration in response to fungal invasion. Although there is increasing evidence that phytoalexins are of importance in the protection of higher plants against fungal colonization, their production does not necessarily limit the invasion of every pathogenic organism. For example, some fungi have the ability to metabolize them further to harmless products. Furthermore, phytoalexin synthesis is only one of a number of barriers to microbial attack present in higher plant tissues so that their importance in controlling disease varies from species to species (Bailey and Mansfield, 1982).

Although most studies of phytoalexins subsequent to the isolation of pisatin concentrated on crop plants, it soon became evident that a taxonomic element was present in the type of phytoalexin produced by a plant. It is now abundantly clear that different families tend to accumulate their own distinctive class of phytoalexin molecule. Thus, the Leguminosae in general produce isoflavonoids, the Solanaceae diterpenoids, the Compositae polyacetylenes, the Orchidaceae dihydrophenanthrenes and so on (Ingham, 1972; see Fig. 15.1). Anomalies are rare; for example, the furanoacetylene wyerone is produced atypically in *Vicia faba* (Leguminosae), a member of a family which predominantly produces isoflavonoids as phytoalexins.

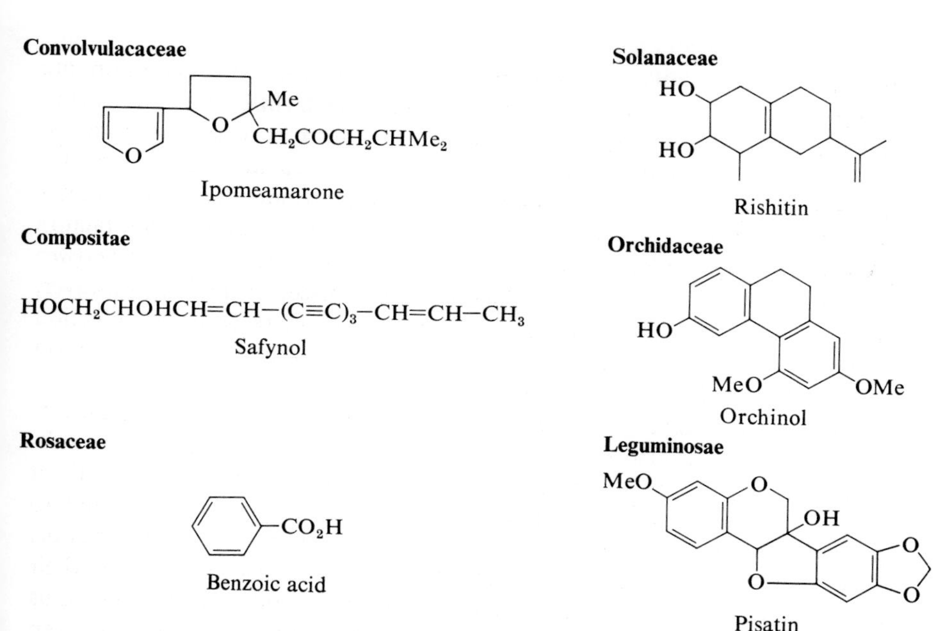

Fig. 15.1 Phytoalexins of various plant families.

The possibility of actually turning these disease resistance investigations to taxonomic advantage was first considered with different species of the genus *Trigonella* (Ingham and Harborne, 1976). The results were successful in that three clear-cut patterns were apparent in the responses shown by some 35 species surveyed in this legume genus. Further work at several different levels of classification suggest that this method represents a valid new approach to plant relationships. The approach differs from conventional chemotaxonomic practice, since it specifically uncovers, for comparative purposes, the synthesis of organic molecules, which cannot be detected in the normal pool of secondary metabolites in the plants under study.

A few examples of the chemotaxonomic results achieved with this phytoalexin approach will be described here. The data available at the family level will be briefly reviewed and the results obtained at tribal and generic level will be discussed within the family Leguminosae, since most effort has been devoted to this plant group (Harborne and Ingham, 1978; Ingham, 1981). First, however, some mention must be made of the experimental technique.

II. The experimental approach

The practical procedures for phytoalexin induction are very simple, but they do require living plants, usually as leaf tissue. Phytoalexins can generally be obtained by the drop diffusate technique (Higgins and Miller. 1968), in which droplets containing a non-pathogenic fungus (such as *Helminthosporium carbonum*) are placed on the surface of the leaves, which are floated on water in a suitable container and left for 48 hours in diffuse light (Fig. 15.2). After a few hours incubation, the spores in the droplets germinate and the fungus starts penetrating the leaf surface. This triggers phytoalexin synthesis in the tissues at the site of the invasion. Quite massive amounts may be produced

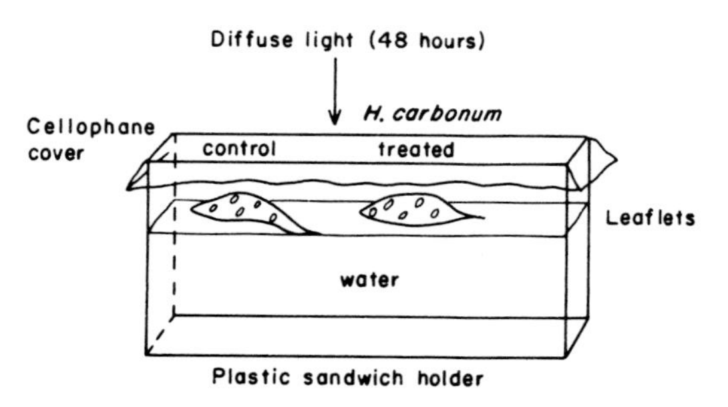

Fig. 15.2 Phytoalexin induction in the drop diffusate technique. For details, see the text.

and most is exuded from the leaf surface and is pushed into the droplets. After 48 hours, the phytoalexins, which are now present in the droplets in quantity, can be collected, isolated and identified by standard phytochemical techniques. Bioassays for antifungal activity are carried out at the same time, since phytoalexins by definition are significantly fungitoxic (Fig. 15.3). It is also essential to run water controls in parallel and demonstrate that no phytoalexin is produced in the absence of the microbial trigger.

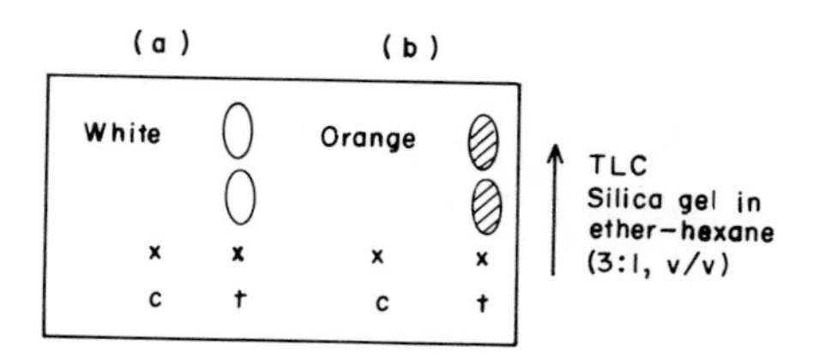

Fig. 15.3 Phytoalexin detection. *Key*: (a) bioassay after spraying with spores of *Cladiosporum herbarum* and incubating for 3 days, white inhibitory zones on a grey background; (b) chemical identification of phenolics with diazotized *p*-nitroaniline c, control droplets; t, treated droplets.

The simple elegance of this technique lies in the fact that it yields a concentrated solution of phytoalexin which is free from contamination by the normal cellular constituents of the plant. Most other techniques that can be used lack this refinement. If negative results are obtained in the drop diffusate technique, it should not be assumed that the phytoalexin response is absent. Other procedures must be tried with other organs of the plant before assuming that phytoalexins are never formed in a given taxon. For example, fungal infection can be applied to plant stems, hypocotyls or germinating seeds (Keen, 1975), positive results often being obtained where leaves have failed to respond to the drop diffusate method.

In general, more than one phytoalexin is likely to be produced by a plant. In some cases, five or more substances may be formed, although only one or two are major components. Substances produced are almost invariably of the same chemical type and tend to be closely related biosynthetically. For example, in the potato, seven phytoalexins are formed; they are all, however, sesquiterpenoids biosynthetically related to the major compound, rishitin (see Fig. 15.1; Stoessl *et al.*, 1976). Very rarely, compounds of more than one type may be detected. Thus, in *Vicia faba*, the major phytoalexins are six furanoacetylenes, including wyerone. Careful scrutiny of the phytoalexin response in this plant has also revealed the presence in trace amounts of medicarpin, an isoflavonoid commonly produced very widely elsewhere in the Leguminosae (Hargreaves *et al.*, 1976).

III. Phytoalexin variation at the family level

The variation in phytoalexin structures produced in different angiospermous families is summarized in Table 15.1. It must be emphasized that the data are still very incomplete in that most families have yet to be systematically surveyed. However, the results to date establish that 17 families widely differing in their taxonomic position are capable of responding to fungal infection by synthesizing phytoalexins. There is thus good reason for believing that

Table 15.1 Chemical variation in phytoalexins in families of the angiosperms

Family[a]	Genus	Chemical type	Structural example
Monocotyledons			
Amaryllidaceae	*Narcissus*	Flavan	7-Hydroxyflavan
Gramineae	*Avena*	Benzoxazin-4-one	Avenalumin I
	Oryza	Diterpene	Momilactone A
Orchidaceae	*Orchis*	Phenanthrene	Orchinol
Dicotyledons			
Caryophyllaceae	*Dianthus*	Benzoxazin-4-one	Dianthalexin
Chenopodiaceae	*Beta*	Isoflavone	Betavulgarin
Compositae	*Carthamus*	Polyacetylene	Safynol
Convolvulaceae	*Ipomoea*	Furanoterpene	Ipomeamarone
Euphorbiaceae	*Ricinus*	Diterpene	Casbene
Leguminosae	*Pisum*, etc.	Isoflavonoid	
		Isoflavone	Wighteone
		Isoflavanone	Kievitone
		Pterocarpan	Medicarpin
		Isoflavan	Vestitol
	Arachis	Stilbene	Resveratrol
	Vigna	Benzofuran	Vignafuran
	Vicia	Furanoacetylene	Wyerone
Linaceae	*Linum*	Phenylpropanoid	Coniferyl alcohol
Malvaceae	*Gossypium*	Naphthaldehyde	Gossypol
		Naphthafuran	Vergosin
Moraceae	*Morus*	Benzofuran	Moracin-C
		Stilbene	Oxyresveratrol
Rosaceae	*Eriobotrya*	Biphenyl	Aucuparin
	Malus	Phenolic acid	Benzoic acid
Solanaceae	*Lycopersicon*	Polyacetylene	Falcarinol
	Solanum, etc.	Sesquiterpene	Rishitin
Tiliaceae	*Tilia*	Sesquiterpene	7-Hydroxycalamenene
Umbelliferae	*Daucus*	Chromone	Eugenin
		Dihydroisocoumarin	6-Methoxymellein
	Pastinaca	Furanocoumarin	Xanthotoxin
Vitaceae	*Vitis*	Stilbene oligomer	α-Viniferin

[a]For references, see Harborne and Ingham (1978) and Bailey and Mansfield (1982).

phytoalexin induction is a general property of higher plants. Some families may well be negative and there is some evidence that the Cucurbitaceae is among these. Most families that have been deliberately tested have responded positively so that it is probable that negative families are in a minority.

In general, it is apparent (Table 15.1) that each family has a characteristic phytoalexin response. There is very little overlap between families and where the same chemical type is found in two families, the actual structures produced are usually different. Most families appear to produce one main type of phytoalexin. Where wider investigations have been carried out (e.g. in the Leguminosae), several minor structural types often appear. In many families, too few taxa have been investigated to be able to discern the main type. In Umbelliferae, for example, isocoumarins, chromones and furanocoumarins have been variously detected in *Daucus* and *Pastinaca*, but it is too early to be sure which of these structures are commonly formed. Preliminary experiments with several other umbellifers at Reading suggest that furanocoumarin synthesis may be the most usual response in this family. The most consistent data obtained so far within a family are from the Solanaceae, where 20–30 species from five genera (*Capsicum*, *Datura*, *Lycopersicon*, *Nicotiana*, *Solanum*) have all been shown to produce sesquiterpenoids of the same general type as rishitin (Stoessl *et al.*, 1976).

The utility of the phytoalexin data for chemically relating families is not yet clear. One may note, however, that the closely related Convolvulaceae and Solanaceae are chemically related in their phytoalexins, since the former produces the sesquiterpene furanolactone ipomeamarone, which is not too different in structure from the sesquiterpene alcohols of the Solanaceae. Again, it is interesting that the Moraceae produce benzofurans and stilbenes, two types found elsewhere together in the Leguminosae, a family that is not too distant morphologically from it. These data (Table 15.1) at least suggest that phytoalexin comparisons between families may be worth pursuing to yield new information for systematic purposes.

IV. Phytoalexin variation below the family level

Although it is too early yet to draw taxonomic conclusions from the phytoalexin data available at the family level, the same is not true for the classification within genera and tribes, since much more consistent data are available, at least within the family Leguminosae. Here, over 400 taxa have been investigated and the results are producing a very interesting pattern in relation to the systematic classification, particularly of temperate members of the family (Ingham, 1981). Here, a few of the results that have been obtained with *Trigonella*, *Lathyrus* and *Trifolium* will be mentioned to illustrate the diversity of responses that are present in these plants.

One of the first genera to be properly surveyed was *Trigonella*, a group of some 70 annual species, the best known member being the spice plant, *T. foenum-graecum*. *Trigonella* is taxonomically difficult because it merges into the closely related *Medicago* or *Melilotus*; some of its species at one time or another have been classified under these other genera. Results were obtained with 35 species (Table 15.2), which variously produced one or more of five phytoalexins (Ingham and Harborne, 1976). Two of these are pterocarpans, medicarpin and maackiain and two are isoflavans, vestitol and sativan (Fig. 15.4). The results of the survey indicate that *Trigonella* can be divided into three groups: group 1, with medicarpin and maackiain; group 2, with medicarpin alone; and group 3 with medicarpin and vestitol. These groups are further divisible into subgroups on the basis of further minor variations in phytoalexin production (Table 15.2).

Table 15.2 Groupings of *Trigonella* species based on patterns of phytoalexin production

Group 1 Medicarpin + maackiain	Group 2 Medicarpin	Group 3 Medicarpin + vestitol
Group 1A (ratio 1: 1)[a]	Group 2A (in quantity)[a]	Group 3A
T. berythaea	*T. anguina*[b]	*T. brachycarpa*
T. foenum-graecum	*T. arabica*[b]	*T. noëana*
T. gladiate	*T. balansae*[b]	*T. radiata*
Group 1B (ratio 10 : 1)[a]	*T. caelsyriaca*[b]	Group 3B (with sativan as well)
T. caerulea	*T. corniculata*[b]	*T. arcuata*
T. kotschyi	*T. cretica*[b]	*T. fischerina*
T. melilotus-caerulea	*T. hamosa*	*T. geminiflora*
T. procumbens	*T. rigida*[b]	*T. incisa*
T. sibthorpii	*T. schlumbergeri*[b]	*T. monantha*
	T. spikata	*T. orthoceras*
	T. stellata[b]	*T. platycarpos*
	T. suavissima[b]	*T. polycerata*
	T. uncata[b]	
	Group 2B (in traces)[a]	
	T. lilacina	
	T. monspeliaca	
	Group 2C (with unknown pterocarpan)	
	T. calliceras[b]	

[a]Quantitative differences in response.
[b]Coumarin producer; all others negative for this character.

That these phytoalexin studies divide *Trigonella* species into systematically meaningful groupings is apparent from supporting morphological and chemical evidence. In particular, many species from group 2 show some morphological resemblance to *Melilotus*, a genus characterized by the uniform synthesis of medicarpin. Further, a special chemical feature of *Melilotus*, namely the ability to release the volatile sweet-smelling coumarin when tissue is macerated, is present in most *Trigonella* species of group 2, but is absent from members of groups 1 and 3. Again the 11 species of group 3 have morphological affinities with members of *Medicago*. It is significant here that several attempts have been made to move one member of group 3, *T. platycarpos*, into this latter genus.

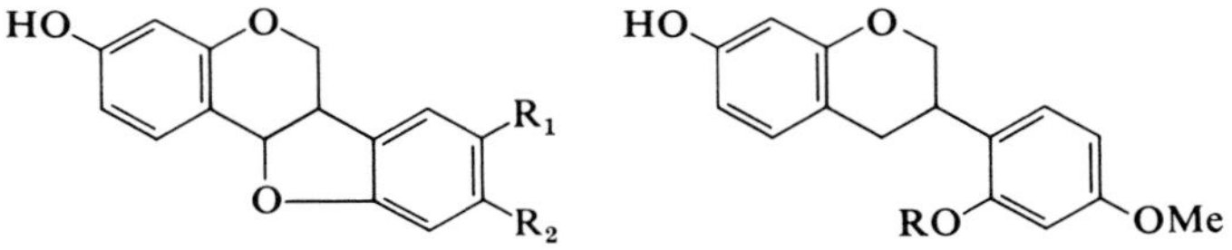

Medicarpin (R_1 = H, R_2 = OMe)
Maackiain (R_1 = R_2 = OCH_2O)

Vestitol (R = H)
Sativan (R = Me)

Fig. 15.4 Phytoalexins of *Trigonella*.

Similar surveys in other legume genera, while not showing such clear-cut groupings as in *Trigonella*, have revealed some interesting correlations between phytoalexin structures and taxonomic groupings. A study of 55 species in *Trifolium* showed that most produce pterocarpans and isoflavans, with several structures being unique to the genus (Ingham, 1978). Exceptionally, the stilbene resveratrol is the only phytoalexin formed in two species, *T. campestre* and *T. dubium*; significantly these both belong in the same section *Chronosemium* of the genus. Again, a survey of 29 species belonging to 10 sections of *Lathyrus* revealed the pterocarpan pisatin in all but one taxon (Robeson and Harborne, 1977). The exception was *L. nissolia*, section *Nissolia*, which is unlike any other *Lathyrus* species in its superficially grass-like appearance due to the elongated leaf phyllodes. Pisatin is replaced in this morphologically distinctive species by two new pterocarpans, not previously recorded in the Leguminosae (Robeson and Ingham, 1979).

The taxonomic implications of phytoalexin induction in the Leguminosae can perhaps be better appreciated when the results obtained are analysed at the tribal level. One tribe where representative surveys have been accomplished is the Vicieae (Table 15.3), in which plants pterocarpans, 6a-hydroxypterocarpans and furanoacetylenes (Fig. 15.5) are variously produced. Here, the most striking result from phytoalexin comparisons is the dichotomy in

Table 15.3 Phytoalexin differences within the tribe Vicieae

Genus	Number of species studied	Phytoalexin class	Compounds identified
Vicia	27	Furanoacetylene	Wyerone and wyerone epoxide widespread
Lens	2		
Pisum	2	6a-Hydroxypterocarpan	Pisatin in all *Pisum* and in 29 of 31 *Lathyrus* spp. representing 10 sections
Lathyrus	31		
Cicer	1	Pterocarpan	Medicarpin and maackiain

Data mainly from Robeson and Harborne (1980) but for *Cicer*, see Ingham (1976). Other compounds besides those mentioned were found in individual species. Traces of medicarpin have been found in *Vicia faba*, but the major response is furanoacetylene production (Hargreaves *et al.*, 1976). The genus *Vicia* was screened mostly using cotyledon rather than leaf tissue.

the tribe, which cuts across previous views of the relationships between the four main genera. Thus *Vicia* and *Lens* are grouped together, since they both produce furanoacetylenes (Robeson and Harborne, 1980) a type not known anywhere else in the Leguminosae, in spite of a wide search (Ingham, 1980).

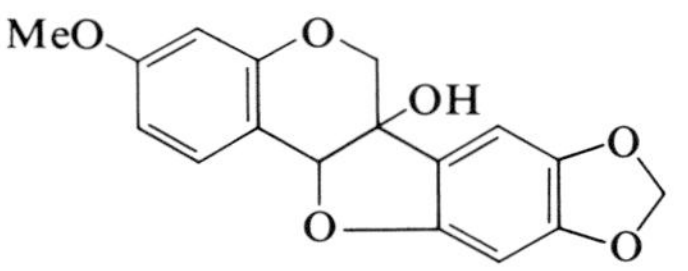

Pisatin
(*Pisum, Lathyrus*)

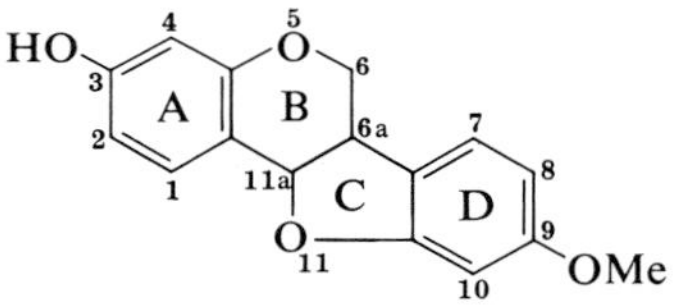

Medicarpin
(*Cicer*)

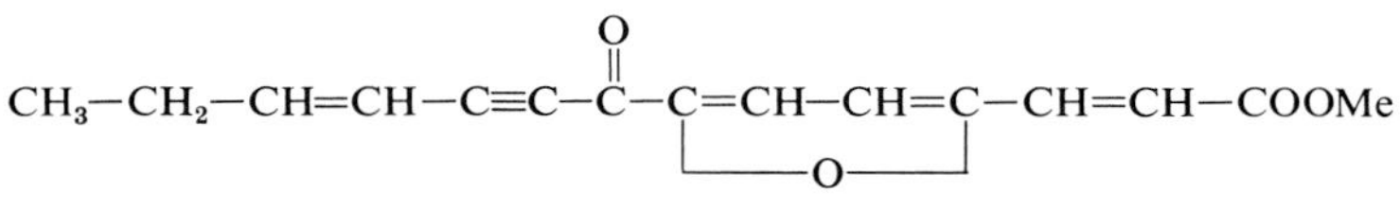

$$CH_3-CH_2-CH{=}CH-C{\equiv}C-\overset{\overset{O}{\|}}{C}-C{=}CH-CH{=}C-CH{=}CH-COOMe$$

Wyerone (*Vicia, Lens*)

$$CH_3-CH_2-\overset{O}{CH-CH}-C{\equiv}C-\overset{\overset{O}{\|}}{C}-C{=}CH-CH{=}C-CH{=}CH-COOMe$$

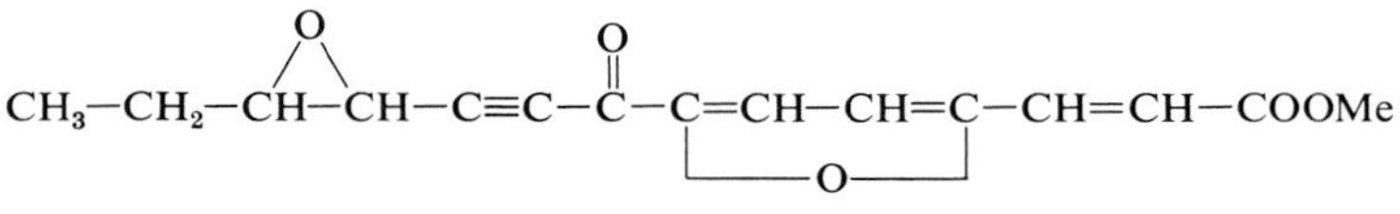

Wyerone epoxide (*Vicia, Lens*)

Fig. 15.5 Phytoalexin structural variation in the Vicieae.

By contrast, *Pisum* and *Lathyrus* are united in forming the usual legume-type phytoalexin, that based on pterocarpan (Robeson and Harborne, 1977). Species of these latter genera regularly produce pisatin as well as several other related isoflavonoids. This chemical separation of *Vicia* and *Lathyrus* thus disclosed is unexpected, since these two genera are morphologically close; they are also similar in this ability to accumulate large amounts of non-protein amino acids in the seeds (Bell, 1966).

As indicated in Table 15.3, *Cicer* is somewhat distinct from *Pisum* or *Lathyrus* in making a simple pterocarpan lacking the 6a-hydroxyl present in pisatin, namely medicarpin. This chemical difference supports the separation of *Cicer* into a separate tribe, the Cicereae, as recently proposed by Kupicha (1977) on the grounds of serology and pollen morphology.

Table 15.4 Distribution of isoflavonoid phytoalexins in seven tribes of the Leguminosae

Tribe	Presence or absence of: Isoflavone	Isoflavanone	Pterocarpan	Isoflavan
Cajaneae	+	+	−	−
Diocleae	−	−	+	−
Glycineae	+	−	+	−
Phaseoleae[a]	+	+	+	+
Vicieae[a]	−	−	+	−
Trifolieae[a]	−	−	+	+
Loteae [a]	−	−	−	+

[a]Members of these tribes also contain non-isoflavonoid phytoalexins.

The results from phytoalexin induction are also meaningful above the tribal level, since different legume tribes within the subfamily Papilionoideae have different patterns (Table 15.4). These data have evolutionary significance, since the different types of isoflavonoid phytoalexin can be placed in a biogenetically developing series. This is shown in Fig. 15.6 and is based on the latest available data from biosynthetic studies on isoflavonoids. It is significant that the efficiency of the various isoflavonoid types as fungitoxins follows the same progression, isoflavones and isoflavanones being the least effective and isoflavans the most. According to this scheme, the Cajaneae are a primitive group, the Phaseoleae intermediate and the Loteae advanced. Such a phylogenetic series is not seriously at variance with ideas on tribal relationships derived from other sources. Many critical groups, however, remain to be examined, so that the present data are only a pointer to the possible benefits that may arise from more intensive investigations of phytoalexin synthesis among these plants.

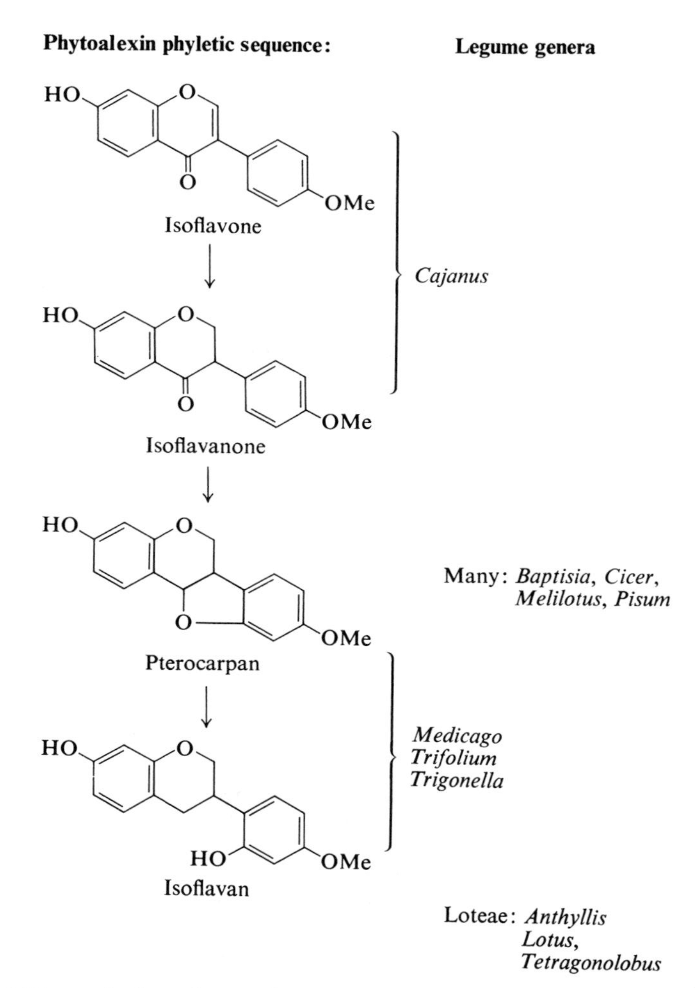

Fig. 15.6 Biosynthetic pathway of phytoalexins linking increasing fungitoxicity with evolutionary advancement.

V. Conclusion

It has been proposed in this chapter that one class of disease-resistant factors, the phytoalexins, offer a new line of enquiry for systematic purposes in plants. These substances are particularly promising because they are specifically elicited by fungal pathogens in a convenient and simple test system—the leaf drop-diffusate technique. They have been shown to be induced in at least 17 families and within one of these, the Leguminosae, they are clearly taxonomically useful. Chemotaxonomic studies of phytoalexin induction would certainly be applicable in the future to other plant groups.

Other disease-resistant factors present in plant tissues might also be worth surveying, in a taxonomic context. Fungitoxic substances are known to occur sometimes on the leaf surface: thus isopentenyl isoflavones have been so detected on *Lupinus* leaves (Harborne *et al.*, 1976) and diterpenes on *Nicotiana* leaves (Bailey *et al.*, 1974). The screening of leaf washings for petroleum-soluble fungitoxic substances thus presents another feasible approach to plant chemotaxonomy. In addition, preformed bound toxins have been recorded in a variety of plant species. The free toxins are released by enzymic hydrolysis during microbial attack. Again, there is evidence of considerable chemical variation in such bound toxins, many of them being saponin in nature (see Harborne and Ingham, 1978). The taxonomic possibilities of such toxins have already been explored in the case of the lactone glucosides, tuliposides A and B, which occur exclusively in five related genera within the Liliaceae and Alstroemeriaceae (Slob *et al.*, 1975). Comparative studies of the saponin toxins, known to be present for example in Gramineae, Araliaceae and Primulaceae, would seem to be particularly worthwhile.

16
The Handling of Chemical Data

I. Introduction 374
II. Infraspecific treatment of chemical data 376
III. Specific and generic treatment of chemical data 377
IV. Cladistic analyses 381
V. Conclusion 385

I. Introduction

In the final paragraph of the concluding chapter of Alston and Turner's text, Biochemical Systematics (Alston and Turner, 1963a), it was stated that '. . . the art of assessing the phylogenetic value of morphological data is further advanced than the art of assessing the value of biochemical data' and although '. . . we know far less at this time about variation in the chemistry of the plant, it is probable that in fifty years this situation will be reversed'. Indeed, this has already proven to be so. The score of years since 1963 has seen the development of powerful tools for the isolation and identification of micromolecular components; while the sophisticated instrumentation and techniques that have evolved out of macromolecular studies have permitted an awesome assemblage of data which has provided an insight into evolutionary mechanisms that would have been difficult, if not impossible, to predict from the state of the art at that time.

This advance in methodology has been accompanied by equally potent conceptual insights and developments such as, for example, numerical programs for the analysis of both phylogenetic and phenetic data. This includes statistical treatment of population data, automatic weighting schemes based on such information and yet other mathematical treatments too numerous to discuss in detail.

Macromolecular developments, in particular, have seen the introduction of concepts that have permitted the erection of 'molecular clocks' for phyletic purposes at the higher categorical levels as well as population theory, using isozymic data, for the calculation of 'genetic distance' among more closely related taxa. These developments, as well as others, will be touched on in the next two chapters and will not be discussed further here.

In the early period of chemosystematics most data were treated phenetically. That is, they were considered as mere 'bits' of information that might be added up in columnar fashion to support this or that comparative relationship. In fact, many, if not most, of these early studies were based on mere 'spot data', especially in studies by plant systematists. Ellison *et al.* (1962) reviewed problems relating to the treatment of chromatographic spots and adapted polygonal methods for the presentation of such data. Although very 'crude' by current standards, their technique provided a visual configuration of the chemical data that was readily understood by morphological workers since such graphs were also used in a similar fashion for the presentation of morphological data.

A number of workers adopted this procedure for the presentation of mostly chromatographic profiles (e.g. Ellison, 1964; Harney and Grant, 1965; Tilney and Lubke, 1974). While acceptable in a phenetic sense, the polygonal graphs are only as valid as the spot-data is truly reflective of the secondary chemistry of the taxa concerned. However, the enthusiasm of early workers to compile spot-data without sound training in chromatographic methods, and care in the collection of appropriate plant organs for analysis (Chapter 9), undoubtedly introduced much unneeded, if not artifactual, variation into such presentations.

Other workers attempted to treat spot-data in a more numerical vein. Weimark (1972) commented critically on the errors inherent in such treatments, a paper which Adams (1974) subsequently criticised largely on the grounds that criteria for good chemotaxonomy is no different from the criteria for good morphotaxonomy, whether the compounds are represented as chromatographic spots or identified chemically. Weimark (1974) again reviewed the problems inherent in the numerical treatment of 'spot-data', but most subsequent chemosystematics have dealt mainly with *identified* compounds and this has changed considerably the methods by which such data might be evaluated.

Of course, numerical treatment of *identified* chemical components is also possible (Parker, 1976) but prior to about 1974 such data was mostly treated as essentially phenetic data. For example, Challice and Westwood (1973) used 22 morphological characters and 29 flavonoids (all identified) to classify species of the genus *Pyrus* (pears); their resulting phenetic treatment was based in part on the presence or absence of individual compounds and no attempt was made to provide a *numerical* phylogenetic treatment. However, they did construct a hypothetical phylogenetic tree (Fig. 16.1) whose branches

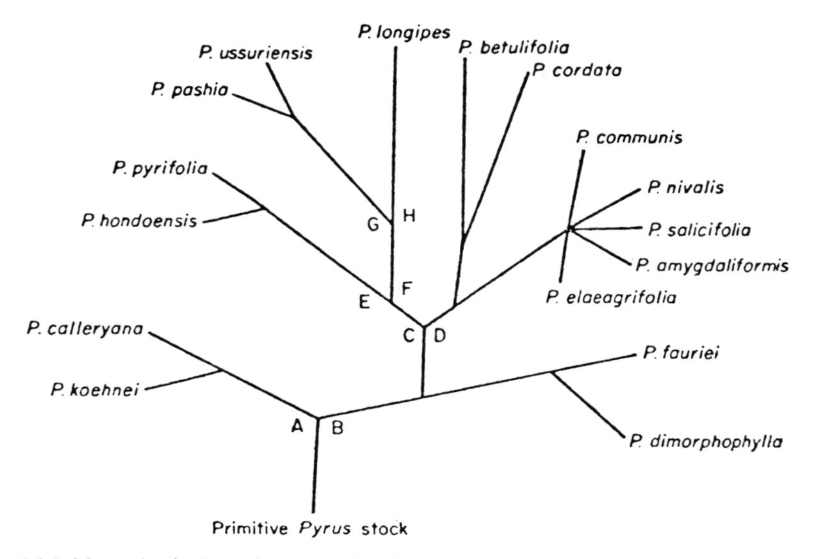

Fig. 16.1 Hypothetical phyletic relationships among *Pyrus* species as proposed by Challice and Westwood (1973). Letters refer to the following: A, retention of C_6–C_1 phenolic acid esters of calleryanin from original primitive *Pyrus* stock; B, loss of above; C, retention of flavone glycosides from original primitive *Pyrus* stock and gain of ability to synthesize flavone FS; D, loss of flavone glycosides; E, continuation of line (C); F, gain of ability to *O*-glucosylate flavones at the 4′-hydroxyl; G, retention of luteolin 7-rhamnosylglucoside; H, loss of above.

were largely determined by assumed biosynthetic pathways based on a 'primitive' stock with gain or loss of various substitutions on particular flavonoid types. The latter is perhaps a better utilization of the chemical data than their use in phenetic clusterings. At least a tree erected from these data is perhaps more convincing than one erected from purely morphological data, largely because chemical homologies and their transmutations along metabolic pathways are more readily perceived by the systematist than are 'morphobolic' pathways.

II. Infraspecific treatment of chemical data

Because of the populational complexity of interspecific chemical variation in plant taxa we tend to believe that the most difficult problem involved in such analyses is that of sampling. It is a rare situation where closely related allopatric infraspecific taxa possess absolutely unique compounds. Rather, because of their recent divergence and/or occasional gene exchange among peripheral populations, such categories or populations are characterized by statistical differences in the frequence of compounds or else are characterized by suites of cohering quantitative variables that are perhaps best portrayed in contour form after detailed numerical-computer analysis (Adams, 1970b; Flake and Turner, 1968, 1974; cf. also chapter 10). Such methods require well-organized sampling schedules, and consume much time and energy, so much, in fact, that some workers feel that the results are not worth the effort and despair at attempting such treatments. But for a broad-ranging taxon which occupies several or more physiognomic provinces, chemosystematic analyses of this nature can be very rewarding, especially if the taxa concerned are important to agronomic research or might serve as genome donors to established crop plants.

III. Specific and generic treatment of chemical data

On the surface it would appear that the treatment of chemical data at the specific or generic level would be similar to that at the infraspecific level. In as much as the latter might ultimately be expected to evolve into good biological species (cf. chapter 11), numerous intermediate situations undoubtedly occur in nature and the chemosystematist must judge treatment of his data according to the variation exposed.

At the specific level and higher, most taxa which will have been sampled carefully and thoroughly are likely to show both quantitative and qualitative differences in their micromolecular components. The latter are difficult to treat on a par with the former; at least most numerical schemes have not provided adequate methods for their treatment together in a convincing or meaningful way. That is, how does one evaluate missing data among taxa? If contour lines based on population or statistical data are drawn to delineate chemical relationships such as provided by Adams and his workers (see above and chapter 10) does one equate absence of one or more compounds as mere quantitative variation (i.e. 0.1% versus 99.9%), or how?

Problems of the above nature have led most chemosystematists to treat specific chemical data as either 'present' or 'absent' (although trace amounts might be categorized as either one or the other). Because of this, chromatographic components in the period 1950–1970 were generally treated in a phenetic way. Such treatments are valid and serve a purpose if one wishes merely to cluster taxa without insight into their likely phyletic relationships (for an excellent *introductory* account of such numerical treatments see Dunn and Everitt, 1982).

As indicated above, that chemical characters might contain 'directional' phyletic information was noted early on by both enzymologists as well as by comparative phytochemists (Bate-Smith, 1962; Harborne, 1967a), including those of a much earlier generation (cf. Alston and Turner, 1963a).

Bate-Smith (1973), in an attempt to use flavonoid data in a more meaningful, if not more rigorous way, introduced a 'point-method' or numerical score for each taxon based on the number of 'primitive' or 'advanced' chemical components each possessed. A 'primitive' component was said to be one whose formation preceded the more 'advanced' component in a metabolic pathway. Earlier phytochemical workers espoused such concepts, but Bate-Smith was apparently among the first to apply a points system (e.g. a zero or less for a primitive or basal compound; higher scores for more advanced compounds). His first effort in this direction dealt with species of *Geranium* but in a later paper he applied the method to *Ulmus* (Bate-Smith and Richens, 1973).

Such ideas are not new, however. Morphological systematists have long attempted to assess the 'primitive' or 'advanced' state of a given character or group of characters to ascertain which species among several or more, might possibly have been the progenitor of yet other more derived populations or species. To do this they merely surmise from morphogenetic theory, character distributions among related taxa, or from biogeographic theory, what character states might or might not be primitive. Numerical states were assigned to such characters (usually zero for 'primitive' and one for 'advanced') and these would then be added up to achieve some sort of relative position, one to the other, in some imaginary tree diagram. The branching pattern of the hypothetical tree would normally be constructed according to character-paths; i.e. it would be assumed that a number of different states of the same character suggested a lineage. In such a fashion the junior author erected his hypothetical phylogeny of *Hymenopappus* (cf. Fig. 2.2, chapter 2).

This approach to biological classification was formalized early on by Prof. W. H. Wagner, fern systematist at the University of Michigan, and a considerable cult has developed around its use (cf. Wagner, 1980). Such diagrams are often referred to as Wagner diagrams. It is to be noted that Wagner, himself, modestly refers to the use of such diagrams as 'the groundplan-divergence method'. Such diagrams are projected as targets with radiating lines from a zero (central or primitive) point to an ever-expanding circular periphery where the more advanced taxa are positioned by numerical scores.

It should have been obvious that the use of such techniques was ideal for chemical characters, especially since the concept of biosynthetic pathways has long been known to biochemists. Nevertheless, relatively few applications using Wagner diagrams have been applied to chemosystematic data. This is probably because relatively few genera were sufficiently investigated to warrant such a treatment and early workers were not really certain of the directionality of a given hypothetical pathway.

One of the better chemosystematic studies using a Wagner diagram is that of Bacon (1978) who investigated for flavonoids all ten species of the largely Mexican genus *Nerisyrenia* (Cruciferae). From among these he detected 40 flavonoids, mostly ether-linked glycosides of the flavonols kaempferol, isorhamnetin and quercetin, of which 31 were isolated and fully characterized. Following the views of Harborne (1967a), that increasing complexity of glycosylation patterns is indicative of evolutionary advancement, such as occurs in *Baptisia* (Harborne, 1971; Turner, 1969), Bacon was able to construct the hypothetical phyletic diagram shown in Fig. 16.2.

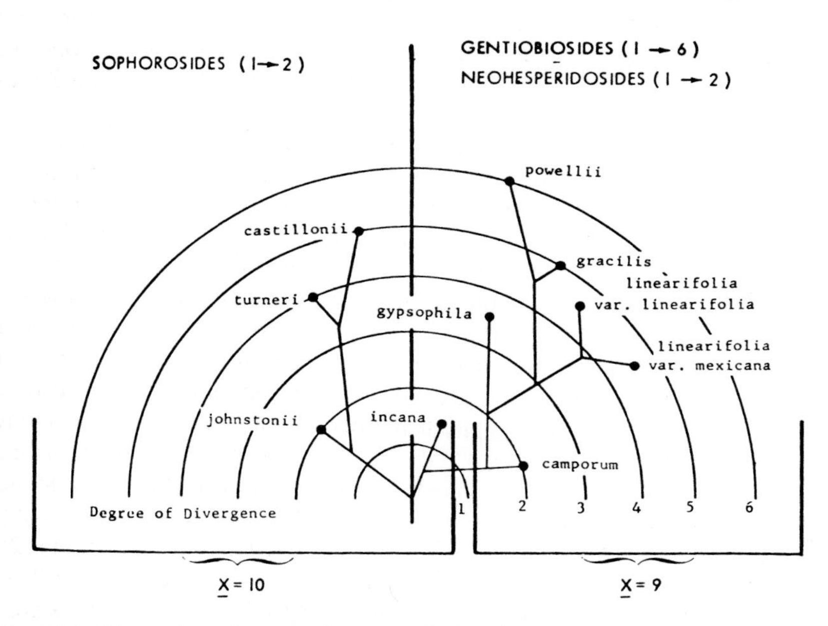

Fig. 16.2 Wagnerian diagram showing relationship among species of *Nerisyrenia*. The right side of the figure shows those species with a base chromosome number of $x = 9$; those to the left have $x = 10$. The concentric position of taxa was determined largely from morphological data. The major flavonoid chemistry of the two groups is indicated along the top. Positioning of the various taxa, as shown, could not have been deduced from purely morphological or chemical data. (From Bacon, 1978.)

From a systematic viewpoint the really remarkable aspect of Bacon's study is that it combines a synthesis of information from cytology, morphology *and* chemistry. Thus, he notes that his chemical data alone might have led to one classification and the morphological data to quite another:

> The relationships illustrated in Figure [16] 2 are based largely upon morphology, although chemical evidence has been utilized in the recognition of this

> dichotomy. Some minor chemical incongruities exist in the line in which *Nerisyrenia incana* is basal; indeed, a better "chemical phylogeny" would result by assuming that *N. powellii* and *N. castillonii* are the more "primitive" taxa, a reversal of that proposed on morphology. This suggestion arises from the fact that *N. johnstonii*, *N. castillonii* and *N. turneri* all utilize glucose and galactose as glycosidic constituents. In the other grouping, the species considered "primitive" on morphological grounds do not utilize galactose, whereas galactosides occur sporadically in the more advanced taxa and become quite frequent in the most advanced species. If it is assumed that the original ancestor of *Nerisyrenia* had the ability to produce both types of glycosides, then one might expect this ability to be retained in the more primitive species. This assumption would favor *N. powellii* and *N. castillonii* as the most primitive species. Furthermore, by assuming that the former species are primitive, one finds a progressive decline in the ability to produce galactosides in the "powellii" line, resulting in a decrease in the complexity of the flavonoid profile, while a similar trend is seen in the "castillonii" line, since *N. johnstonii* would then be an advanced taxon, and it lacks the ability to produce the triglycosides characteristic of *N. turneri* and *N. castillonii*. Under this scheme, the trend in flavonoid chemistry would support the generality proposed by Mabry (1973) that within a genus reduction in the complexity of the secondary products chemistry, reflecting a loss of biosynthetic ability, is indicative of evolutionary advancement.

Bacon then goes on to state his reasons for accepting largely a morphological and cytological bias for his arrangement which is essentially that *Nerisyrenia incana* and *N. johnstonii* share morphological, cytological (and chemical) relationships with the related genus *Synthlipsis*; and such considerations lead to the cladistic analysis of chemical characters, which will be discussed in more detail below.

Inclusive chemosystematic treatments of the type rendered by Bacon are unfortunately rare. Most studies are based on small samples and relatively few attempts are made to relate the chemical data to yet other character sets. Nevertheless, phytochemists have not been reluctant to use their data in as vigorous a fashion as have 'morphogenetic' phylogenists or chromosomal enthusiasts. The recent book by Gottlieb (1982) attests to the zealous enthusiasm with which micromolecular components ('allelochemics') might be applied to phyletic problems at whatever taxonomic level. In this he has proposed a number of point methods for the evaluation of chemical data, both at the generic and familial level, or higher.

Richardson and Young (1982) have reviewed much of the literature relating to the phylogenetic content of 'flavonoid point scores'. They point out that phylogenetic arrangements based on numerical scores, such as championed by Gottlieb (1982) and yet other workers, have serious limitations. Richardson and Young rightly note that:

> In all the methods used to obtain flavonoid scores, a particular species is given a final score. This is obtained by either adding up the scores for individual

compounds present in the species or adding up the points accumulated for the particular molecular modifications occurring in the biosynthesis of individual compounds, or instead by using an average value obtained by dividing the total score by the number of compounds in the species. This final score is then used to compare species (or groups of species) in terms of their relative evolutionary advancement. However, the use of these final scores for comparisons conceals much of the information which might otherwise be used to deduce the phylogenetic relationships among the taxa. For example, two species with the same final flavonoid score may either contain the same compounds or they may contain different compounds with an equivalent point value, or a combination of both. This distinction is very important if one wishes to determine the phylogenetic relationships of the species.

Crawford and Levy (1978) presented a perceptive analysis of the problems involved in the comparison of flavonoid profiles noting that one should distinguish between 'biosynthetic affinities' and 'genetic affinities'. Thus, the fact that two plants have identical flavonoid profiles does not necessarily mean that they are identical genetically at those loci involved in the biosynthesis of the component compounds. They go on to document this statement with Brederode's and co-workers data on the genetic control of flavonoids in the genus *Melandrium*, a series of publications which are notably lacking in the ambitious presentation of Gottlieb (1982) and which might account in part for his misunderstanding (p. 21) of Crawford's (1978) excellent review of the systematic significance of flavonoids among angiosperms generally:

> [Crawford advises us] . . . to forget about micromolecular systematics and to embark on the study of macromolecules in spite of this author's admonition that the time and labor involved are immense and that such investigations will not become common place in the near future.'

Knowing Crawford as we do it is unlikely that he intended such an admonition. What we understand him to say is that much more needs to be known about the enzymological control of 'allelochemics' before we can attribute to them the degree of homology among distant (or even closely) related taxa before any reasonably accurate phylogenetic insights can be forthcoming.

IV. Cladistic analyses

Within the past decade or so a 'new breed' of plant systematists has emerged from the heretofore largely classically oriented field in which morphological data were mostly treated phenetically. The subdiscipline arose first in zoology, primarily because animals have a relatively good fossil record which provides a background fabric of phylogenetic theory. As noted in chapter 3, it was inevitable that cladistic (or phylogenetic) treatments become the favoured approach to systematic monography by plant workers generally, for if plants evolved via branching patterns through time, then the concerned intellectual is wont at least to speculate on the make-up of such branches and the relative age of their divergence, regardless of the rate of morphological differentiation after such branching. In other words, plant systematists, like animal systematists, are intrigued by the two phenomena: divergence as a reflection of speciation ('taxation') and the relative age of branching patterns as a reflection of phylogeny.

Among plant workers, proceedings from several recent symposia have been published (e.g. Stuessy, 1980; Funk and Brooks, 1981) which cover the concepts and methods of cladistic treatments in considerable detail. Unfortunately, relatively few presentations have dealt with chemical data, in spite of the fact that such data provide some of the better character suites for cladistic analysis.

As to a relatively brief simple explanation of what cladistics is all about we can do no better than refer the interested worker to the very short readable accounts by Patterson (1980) and Charig (1981). But in order to present a description of its use with chemical data, we have drawn on the treatment of the Imaginaceae by Richardson and Young (1982). Although this is an idealized imaginary problem it can be readily understood as to purpose and easily visualized in its application. Their account follows (Table and Figure numbers have been changed to conform with those of this chapter):

> The family Imaginaceae contains four taxa and five flavonoid features. These five flavonoid features are not present in the hypothetical outgroup, the family most closely related to the Imaginaceae. This outgroup, or sister group, to the Imaginaceae contains only flavonoid-3-*O*-monoglycosides. So in the context of the Imaginaceae the five flavonoid features in Table 16.1 (5-*O*-glycosylation, *O*-methylation, acylation, complex glycosylation and sulphation) are considered to be advanced flavonoid features. The distribution of these

Table 16.1 The flavonoid features of the Imaginaceae and their primitive and derived (advanced) states. (From Richardson and Young, 1982.)

Character	Primitive state (−)	Derived state (+)
A 5-*O*-Glycosylation	Absent	Present
B *O*-Methylation	Absent	Present
C Acylation	Absent	Present
D Complex glycosylation	Absent	Present
E Sulphation	Absent	Present

characters in each species of the Imaginaceae is shown in Table 16.2. The presence of any advanced flavonoid feature is given a score of 1, and a total flavonoid score for each species is shown. Note that *Imagina compounda* and *Makeupa featura* both have a flavonoid score of three but they have quite different flavonoid profiles. The cladogram in Fig. 16.3 is constructed directly from the data in Table 16.2. Character D (complex glycosylation) is present in all four species and leads to divergence level 1. Here the cladogram branches depending on the presence of character C (acylation) or A (5-*O*-glycosylation) or B (*O*-methylation). Character C leads to divergence level 2 and the species *Thinkova structura* with the divergence formula CD. Characters A and B (in either order) lead to divergence level 3 and the species *Imagina compounda* with the formula ABD. Character E (sulphation) occurs in two species on lines that have already diverged on the cladogram and we may consider the occurrence of sulphation in these species as being parallel developments. This is also the most parsimonious explanation of the data, i.e. any other arrangement of the characters on the cladogram would result in a tree containing more branches (or reversals). Thus the cladogram (Fig. 16.3) indicates that all four species have one advanced flavonoid feature in common and then a divergence occurs to form two branches, each with two species. Both branches share one character and this character is interpreted as being independently derived on two occasions. Note that *Makeupa featura* and *Imagina compounda* both lie on level 3 but on divergent lines!

Table 16.2 Occurrence of derived flavonoid characters in the Imaginaceae. (From Richardson and Young, 1982.)

	Character states					Flavonoid
	A	B	C	D	E	score
Imagina structura	+	+		+	+	4
I. compounda	+	+		+		3
Thinkova structura			+	+		2
Makeupa featura			+	+	+	3

They follow up this simple presentation with an analysis of real data from the Campbell *et al.* (1979) treatment of flavonoids among *Marchantia* and its allies. The latter authors used a point-count method for their determination of primitive versus advanced taxa, 'averaging' such counts for the positioning of a given taxon. Richardson and Young's (1982) system does not average or calculate mean values; rather, as in most cladistic analyses, they note that advanced or derived compounds provide the basis for branching lengths and hence degree of divergence, while the presence of primitive compounds provide little or no useful information for phylogenetic reconstruction. Taken together, however, shared characters, evaluated at their full numerical value, can be used to construct a readily comprehensible hypothetical phylogenetic arrangement or cladogram (Fig. 16.4). It will be noted that they superimpose their arrangement upon a Wagnerian diagram much in the manner of Bacon's arrangement of *Nerisyrenia* (see Fig. 16.2) except that the latter utilizes, in addition to chemical data, both morphological and chromosomal data.

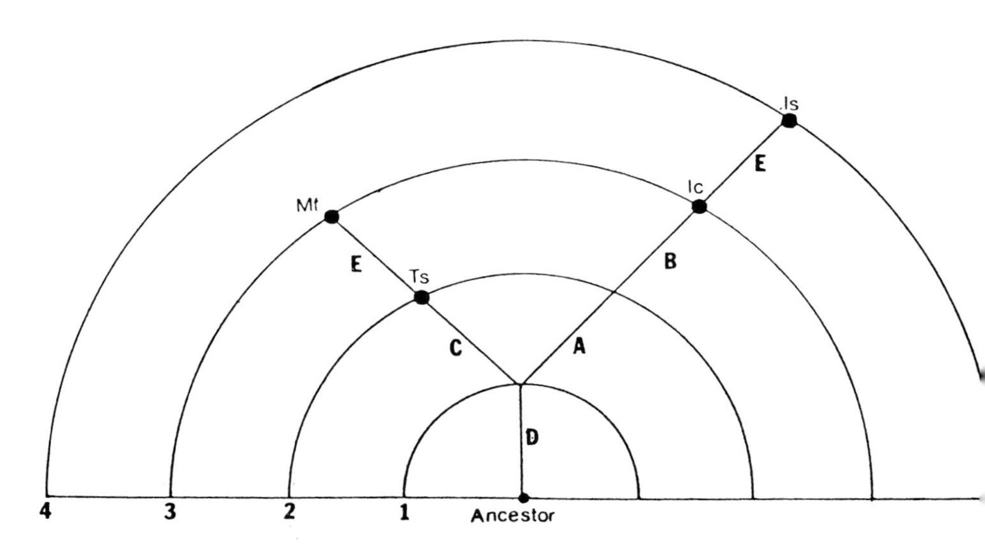

Fig. 16.3 Cladogram of the four species of the Imaginaceae. For abbreviations see Tables 16.1 and 16.2. (From Richardson and Young, 1982.)

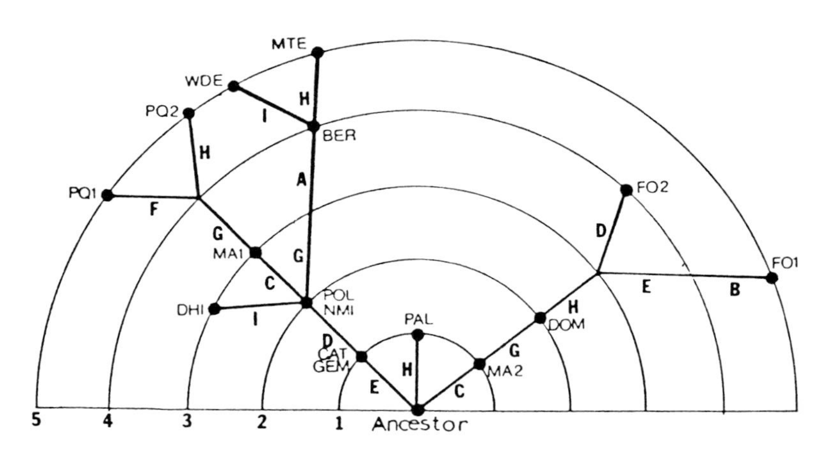

Fig. 16.4 Cladogram of relationships in the Marchantiaceae based on shared advanced flavonoid compounds. (From Richardson and Young, 1982.)

Regarding *Nerisyrenia*, it should be emphasized that Bacon did not perform a purely cladistic treatment of the chemical data as espoused by the Hennigian School (cf. Charig, 1981). Had he done so (personal communication), comparison of his flavonoid data with that from his outgroup would have yielded essentially the arrangement he projected from ingroup data (albeit three sets, including flavonoids, morphology and chromosomal) in which *N. incana*, *N. camporum* and *N. johnstonii* are viewed as the basal species. As noted in the discussion above, consideration of the ingroup flavonoid data, taken alone, would be open to several interpretations of parsimony and hence 'phyletic validity'.

Seaman and Funk (1983), using sesquiterpene lactone data, have presented somewhat similar views to those of Richardson and Young in their analysis of these compounds among the Compositae, especially the genera *Iva* and *Tetragonotheca*. Like the latter authors, they also deplore the numerical averaging of chemical data and emphasize the need to consider foremost metabolic pathways, successive modifications among these, and the search for comparative pathways among outgroups so that more reliable phyletic insights might be had.

Finally, it should be noted that in all of the above diagrams no account is given of the consequences of hybridization among specific taxa within these groups and the reticulate phylogenetic web that such co-mingling might produce. Purely cladistic approaches do not resolve, nor do they reveal, such webs. Rather, the experienced taxonomist must try to evaluate his chemical and other data sets against a broad spectrum of possible phyletic events, including hybridization among taxa, both gradual and rapid speciation, abrupt speciation, to say nothing of parallelisms, character reversals, and yet other pitfalls of the practising cladistist.

That hybridization is so rampant among higher plants, as opposed to the higher animals, must impose a higher degree of restraint on the phytochemists in their phyletic considerations of chemical data. But it should not deter the concerned from attempting to place their data in some such perspective.

V. Conclusion

In summary, various methods have been proposed to treat micromolecular data. Early workers tended to use these in a purely phenetic way. But nearly all recent workers have tended to treat their data in a phyletic way, either applying numerical point-scores to determine degree of advancement of a given taxon, or by attempting cladistic analyses to ascertain branching patterns. The ideal treatment is one that attempts to do both, but at the same time making an effort to incorporate data from yet other disciplines, and especially combined with field studies to ascertain if hybridization is a factor in the taxa concerned.

Part III
Macromolecular Approaches

17

Proteins

I.	Introduction	389
II.	Amino acid sequence analysis	395
III.	Storage protein analysis	408
IV.	Fraction I protein analysis	420
V.	Isozyme analysis	424
VI.	Serological analysis	429
VII.	Summary	437

I. Introduction

No other characters in biochemical systematics have generated more controversy and discussion during the long period of their application to plants than have the proteins. Thus, the early work on serological comparisons of plant protein extracts by Mez and his associates (see e.g. Mez and Ziegenspech, 1926) was subjected to both valid technical and prejudiced irrational criticism. As a result, their findings have not stood the test of time. More recently, as noted below, developments of powerful chemical techniques have permitted the direct comparison of proteins, particularly that of amino acid sequencing; consequently, as noted by Turner (1967), biochemists interested in the evolution of proteins have developed phyletic concepts and ideas quite similar to those propounded by organismally oriented workers of an earlier century. Thus Pauling and Zuckerkandl's (1963) idea for the 'molecular restoration' of extinct proteins via extrapolation from what is known of extant proteins (by using comparative sequence data) is nothing more than the plant taxonomists' attempt to reconstruct a 'primitive flower' from comparative studies among flowering plants generally. But oh the difference in sophistication! The organismal taxonomist, especially at the familial level and higher, is hounded by a complex set of organs, the structural or genetic basis of which is not known; but the molecular worker using sequence data is presented with a relatively simple set of variables (about

20 amino acids) the sequential base of which rests on a yet simpler genetic code (four nucleotides in sets of three).

Indeed, it is this difference in sophistication of data that led the junior author (1969) to his somewhat audacious remark (at the time) that:

> *If* I were interested in obtaining the most meaningful arrangement of present-day angiosperm families, *phylogenetically speaking*, I would rather have available to me the primary structure (amino acid sequence) of ten *metabolically important* enzymes (such as cytochrome *c*) of all of the taxa which comprise these groups than have a detailed listing of all of the exomorphic features which characterize the groups.

When the above was composed it was largely unappreciated by the systematic community that certain protein molecules, mainly enzymes involved in metabolic processes, might have incorporated into them random mutational events which might permit them to serve as 'molecular clocks' (discussed below) so that the relative time of divergencies among taxa might be read.

The existence of a 'molecular clock', at least for plants, is variously disputed. Thus Cronquist (1976) doubts, if not deplores, the existence of such information molecules among higher plants, a view and point of logic strongly refuted by Turner (1977). As noted by Wilson *et al.* (1977):

> The molecular clock came as a surprise and is having a major impact on evolutionary biology. It allows the properties of organisms, even those with a poor fossil record, to be viewed easily from a time perspective. That is, rates of evolutionary change can be calculated whether the properties are chromosomal, morphological or behavioral. By comparing these various rates, one can identify important evolutionary parameters at different levels of organization. It is on this basis that regulatory evolution is postulated as being at the basis of morphological evolution.

Clearly the issue of molecular clocks is not a dead one as implied by Cronquist (1976). It has a rational theoretical base and its accuracy as a phylogenetic chronometer fits what is known of the fossil record, albeit largely animal (Fitch, 1976). Although the reality of 'neutral' mutations is a continuing controversy among evolutionary biologists and its implications for molecular evolution is a continuing major issue in the literature (compare for example the many articles in the Journal of Molecular Evolution, 1977–1983), the potential efficacy of comparative enzymology as a major phyletic tool for systematic purposes is difficult to deny. In fact, the use of comparative amino acid sequence data (like comparative DNA data, see chapter 18) is certain to be one of the major underpinnings of any ultimate biological classification.

This highly optimistic portrayal of the *potential* of comparative enzymology for systematic purposes should not prejudice the chemosystematist towards the seemingly more mundane approaches largely espoused in the present text. The fact is, enzymes are difficult and expensive to isolate, time-consuming to sequence and, in general, at the present state of the art, do not provide as much useful systematic data per unit input of time, or effort.

Nevertheless, in spite of the above controversy, more experiments have been done with proteins in a systematic context than with any other type of biochemical character. The main stimulation to employ proteins lies in the enormous number of bits of information hidden within their structures. Furthermore, the considerable variety of proteins (and enzymes) present universally in plant tissues means that many different taxonomic comparisons are possible. An important practical stimulus in recent times has been the elegance and ease of displaying protein and enzyme variation between organisms through the employment of gel electrophoretic techniques.

From the evolutionary viewpoint, too, proteins have certain advantages over other macromolecules in that they more clearly reflect in their structures the processes of natural selection which operate on the phenotype rather than on the DNA code itself in the genotype. The environment and the cytoplasm are known to exert their effects on the primary products of the genome, the proteins that are produced within the living cell (Watts, 1971). Comparative studies of amino acid sequences in functional proteins also have an important biochemical rationale. Knowledge of which amino acids in a particular sequence are invariant or fixed and which can vary at will can indicate to the biochemist which portions of the enzyme are essential for functional activity.

In taxonomic surveys, single proteins can be compared or the whole range of proteins in a given tissue can be compared. For pure proteins, the most information can be obtained by determining differences in the primary amino acid sequence. Much less satisfactory information on amino acid differences can be produced by 'peptide mapping'; the protein is subjected to partial hydrolysis and the numerous peptides formed are separated two-dimensionally to produce a peptide 'fingerprint'. Other properties of pure proteins (molecular weight, isoelectric point etc.) may be found to vary from species to species and produce on measurement further characters for taxonomic assessment.

When comparing the whole spectrum of proteins present in a given tissue, e.g. by gel electrophoresis of seed extracts, the information content is more restricted and has relatively little phyletic input. Each protein band displayed on an electrophoretogram may be treated as a simple +/— character, or it can be given a quantitative value, depending on its relative contribution to the total protein in the extract.

Before further discussing the application of proteins to taxonomic studies of plants, it is necessary to outline briefly their nature and properties. The proteins of plants, as in other organisms, are high molecular-weight polymers of amino acids. Twenty different amino acids are normally found in peptide linkage, the proportions of these different 'building blocks' varying from protein to protein. These amino acids are arranged in a specific linear order, as determined by the triplet base code of the DNA in the nucleus, and each protein has a well defined amino acid sequence. That of cytochrome *c* of wheat germ is illustrated in Fig. 17.1. The two ends of the sequence can be distinguished by the fact that one of the terminal amino acids has a free amino (NH_2) group and the other has a free carboxyl (CO_2H) group. The amino acids in the sequence are numbered from the amino terminus, so that in the cytochrome *c* of wheat the first position is occupied by alanine, the second by serine, the third by phenylalanine and so on (Fig. 17.2).

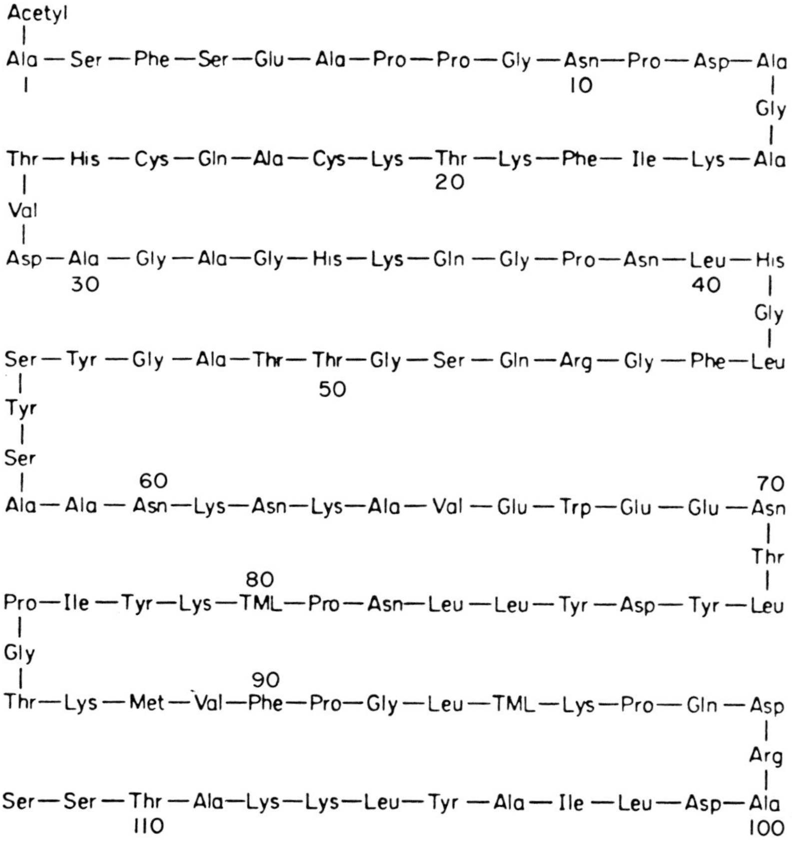

Acetyl
Ala—Ser—Phe—Ser—Glu—Ala—Pro—Pro—Gly—Asn(10)—Pro—Asp—Ala—Gly—
Ala—Lys—Ile—Phe—Lys—Thr(20)—Lys—Cys—Ala—Gln—Cys—His—Thr—Val—
Asp—Ala(30)—Gly—Ala—Gly—His—Lys—Gln—Gly—Pro—Asn—Leu(40)—His—Gly—
Leu—Phe—Gly—Arg—Gln—Ser—Gly—Thr(50)—Thr—Ala—Gly—Tyr—Ser—Tyr—Ser—
Ala—Ala—Asn(60)—Lys—Asn—Lys—Ala—Val—Glu—Trp—Glu—Glu—Asn(70)—Thr—
Leu—Tyr—Asp—Tyr—Leu—Leu—Asn—Pro—TML(80)—Lys—Tyr—Ile—Pro—Gly—
Thr—Lys—Met—Val(90)—Phe—Pro—Gly—Leu—TML—Lys—Pro—Gln—Asp—Arg—
Ala(100)—Asp—Leu—Ile—Ala—Tyr—Leu—Lys—Lys—Ala—Thr(110)—Ser—Ser

Fig. 17.1 Amino acid sequence of the protein of cytochrome *c* from wheat germ. TML = E-*N*-trimethyl-lysine

```
  1           2                3               110              111            112
NH2        CH2OH                                               CH2OH
  \          |                                                   |
   CH-CO-:-NH-CH-CO-:-NH-CH-CO····NH-CH-CO-:-NH-CH-CO-:-NH-CH-CO2H
  /                       |              |                           |
CH3                      CH2            CHOH                       CH2OH
                          |              |
                          Ph            CH3
Alanine      Serine    Phenylalanine   Threonine        Serine      Serine
AMINO                                                               CARBOXYL
TERMINUS                                                            TERMINUS
```

Fig. 17.2 Numbering of the amino acid sequence of wheat germ cytochrome *c*. (Note: the *N*-terminus, alanine, carries an *N*-acetyl substitution: this is absent from animal sequences.)

Several identical chains (or subunits) may form a higher molecular-weight hydrogen-bonded aggregate. Most plant enzymes contain two identical subunits and are called dimeric, but trimeric and tetrameric proteins are also possible. Because of the stereochemistry of the peptide bond, a polypeptide chain does not usually exist as a random three-dimensional structure. The chain is more usually coiled in the form of an α-helix and this, in turn, can fold in on itself and adopt a particular shape. Many plant proteins are, in fact, roughly rounded in appearance and hence are called globular proteins.

Proteins which contain other structural elements besides amino acids are classified as conjugated proteins. The linkage between the polypeptide chain and the other structural moiety may be ionic or covalent; in the latter case, the side chains of amino acids such as serine and cysteine may be involved. A simple conjugated protein is one that contains a metal such as iron, copper or molybdenum complexed into its structure. Examples are ferredoxin, an iron–protein complex, and plastocyanin, a copper–protein complex; the amino acid sequences of these two electron-carrier proteins will be described in the next section.

Other conjugated proteins contain lipid, phosphate, sulphate, carbohydrate or nucleic acid. Proteins with carbohydrate attachment are called glycoproteins; within this class are the lectins, special proteins involved in recognition phenomena in plants. Chromo-proteins have a chromophoric group attached and are coloured. The respiratory enzyme cytochrome *c*, for example, is yellow and consists of a polypeptide chain (see Fig. 17.1) linked to a haem (iron–porphyrin) chromophore. Such a chromophore has to be detached from the protein before sequence studies can be carried out.

Plant proteins were at one time specifically classified on the basis of solubility properties and this is still a useful criterion for distinguishing different types of storage protein, as in seeds. The classes are separated according to their solubility, or lack of solubility, in water, aqueous acid and alkali and 70% alcohol. Thus, they can be fractionated into albumins (soluble in water), globulins (soluble in salt), glutelins (soluble in dilute alkali) and prolamins (soluble in 70% alcohol). Further purification steps are usually required to completely separate the different protein classes.

An alternative system of classification is according to molecular weight. This can vary from 12 000 in cytochrome *c* and 60 000 in alcohol dehydrogenase to 500 000 in urease. Molecular weights of over 1 000 000 are also known. Proteins separate according to molecular weight when they are subjected to gel filtration on a column of Sephadex, a cross-linked dextran. The higher the molecular weight, the faster the protein moves down the column. Separation of proteins by gel electrophoresis (see section III) is also partly determined by their molecular size, although the charge properties of the protein are equally important in determining their mobility on the gel. The net charge on a protein depends on the balance of basic and acidic amino acids present in the polypeptide chain. Most plant proteins have more acidic than basic amino acids and thus have a net negative charge and move towards the anode.

A third method of protein classification is on the basis of function. Many proteins are also enzymes, catalysing particular steps in either primary or secondary metabolism, and can clearly be distinguished from other proteins by their enzymic properties. All enzymes are, of course, not necessarily functional and the purpose of some of the enzymes most easily detected in plant tissues and hence commonly used for comparative purposes, the peroxidases and esterases, is still not entirely clear. On the other hand, non-enzymic protein is definitely known to be used either for storage purposes, particularly in seeds and roots, or as part of the internal structure of the cell, in the membrane or in the cell wall.

Plant proteins have a few special features which distinguish them from proteins in animals. They are sometimes relatively low by comparison in their content of the sulphur amino acids, methionine and cysteine. Occasionally, as in cereal crops, certain protein classes are unusually low in lysine content. They also sometimes contain amino acids other than the so-called 20 protein amino acids. Plant (but not animal) cytochrome *c* has the compound trimethyl-lysine present as one of its amino acid moieties.

Proteins occur throughout the plant in all types of tissue and even a simple organ like the leaf may contain several thousand, mainly enzymic, proteins. The problem of separation and isolation of a particular individual protein can therefore be a formidable one. It is often preferable to work with plant tissue rich in protein, and this is the reason why many protein studies have been based on the seeds of legume plants such as beans and peas. For isolating enzymes, it is wise to employ very young seedling tissue, since this tends to have a particularly active enzyme complement. For this reason, isozyme studies (section V) are normally carried out on young seedling tissues.

Although the amino acid composition of the total leaf protein is practically identical, whatever the plant source, there may be very small measurable differences which can be utilized in a taxonomic context. Variations in the amino acid composition of total leaf and seed (caryopsis) protein have been detected between different grasses and the differences observed correlate fairly well with tribal and subfamily classification (Yeoh and Watson, 1981). Such analyses are relatively rare outside the Gramineae (cf. chapter 11).

A comprehensive and up-to-date handbook on plant proteins has recently appeared (Boulter and Parthier, 1982) and this should be consulted for further details. In recording the contribution of proteins to plant systematic studies, emphasis will be given first to analysis of amino acid sequences (section II). Electrophoretic studies of storage proteins from seed or tuber will then be described (section III); this will be followed by an account of the isoelectric focusing of fraction I protein, the major leaf protein of green plants (section IV). Then, isozyme studies will be outlined and finally the contribution of serological studies will be assessed (section V).

II. Amino acid sequence analysis

A. Cytochrome *c*

The respiratory enzyme cytochrome *c* has a special place in the history of the biochemical systematics of plants. It was the first higher plant protein to be sequenced, that extracted from wheat germ (Stevens *et al.*, 1967). It was also the first plant protein for which a range of sequences from different plants were determined (Boulter *et al.*, 1972). The main incentive for studying this particular protein in plants was the fact that cytochrome *c* had earlier been sequenced from over 200 animal species and the phylogenetic tree derived from a computer analysis of the resulting data fitted in remarkably well, apart from some minor discrepancies, with that derived from the vertebrate fossil record (Fitch and Margoliash, 1967; Dayhoff, 1972).

From the data derived from the animal sequence, it appeared that cytochrome *c* was evolving very slowly in evolutionary time, with approximately one substitution in the sequence occurring on average every 20 million years. Because of the poor fossil record of plants, most speculations about the phylogeny of the angiosperms have of necessity been highly speculative. The particular attraction of looking at the amino acid sequences of cytochrome *c* in representative plants lay in the possibility of constructing a phylogeny of the angiosperms which for the first time had an objective basis.

In choosing a protein for comparative sequence analysis, a number of criteria have to be met (Table 17.1). It will be seen that cytochrome *c* meets these requirements in almost all respects. The amino acid sequence of this enzyme has undergone considerable change in the course of evolution. In terms of mutability, it lies somewhere between certain histone proteins, which show almost no change in amino acid substitution from pea to calf liver, and proteins such as plastocyanin and ferredoxin, which can vary in sequence from species to species (see below).

The only serious disadvantage of cytochrome *c* is the fact that the content in plant tissues is relatively low (Table 17.1). This factor was only realized after the programme of sequencing this protein was launched but it has significantly limited the number of taxa for which analyses have been achieved. At least a kilogram of seed material is essential, from which sufficient seedlings can be grown for isolation purposes.

Table 17.1 Cytochrome *c* as a model protein for phylogenetic studies in plants

Criteria for selecting a protein for sequence analysis	Properties of cytochrome *c*
1. Must show variation in amino acid sequence	Less than one-third of amino acid positions are invariant.
2. Must not be too long, so that it can be readily sequenced	It is fairly short, ranging from 104 amino acids in animals to 112 in plants.
3. Must be easy to isolate and purify	It is soluble, has a yellow chromophore and a distinctively highly basic isoelectric point.
4. Must occur in reasonable quantities	Occurs to the extent of 100 mg kg^{-1} in some vertebrates. In plants, young dark-grown seedlings may have between 1 and 4 mg kg^{-1}.
5. Must be homologous in function	It is an essential mitochondrial enzyme of terminal respiration in all living organisms. Has same overall shape, independent of source.

The processes of protein purification and sequence analysis for cytochrome *c* are outlined in Table 17.2. For further details of amino acid sequence analysis, see Boulter and Ramshaw (1975), Croft (1980) and Walsh *et al.* (1981). In recent years, the laborious and repetitive procedures of sequence analysis of peptides has been automated. Sequenators or sequencers are now available which are capable of analysing pure peptides containing between 40 and 50 amino acid residues.

Following the critical first sequence determination of a higher plant cytochrome *c*, from wheat germ, by Stevens *et al.* (1967), Boulter and his co-workers (1972) determined sequences for 24 other plants (19 dicotyledons, four monocotyledons and one gymnosperm). In three specific features, the plant cytochrome *c*, differ from those determined in other phyla. Thus higher plant cytochrome *c* is unique in that the *N*-terminal amino acid is not free but carries an *N*-acetyl blocking substituent. Its presence suggests that the initiation of this protein synthesis in higher plants may differ from that in other organisms. Secondly, the sequence in higher plants contains 111–112 residues, as compared to 104 in animals and 106–108 in micro-organisms. These extra eight amino acids, when compared to the sequence in animals, can be seen to be added at the *N*-terminal end. Thirdly, there is an unusual amino acid, trimethyl-lysine, which is present in addition to the normal 20 protein amino acids. It occurs at positions 80 and 94 (see Fig. 17.1) invariably in all higher plant sequences so far determined. Trimethyl-lysine is not found in animal cytochrome *c* but has been detected as a constituent of this protein in *Neurospora crassa*.

Table 17.2 Procedures required for purifying and sequence analysing cytochrome *c* in plants

Enzyme Purification

1. Macerate fresh tissue in buffer, adjust to pH 4.6
2. Filter, adjust to pH 8.0
3. Elute through Amberlite CG 50 ion-exchange resin
4. Elute through CM-Sephadex column
5. Elute through G-Sephadex column *or* subject to isoelectric focusing
6. Check purity by spectral characteristics, λ_{max} 280, 410, 550 nm

Sequence Analysis

1. Determine amino acid composition by automatic analysis, after hydrolysis with hot 6 M acid
2. Digest protein separately with chymotrypsin and trypsin
3. Separate each set of peptides by TLC/electrophoresis
4. Determine sequence of peptides by Dansyl/Edman procedure, identifying *N*-dansylated amino acids formed by polyamide chromatography
5. Check identity of peptides by colour reactions and TLC/electrophoresis
6. Determine sequence from above data

Remarkably little molecular heterogeneity has been observed during the determination of sequence of plant cytochromes, so that it appears to be a monomorphic protein. Among vascular plants only in pumpkin, *Cucurbita maxima*, was evidence obtained that more than one form of the enzyme might be present (Ramshaw, 1982).

The results of sequence analysis can be expressed in a simple way as the number of substitutions that separate the cytochrome *c* of different organisms (Table 17.3). In this analysis, one would expect that plants that are most closely related would have fewer differences than those that are unrelated. This is borne out in Table 17.3 where the four dicotyledons that are included differ among themselves by between 9 and 13 substitutions, whereas they differ from a monocotyledonous representative, namely wheat (*Triticum aestivum*), by between 14 and 16 substitutions. The cytochrome *c* of *Ginkgo biloba* (a primitive gymnosperm) expectably is more widely divergent with between 17 and 20 substitutions being different. Such an analysis, of course, does not make the most of the informational content present in each sequence.

Table 17.3 Matrix of amino acid differences in cytochrome *c* from six higher plant species. (From Boulter *et al.*, 1972.)

Species	*Ginkgo*	*Triticum*	*Helianthus*	*Vigna*	*Ricinus*	*Sesamum*
Ginkgo biloba	0					
Triticum aestivum	17	0				
Helianthus annuus	20	15	0			
Vigna radiata	20	15	12	0		
Ricinus communis	17	14	13	10	0	
Sesamum indicum	18	16	11	9	9	0

Another way of considering the cytochrome *c* data is to analyse the substitutions that have occurred at particular mutable sites. For example, the amino acids that occupy position 65 in the sequence in different phyla are indicated in Table 17.4. This shows that cytochrome *c* from higher plants can be separated from those of other phyla on the basis of particular insertions into the sequence, presumably reflecting the enormous distances in time since the ancestral taxa diverged. It is interesting that such mutations appear to be largely retained within a division or phylum, in spite of the apparent lack of functional significance.

Table 17.4 Amino acid substitutions at position 65 in cytochrome *c* of different phyla

Amino acid 65	Tyrosine →	Serine →	Phenylalanine →	Methionine
phylum	higher plants	fungi	insects	vertebrates
mutational changes needed	one	one	two	

The number of mutations necessary at the DNA level to change tyrosine to serine or serine to phenylalanine can be determined from a knowledge of the base triplet code. In these two cases, single mutations separate these amino acids; by contrast, the change from phenylalanine to methionine (Table 17.4) requires at least two mutations, a change which could have occurred via an intermediate cytochrome *c* with leucine present at position 65. These calculations assume that only the minimal number of mutations have taken place in the nucleic acid base sequence of DNA to produce the changes observed in the protein amino acids. Indeed, this principle of 'maximum parsimony' is usually implicit in any calculations associated with the determination of evolutionary significance in amino acid substitutions.

In order to extract the maximum taxonomic information from the sequence differences in cytochrome *c* or any other protein, it is normal practice to compare the sequences in every possible combination, noting all substitutions that occur in terms of minimal numbers of mutations. This requires a computer analysis of the data and the eventual construction of a phylogenetic tree, using the 'ancestral sequence' (Dayhoff and Eck, 1966) or other procedures (Fitch and Margoliash, 1968; Fitch and Yasunobo, 1975). One such tree derived from the cytochrome *c* data for plants is shown in Fig. 17.3 (Boulter *et al.*, 1972). This one is based on using *Ginkgo biloba* as the ancestral sequence. However, by using the 'flexible numerical matrix' method of Lance and Williams (1967), it is possible to construct a phylogenetic tree without recourse to an actual ancestral sequence. This procedure gives a fairly similar result (Fig. 17.4). In this latter tree, the five monocotylodonous species that have been analysed (representing five genera among three families) have been included. It will be noted that these taxa all cluster together and emerge from the dicotyledon members near *Pastinaca sativa*.

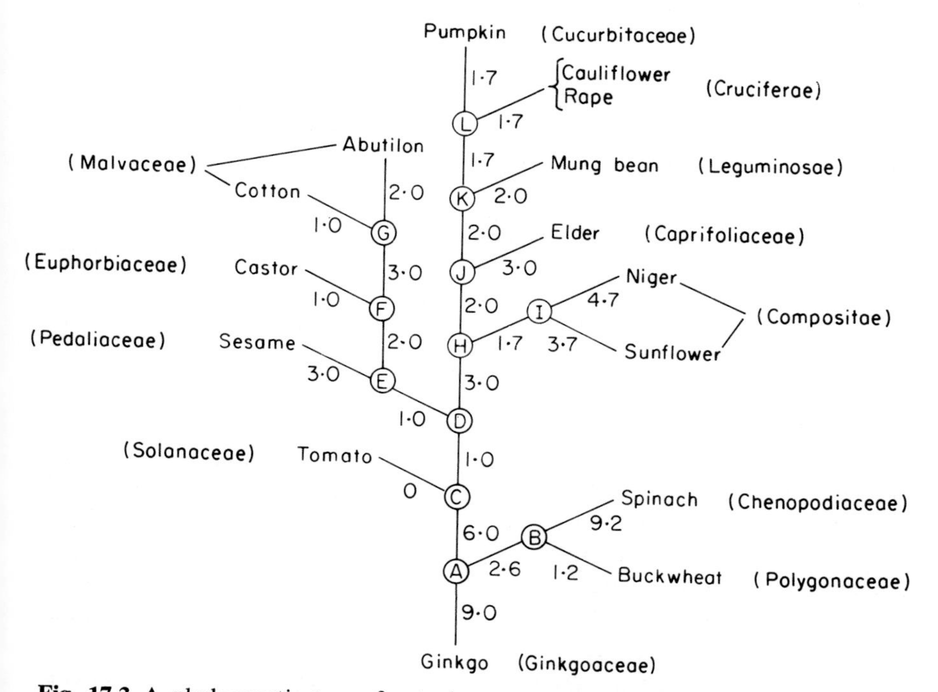

Fig. 17.3 A phylogenetic tree of cytochrome *c* sequences of fifteen plant species derived from the ancestral sequence method.

Fig. 17.4 A phylogenetic tree of cytochrome *c* sequences of plants, derived from the flexible numerical method.

There has been much argument, misunderstanding and critical comment (cf. Cronquist, 1976) about the interpretation of this cytochrome *c* tree. It is essential to realize that this tree does *not* represent the phylogeny of the organisms indicated in the figure but is of the cytochrome *c* sequences extracted from them. Thus, it is one line of evidence among many for relationships and it could not be expected to relate precisely with phylogenies based on morphological data. The fact that the tree actually corresponds in certain features (e.g. *Spinacia*, Chenopodiaceae, is generally regarded as a relatively primitive taxon) with evolutionary views of plant taxonomists is really quite remarkable and at least argues in favour of taking this evidence into account in phylogenetic speculation.

Since the present results represent a very low sampling of angiosperms

(0.001% at the species level), it is a mistake to draw any firm conclusions from such a limited base. Another mistake is to assume that each species sampled necessarily represents the family and order to which it belongs. It is true that two *Brassica* species studied, *B. napus* and *B. oleracea*, have identical sequences. However, in another example, in the Compositae, two species in the same tribe, namely *Guizotia abyssinica* and *Helianthus annus* differ by eight substitutions. Interestingly, in spite of these differences, these two cytochrome *c* sequences branch off close together in the tree (Fig. 17.4). More sampling would clearly be necessary before the representative cytochrome *c* of a particular family could be arrived at.

Among those assumptions needed to produce phylogenetic trees from cytochrome *c* data are (1) that convergence does not take place, and (2) that there is a constant rate of mutation with time, i.e. a constant time clock. The first assumption means that the tree truly represents common ancestral descent and that random chance or unidentified selective pressures do not affect the production of the same residue at a given position. Subsequent reconsideration of this matter by Peacock and Boulter (1975) has indicated that some 33% of all amino acid substitutions within the cytochrome *c* data set are due to convergence and *not* to common ancestry. Such problems are especially acute when the sequences being compared fall between 15 and 25% identity where, as noted by Doolittle (1981), 'it is difficult to decide between chance similarity and genuine common ancestry'.

The second assumption of a common evolutionary time clock for all proteins was seemingly rudely shattered when it was discovered that plastocyanin sequences (see below) showed many more changes between taxa than was apparent in the cytochrome *c* results. Presumably, it is still possible that the genes coding for proteins, because of differences in their location within the chromosome, vary in their susceptibility to natural radiation and hence to mutational change.

The question of whether differences in sequences are due to 'neutral' changes or to Darwinian selective pressures is one that has been debated continuously ever since the first amino acid sequence became available. One view is that each amino acid substitution must have a unique survival value in the phenotype of the organism—the phenotype being manifested in the structure of the proteins (Smith, 1967). The other view is that although certain positions along the polypeptide chain have specific functions in relation to enzymic 'active sites', most of the remainder of the protein molecule can change freely in evolutionary time by random drift (Jukes, 1980). It seems that both apparently contradictory viewpoints may contain an element of the truth so that neither one should be abandoned in favour of the other at the present time.

B. Plastocyanin

The practical difficulties associated with accumulating sufficient protein for cytochrome *c* sequence studies from further plant species led Boulter and his co-workers to turn to another more abundant material. A second reason for analysing a further protein was the hope that the results obtained might coincide with those from cytochrome *c* and thus establish the validity of protein sequence analysis for phylogenetic reconstruction. These workers chose to examine plastocyanin, a copper–protein which is an essential component of electron transfer in the plant chloroplast. It can be obtained in some quantity simply by extracting mature green leaf tissue, the yield being of the order of 10 mg per kg fresh leaf (Boulter *et al.*, 1977). Plastocyanin is blue in the oxidized state and hence, like cytochrome *c*, can be monitored during purification by spectral measurements in the visible region (at 597 nm). In addition, because it has associated copper ions, it can be detected by electron-spin-resonance spectroscopy. Like cytochrome *c*, it is a conveniently small protein of molecular weight around 10 500 and a sequence of 99 amino acid residues. Unlike cytochrome *c*, it lacks an *N*-terminal substituent and hence is more readily sequenced automatically than that protein.

Indeed, most of the 63 higher plant sequences available for plastocyanin have been determined automatically. When some 13 plant species had been completely sequenced, it was realized that most amino acid substitutions occur in the first 50 positions, the rest of the sequence being relatively invariant. In fact, 32 residues are invariant, with highly conserved regions occurring at positions 31–45 and 82–94. Subsequent analysis of a further 50 species has been confined to the *N*-terminal sequence of the first 40 positions. Such analysis represents the limit of accuracy of present day automatic sequencers (Croft, 1980), into which the pure polypeptides are fed for amino acid sequence determination.

It is clear from the available data on plastocyanin, which has been collected together by Ramshaw (1982), that the rate of mutation is at least twice as fast as that in cytochrome *c*. Furthermore, significant heterogeneity has been observed in a number of the species examined. From the affinity trees that have been derived from the plastocyanin data (Fig. 17.5) (Boulter *et al.*, 1979), it is already apparent that the determination of interfamily relationships will be difficult. Thus, significant overlap at the family level occurs between members of the Solanaceae and the Caprifoliaceae. The disparate separation in this tree of *Vigna* and *Phaseolus* (both Phaseoleae) from other members of the Leguminosae is also unacceptable taxonomically. Even at the species level, there may be surprisingly large differences between taxa. On the one hand, two species of *Heracleum*, *H. sphondylium* and *H. mantegazzianum*, have identical partial sequences while on the other, two species

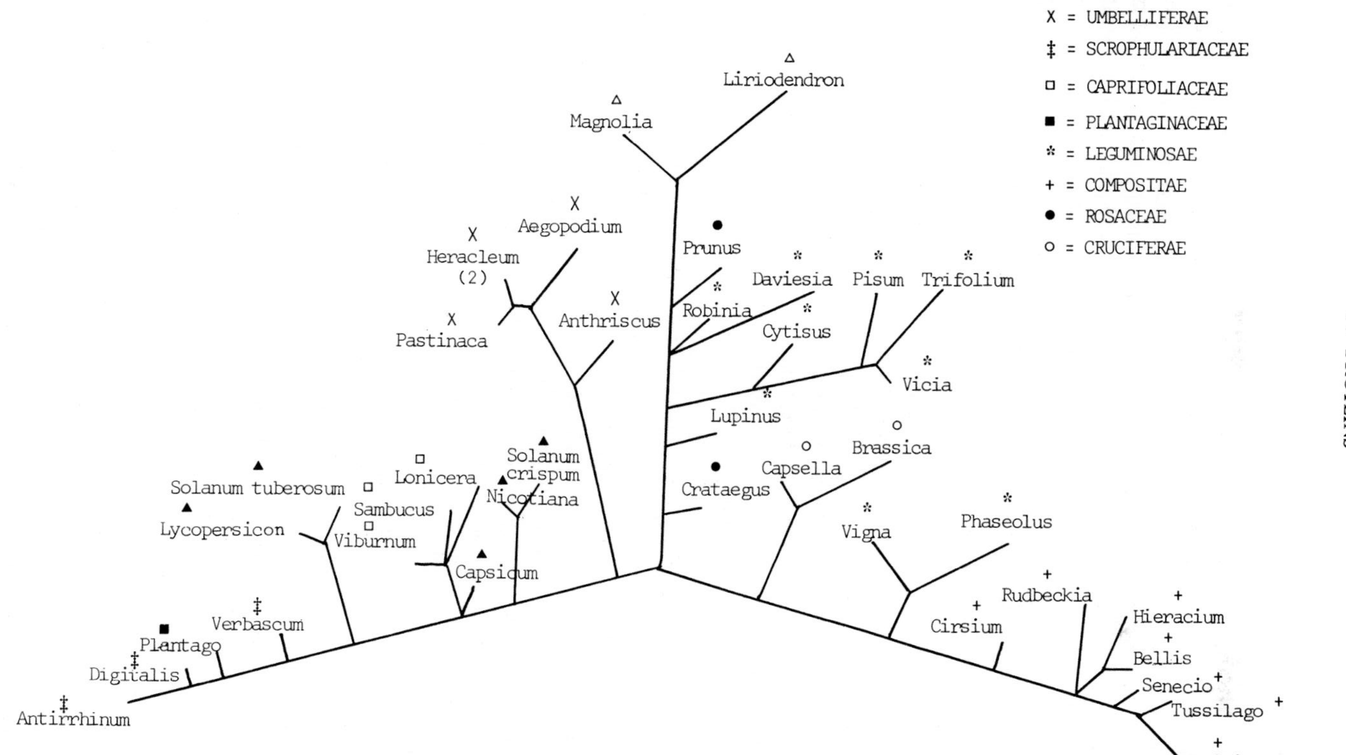

Fig. 17.5 Affinity tree of plastocyanin amino acid sequence data constructed by the compatibility method. In general, the species studied were the most common members of a given genus, e.g. *Bellis perennis* for *Bellis*, *Lycopersicon esculentum* for *Lycopersicon*. (From Boulter *et al.*, 1979.)

of *Solanum*, *S. tuberosum* and *S. crispum*, are very different in partial sequence (Fig. 17.5).

More recently, partial plastocyanin sequences have been determined for five representative members of the Ranunculaceae and the results have been compared, by constructing dendrograms, with similar data from 28 other higher plant families (Grund *et al.*, 1981). On this basis, the rather surprising result emerges that the Ranunculaceae is more closely related to the Rosaceae and the Leguminosae than to any of the other families investigated. These data are, however, in accord with Wettstein (1935) who derived the Rosales from the Polycarpicae (which includes the Ranunculaceae) and with the system of Meeuse (1970) who includes the Leguminosae in his Ranunculacean–Berbidalean–Rutalean line on the basis of common alkaloid types. Most other taxonomists, however, do not regard the Ranunculaceae as having much affinity with the Leguminosae or Rosaceae and serological comparisons have failed to indicate any determinant similarities (see section VI). Since the present plastocyanin data are derived from only a few species, it should be a matter of some priority to extend the species sampling; certainly this would be needed for confirmation of the unexpected family interrelationships thrown up by these results.

The phyletic import of sequence data from the plastocyanins examined to date is difficult to assess. It is widely believed that chloroplasts in eukaryotes arose through the endosymbiotic incorporation of a cyanobacterium into a more ancestral eukaryotic cytoplast, much as discussed by Fox *et al.* (1980). At least the phylogenetic relationships of chloroplasts would appear to be with the DNA of cyanobacteria rather than with nuclear DNA of the eukaryote which possesses the chloroplast. Little is known about the interplay of nuclear versus chloroplastic DNA in the developmental control of plastocyanin among eukaryotic plants (Weeden *et al.*, 1982). Incorporation of neutral or near-neutral mutations into codons responsible for the first 35 or so amino acids in plastocyanins may be so readily accomplished and/or erratic that their use as a molecular chronometer, especially at the familial level or higher, is rendered moot. Perhaps complete sequence data for the more interior amino acids (e.g. those at position 46 to 81) will prove more reliable for systematic comparisons.

C. Ferredoxin

Ferredoxin is similar to plastocyanin in that it is located in the chloroplast and is involved in the electron transport process of photosynthesis. Unlike plastocyanin, it is also an electron carrier for other processes, such as nitrogen fixation, and so has a wider distribution, commonly being present in bacteria as well as in algae and in higher plants. The copper of plastocyanin is replaced by iron in ferredoxin and the protein of higher plants has two ferrous ions at the active centre. It is similar in size to plastocyanin, having between 93 and 98 residues.

Ferredoxin can be isolated pure in quantity from green tissues and thus has similar advantages to plastocyanin in the determination of higher plant affinities through protein sequence analysis. Remarkably, the rate of mutation in ferredoxin parallels that in plastocyanin and differences in substitution can occur between species. In the genera where more than one species has been examined, namely in *Equisetum* and *Phytolacca*, the sequences vary by between one and six substitutions.

An additional complication with this protein is that several species examined have been found to have two quite different ferredoxins. Gene duplication, presumably responsible for this phenomenon, appears to have occurred prior to speciation. In the three genera where pairs of ferredoxins occur (in *Phytolacca*, *Equisetum*, *Dunalliela*) all the species examined have both sets. The two ferredoxins in a given species differ markedly in amino acid substitution so that, in making taxonomic comparisons, it is necessary to match up the homologous sequences.

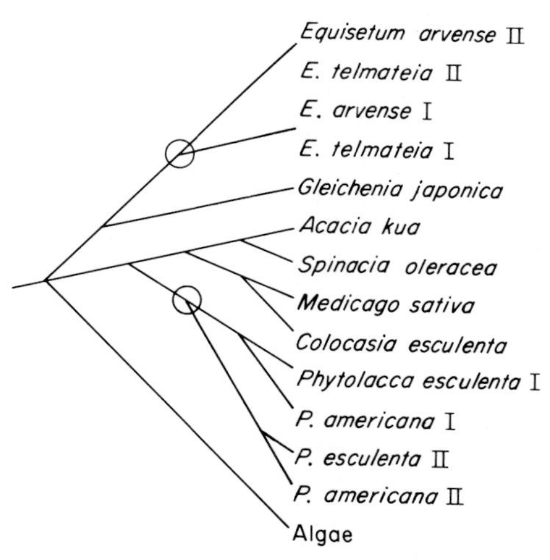

Fig. 17.6 Affinity tree of ferredoxin sequences, based on the matrix method. The circles on the lines indicate gene duplication

The sequence data for higher plants is displayed in the ferredoxin tree of Fig. 17.6; this has been prepared by the matrix method. As will be seen, the data are far too sparse for any taxonomic conclusions to be drawn. All that can be deduced is that there are clear separations at the phylum level, the horsetails and single fern (*Gleichenia*) being clearly separated from the six angiosperm representatives (Matsubara *et al.*, 1980; Ramshaw, 1982). Most phylogenetic interest in the ferredoxin data at present centres on improving our understanding of the early evolution of life forms. Some bacterial ferredoxins (e.g. of *Clostridium*) are much simpler in structure and shorter than the others and it is possible to relate the lengthening of the polypeptide chain with evolutionary advancement among these organisms (Matsubara *et al.*, 1980).

A number of other blue copper proteins are known from both higher plants but as yet they have not been shown to have biological functions, as do plastocyanins (Murata *et al.*, 1982). These include proteins from *Populus* spp., rice, mung bean and cucumbers. Amino acid sequence data are available for only one of these (that from cucumber) and it has a 'remarkable similarity' to that of stellacyanin, a glycoprotein obtained from the Japanese lacquer tree, *Rhus vernicifera* (Ryden and Lundgren, 1979). It would appear that sequence comparisons among the various pigments of the cellular plastids, for systematic purposes among higher plants, must await knowledge of their phyletic origin, whether controlled by chloroplastic or nuclear DNA, and the role each might play in metabolic processes.

Some of the problems involved in establishing protein homology among cellular plastids is exemplified by the study of Weeden *et al.* (1982) who showed, using three-way immunochemical tests, that isozymes of phosphoglucose isomerase from spinach plastids were more similar to those of the prokaryote, *Synechococcus*, than they were to the spinach cytosolic isozyme itself. They conclude that, in spinach, the plant nuclear genome may consist of genes derived from both eukaryotic and prokaryotic sources and that a knowledge of such genomic heterogeneity is important to an understanding of cellular evolution.

D. Other proteins

Other plant proteins that have been sequenced are listed in Table 17.5. They range from comparatively short polypeptides, such as the mistletoe toxin viscotoxin, to relatively large enzymes, such as ribulose bisphosphate carboxylase (RuBP carboxylase). The large subunit of the latter was sequenced indirectly, by sequencing the DNA in the gene responsible for coding its synthesis in *Zea mays* (McIntosh *et al.*, 1980). This is an alternative method of sequencing proteins which may be more widely adopted in the future. There is very little comparative sequence data for RuBP carboxylase, but that which is available from partial analyses of barley and spinach indicate that the large subunit enzyme is highly conserved and hence it may not be of much interest to taxonomists as such. Comparative studies, however, have been carried out at the whole protein level and these are discussed below in section IV (p. 420). In addition, RuBP carboxylases have been compared in total amino acid composition when isolated from 38 grass species representing 26 genera (Yeoh *et al.*, 1981). Some small differences at subfamily level were apparent, e.g. the pooids were distinguishable from all other grass groups.

Many of the proteins listed in Table 17.5 are of limited distribution and hence do not lend themselves to taxonomic comparisons. The choice of further proteins for sequence surveys would seem to be limited. The storage proteins of seeds are an obvious target for comparative studies and some partial sequence data are already available for cereal and legume sources (see Ramshaw, 1982). However, these particular proteins are also open to comparative investigation by simpler analytical techniques, as will be discussed in section III (p. 408).

Table 17.5 Plant proteins (and enzymes) that have been sequenced

Class	Protein	Sources	Number of residues
Enzymes	Ribulose 1,5-bisphosphate carboxylase: small subunit	*Spinacia*	120
	large subunit[a]	*Zea mays*	475
	Protease	6 spp.	ca. 220
	Peroxidase	Horseradish and turnip	ca. 300
	ATP synthase	*Spinacia*	81
Toxins	Phytohaemagglutinin	7 legumes	ca. 237
	Protease inhibitor	2 legumes	ca. 181
	Viscotoxin	*Viscum album*	46
	Purothionin	Wheat	45
Storage	Globulin: small subunit	2 legumes	135–154
Histones	Histone H3	Pea	135
	Histone H4	Pea	102
Symbiotic	Leghaemoglobin	5 legumes	ca. 150
Sweet proteins	Monellin (two chains)	*Dioscoreophyllum*	94
	Thaumatin I	*Thaumatococcus*	207

For further details, see Ramshaw (1982)
[a]Sequenced *via* the DNA

In spite of the difficulties encountered so far in protein sequence studies among higher plants, it is still possible that, by choosing a range of suitable proteins for analysis, meaningful results will emerge for evolutionary comparison. Penny *et al.* (1982) have recently tested the prediction that similar phylogenetic trees should be obtainable from different sets of character data in the animal kingdom. They employed sequence data for cytochrome *c*, fibrinopeptides A and B, and haemoglobins A and B for 11 vertebrate species and showed that from the 39 minimal and near-minimal trees produced from the data, a 'consensus' tree could be selected to best express the interrelationships of the different taxa. The present requirement in the plant kingdom is for further sequence data on cytochrome *c*, especially among ferns and gymnosperms, where a reasonable fossil overview might be had, coupled with similar determinations on one or more completely new metabolically critical enzymes.

III. Storage protein analysis

A. Analytical procedures

It will be apparent from the previous section that the determination of the amino acid sequence of a protein, in spite of the development of automatic sequencers, is still a relatively complex procedure. It is time consuming, requires large plant samples and very expensive equipment and is best operated by a scientist with specialized biochemical training. In complete contrast, the analysis of proteins by gel electrophoresis is rapid and easy and requires only small plant samples and cheap apparatus. It is not surprising that this method is one that has been extremely widely applied and has been used equally by both biochemists and taxonomists. It remains the most popular of all techniques in the field of biochemical systematics.

In its simplest form, the analysis begins with mature seeds of the species to be compared. The material is crushed in a suitable buffer, containing additives to prevent protein denaturation, and, after centrifuging, the supernatant containing all the soluble proteins is subjected to electrophoretic separation in a starch or polyacrylamide gel. Preferably, the gel is in a form of a rectangular slab and the various samples to be analysed are placed side by side in wells or slots in the gel. They are run alongside each other so that the results are directly comparable. After subjecting the gel to horizontal low-voltage electrophoresis for 3–4 hours in a suitable buffer, it is removed and stained for protein with either amidoblack or Coomassie blue. After washing out the excess stain, the proteins appear on the slab as a series of dark bands, as shown in Fig. 17.7. The presence and absence of individual proteins in the different samples can then be noted and analysed.

Further information can be elicited from the same protein extract by detecting enzyme activities through appropriate stains. In this case, the slab, before protein staining, is cut transversely into up to four slices. Only the top slice is then stained for protein, the other three being tested for peroxidase, esterase and other enzyme activity. The number of bands formed after staining will be relatively few compared to the number of protein bands.

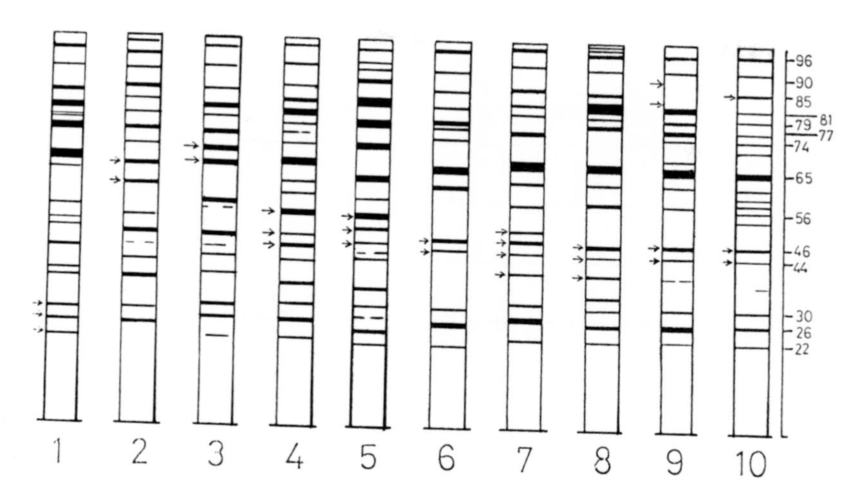

Fig. 17.7 Gliadin protein gel electrophoresis of ten cultivars of wheat: 1, Maris Dove; 2, Highbury; 3, Bouguet; 4, Clement; 5, Maris Ranger; 6, Maris Nimrod; 7, Cappalle-Desprez; 8, Maris Freeman; 9, Aton; 10, Maris Huntsman. Arrows indicate cultivar-specific bands. (From Ellis, 1979.)

The preferred tissue for comparative purposes is seed (or fruit) because the storage proteins laid down here are stable, their composition being little affected by environmental conditions or seasonal factors. Also, seeds are rich in protein so that it may be possible to carry out an analysis on a single seed, e.g. a single grain of wheat. Leaf tissue is to be avoided since the protein composition is generally much more variable. Besides the effect of physiological and ecological factors, leaf protein may show significant population variation due to genetic factors. It is only in comparative isozyme studies, where genetic variation is specifically of interest, that young leaf tissue becomes the material of choice (see section V, p. 424).

Plant pollen is also a rich and stable source of protein and pollen has been occasionally used in comparative protein analysis. The protein content may not be as stable with storage as that of seed. For example, Johnson and Fairbrother (1975) found that in *Betula populifolia* some of the protein bands in fresh pollen disappear from pollen that has been stored for several months. Thus, freshly collected pollen should be used whenever possible. An example of the successful application of protein analysis using fresh pollen is the work of Krattinger *et al.* (1979) on four European *Typha* species and the hybrids thereof. In this case, each species has up to five characteristic pollen protein bands (Fig. 17.8). These always appear without fail in the hybrid pollen, so that the parental origin of a particular hybrid plant can be readily delineated. Since morphological characters are insufficient for identifying such hybrids

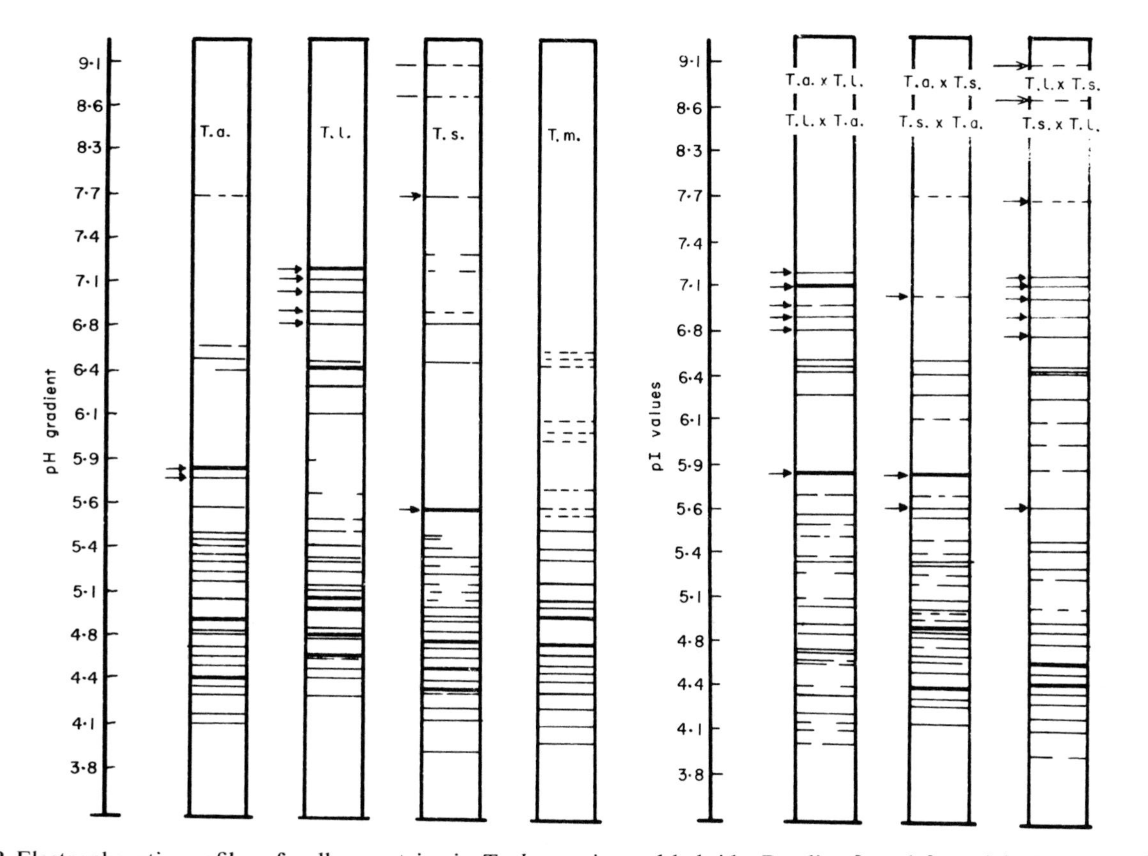

Fig. 17.8 Electrophoretic profiles of pollen proteins in *Typha* species and hybrids. Reading from left to right: *T. angustifolia*, *T. latifolia*, *T. shuttleworthii*, *T. minima*, *T. angustifolia* × *T. latifolia*, *T. angustifolia* × *T. shuttleworthii*, *T. latifolia* × *T. shuttleworthii*.

in *Typha*, the method has direct application to the unequivocal recognition of natural hybrid swarms among these particular taxa. Similar, but somewhat more controversial work, has been performed on the North American species of *Typha* (Sharitz *et al.*, 1980).

In some plants, e.g. tree species, seeds may not always be available for analysis and, in such cases, it is possible to resort to root tissue. Thus, Huang *et al.* (1975) obtained clear protein banding patterns from root extracts of 41 trees of *Robinia pseudacacia*. Comparison of the protein bands of a dominant stem variant with those of the typical tree showed no protein differences and the authors concluded that the dominant stem form was only an ecotype not worthy of varietal status. In some cultivated plants, other storage tissues besides seed or root may be used to yield protein profiles. In the case of the potato, *Solanum tuberosum*, tuber sap has been extensively analysed. Different cultivars can be identified and typified on the basis of tuber protein patterns and of patterns of the esterase isozymes (Stegemann, 1979).

The number of protein bands detectable by electrophoresis in storage tissue is relatively unpredictable, but may average between 10 and 30. As many as 84 bands, however, were detected in seed tissue of *Bulnesia* species (Comas *et al.*, 1979). Cereal seeds also yield many electrophoretic bands (Miege, 1975). In such cases, sufficient taxonomic information may be obtained by analysing only a small fraction of the total protein. The gliadin fraction of cereal seeds has been found to yield between 30 and 60 individual components (Ellis, 1979). In such cases, some simple fractionation of the seed protein is a necessary preliminary before electrophoresis is applied.

B. Interpretation of protein band patterning

The bands on a developed, stained gel are assigned R_p values accordingly to their mobility relative to the buffer front. On the assumption that a band of identical mobility in two related taxa is the same protein, the results of electrophoretic analysis can be presented as the number of bands that are common to two or more species and the number which are species-specific. Each band then represents a single taxonomic character. In species comparisons, it is important to test for population variation and exclude from comparison any band that is not regular in its occurrence. A 70% or better frequency is desirable in a band used for taxonomic purposes; it is preferable, if possible, to employ only bands with a 95–100% frequency. When comparing a number of species in the same genus, the results can be simply analysed by assigning a similarity index to each interspecies comparison; the resulting data can then be compared with the species relationships that have been developed from the morphological, anatomical and cytological features.

Some idea of the procedures to be used in species comparisons by gel electrophoresis can be gained from a study of the investigations by Vaughan and his associates on seed proteins of *Brassica* and *Sinapis* (see e.g. Vaughan, 1975). These two allied genera in the Cruciferae include such well known crop plants as cauliflower, turnip, swede, rape and mustard. Since *Sinapis*, the mustard genus, is capable of hybridization with *Brassica*, the cabbage genus, it is not sure whether *Sinapis* is sufficiently distinct from *Brassica* to be treated as a separate genus; in this respect, the morphological and cytological evidence are equivocal.

In this situation, the albumin proteins in seeds of representative taxa of the two groups were analysed by gel electrophoresis. From the resulting band patterns (Fig. 17.9), it will be seen that the two *Sinapis* species differ

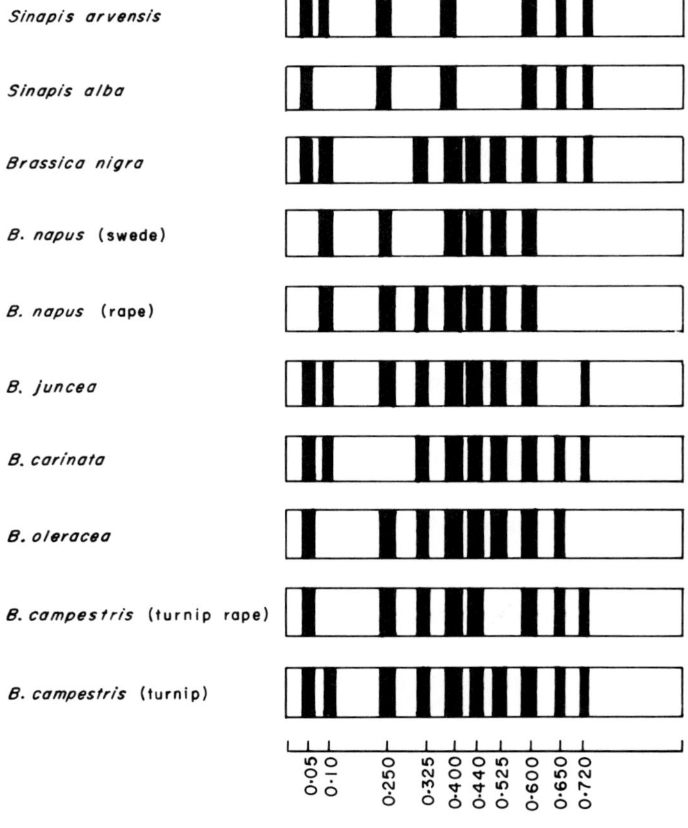

Fig. 17.9 Electrophoretic patterns of seed albumin proteins of the *Brassica–Sinapis* complex. (From Vaughan and Denford, 1968.)

in lacking three major bands (with R_p values 0.33, 0.44 and 0.53) which are usually present in the eight *Brassica* species analysed. These differences can be expressed as a three-dimensional model (Fig. 17.10) in which the lengths of the lines which join the different taxa are calculated from the similarity coefficients. One can conclude from these results that *Sinapis* is sufficiently different in its proteins to separate it from *Brassica*.

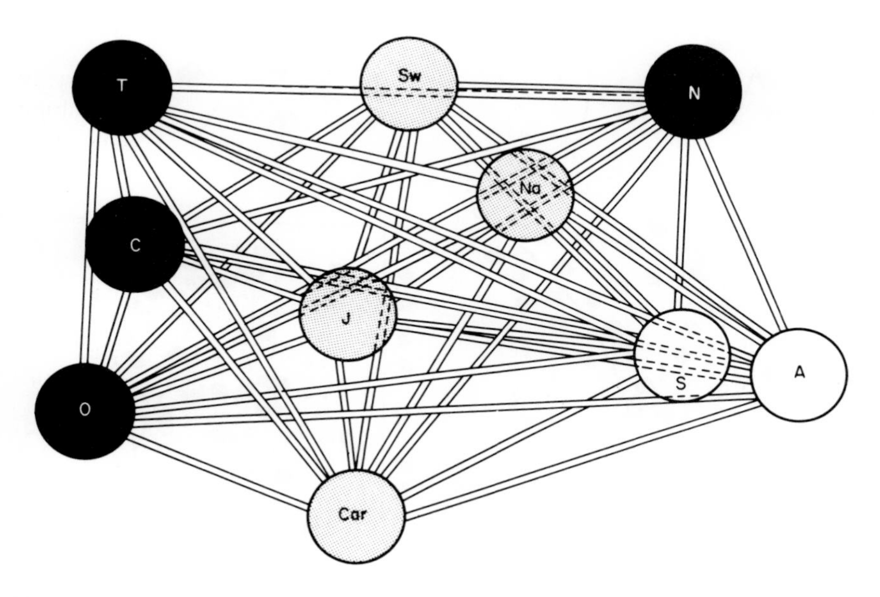

Fig. 17.10 Three-dimensional representation of relationships within the *Brassica–Sinapis* complex based on protein data. For details, see the text. (From Vaughan and Denford, 1968.)

The results of the analysis also have some bearing on the supposed hybrid origin of four of the *Brassica* taxa, namely the swede and the rape plants (both *B. napus*), *B. carinata* and *B. juncea*. The fact that these four taxa show no species-specific bands but instead have various combinations of bands present in the four true *Brassica* species indicates that they are indeed of hybrid origin. This point is also underlined in the three-dimensional diagram (Fig. 17.10), where the four hybrid taxa (grey circles) separate on a different plane from the true species (black circles). Finally, the fact that the turnip *B. campestris* uniquely possesses all the bands variously but incompletely present in the other three *Brassica* species indicates that it could be regarded as the ancestral taxon, the others being derived from it by loss of one or more protein bands (Vaughan and Denford, 1968).

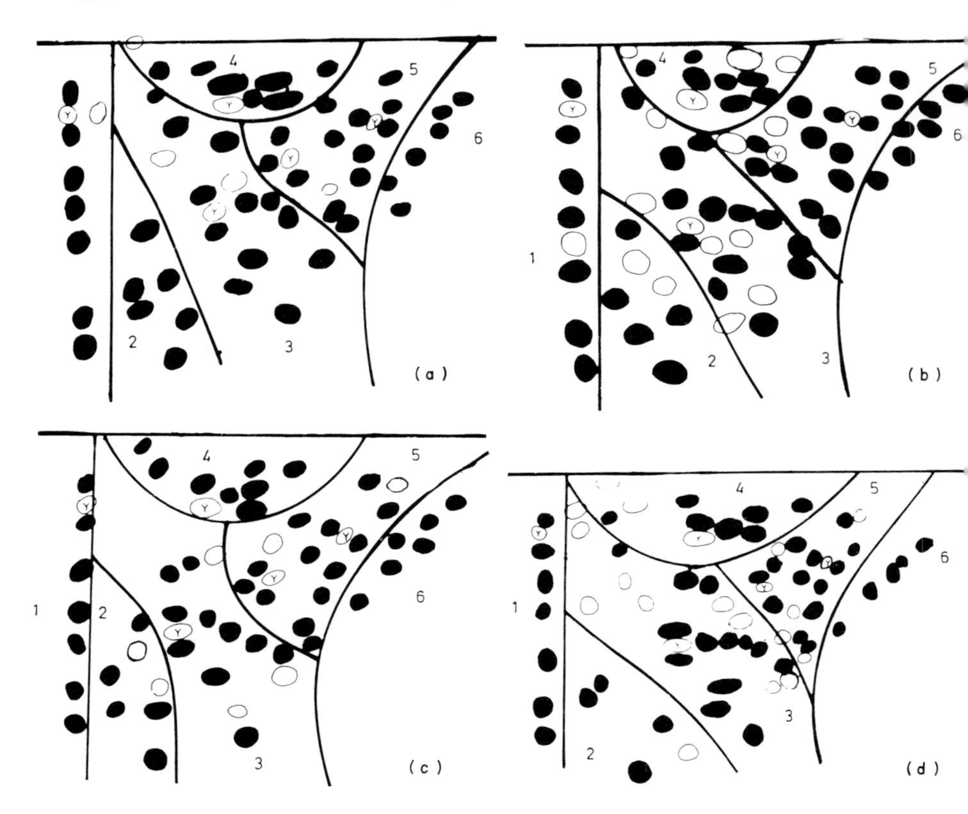

Fig. 17.11 Peptide fingerprints of globulins from legume seeds. Sample spotted TL corner, electrophoresis L → R, chromatography T → B. (a) *Vicia dumetorum*, (b) *Lathyrus sylvestris*, (c) *Vicia angustifolia*, (d) *Lens esculenta*. (From Jackson *et al.*, 1967.)

Within a closely knit group of related plants as with the *Brassica–Sinapis* complex, it is a reasonable assumption that protein bands of identical R_p value are homologous. Such homology could be further tested by separating the individual bands and comparing them for similar behaviour in other electrophoretic systems. Regrettably such tests are not often employed because of the additional labour involved. Another method of testing for protein homology is to use peptide fingerprinting. This procedure was adopted by Jackson *et al.* (1967) in a comparison of seed globulins in the Leguminosae. In this method, the major globulin fraction is subjected to tryptic digestion and the resulting complex mixture of peptides is separated two-dimensionally, by electrophoresis in one direction and by chromatography in the other. Some typical patterns are illustrated in Fig. 17.11. A study of the fingerprint patterns then indicate the number of spots that are

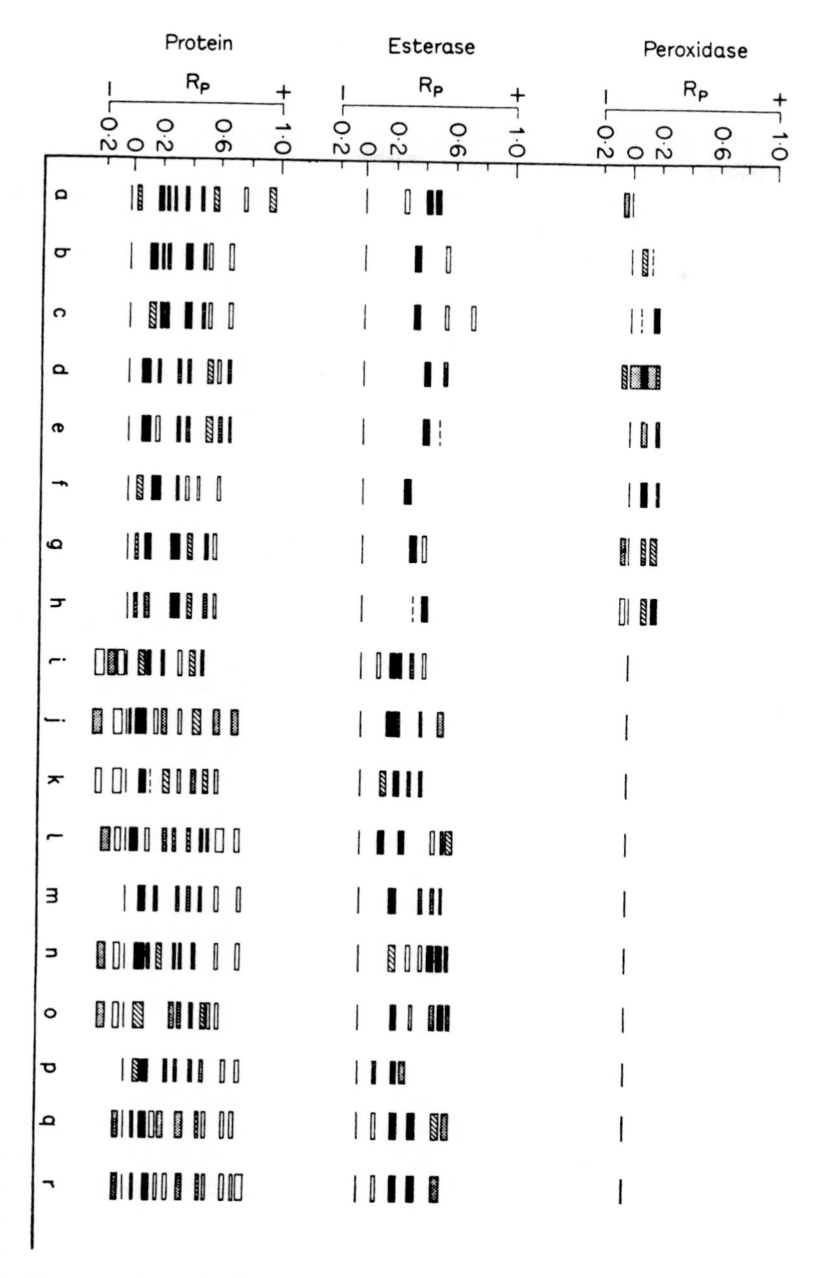

Fig. 17.12 Gel electrophoretic profiles of proteins, esterases and peroxidases of umbellifer seeds. (From Harborne, 1971.)

Key: Echinophoreae, species a; Scandiceae, subtribe Scandicineae, species b-h; Scandiceae, subtribe Caucalineae, species i-t.

common to a given pair of taxa and the number that are different. The actual results obtained fitted in well with accepted generic and tribal boundaries within the Leguminosae. For example, the representative species of five genera in the tribe Vicieae gave almost identical peptide fingerprints for their globulins. By contrast, *Phaseolus vulgaris* (tribe Phaseoleae) showed many differences, as did *Abrus precatorius* (tribe Abreae). More detailed comparisons, using e.g. immunological tests, have subsequently been conducted of the major globulins in legume seeds and the distribution in the family of the two major proteins, legumin and vicilin, has been enumerated (Derbyshire *et al.*, 1976).

An alternative approach to circumventing the problem of doubts about protein homology in seed extracts is to include in the analysis a survey of several enzyme activities. This has been done, for example, in a study of umbellifer seed extracts (Crowden *et al.*, 1969) where differences in protein band patterns could be correlated with variations in the position and number of esterase and peroxidase bands, which were determined at the same time in the same separation by slicing the gel (see above). Some typical results for umbellifer seeds are illustrated in Fig. 17.12. Here, a distinct difference in protein characters is observable at the subtribal level within the tribe Scandiceae. In the subtribe Scandicineae, there are two major proteins with R_p values 0.2 and 0.4, while in the Caucalineae, there is a single major band at R_p 0.10–0.20 and several minor bands; another difference in the latter subtribe is the additional presence of cationic bands. These differences in proteins are also reflected in differences in peroxidase and esterase activity. The data are of some taxonomic significance in that while Drude (1897–8) places these taxa together in one tribe, Bentham (1867) separate them, treating the Scandicineae as the tribe Scandiceae and placing the Caucalineae with the Dauceae into another tribe. The macromolecular data, as will be seen, are more closely in accord with the latter than the former system of classification.

C. Taxonomic applications

The electrophoretic separation of storage proteins in plants now has an established place in modern chemotaxonomic practice. Most applications have been within groups of closely related taxa, either at the population level or as a comparison of sympatric species within the same genus (Table 17.6). In some cases, other macromolecular approaches are included, such as the comparison of isozymes in seeds or seedling tissues (see section V). Interestingly, direct comparison of seed protein data with leaf isozyme data in *Zea* species showed that the seed protein data correlated more closely with conventional taxonomy of the genus than did the isozyme results

(Mastenbroek *et al.*, 1981). Indeed, the results of seed protein analysis very generally accord with classifications based on morphological and anatomical criteria. Occasionally, they may indicate new interrelationships not apparent in the formal herbarium comparison of plant taxa. Some idea of the kinds of plants so far studied can be gleaned from Table 17.6.

The general conclusions of storage protein studies are fourfold. (1) The seed protein profile is usually species-specific and categories below the species level tend to share the same profile. In favourable cases, the results can be used to circumscribe the species accurately. (2) Closely related species have more protein bands in common than remote species. Most often the data can then be used to clarify sectional relationships within a genus. (3) Hybridity of a taxon is usually revealed by additive inheritance of seed proteins from the parents; occasionally, however, one or more parental bands may be missing. (4) The wild ancestor(s) of a cultivar may be revealed through protein patterns. If directly related, the wild ancestor may share the protein profile of the cultivar. If the cultivar is a product of hybridization, then bands from several ancestral taxa may be discernible.

One taxonomic problem where seed protein analysis has been extensively applied is to the origin of the cultivated hexaploid bread wheat *Triticum aestivum*. The origin of this wheat plant is still a matter of some uncertainty, although the now well known electrophoretic experiments of Johnson and Hall (1967) and Johnson (1972) have provided valuable support for the theory that it originated through hybridization between the diploid *Aegilops squarrosa* of genome DD and the ancient cultivated tetraploid *T. dicoccum* (genome AABB). The resulting triploid is assumed to have given rise, by doubling of the chromosomes, to the fertile hexaploid *T. aestivum* of known genome AABBDD. At least, the results of numerous electrophoretic scans of seed proteins are not incompatible with this view. In particular, Johnson (1972) was able to simulate closely the pattern in different introductions of *T. aestivum* with a mixture (2 : 1, v/v) of the protein extracts of *T. dicoccum* and of *A. squarrosa*. More recently, various diploid species (e.g. *T. urartu*) have been examined as possible ancestors of the tetraploid *T. dicoccum* and work is continuing to establish species relationships among the ten or more known diploid taxa (Brody and Mendlinger, 1980). Undoubtedly, protein electrophoresis remains a powerful and elegant tool for investigating this complex and fascinating evolutionary problem.

This technique has yet to be applied successfully to the origin of the cultivated barley, *Hordeum vulgare*, although attempts have been made in this direction. Some experiments have been applied to the origin of polyploids in the wild barley species *H. murinum* and gel electrophoresis has been shown to differentiate diploids, tetraploids and hexaploids among these taxa (Booth and Richard, 1978). The origins of other cereal crops have also been similarly investigated through storage protein patterns (Table 17.6); similar studies using protein bands generally have also been used in the documentation of natural hybrids among closely related feral species (Table 17.7).

Table 17.6 Selected examples of seed storage protein surveys at the generic level

Family	Genera	References
Chenopodiaceae	*Chenopodium*	Crawford and Julian (1976)
	Suaeda	Unger and Boucaud (1974)
Compositae	*Lasthenia*	Altosaar *et al.* (1974)
Cruciferae	*Brassica* and *Sinapis*	Vaughan and Denford (1968), Vaughan (1975)
Gramineae	*Hordeum*	Booth and Richard (1978)
	Oryza	Monod *et al.* (1972)
	Tripsacum	de Wet *et al.* (1981)
	Triticum	Johnson (1972)
	Zea	Mastenbroek *et al.* (1981)
Leguminosae	*Phaseolus* and *Vigna*	Sahai and Rana (1977)
	Trigonella	Ladizinsky (1979)
	Vicia	Ladizinsky (1975)
Loasaceae	*Mentzelia*	Hill (1977)
Pinaceae	*Abies*	Clarkson and Fairbrother (1970)
Solanaceae	*Solanum*	Edmonds and Glidewell (1977)
Zygophyllaceae	*Bulnesia*	Comas *et al.* (1979)
	Larrea	Hunziker *et al.* (1972)

Table 17.7 Selected studies in which additive protein bands have been used to document hybridization

Family	Genus	References
Compositae	*Carthamus*	Efron *et al.* (1973)
	Senecio	Hull (1974)
	Stephanomeria	Gottlieb, 1973a; Gallez and Gottlieb (1982)
	Tragopogon	Roose and Gottlieb (1976)
Leguminosae	*Phaseolus*	Garber (1974)
Labiatae	*Galeopsis*	Houts and Hillebrand (1976)
Malvaceae	*Gossypium*	Cherry *et al.* (1971), Hancock (1982)
	Hibiscus	Hoisington and Hancock (1981)
Gramineae	*Agropyron*	Hunziker (1967)
	Avena	Murray *et al.* (1970)
Polemoniaceae	*Phlox*	Levin and Schaal (1972)
Solanaceae	*Nicotiana*	Sheen (1970)
	Solanum	Desborough and Peloquin (1966)
Sterculiaceae	*Fremontia*	Scogin (1979)
Ulmaceae	*Ulmus*	Feret (1972)

More obvious taxonomic success with cereals has been achieved in the area of cultivar identification and gel electrophoresis has been extensively applied to all the common cereal crops (Ladizinsky and Hymowitz, 1979). There are often no visible morphological differences between the grains of different varieties so that the tedious process of growing plants to maturity is necessary before identification by morphological characters becomes possible. By contrast, electrophoretic analyses can be carried out on a single grain and give an answer within 24–48 hours.

The regular monitoring by the miller of wheat varieties supplied by the farmer is of utmost importance, since only certain varieties have the right quality for making into bread (Ellis, 1979). Much attention has been given, therefore, to the starch-gel electrophoresis of wheat proteins to provide an answer to this need. Analysis can be limited to the proteins in the gliadin fraction of the seed protein, i.e. that fraction soluble in dilute acetic acid or in solvents containing sodium dodecyl sulphate (SDS). Forty three gliadins can be recognized by gel electrophoresis (Ellis and Beminster, 1977) and any one variety has about 20 gliadin bands.

An analysis of 29 wheat cultivars grown in the U.K. provided unique electropherograms for each, apart from three which were closely related ancestrally. Some idea of the variety patterns can be seen in Fig. 17.7 (p. 409). The patterns were found to be consistent in any one cultivar, irrespective of the normal variations that occur in crop husbandry. Furthermore, individual grains of a given cultivar gave identical bands in all but three cases. In three varieties, 'biotypes' were encountered where one or two of the characteristic bands of that particular variety were missing. The presence of these biotypes, however, provided no serious impediment to correct cereal identification.

An alternative approach to studying wheat storage proteins is to determine isozyme patterns (see section V) in 15-day-old seedlings. A study of isozymes in hexaploid wheat cultivars by Salinas *et al.* (1982) offers an interesting comparison. In this work, 38 cultivars of *T. aestivum* and one of *T. spelta* were analysed using both starch and polyacrylamide gels. Some 15 enzyme activities were variously assayed and the patterns obtained were distinctive for 26 of the 39 cultivars. The remaining 13 cultivars fell into five pairs, with one group of three. Thus, the more complex procedure of isozyme separation provides a poorer key to cultivar identification than gliadin analysis.

Gel electrophoretic procedures have been applied to cultivar identification in barley. Here, Shewry *et al.* (1978) have used SDS–polyacrylamide-gel electrophoresis at pH 8.9 to analyse rapidly the hordein fractions of 88 varieties. Twenty nine patterns were obtained, with one to 25 cultivars in each group. Further subdivision of the largest group was achieved by a second electrophoretic run with SDS being replaced by urea and the pH changed to 4.6. In this case, the method does not give a complete separation of all varieties, but most can be identified by one or other procedure.

Yet other crop plants where protein electrophoresis has been applied to cultivar identification are those belonging to the genus *Brassica*. The earlier work of Vaughan and his co-workers on the gel electrophoresis of albumin fractions of the seeds of cultivated and wild *Brassica* species has already been described. More recent protein analyses in *Brassica* cultivars have been accomplished by using both the soluble seed proteins and certain isozymes, especially the esterases (Phelan and Vaughan, 1976). The results showed *inter alia* that a taxon *B. alboglabra* of disputed specific rank is actually better treated as a member of the *B. oleracea* complex, at least from the chemotaxonomic viewpoint. The more general application of protein electrophoresis to the taxonomy of both wild and cultivated plants is reviewed by Vaughan (1983).

IV. Fraction I protein analysis

A. Subunit variation

Fraction I protein constitutes up to half the total leaf protein in all green plants and has been claimed to be the most abundant protein on this planet. It has long been studied biochemically because of its enzymic activity. It is the key catalyst in the carbon dioxide fixation step in the Calvin carbon cycle of photosynthesis and is known alternately as ribulose 1,5-bisphosphate carboxylase (RuBP carboxylase). It also has oxygenase activity and is responsible in part for the phenomenon of photorespiration. Comparative studies at the level of amino acid composition and sequence on this protein have already been mentioned (p. 406).

The high concentration of fraction I protein in the chloroplast has long puzzled plant biochemists, particularly since all the other enzymes of photosynthesis are present in much lower amounts. Its abundance is in fact due in part to its relative inefficiency as a catalyst. Its production in large quantities means that it is readily obtained in pure form from green plants. Indeed, it can be crystallized directly from clarified leaf extracts of tobacco and other *Nicotiana* species (Chan *et al.*, 1972). It was as a result of studies of its properties in the tobacco plant that its value as a new chemotaxonomic tool in higher plants became apparent (Wildman *et al.*, 1975).

Its use in chemotaxonomy derives from the fact that in any one plant species, fraction I protein is composed of several subunits and the number and nature of these subunits can vary from species to species. In fact, FI protein is usually made up of three different large subunits (of molecular weight 55 000) and of one to four small subunits (molecular weights 12 000–15 000). A further bonus of studying this protein is the fact that the coding for the large subunits is extrachromosomal and is contained in the DNA of the plant chloroplasts, whereas that for the small subunits is present in the DNA nucleus. The consequence is that any hybrid resulting from the cross between two different plant species will contain the large subunits of the maternal parent, while the small subunits present will be the sum of those present in both male and female parents. It may thus be possible to determine the direction of that original cross, i.e. which was the female and which the male parent.

Since the nature and number of subunits of fraction I protein vary at the specific level in most plant genera so far examined (Gray, 1980), examination of these subunits can provide a new way of identifying the hybrid origin of both wild and cultivated plants. The method is particularly reliable, since no variation in the subunits of this protein has yet been encountered below the specific level. Although it is possible to separate the subunits by simple gel electrophoresis, the preferred technique, giving better resolution of the bands, is isoelectric focusing on polyacrylamide gel in the presence of 8 M urea.

B. Applications

One of the first crops to be examined with a view to determining its origin was the commercial tobacco plant, *Nicotiana tabacum*, a member of the Solanaceae. A tetraploid, it is believed to have arisen by doubling of the chromosomes from a hybrid between the diploid wild species *N. sylvestris* and either *N. tomentosiformis* or *N. otophora*. Of these two options, the latter is favoured on geographical grounds since present day populations of *N. otophora* are found growing with *N. sylvestris* on eastern slopes of the Andes in South America. By contrast, plants of *N. tomentosiformis* grow on their own in a region further north from this.

Examination of the fraction I protein subunits of the above three diploid species and comparison with the tobacco plant (Fig. 17.13) clearly showed that only *N. sylvestris* could have provided the large subunits so that this must have been the maternal parent (Gray *et al.*, 1974). Furthermore, the small subunit patterns (Fig. 17.13) indicate that *N. tomentosiformis*, and not *N. otophora*, was the male parent. Thus, this critical analysis shows unambiguously that the tobacco originated from the diploid hybrid *N. sylvestris* ♀ × *N. tomentosiformis* ♂. This conclusion has been further substantiated by the production of the synthetic hybrid tetraploid and the demonstration that its fraction I protein is identical to that of the cultivated tobacco. Thus, this analysis of a single leaf protein has provided a clear answer to the origin of today's tobacco plant.

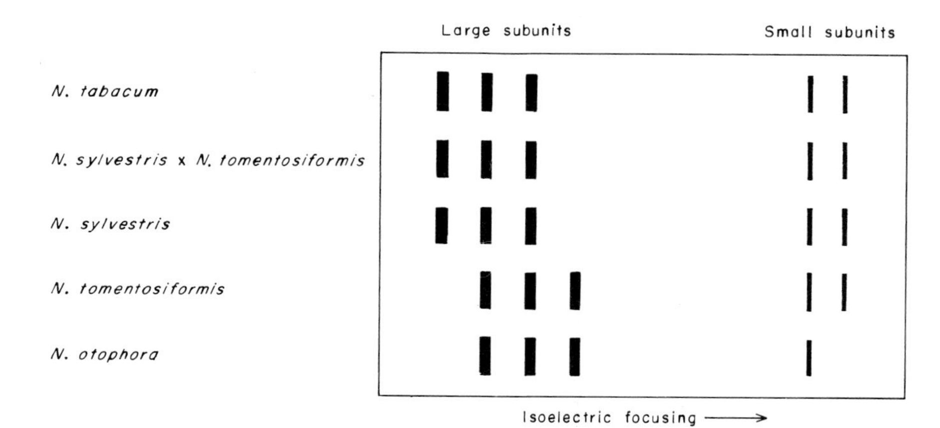

Fig. 17.13 Subunit patterns of fraction I protein from tobacco: the solution to the hybrid origin of *Nicotiana tabacum*.

Fraction I protein subunits have also been surveyed in several other crop plants and their wild relatives, notably cotton, rape, wheat, barley and potato. In general, the results have supported conclusions about parental origin derived from other sources; new information about these crops has also accrued. In the case of wheat, for example, it has already been mentioned that seed storage protein profiles support the hypothesis that the hexaploid *T. aestivum* was derived by hybridization between the tetraploid *T. dicoccum* and the diploid *A. squarrosa* (p. 417). The results of FI protein studies have indicated that such a hexaploid could only have been produced if the direction of the cross was *T. dicoccum* ♀ × *A. squarrosa* ♂. Additionally, these data have indicated that if *T. monococcum* is the diploid donor of the AA genome in *T. dicoccum*, as appears from other studies, then it must have been the male parent in a cross with a second, so far unidentified diploid species, the source of the B genome (Chen *et al.*, 1975).

FI protein subunits have been determined so far in only a relatively small number of plants and much yet remains to be learnt about its natural distribution. Variations in the small subunits are most meaningful taxonomically, since they are probably related to amino acid sequence differences. Some of the multiplicity in large subunits has been ascribed to artifactual changes caused by the reagents employed in preparing the enzyme for isoelectric focusing.

Up to the present, FI proteins have been determined in some 173 species belonging to seven plant families (Table 17.8). Two distinct patterns are apparent: plant groups where variation is very marked, such as *Nicotiana*, *Gossypium* and *Spirodela*; and plant groups where there is little variation, e.g. *Beta*, *Oenothera* and *Triticum*. These data have been interpreted by Chen and Wildman (1981) to suggest that the former taxa are relatively ancient in evolutionary terms, i.e. they have had more time to change in the polypeptide composition of this essential plant enzyme. Support for such a view stems from the fact that species of *Spirodela* and other Lemnaceae can be dated to at least 50 million years, since their fossils have been identified in strata of the Upper Cretaceous.

Table 17.8 Polypeptide composition of large and small subunits of fraction I protein of angiosperms

Family	Genus	Number of species analysed	Kinds of polypeptides LS	SS	Range in number of small subunits per species
Dicotyledons					
Solanaceae	*Nicotiana*	63	4	13	1–4
	Lycopersicon	8	1	3	1–3
	Solanum	7	2	3	1–3
Cruciferae	*Brassica*, *Sinapis*, *Raphanus*	8	2	4	1–2
Chenopodiaceae	*Beta*, *Spinacia*	2	2	2	1–2
Onagraceae	*Oenothera*	12	1	1	1
Malvaceae	*Gossypium*	19	4	8	2–4
Monocotyledons					
Gramineae	*Zea*	3	1	2	1–2
	Sorghum	7	1	1	1
	Hordeum	4	1	1	1
	Triticum, *Aegilops*	8	2	1	1
	Avena	7	3	1	1
	Oryza	14	3	6	1–4
Lemnaceae	*Lemna*, *Spirodela*, *Wolffiella*, *Wolffia*	11	4	8	1–4

Data from Chen and Wildman (1981). Abbreviations: LS, large subunit; SS, small subunit.

By contrast, plant genera which show little variation in FI protein are assumed to have evolved more recently, and this seems reasonable, at least in the case of the Gramineae taxa examined (Table 17.8). Truly, more taxa need to be investigated before this idea is substantiated. Nevertheless, this evolutionary interest should be a spur to further taxonomic investigations of subunit variations in such an interesting and all important plant protein.

V. Isozyme analysis

A. Introduction

Isozymes or isoenzymes are different molecular forms of an enzyme which are all capable of catalysing the same metabolic reaction (Rider and Taylor, 1980). Strictly speaking, isozymes are polypeptide constituents which are coded for at more than one gene locus, whereas allozymes are different forms of enzymes coded for by different alleles at the same locus. However, the term isozyme is often used loosely to embrace both genetic situations. The fact that isozyme variation is common in nature, both in animals and plants, was only realized when gel electrophoresis was widely applied to crude protein extracts during the 1950s and 1960s. This discovery of isozyme variation led to an enormous upsurge of interest in population genetics and the major applications of isozyme studies remain in this field.

Isozyme differences can be characteristically picked up by surveying natural populations of a given plant species. Indeed, isozymes provide a unique set of inherited characters, hidden within the leaf but readily revealed by electrophoresis; previously, the study of genetic variation was limited to relatively rare morphological mutations or to flower colour forms. From the purely taxonomic viewpoint, characters that are variable within populations are normally avoided. On the face of it, isozyme analysis would seem to have little systematic relevance. In fact, isozymes have been used to significant purpose in a number of systematic studies over the last decade. The point is that isozymes provide a direct measurement of the degree of genetic identity (or divergence) that exists between species, so that they provide fresh insight into relationships between taxa within the same section or genus.

One or two examples of the taxonomic value of isozyme studies will be mentioned below. Before this, some mention must be made of the experimental procedures to be employed. Isozyme analysis in animal populations has been well covered by Ferguson (1980) and that in plants has been reviewed by Gottlieb (1981) and by Hurka (1980). Some of the problems involved in the assessment of isozymic data and the detection of so-called 'cryptic alleles' are covered by Thorpe (1979) and Ayala (1982), respectively.

B. Experimental procedures

The electrophoretic technique for isozymes is essentially the same as that used for storage protein analysis (see p. 408). Small amounts of plant tissue (usually fresh leaves, but stems, flower petals, endosperm and pollen may also do) are extracted into a suitable buffer, containing a reducing agent and polyvinylpyrrolidone as protein protectants. After centrifugation, the supernatants are loaded on paper wicks which are then placed in adjacent slots cut in a buffered gel and then the gel is developed in the usual way. After electrophoresis, the slab is sliced and each slice incubated with the substrate of an enzyme of interest, together with necessary cofactors such as $NADP^+$ or NAD^+, and a reagent that gives a coloured reaction with the product of the enzymic activity. This results in coloured bands forming where the enzyme is located and the number and position of such bands are then recorded for each sample.

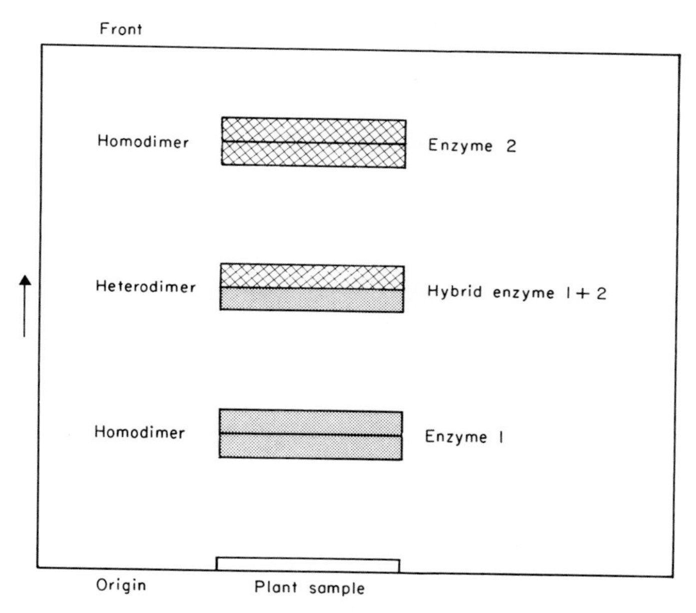

Fig. 17.14 Gel banding pattern for a dimeric enzyme, when two isozymes are present.

The pattern, or zymogram, of enzyme bands after gel electrophoresis depends on the particular enzyme assayed, its quaternary structure, its mode of inheritance and the genotype and ploidy of the individual being examined. Most enzymes have a quaternary structure such that the active catalyst consists of two (or occasionally four) identical subunits. The enzyme is thus designated as dimeric (or tetrameric). In the case of a dimeric enzyme, a plant heterozygous at a given locus will show the presence of three bands, two due to the homodimers and one due to the heterodimer made up of subunits derived from the two homodimers. This third band is usually of intermediate mobility and is more intensely stained (Fig. 17.14). With tetrameric enzymes, five bands will appear when the coding gene is heterozygous.

The complexities of interpretation of these gels are therefore considerable and it may be necessary to sort out the situation by genetic analysis. In the standard case, this simply means crossing individuals with different banding patterns and scoring the number of individuals of different patterns in the F_1 generation; normal Mendelian inheritance will be exhibited. Another technique that has been developed for this purpose is the comparison of leaf or stem extracts with those of the pollen. Since pollen is haploid, the grains will contain only one allele at each locus. Thus, the pollen of a heterozygote will produce only one or other subunit; hence no 'hybrid' heterodimers will appear on the zymograms.

In an isozyme analysis that is truly representative of the genetic variation within a plant population, it is essential to sample as many enzymes as possible. This avoids any bias due to the variable nature of inheritance or forms of any individual enzyme. In practice, it is possible to assay between 10 and 20 enzyme activities. The difficulties of extending the survey beyond this number lie in part in the need to find suitable reagents which will give coloured products where the enzyme bands are located on the gels.

Table 17.9 Enzymes most frequently studied by gel electrophoresis in plant populations

Enzyme	Abbreviation	EC number	No. of isozymes usually present
Glutamate oxaloacetate transaminase	GOT	2.6.1.1.	2
Peptidase	PEP	3.4.1.–	Variable
Phosphoglucomutase	PGM	2.7.5.1.	2
Malate dehydrogenase	MDH	1.1.1.37.	3–4
Acid phosphatase	Acph	3.1.3.2.	Variable
Esterase	EST	3.1.1.–	Variable
Glutamate dehydrogenase	GDH	1.4.1.3.	1
Alcohol dehydrogenase	ADH	1.1.1.1.	2–3
Peroxidase	PER	1.11.1.7.	Variable
Superoxide dismutase	SoD	1.15.1.1.	Variable
Isocitrate dehydrogenase	ICD	1.1.1.42.	2
Phosphoglucose isomerase	PGI	5.3.1.9.	2

The most commonly studied enzymes are listed in Table 17.9. Most of these are metabolically important; e.g. phosphoglucomutase is a key enzyme of the glycolytic pathway of carbohydrate utilization. A few, such as peroxidase, are of as yet ill-defined function in plant metabolism. For the enzymes of primary metabolism, the isozymic variation is primarily due to the fact that the same reaction has to be catalysed in different subcellular compartments within the cell, e.g. the plastids and the cytoplasm; each organelle tends to have its own distinct isozyme. The number and subcellular location of these isozymes appear to be highly conserved during plant evolution. However, gene duplication in diploid species and the addition of new genomes in polyploid species have increased the number of isozymes that can be detected (L. D. Gottlieb, 1982). The advantages, or otherwise, to an organism of having several different forms of the same enzyme within the same organelle is not yet at all clear, but multiplicity in an essential enzyme may be an important safety factor under conditions of stress.

C. Taxonomic applications

The special contribution of isozyme analysis to plant taxonomy is that it provides the means of calculating the degree of genetic similarity between species. This can be expressed as the mean genetic identity, $\bar{I}$, which is based on comparing for as many enzymes as possible all pairs of populations representing the two given species. The genetic identities of species pairs in a few selected cases are shown in Table 17.10 (from Gottlieb, 1981). The average value of such comparisons that have now been made for 21 pairs of related species is 0.67. Taking the negative logarithm, this converts to a genetic distance estimate $\bar{D}$ of 0.40. This represents about 40 detectable changes per 100 loci having occurred during the separate evolution of an average pair of plant species. Such a genetic distance represents more than a tenfold increase over the average distance between conspecific members of the same population. In a sense it is a measure of the 'barriers' that exist between one species and another in natural plant populations.

Table 17.10 Mean genetic identities between pairs of plant species based on isozyme comparisons

Species pair	Genetic identity
Clarkia biloba–C. lingulata	0.88
C. rubicunda–C. franciscana	0.28
Phlox drummondii–P. cuspidata	0.84
P. drummondii–P. roemariana	0.70
P. cuspidata–P. roemariana	0.59
Tragopogon dubius–T. porrifolius	0.50
T. dubius–T. porrifolius	0.62
T. porrifolius–T. pratensis	0.53

For references, see Gottlieb (1981)

These calculations of genetic identity have immediate taxonomic relevance in indicating how closely related, or otherwise, are species within the same genus. They may also help to indicate phylogenetic trends at the species level.

They may confirm, or otherwise, particular progenitor-derivative species relationships within actively evolving plant genera. One interesting example of the use of genetic identity is in the genus *Clarkia* of the Onagraceae.

One *Clarkia* species, *C. franciscana*, a predominantly self-fertile annual plant, was found to occur in a single locality in San Francisco, California. Lewis and Raven (1958) suggested that it might have evolved very recently from within a population of *C. rubicunda* which grows in the same region. The two species resemble each other in a number of morphological traits and they will hybridize, although the hybrid is almost completely sterile. Electrophoretic studies by Gottlieb (1981) for eight enzyme activities controlled by about 14 gene loci indicate a genetic identity between *C. franciscana* and *C. rubicunda* of only 0.28. This suggests that if the two species are indeed related as progenitor-derived species, then the separation must have occurred much earlier in time than that suggested by Lewis and Raven (1958).

It may be noted that in the genus *Clarkia*, another pair of species, *C. bilobata* and *C. lingulata*, show a much closer genetic identity ($\bar{I} = 0.88$, Table 17.9). This value fits in with the morphological situation in that these two taxa are nearly identical, only differing from each other in petal shape. By contrast, *C. franciscana* and *C. rubicunda* are readily separated morphologically and can even be distinguished from each other at the seedling stage.

<table>
<tr><th></th><th>1</th><th>2</th><th>3</th><th>4</th><th>5</th><th>6</th><th>7</th><th>8</th><th>9</th><th>10</th><th>11</th><th>12</th><th>13</th></tr>
<tr><td>T. pratensis</td><td>|</td><td></td><td></td><td>|</td><td></td><td>|</td><td></td><td></td><td></td><td>|</td><td>|</td><td></td><td></td></tr>
<tr><td>T. miscellus</td><td>|</td><td>|</td><td>|</td><td>|</td><td>|</td><td>|</td><td>|</td><td>|</td><td>|</td><td>|</td><td>|</td><td>|</td><td>|</td></tr>
<tr><td>T. dubius</td><td></td><td></td><td>|</td><td></td><td></td><td></td><td></td><td>|</td><td></td><td></td><td></td><td></td><td>|</td></tr>
</table>

Fig. 17.15 Alcohol dehydrogenase isozymes in *Tragopogon* parental species and the hybrid *T. miscellus*. Bands 2,5,7,8 and 12 are due to hybrid enzymes.

The mean genetic values (Table 17.9) do not, of themselves, provide a complete index of phylogenetic relatedness and may work in some cases (e.g. *Phlox*, see Table) but may fail to be of much value in assessing species relationships in others. One may note that the three pairs of diploid species in the genus *Tragopogon* (Compositae) have very similar I values, so their relative taxonomic positions are not clear from these data. In the case of *Tragopogon*, however, isozyme studies have been useful for another reason, for confirming the parentage of two tetraploid taxa: *T. mirus* (*T. dubius* × *T. porrifolius*) and *T. miscellus* (*T. dubius* × *T. pratensis*). In *T. miscellus* for example, the alcohol dehydrogenase profile was composed of 13 distinguishable isozymes (Fig. 17.15). Each of the 13 isozymes was fully accounted for by simple additivity of the polypeptides specified by genes inherited from the two diploid parents; five of the bands were simply of hybrid origin. Similar patterns occurred with other enzyme activities, so that in this case, there was overwhelming confirmation of the precise parental origin of *T. miscellus* (Gottlieb, 1981). It is interesting in this case that two-dimensional chromatography of the leaf flavonoids in these plants was considerably less successful in establishing the parentage of the two tetraploid hybrids (Brehm and Ownbey, 1965).

VI. Serological analysis

A. Introduction

No mention of protein comparisons in a taxonomic setting would be complete without reference to serology. The major applications of serology have always been in the medical field, e.g. in the development of immunization against infectious diseases, but there has been a connection with plant proteins ever since the discovery in 1888 by Stilman that certain legume seed extracts have the ability to agglutinate erythrocytes of human blood corpuscles. Serological techniques were developed early this century on an empirical basis before proteins had been properly purified and characterized. They have been employed ever since in both plant and animal systematics, with continuing refinements of the original techniques. Although used widely in plant studies, their most fruitful application has probably been at the higher levels of classification, largely because other biochemical characters are relatively scarce here.

In its simplest form, serological comparison is no more difficult than any other type of protein comparison. It is based on the fact that plant proteins, either singly or as mixtures, are antigenic. Thus, their injection into animals will induce the formation of specific antibodies, called immunoglobulins, in the blood of these vertebrates. After a few weeks, the blood of the animal can be sampled and the antisera containing these antibodies can be separated from the erythrocytes. If these antibodies are then allowed to mix with the original plant antigens in a test tube, there will be a complementary interaction with a precipitate being formed (the so-called precipitin reaction). If the same antibodies are mixed with antigens isolated from a second plant species, then the degree of precipitation will be less. The extent of such a precipitation will be a direct measure of the relatedness of the two taxa being investigated.

Serological tests are thus concerned with matching the proteins of one species against those of another in a lock and key approach. Normally, one test species is chosen, the antisera developed in rabbits and this is matched against the proteins of a range of other species within the group. Ideally, a second test species should be chosen to produce antisera and this is then again tested against all other taxa. In theory, all reciprocal tests should be carried out but in practice this is rarely achieved.

There is a limited factor in such serological experiments in that a separate series of immunizations with rabbits are needed for every test species. Rabbits, incidentally, are used almost exclusively in plant serology. Since the individual reaction of a rabbit varies with its physiological state and its genetic constitution, one batch of antisera may differ slightly from another. In order to avoid loss of the antigenic protein via excretion or metabolism in the animal body, certain adjuvants are added to prolong the life of the injected antigen. A further complication in serological analysis is the fact that the results are based on serological cross-reactivity and they require interpretation before they can be applied to data processing with other biochemical information (Cristofolini, 1980).

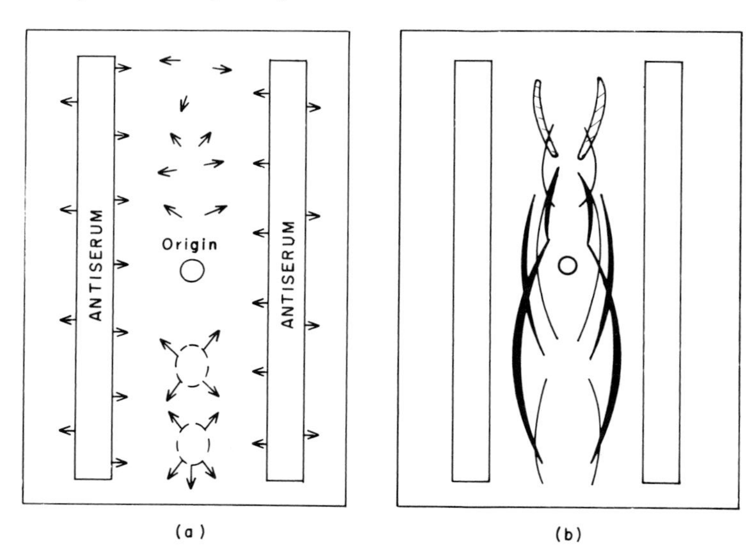

Fig. 17.16 Immunoelectrophoresis of plant proteins (a) after gel electrophoretic separation, antisera are placed in troughs at side. (b) When antigens and antibodies diffuse towards each other, they meet and precipitation occurs along an arc of optimal antigen–antibody proportion. (From Smith, 1976.)

In the period of classical plant serology, the results were derived from direct observation in the test tube of the extent of the precipitin reaction which varied with time. Serial dilution was employed, since there is an optimal point in the ratio of antigen to antibody which yields the greatest cross-reaction. A significant advance in technique came when it was found that the precipitin reaction could be quantified by determining the turbidity of the solution with a densitometer or reflectometer (Boyden, 1954). Two even more important developments were the introduction of gels, through which the protein extracts could diffuse, for studying the immunological reaction (Ouchterlony, 1948), and the application of electrophoresis to the process (Grabar and Williams, 1953). In immunoelectrophoresis, the antigenic proteins are separated by gel electrophoresis and each protein is then tested against the antiserum, the intensity and position of the precipitin lines being an indication of the serological reaction (Fig. 17.16). These new techniques have undergone subsequent modification (see Daussant, 1975), one of the most widely used modifications being double-diffusion experiments (Fig. 17.17).

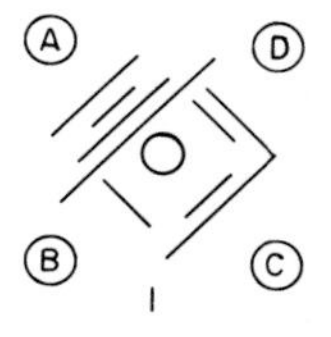

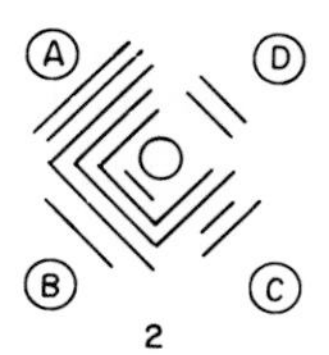

Fig. 17.17 Diagram of Ouchterlony double-diffusion experiment. (1) The centre well contains antisera of *Sambucus canadensis* and is being tested against antigens of *S. canadensis* (in well A), *Viburnum trilobum* (well B), *Cornus canadensis* (well C) and *Nyssa sylvatica* (well D). There is little correspondence with these taxa. (2) The centre well contains antisera of *Cornus racemosa* and is being tested against antigens of *C. racemosa* (well A), *C. amomum* (well B), *Nyssa sylvatica* (well C) and *Sambucus canadensis* (well D). There is correspondence between all taxa except *S. canadensis*. (From Hillebrand and Fairbrother, 1970.)

In botanical serology, proteins have been extracted from leaves, stems, pollen, root and tuber. However, most attention has been given to seeds and the storage proteins thereof and in almost all analyses mentioned below, seeds have been the chosen source, because of their accessibility and stability.

Usually, the total soluble fraction of the seed is used as antigen. Various purified protein fractions have also been employed. For broad comparisons above the family level, Jensen and his co-workers have concentrated on the major storage protein after it has been purified on columns of Sephadex G-200 and DEAE-Sephadex A50 (Jensen and Penner, 1980). Particular enzyme activities in the seeds, such as the proteases, have also been used as antigens (Chrispeels and Baumgartner, 1978).

The biochemical basis of the antigen–antibody reaction is still being actively investigated in a number of laboratories and has not yet been fully elucidated. The ability to produce antibodies is related in some way to the amino acid sequence of the protein antigen. Thus, although all plant proteins are antigenic, the larger the protein the greater the antigenicity. This can be explained by assuming that the larger the protein is, the greater the number of antigenic sites or 'determinants'. In the case of the small protein cytochrome *c*, three or four antigenic sites are known to be present. A large globular protein of seeds may have many more.

A single determinant represents a well-defined site on the protein surface and hence is coded for by the DNA. The size of a determinant sequence is not known exactly, but may be from six to twelve amino acid residues. In the case of cytochrome *c*, it is possible that nearly half the protein could be involved in determinant sites (Margoliash *et al.*, 1970). It is clear that the tertiary and quaternary structure of a protein is also involved in some way in the serological reaction.

During the process of immunization, the rabbits, after injection, produce specific antideterminants on protein antibodies, often referred to as immunoglobulins. The antideterminants are apparently always carried in identical pairs on antibody molecules and it is these that are used to analyse the nature and number of determinants in the antigenic protein of the various species being compared.

Serological identity between two proteins can be taken to indicate closely similar or identical molecular sequences. However, serologically dissimilar proteins may or may not be homologous. Thus a single amino acid substitution in an antigenic site may be sufficient to upset the serological reaction.

The literature on plant serology is extensive and only a few applications to taxonomy can be considered here. Serology is an important technique in its own right and is used for purposes other than taxonomy. When used taxonomically, it may be combined with other protein techniques, especially electrophoresis. A general account of plant serological principles is provided by Smith (1976). Serological techniques are extensively reviewed by Daussant (1975) and the problems of interpretation are dealt with by Cristofolini (1980). An extensive review of the application of serological techniques to plant phylogeny is presented elsewhere by Fairbrother *et al.* (1975).

B. Taxonomic applications

Serology has been applied to taxonomic problems in plants at all levels of classification. It has been used to evaluate relationships at the subspecific level in *Phaseolus vulgaris* (Kloz, 1971) and at the specific level, to assign two anomalous taxa, *Phaseolus aureus* and *P. mungo* to another genus, *Vigna* (Chrispeels and Baumgartner, 1978). There has been widespread use at the generic level to study affinities within families and to ascertain relationships within several related families. Some of the more familiar examples from the recent literature are shown in Table 17.11.

Table 17.11 Some plant families which have been investigated serologically

Family	Type of taxonomic information produced	Reference
Caprifoliaceae	Indicates distinctiveness of *Viburnum* from other genera	Hillebrand and Fairbrother (1970)
Gramineae	Of value in revising species classification within genera, e.g. in *Bromus*	Smith (1976)
Leguminosae	Mainly of value in delimiting genera and tribes; separates *Vigna* from *Phaseolus* in the Phaseoleae	Kloz (1971), Cristofolini (1981)
Primulaceae	Upholds subdivisions based on morphology; confirms *Lysimachia* as a family member	John (1978)
Ranunculaceae	Clarifies generic relationship and correlates with karyotype characters	Jensen (1968)
Rubiaceae	Separates *Asperula* and *Galium* from other genera	Lee and Fairbrother (1978)
Solanaceae	Clarifies species relationships at sectional level in *Solanum*	Gell *et al.* (1960), Lester (1965)
Umbelliferae	Confirms tribal groupings in various subfamilies	Pickering and Fairbrother (1971)

One of the most widely quoted family surveys is that of the Ranunculaceae, where Jensen (1968) used proteins of mature seed and a variety of standard techniques, including turbidometry of the precipitin reaction (the Boyden procedure), immunodiffusion and absorption. Jensen obtained antisera from ten key taxa, including *Adonis vernalis* and *Hydrastis canadensis*, and tested them reciprocally against generic representatives throughout the family. The integrated results are indicated in a two-dimensional scheme (Fig. 17.18) of generic affinities. The most striking correlation is with chromosome number and size (Gregory, 1941), but this scheme also fits in with the distribution pattern of secondary chemicals, e.g. the lactone ranunculin (Ruijgrok, 1968), in the family.

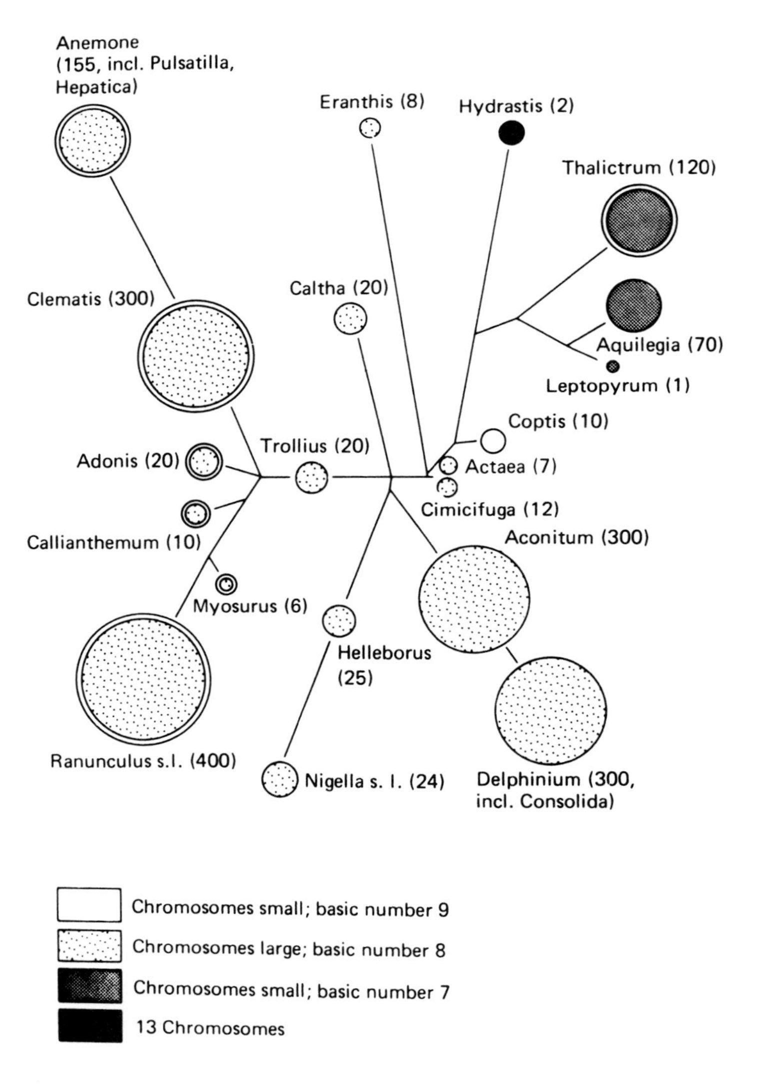

Fig. 17.18 An arrangement of genera in the Ranunculaceae based on serological evidence. (From Jensen, 1968.)

Some of the results indicated in the scheme do not correspond with generic relationships indicated by the morphological evidence. For example, *Hydrastis* has been classified with the neighbouring Berberidaceae in some systems and when included in the Ranunculaceae is regarded as a link taxon. However, the serological results clearly tie in *Hydrastis* with the Ranunculaceae, since there was little correspondence in cross-reactions when it was tested against four Berberidaceous taxa, which were included especially for this comparison. In spite of its distinctive chromosome number, it seems therefore that *Hydrastis* should be retained in the Ranunculaceae.

In most other ways, the serological survey of the Ranunculaceae by Jensen agreed with orthodox classifications of the family. Thus the serologically related pairs of genera, *Aconitum–Delphinium*, *Anemone–Clematis* and *Ranunculus–Myosurus*, are morphologically similar. The serological isolation of *Nigella* and of *Eranthis* from the rest of the family also agreed with the evidence from other characters.

Systematic serology has also been applied above the family level to test out various theories of ordinal classification. Sampling of families is usually limited because of practical limitations on the number of serological tests that can be achieved. The reasonable assumption is made that serological similarity is likely at the generic level and single representatives of genera are sufficient at this level of comparison. Much of these data have been thoroughly reviewed by Fairbrother *et al.* (1975). One example must suffice to indicate the usefulness of the serological approach. The systematic position of the Didiereaceae, a small family of cactus-like thorny shrubs indigenous to Madagascar, was for a long time controversial and Cronquist, for example, in 1957 placed them in the Euphorbiales. The discovery of betalain pigments in these plants (see p. 306) firmly placed them on chemical grounds within the Centrospermae (or Caryophyllales). Other evidence was needed to confirm this reassignment. This was provided by Jensen (1965) who obtained antisera from the genus *Alluaudia* (Didiereaceae) and tested it against antigens from 23 taxa of the Centrospermae and 11 taxa outside the Centrospermae. The serological results strongly supported the association with the Centrospermae and this conclusion is now widely accepted in most modern treatments of angiosperm classification.

The most recent serological experiments of taxonomic relationships at the higher categories have employed purified storage proteins, rather than whole seed protein fractions (Jensen and Penner, 1980; Jensen and Buttner, 1981). Two similar major storage proteins appear to occur very widely in angiosperm seeds, so that serological comparison is possible with antisera raised against purified proteins of single taxa to represent whole families. In one experiment, the two reference proteins chosen were aquilegin, isolated from *Aquilegia vulgaris*, to represent the Ranunculaceae and tubiflorin from *Digitalis purpurea* to represent the Scrophulariaceae. Serological tests at the family level showed that taxa of different genera within a family were only distinguishable serologically when the reference antisera had been obtained from within that family; otherwise, all genera within a given family responded similarly. In the case of proteins isolated from different species within a genus, there was great serological uniformity.

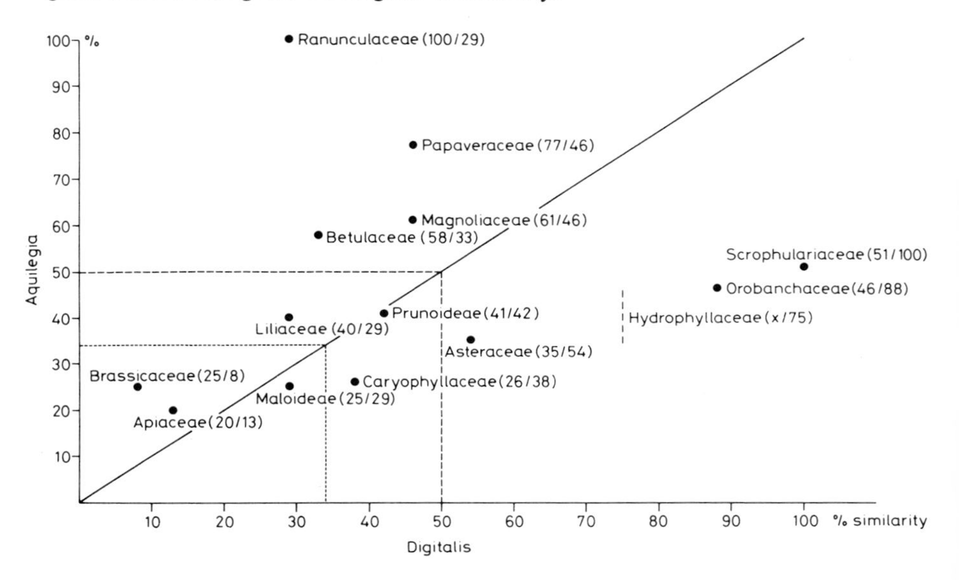

Fig. 17.19 Serological similarities of 14 families of subfamilies when tested against purified storage protein of *Aquilegia* (Ranunculaceae) and *Digitalis* (Scrophulariaceae). (From Jensen and Buttner, 1981.)

By testing antisera, produced from injecting aquilegin and tubiflorin into rabbits, against representatives of other angiosperm families, it was possible to map out the serological similarities in the form of a two-dimensional plot, as indicated in Fig. 17.19. Such a diagram necessarily only presents an incomplete picture of the total storage protein comparisons that are possible. Nevertheless, it is interesting to see that a number of established taxonomic relationships are confirmed, e.g. the closeness of the Scrophulariaceae to the Orobanchaceae. The most striking feature of this diagram is the separation of the Prunoideae and the Maloideae within the same family, the Rosaceae. This separation is apparent when other reference antisera have been used. This conclusion does not, of course, require that the Prunoideae should be raised to family rank and moved away from other Rosaceous members. The authors conclude that 'when evaluated with all other taxonomic characters, the position of both subfamilies within the Rosaceae should be maintained. *Nevertheless*, the serological characters indicate that the differences between these subfamilies are greater than has been proposed before by other researchers.' This approach through purified storage proteins would seem to be a particularly fruitful one for further investigation of taxonomic relationships and indicates one of the new directions that plant serology is taking during the present decade.

VII. Summary

Plant proteins can be studied in a variety of different ways to yield useful and interesting comparative data (Table 17.12). So far the simplest of all experimental procedures—gel electrophoresis—has been the most successful in yielding results of taxonomic value. This is in spite of the fact that seed storage protein studies often only measure the number of proteins present and in such studies, there is a real problem of defining protein homology.

Table 17.12 Protein characters and their utilization in plant taxonomy

Properties	Proteins studied	Taxonomic application
Amino acid sequence	Cytochrome *c*, plastocyanin and ferredoxin of fresh leaves	Not yet of established utility; but of potential phyletic importance
Subunit variation	Fraction I protein of leaves	Valuable for identifying species hybrids
Enzyme types (isozymes)	Enzyme extracts of young seedlings	Chiefly useful at subspecific and species levels
Serological interactions	Total protein fraction or major storage protein of seed	Most useful at family or order level
Storage protein complexity	Seed or pollen proteins	Useful at species and generic levels

There is a clear trend in recent years towards analysing variation within a single class of protein or even a single protein, after suitable separation. Amino acid sequences of proteins, which are regarded by many as the most important of all characters, have not yet proven of direct taxonomic application. However, the pioneering work has now been done and it will be interesting to see amino acid sequencing being applied to real taxonomic problems among the angiosperms. So far, among flowering plants, most studies have simply displayed the vast amount of information that can be obtained by such an approach.

Subunit variation in fraction I protein has proven to be an interesting new source of taxonomic and possibly phyletic information. Although the number of subunits present is relatively small, further surveys of such variation should be of value.

Isozyme variation in plants remains primarily of genetic interest, although there are taxonomic benefits to be derived from such studies. Serology continues to be of importance, largely because it allows wide comparisons in protein types between genera, families and orders. The actual biochemical basis of serological cross-reactions is unfortunately still not completely sure.

18

Nucleic Acids

I.	Introduction	439
II.	Nucleic acid base ratios	443
III.	DNA–DNA hybridization	448
IV.	Sequence studies	456
V.	Conclusion	459

I. Introduction

The nucleic acids of the plant cell have unrivalled taxonomic importance when it is considered that all the information which determines what an organism is going to be is tied up in the triplet code of the deoxyribonucleic acid (DNA) in the nuclear chromosomes. As Smith (1976) has expressed it:

> It is clear that the nucleic acids of different organisms are treasuries of taxonomic evidence. Taxonomists operate by compiling and then classifying specifications of organisms. In the nucleic acids the natural specifications are available for the asking.

Within the DNA base sequences of an organism, there could be data which would provide a definitive measure of the relative similarities and differences that exist in form and function between related plant species.

The possibilities of applying DNA data to problems of constructing plant phylogenies are even more promising. Belford *et al.* (1981) have detailed some of the advantages of DNA sequence comparisons over classical taxonomic approaches. They write:

> with hybridisation, the complete spectrum of DNA sequences present in a species is compared with that of other species, thus avoiding subjective decisions on which characters to use. Also, the hazard of basing a phylogeny on characters which might have been subject to great fluctuations in selective pressures is avoided. And finally, molecular approaches by-pass the difficulty of prezygotic and postzygotic mating barriers.

This latter advantage means that comparisons are not limited, as at present, to those species or genera that hybridize in nature or which can be induced to hybridize with each other in the glasshouse.

There are two main reasons why nucleic acid data have not been as widely applied in plant taxonomy as might have been expected from the above statements. The first simply relates to the considerable technical problems associated with studying these macromolecules, their isolation in pure form without damage and the analysis of the information they contain. The second reason is more fundamental. In eukaryotes, the nucleic acid that codes for protein is only a small part of the total DNA. Both DNA and RNA exist in many different forms and lengths and, until the complexity of their organization in the cell is understood, nucleic acid comparisons between organisms are difficult to assess.

In the case of prokaryotes (e.g. the bacteria and cyanobacteria), the nucleic acid exists in a relatively simple form as a circular thread of double helix. Only technical problems delayed the rapid utilization of DNA and also RNA (ribonucleic acid) data to classification in these organisms. DNA base ratios and DNA homologies have now been measured in most bacterial species and the results applied most successfully to problems of generic classification (de Ley, 1968). An interesting recent application of DNA data is that differences in base composition and homology allowed the recognition of the organism responsible for Legionnaire's disease as a new genus, called *Legionnella pneumophila* (Lattimer and Ormsbee, 1981). The application of ribosomal RNA comparisons to bacterial phylogeny has also been developed in recent times (Woese *et al.*, 1978; Kandler and Schleifer, 1980).

For some time after the original molecular biology 'explosion' initiated in 1953 by the discovery of Watson and Crick that DNA exists as a simple base-paired double helix, the nucleic acids of higher plants were ignored in favour of those from microbial or animal cells. However, in the last decade, this neglect has been remedied in part and our information about plant nucleic acids has expanded considerably. Recent reviews include those of Hall and Davies (1979) and of Marcus (1981). Two general points emphasize the difficulties associated with DNA comparisons between different species. The first is that the total amount of DNA present in a plant cell varies between wide limits in a way quite unrelated to taxonomy or chromosome number. The second is that within the higher plant cell, there are three sets of DNA, in the nucleus, chloroplast and mitochondria, each of which carries genetic information and which may be translated via RNA into protein synthesis.

Fairly sophisticated measurements have now been made of the total DNA of plant cell nuclei; the amounts can be estimated by nuclear isolation followed by extraction and direct chemical measurement. It has been found that the content can vary by more than a hundredfold between plants (Table 18.1). At one extreme, the flax plant, *Linum usitatissimum*, contains 1.4 pg (10^{-12} g) per 2 *C* nucleus, while at the other the monocotyledon *Tradescantia ohioensis* has 89.0 pg. Significant variations can occur within quite restricted taxonomic groupings. The considerable range existing within the onion, *Allium cepa*, and related *Allium* species is illustrated in part in Table 18.1. Again, in the genus *Vicia*, the DNA content may differ by up to sevenfold even among diploid species (Bennett and Smith, 1976) (compare values for *V. faba* and *V. angustifolia* in Table 18.1). Yet again, among *Lathyrus* species, the content may vary from 6.9 to 29.2 pg (2 *C* values) (Narayan, 1982).

Table 18.1 Some typical DNA measurements on the nuclei of higher plant cells

Plant (family)	DNA content (pg per 2*C* nucleus)[a]
Tradescantia ohioensis (Commelinaceae)	89.0
Allium globosum (Liliaceae)	75.8
Allium karataviense (Liliaceae)	45.4
Allium cepa (Liliaceae)	33.5
Vicia faba (Leguminosae)	25.8
Secale cereale (Gramineae)	18.9
Allium fuscum (Liliaceae)	18.4
Lolium perenne (Gramineae)	9.9
Helianthus annuus (Compositae)	9.8
Pisum sativum (Leguminosae)	9.1
Vicia angustifolia (Leguminosae)	6.1
Lupinus albus (Leguminosae)	2.3
Linum usitatissimum (Linaceae)	1.4

[a]The actual amount will increase with increasing ploidy levels; thus the tetraploid form of *T. ohioensis* will have 178 pg per nucleus, etc.

Even in a plant with a low nuclear DNA content, a problem exists to explain the role of all the DNA that is present. In such a plant, the size of the genome is in the order of 2×10^{12} daltons, which represents in principle some 2×10^{6} genes. Considering that there is only a finite number (a few 1000) of different proteins present in the cell, this means that only about 1% of this DNA can be directly concerned in protein synthesis. It is, in fact, clear that only a small part of the nuclear DNA is coding DNA, the remainder being either structural or regulatory.

Some idea of the different kinds of DNA and RNA that have been recognized in higher plant cells is indicated in Table 18.2. An important distinction exists between single copy, non-repetitive or unique DNA and repetitive DNA. This last material consists of identical sequences of different lengths which occur in various numbers (from 10 to 10 000 copies of any one sequence may be present). It may be essential to separate these two kinds of DNA before making interspecies comparisons (see section III). One final complication, recognized in recent times, is the presence among the coding genes of higher plants of a few which are termed 'discontinuous genes'. In these genes, the regular triplet code of bases is interrupted by blocks of nucleotides which are not concerned in the transcription of information for protein synthesis. The purpose of these interruptions is still not clear; their presence clearly may somewhat obscure species–species comparisons at the genetic level.

Table 18.2 The different kinds of nucleic acid molecules present in higher plant cells

Nucleic acid	Description
DNA	
Nuclear	The bulk of the DNA in the cell, usually in double-stranded linear sequences
Satellite	A fraction separable by neutral caesium chloride centrifugation, associated with the main nuclear DNA (about 3 to 28% of the total)
Chloroplastic	Circular molecules, some 40 μm in circumference
Mitochondrial	Circular molecules, between 5 and 30 μm in circumference
Single copy	Nuclear sequences of between 1000 and 4000 base pairs
Repetitive	Nuclear sequences of from 200–400 up to 50 000 base pairs
Structural	Required for maintenance of chromosomal morphology
Informational	Concerned in gene regulation
Coding	Carrying the code for individual proteins
RNA	
Nuclear-cytosol	
Chloroplastic	
Mitochondrial	
Transfer (tRNA)	Short 73–90 nucleotide sequences, between 25 and 50 per organelle
Ribosomal (rRNA)	Composed of large subunit of 3000 bases with several small components (about 100 bases) and a smaller subunit (1500–2000 bases). Defined by sedimentation coefficient (70S in chloroplast, 80S in cytoplasm, 78S in mitochondrion).
Messenger (mRNA)	400–4000 bases long, one for each protein synthesized (total number between 10^3 and 10^4).

In the present chapter, the various ways of making nucleic acid comparisons and utilizing the data generated for taxonomic purposes will be considered.

The simplest type of comparison is by means of the base composition but this is only of very limited value in the case of the higher plant. DNA–DNA hybridization, where sequences of two organisms are matched up against each other, is potentially a more informative taxonomic tool. In theory, this is a straight forward procedure but, in practice, it is difficult to achieve. Ultimately, the most valuable comparisons are those that utilize base sequence determinations. As will be made clear below, although sequencing is now a practical objective, only a few sequences have been determined for higher plants so that taxonomic discussion of the data will inevitably be limited.

II. Nucleic acid base ratios

Deoxyribonucleic acid (DNA) is composed of phosphate, the sugar deoxyribose and four organic bases: adenine (A), thymine (T), cytosine (C) and guanine (G). Because of the complementary nature of the base pairing in the double-stranded native DNA (Fig. 18.1), the molar fraction of A is the same as T and that of G is the same as C. Differences in the triplet code of DNA isolated from various organisms are therefore reflected in changing amounts of G + C compared to A + T. For convenience, this is expressed as G + C%, where A + T% = 100 − (G + C). Determination of this G + C ratio thus gives a measure of differences in the cistrons of the DNA and, where significant changes occur, then such measurements may be of taxonomic value.

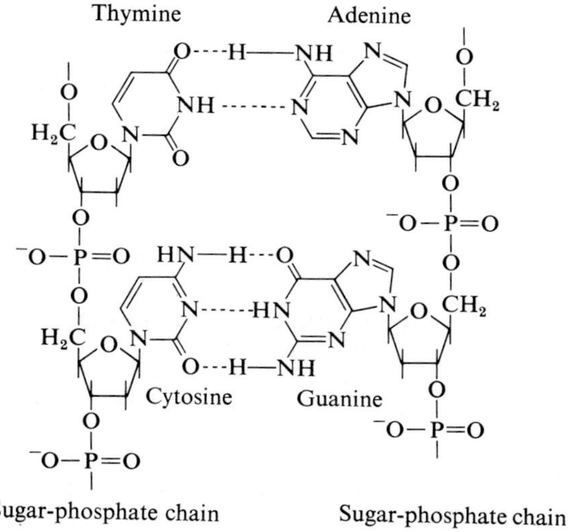

Fig. 18.1 The basic structure of deoxyribonucleic acid, showing the pairing between complementary bases.

The measurement of base ratios is fairly routinely accomplished once the DNA has been isolated and purified. The DNA is hydrolysed with perchloric acid and the amounts of the four bases liberated can be determined spectrally or densitometrically after chromatographic separation (see Harborne, 1973b). It is also possible to determine the G + C ratio directly on the DNA by either measuring thermal denaturation profiles (Mandel *et al.*, 1970) or by centrifugation to equilibrium in the presence of caesium chloride (Schildkraut *et al.*, 1962). Usually both the latter methods are employed, as a useful check on the accuracy of the technique.

Base ratios vary most in the DNA species of lower plants, while in animals and higher plants there are lesser differences. In the case of bacteria, the G + C% may vary from 25 to 75. Because of this very wide variation, such measurements have been made on the majority of bacterial species available. Some typical results are shown in Fig. 18.2 and it will be seen that the data have considerable importance in classifying bacteria at the generic level. Thus, bacteria of the same genus tend to have a closely similar range of G + C ratios, e.g. ± 2–5% as in *Escherichia*, *Proteus* and *Salmonella* (de Ley, 1968). Furthermore, a closely similar G + C ratio at the generic level is well correlated with morphological and physiological features that are shared in common.

In fact, where the G + C ratio of a particular species falls outside the normal range for the genus in which it is usually placed, then it can be indicative of incorrect classification. In many such examples that have been further investigated (de Ley, 1968), phenotypic identification has confirmed the need for reclassification.

Organisms or genera with closely related G + C values may have evolutionary ties, but this does not necessarily follow. Other tests have to be made (e.g. of DNA homology, see section III) to confirm or otherwise such a supposition. On the other hand, organisms that differ significantly (e.g. by 25%) in their G + C ratios must be different and have to be regarded as phylogenetically unrelated. For example, among Gram-positive bacteria, the genus *Clostridium* (G + C 28–38%) has a very different set of cistrons in the DNA from *Streptomyces* (G + C 71–78%). Finally, G + C ratios among bacteria have been used to confirm trends in evolution which are suspected on morphological grounds. One example (de Ley, 1968) where increasing morphological complexity parallels increasing G + C ratios is the series: *Corynebacterium* (48–59%), *Arthrobacter* (63–65%), *Mycobacterium* (60–70%) and *Actinomyces* (68–78%).

There are also wide differences in base ratios among the fungi. Storck and Alexopoulos (1970) examined 322 species from all major classes and found a range from *Pichia kluyverii* at 26% to *Rhodotorula graminis* at 70%. Some typical results obtained in two genera within the Oomycetes are shown in

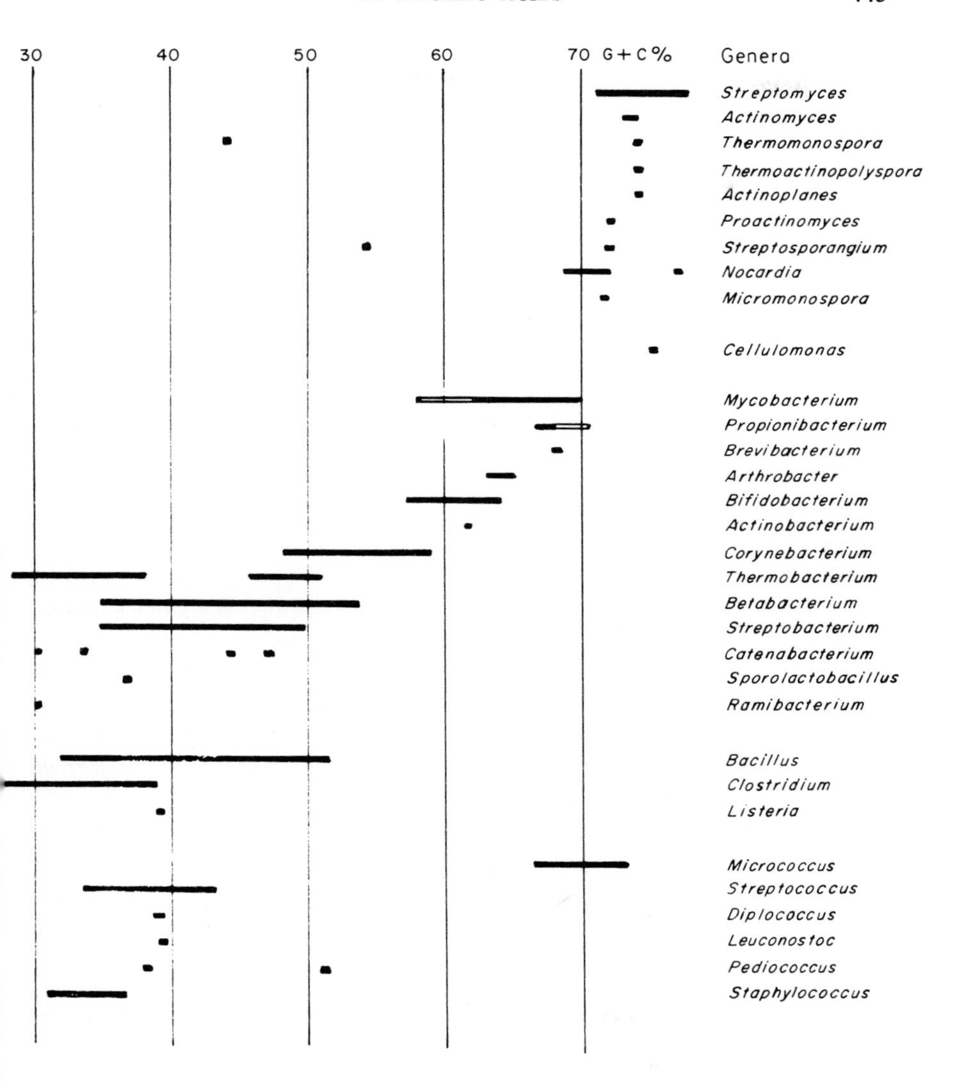

Fig. 18.2 Ranges of G + C% for the DNA of different genera of Gram-positive bacteria. (From de Ley, 1968.)

Table 18.3. These data illustrate the taxonomic potential of such measurements. It will be noted, for example, that *Saprolegnia subterranea* has a value close to most other *Saprolegnia* spp. These data thus support the decision to reclassify a taxon previously included in the genus *Isoachlya*. Additionally, it will be seen that *Achlya inflata* is significantly different from other *Achlya* spp. This fits in with the known morphological heterogeneity within *Achlya*; additionally, differences in chromosome numbers have been detected among these fungal species (Green and Dick, 1972).

Table 18.3 G + C ratios for two fungal genera in the Oomycetes. (From Green and Dick, 1972.)

Taxa	G + C	Taxa	G + C
Saprolegnia ferax	49.5	*Achlya inflata*	42
S. hypogyna	55.5	*A. racemosa*	52
S. diclina	58.5	*A. sparrowii*	51
S. parasitica	59.0	*A. colorata*	52
S. subterranea	61.5	*A. flagellata*	55
Average	56.8	Average	50.4

Unfortunately, among more highly developed plant groups, fewer variations in base composition occur. Thus, in a survey of ferns and fern-allies, Green (1971) found that most plants fell within the narrow range of 37 to 41% (Table 18.4). The only exceptional taxa were the two *Selaginella* species investigated; these had significantly higher values (45 and 46%). Interestingly, these two taxa were the only ones among those examined that contained satellite DNAs. However, without having a lot more data for the Pteridopsida as a whole, it is difficult to know whether *Selaginella* is really distinctive. The range found among ferns is generally comparable with that found in angiosperms (see below). This suggests that the 400 million years since the time of divergence of land plants has been too short for large changes in the DNA base composition to have appeared.

Table 18.4 G + C base ratios for the DNA of ferns. (From Green, 1971.)

Class	Fern or fern-ally	G + C
Psilophyta	*Psilotum nudum*	37
Lycopodophyta	*Selaginella kraussiana*	46
	S. emiliana	45
Arthrophyta	*Equisetum arvense*	39
Pterophyta	*Angiopteris erecta*	38
	Ophioglossum pendulum	39
	O. petiolatum	38
	Ciboteum splendens	41

A complication in the case of DNA base measurements of higher plants is the fact that cytosine may be replaced in part by a modified base, namely 5-methylcytosine (Fig. 18.3). Calculations of the G + C ratio are therefore

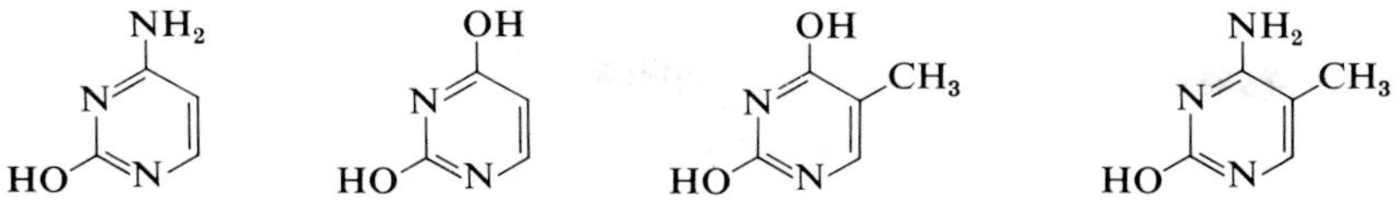

Cytosine Uracil Thymine 5-Methylcytosine

Fig. 18.3 The pyrimidine bases of higher plant nucleic acids.

based on a value where the amounts of cytosine and 5-methylcytosine are added together (Table 18.5). When this is done, it will be seen that G + C ratios in both gymnosperms and angiosperms show little variation and most lie between 36 and 43%. Biswas and Sarkar (1970) surveyed 61 species from 35 families and found a similar range (36–49%), with three members of the Gramineae showing consistently higher values than other taxa. Unfortunately, as will be seen from the data of Table 18.5, other unrelated plants (e.g. *Daucus carota*) may have higher than average values so that the taxonomic value of G + C ratios seem to be quite limited. The data in general reflect the conservative nature of higher plant DNA, at least in terms of base composition. It may be noted (Table 18.5) that the 5-methylcytosine content appears to vary independently of the G + C ratio. Indeed small variations were noted by Ergle *et al.* (1964) at the infraspecific level in species of *Gossypium*. It is therefore possible that measurements of this parameter might on occasion, yield results of taxonomic interest.

Table 18.5 (G + C)% ratios for some higher plant DNAs

Plants	Guanine	Cytosine	5-Methyl-cytosine	(G + C)%[a]
Gymnosperms				
Pinus sibirica (Pinaceae)	20.8	14.6	4.9	40.4
Ginkgo biloba (Ginkgoaceae)	17.2	17.7	—	34.9
Angiosperms				
Daucus carota (Umbelliferae)	23.1	17.3	5.9	46.5
Zea mays (Gramineae)	22.8	17.0	6.2	46.0
Triticum aestivum (Gramineae)	22.7	16.8	6.0	45.5
Cucurbita pepo (Cucurbitaceae)	21.0	16.1	3.7	40.8
Phaseolus vulgaris (Leguminosae)	20.6	14.9	5.2	40.7
Allium cepa (Liliaceae)	18.4	12.8	5.4	36.3
Arachis hypogaea (Leguminosae)	17.6	12.3	5.7	35.6
Gossypium hirsutum (Malvaceae)	16.9	12.7	4.6	34.2
		average value		40.7

[a]Actually the ratio of (guanine + cytosine + 5-methylcytosine) to the (adenine + thymine) contents. (From Grierson, 1977.)

Base compositions may also be measured among the ribonucleic acids (RNA), the only major difference being that thymine is replaced by uracil (Fig. 18.3). Comparison is therefore made between the G + C and A + U contents. Base-pairing is a feature of the large ribosomal RNA macromolecules and Lava-Sanchez *et al.* (1972) have studied the G + C to A + U ratios of the larger (28S) and smaller (18S) subunits from a variety of organisms. The results obtained in respect of the fungi, algae and higher plants studied are illustrated in Fig. 18.4 (from Loening, 1973). These data are expressed as fractions (G + C/A + U) rather than as percentages. Interestingly, the cycads, ferns and gymnosperms all occur in a group, between the fungi and the dicotyledons, with ratios of 1.1 and 1.2 for smaller and larger subunits respectively. The dicotyledons all group together and are apparently well separated from the monocotyledons studied (Fig. 18.4). The G + C/A + U measurements are therefore nicely in accord with the commonly held view that the monocotyledons evolved from within the dicotyledons.

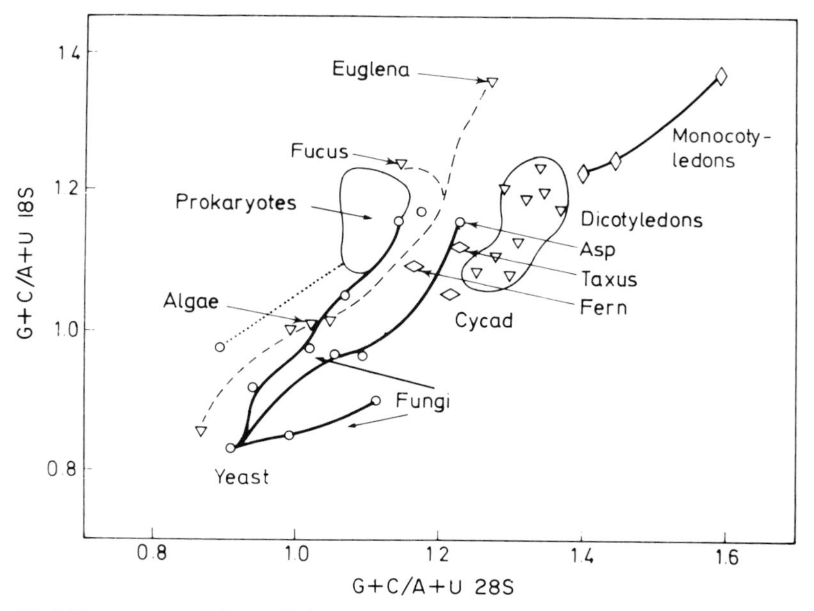

Fig. 18.4 Base compositions of the two ribosomal RNA subunits from fungi, algae and higher plants. (From Loening, 1973.)

III. DNA–DNA hybridization

Although base composition gives only a relatively crude measure of relatedness between the DNA of two taxa, a more informative comparison can be made by directly lining up the sequences of single-stranded DNA preparations of the two species against each other (Fig. 18.5). In principle, this approach to DNA homology comparisons is very simple. It is based on the observation of Doty *et al.* (1960) that the double-stranded DNA molecule, which is loosely held together by hydrogen bonding, can be separated into a single-stranded form by warming in solution. Slow cooling then allows the two complementary strands to come together again, so that the double-stranded duplex can be reformed. The degree of single strandedness in solution can be related to increases that occur in the intensity of absorbance at the ultraviolet maximum of the DNA molecule.

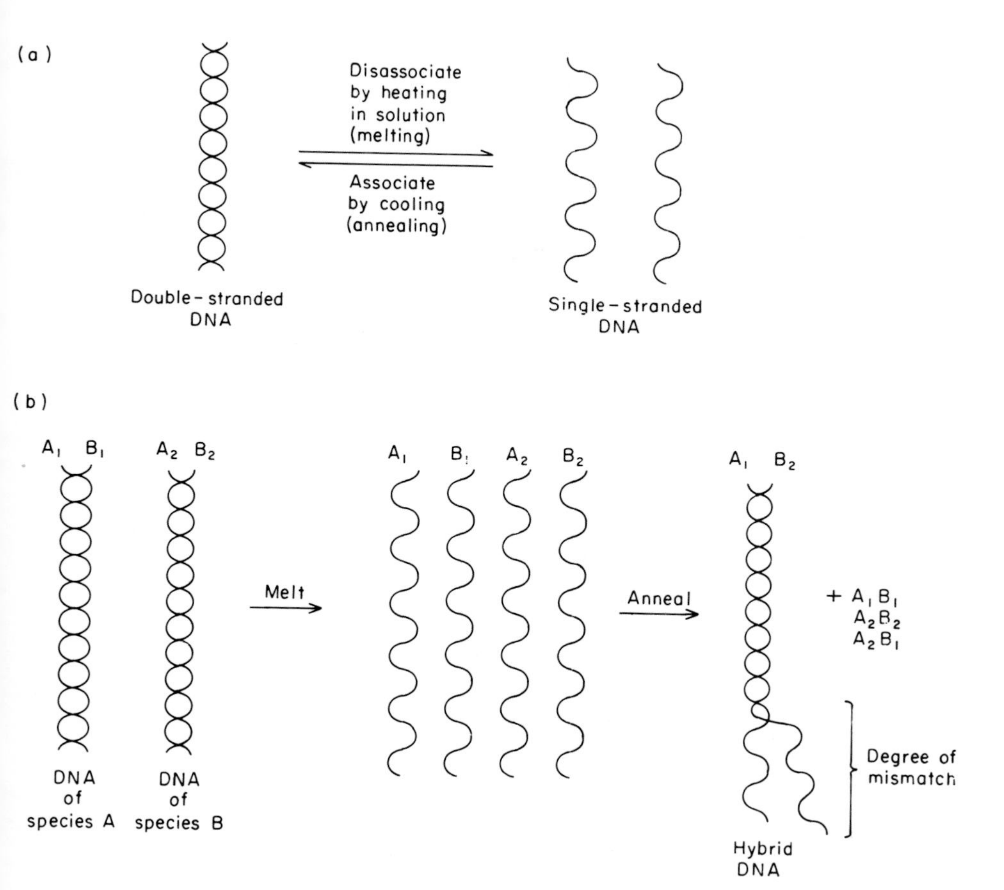

Fig. 18.5 The process of DNA–DNA hybridization. (a) shows the conversion of double to single-stranded DNA, (b) the same process with DNA preparations from two species A and B.

If the above process is applied to DNA preparations from two closely related species that are to be compared, it can be seen (Fig. 18.5) that if such DNA preparations are mixed, then new combinations will occur on cooling such that strand A_1 pairs with B_2 and A_2 with B_1, with the production of hybrid forms of DNA. Such hybrid molecules will differ from the double-stranded DNA of either species A or B in that, because the coding is different in part, the bases will not be completely matched with each other. It will therefore retain some single-strandedness. A measurement of this degree of mismatch in the hybrid DNA preparations will then indicate the degree of relatedness there is between the two species being compared. To avoid the complication of the native DNA of the two species reassociating in the reaction, it is usual to match a high concentration of DNA from species A with a low concentration of DNA of species B which is suitably labelled (e.g. with radioactive atoms).

The experimental problems associated with such DNA comparisons are considerable and, in the case of DNA of higher plants have not yet been fully solved (see below). In principle, similar comparisons can be made between the DNA of one species and the complementary ribosomal RNA of another; such DNA–RNA hybrids have been produced, but the method has not yielded any exemplary taxonomic conclusions (Smith, 1976).

The most successful application of DNA homology experiments has been, once again, with bacterial taxa. Some typical results, obtained by using radioactive labelling of the DNA with tritium, are shown in Table 18.6 (Kohne, 1968). These results indicate that *Shigella flexneri* is the most closely related of the other bacteria to *Escherichia coli* while *Proteus mirabilis* is the least. Although *Salmonella typhimurium* and *Aerobacter aerogenes* are shown by these data to be moderately distant from *E. coli*, they are not *necessarily* closely related to each other. Indeed, they may be quite distinct and other tests would have to be made to determine their relationships. Further details of DNA homology experiments are available in the review of de Ley (1968).

Table 18.6 Relatedness of DNA from *Escherichia coli* with DNA from other bacterial species, measured by DNA–DNA hybridization

Bacterial species	Relatedness to *E. coli* (%)
E. coli	100
Shigella flexneri	85
Salmonella typhimurium	35
Aerobacter aerogenes	35
Proteus mirabilis	2–5

The above results, however, point to at least two obvious limitations to DNA–DNA comparisons. Firstly, as with serological comparisons (see chapter 17, section VI), the method depends for its success on making a multitude of comparisons: testing A against B, B against A, A against C, C against A, B against C, and so on. Reciprocal testing is an essential prerequisite to establish the accuracy of the technique and some early experiments on DNA–DNA hybridization in higher plants by Bendich and Bolton (1967) failed in this respect. Secondly, the experiments can only be conducted between closely related taxa, since the bases along the DNA strands must match to some extent for success. If there is no matching, then the method fails.

Techniques for handling DNA–DNA reassociation experiments have been continually improved since 1960. Procedures that work successfully with DNA preparations from higher plants have been reviewed by Britten *et al.* (1974). The latest methods are illustrated in Fig. 18.6 (from Thompson and Murray, 1981). An important preliminary is the breaking up of the high molecular-weight double-stranded DNA into shorter lengths by sonication or rough handling, a process known as shearing. It then becomes much easier to follow the dissociation and reassociation of the shorter strands so produced. The extent of base pairing in the hybrid DNA can be monitored, as already mentioned, by ultraviolet spectroscopy. A hyperchromic shift (increase in absorbance) occurs on separation and a hypochromic shift (decrease in absorbance) occurs during pairing. Two other important procedures for measuring the degree of association are chromatography on hydroxyapatite (HAP) columns in sodium phosphate buffer (NAPB), which separates double- from single-stranded DNA and the use of SI nuclease, which degrades single-stranded DNA, leaving only duplex regions unattacked. Even with these more refined techniques, the measurement of DNA relatedness requires very considerable skill and absolute control over temperature and salt concentration in the solutions being examined.

From the viewpoint of hybridization experiments, a major complication in higher plant DNA is the presence of both single copy and repetitive sequences; the amounts of these two components can vary quite considerably between one species and another. Most earlier studies of DNA hybridization (reviewed by Smith, 1976) were done on total DNA and hence the results obtained may be of questionable value. This would include the DNA–RNA hybridization studies of Chang and Mabry (1973) in which it was reported that the betalain families within the order Centrospermae were more closely related one to the other than they were to the non-betalain families, both within the Centrospermae and without (Mabry, 1976). Indeed, the considerable experimental error in their results strongly suggests that repetitive DNA and methodological problems render the data as questionable at best. Homology

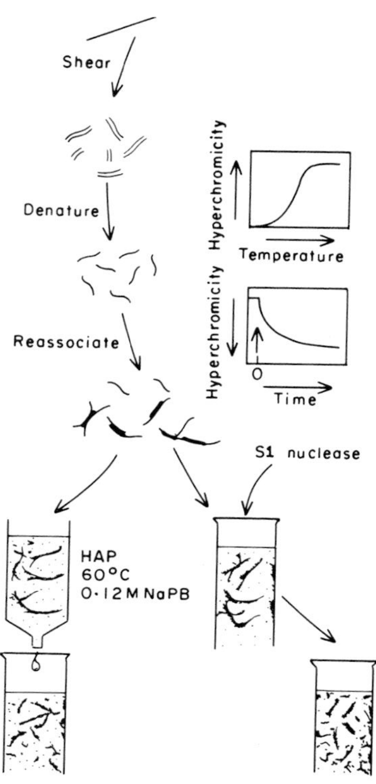

Fig. 18.6 The essential steps in reassociation experiments with plant DNA preparations. (From Thompson and Murray, 1981.)

estimates, which fell between 84 and 100%, reflect an average obtained from only three separate determinations, the standard error of each often being in excess of the differences among the most remote taxa.

Stein *et al.* (1979) were among the first workers to realize that more meaningful results might be obtained by using single-copy DNA for taxonomic comparisons. Their studies of relationships among three fern species of the genus *Osmunda* are illustrated here by the thermal stability profiles of native and reassociated DNA from these plants (Fig. 18.7). In these experiments, single-copy DNA from one species *O. claytoniana* is radioactively labelled with ^{125}I and then allowed to reassociate with the total DNA of the same species and of two other species. The results (Fig. 18.7) clearly show that the base sequences of *O. regalis* and *O. cinnamomea* are equally diverged from the sequence of *O. claytoniana*. The conclusion can be drawn from these data that *O. regalis* and *O. cinnamomea*, assuming they are derived species, became isolated from *O. claytoniana* at about the same time in evolutionary history.

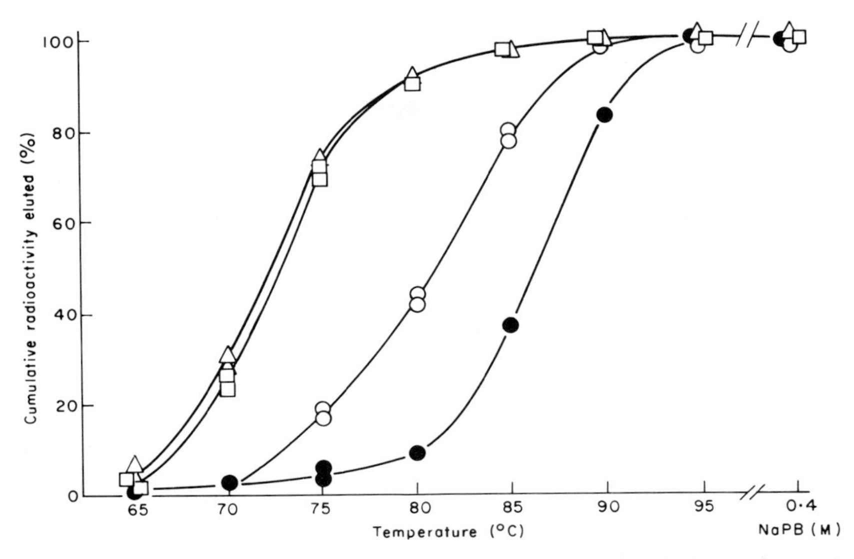

Fig. 18.7 Thermal stability profiles of native and reassociated *Osmunda* species DNA. Native *O. claytoniana* DNA (●), ^{125}I-labelled *O. claytoniana* single copy DNA reassociated with unlabelled total genome DNA of *O. claytoniana* (○), of *O. cinnamomea* (□) and of *O. regalis* (△). (From Stein *et al.*, 1979.)

A more ambitious use of single-copy DNA in DNA homology experiments is that of Belford *et al.* (1981) on eight *Atriplex* species. Here it was of interest to test a phylogenetic scheme for the genus based on morphology (Hall and Clements, 1923) (Fig. 18.8). *Atriplex* (Chemopodiaceae) is one of the few known genera where some species are C_3 and others C_4 in their photosynthetic carbon-fixation pathway (see chapter 14). Thus some species have evolved the ability to operate a pathway that adapts them to subtropical climates and which is not present in the others. According to the morphological data, the C_4 species belong in different subgenera, which would require evolution to the C_4 condition to have occurred on at least two separate occasions.

The results of comparing single-copy DNA isolated from two of the species with the DNA of these and six other species are relatively complicated and need to be studied in detail in order to assess them. However, if accepted at their face value, they indicate that there is a closer affinity (Fig. 18.8b) between all the C_4 plants and between the C_3 plants as a group than between specific C_3 and C_4 plants as suggested by classical treatments. Hence the original subdivision into subgenera appears to be artificial and the relationships indicated by the DNA experiments represent a more natural subdivision. This new scheme also has the great advantage that it is no longer necessary to postulate that evolution to the C_4 condition occurred more than once within the genus.

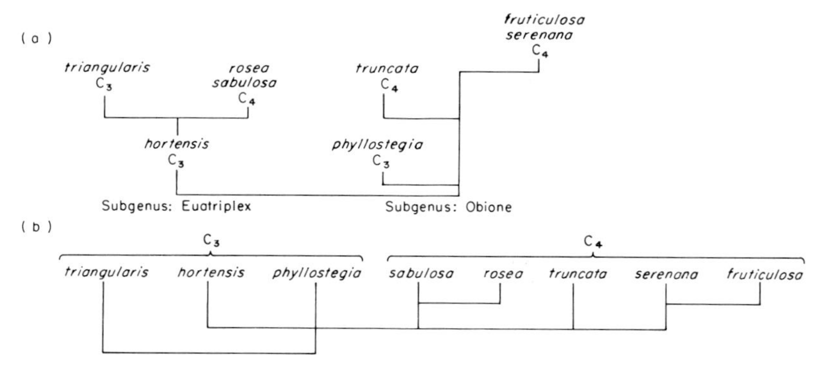

Fig. 18.8 Evolutionary pathways of species within the genus *Atriplex*. (Note: only a fraction of the known species are included.) In (a) the pathway of evolution is based on morphology and in (b) it is based on DNA comparisons.

Belford and Thompson (1981) suggest, from these hybrid thermal stability measurements, 'that lines leading to many present-day *Atriplex* species probably originated during a single period of rapid speciation', negating the view of Hall and Clements (1923) that evolution within *Atriplex* was perhaps gradually reticulate over time. Further, accepting figures calculated from base substitutions among animal genomes, Belford and Thompson (1981) reckon that these major speciation events occurred about 20 million years ago near the Oligocene–Miocene boundary in North America at which time large regions of C_4-type habitats were becoming increasingly available for plant occupancy. Further studies of DNA homologies using single-copy DNA are still needed to confirm the validity of this approach but the results obtained so far seem very encouraging.

A complementary approach to that of Belford and his colleagues with single-copy DNA has been developed by Flavell *et al.* (1977) using variations in repetitive DNA to map out relationships between closely related cereal species. As already mentioned above, the amount and nature of repeating sequences, where they have been studied, vary in a relatively unpredictable manner from taxon to taxon. Nevertheless, amplification of certain sequences occurs frequently enough in evolution that the presence or absence of particular repeated sequence groups in different genomes can be used to trace the pathway of evolution of different species within a given species complex. In this work, it is assumed that an amplification of some sequence or set of sequences in an ancestral taxon will be passed on to all derived lines. However, new amplifications occurring *after* separation of two lineages will involve different sequences in each line, so that each new taxon may have its own catalogue of repeated sequences.

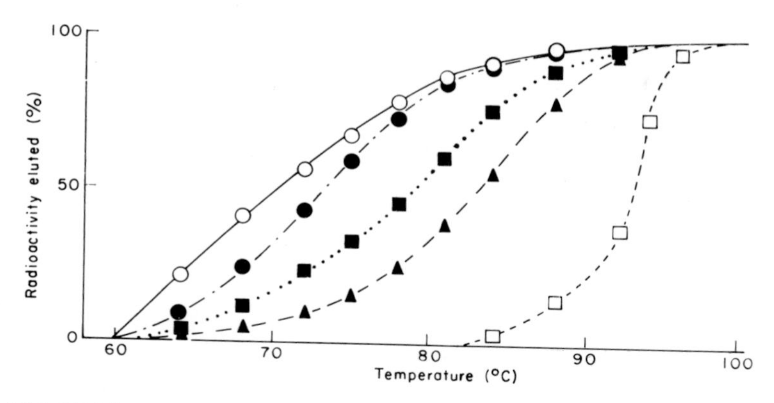

Fig. 18.9 Melting curves of native wheat DNA and duplexes formed between non-labelled wheat DNA and 3H-labelled DNA from each of the four species studied: native wheat, □– – –□; × wheat, ▲- - - -▲; × barley, ●—·—●; × oats, ○——○; × rye, ■ . . . . ■. (From Smith and Flavell, 1974.)

In one set of experiments, Smith and Flavell (1974) studied the order of speciation among oats, barley, wheat and rye. Repeated sequences of DNA were hybridized between the four cereals to yield a matrix of repeated sequence cross-reactivities. The genomes of each taxon contained about 75% repeated sequence and the cross-reaction ranged from 14 to 58%. The thermal stabilities of duplexes formed between unlabelled wheat and labelled rye, barley and oats decreased in that order (Fig. 18.9), indicating that wheat closely resembles rye, whereas barley and oats are less similar. The results were also expressed as estimates of the divergence between any pair of taxa; thus wheat and oats show a divergence of 103, whereas the values for wheat and barley and wheat and rye are 57 and 29 respectively. These and other experiments indicate an evolutionary scheme for these four cereal species (Fig. 18.10), which is in close accord with known taxonomic and genetic considerations.

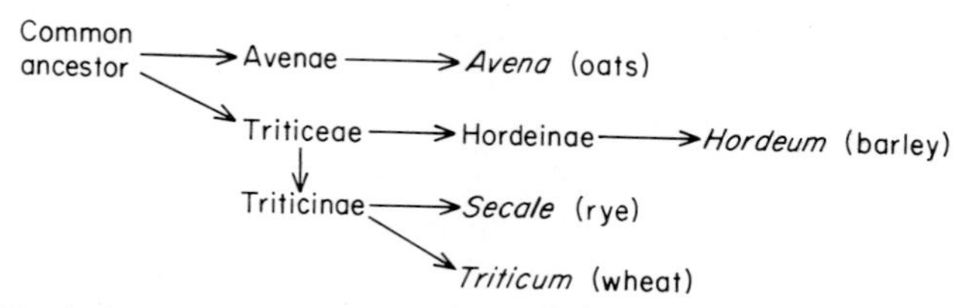

Fig. 18.10 Evolutionary history of cereal crops derived from studies of repeating sequences of nuclear DNA.

Repeat-sequence DNA comparisons have also been made between *Triticum* and *Aegilops* species (Flavell *et al.*, 1979); in general, there are greater similarities here than among the more divergent cereal taxa studied above. Nevertheless, using a DNA probe from *Aegilops speltoides*, it was possible to distinguish diploid *Aegilops* species from diploid *Triticum* species and also different *Aegilops* species from each other. The results confirm the concept that speciation is usually accompanied by quantitative changes in the repeated sequence complements of the genome.

Comparison of the diploid *Aegilops* and *Triticum* species with the hexaploid wheat *T. aestivum* showed that most, if not all, of the groups of repeated sequences in the cultivated wheat could be recognized as being derived from one or other of the putative diploid ancestors. The results also confirm the most likely donor of the B genome is *A. speltoides* and not *T. urartu*. This latter taxon was suggested as a possible B genome donor from protein studies (Johnson, 1975) but it cannot be since it does not possess a highly repeated sequence complement similar to the B genome of wheat. It is valuable to apply data from a new source to the still controversial question of the true ancestral origin of the cultivated wheat plant. Other macromolecular approaches through seed proteins and fraction I protein of the leaf have been mentioned in the previous chapter.

IV. Sequence studies

In taxonomic comparions of plant DNA preparations, studies of base composition (section II) and of DNA homology (section III) are really only essential preliminaries to complete sequence analysis. The ultimate solution to successful DNA comparisons is to be able to compare the order of bases along the chain. Historically, protein sequencing was developed before nucleic acid sequencing became possible so that sequence comparisons in a taxonomic and evolutionary framework have largely concentrated on proteins such as cytochrome *c* (see chapter 17). Clearly, the message is the same whether it is read in the DNA code, the RNA code or in the order of amino acids in the protein. There will be no point now in sequencing the bases in the gene coding for cytochrome *c*; the information is already available.

The difficulty with the DNA, and to a lesser extent with the RNA, is the overwhelming quantity of information involved, much of it being redundant. The enormous complexity of higher plant DNA has already been mentioned (cf. Table 18.2). Additionally, these macromolecules are much longer than proteins and sequences may be many kilobases (1 kilobase = 1000 bases) long. Recently, for example, a comparison of mitochondrial and chloroplastic DNA from *Zea mays* revealed the presence of a 12 kilobase sequence common to both molecules (Stern and Lonsdale, 1982). The problem with DNA sequencing, in taxonomic work, is not now that of determining the sequence but rather that of isolating reasonably sized homologous sections of DNA from different organisms for comparative analysis.

At the present time, it is questionable whether there are any advantages in making comparisons at the DNA level rather than at the protein level. In individual instances, information may come to hand which makes it easier to determine a protein sequence from a DNA sequence rather than vice versa. This has been true in the case of the large subunit of ribulose bisphosphate carboxylase from maize (*Zea mays*) which is 475 amino acids long. This sequence has been obtained by translating the known base sequence of the chloroplast gene coding for this particular protein (McIntosh *et al.*, 1980). Currently, we are witnessing a boom in DNA sequence analysis which is largely directed towards improving our knowledge of genome organization and function and towards the practical benefits that may come from plant genetic engineering. A possible spin-off of this activity may be of new taxonomic data, although at present only relatively few comparative sequences are available. These are largely of RNA sequences, determined in many cases from the ribosomal DNA sequences.

The actual procedures for DNA (and RNA) sequencing are too complicated to describe in simple terms here. They depend on the availability of a range of structurally specific hydrolytic cleaving enzymes, called nucleases. Chemical modification and cleavage using such reagents as hydrazine and dimethyl sulphate are also important. Yet another method of sequencing involves DNA synthesis *in vitro*, using DNA polymerase, a template and primer; the extent of the stepwise synthesis reveals the sequence along the template. In the case of higher plants, the amounts of DNA available for analysis may be insufficient for sequencing but the problem can be overcome in favourable cases by recombinant techniques in bacterial systems. The methods for sequence analysis of plant DNA preparations have been well reviewed by Flashman and Levings (1981).

Not surprisingly, sequence analysis was first applied in the bacterial field because of the relative simplicity of the bacterial chromosome. Catalogues of oligonucleotide sequences of the ribosomal 16S RNA of over 150 prokaryotes have been assembled (Woese *et al.*, 1976). The data have been studied by cluster analysis to produce dendrograms relating the different taxa. In the case of purple bacteria, there is good agreement with the results of cytochrome *c* sequences (Woese *et al.*, 1980). The RNA sequence data have also been incorporated with other chemical characters to produce a new phylogenetic representation of bacterial evolution (Kandler and Schleifer, 1980). In the case of higher plants, ribosomal RNA sequences are only available for a few organisms, notably for the rRNA of maize chloroplasts.

In general, it appears that sequences are conserved during evolution, so that dramatic differences in sequence only occur at the higher levels of classification. A comparison of one to two kilobase sequences in the small ribosomal subunit from a variety of organisms from *E. coli* to *Homo sapiens* (Kuntzel and Kochel, 1981) indicates several regions of conserved primary and potential secondary structures. The only higher plant sequence included, from the maize chloroplast, comes out closest to that of *E. coli* mitochondria. The results of the analysis can also be used to support the theory of endosymbiotic origin for fungal mitochondria.

Transfer RNA preparations of a variety of organisms have also been sequenced. These are much smaller, less than 100 bases, and they take up the now familiar clover-leaf shape. Since the first tRNA determination of yeast (Holley *et al.*, 1965), about 135 other tRNA species have been sequenced (Gauss and Sprinzl, 1981). However, only a few plants have been looked at, and these show few variations (Weil and Parthier, 1981). For example, the transfer RNA for phenylalanine in the cytoplasm has been analysed in wheat, pea, lupin and barley. They are identical, except for a G–C base pair present in the wheat, pea and barley tRNA which is replaced by an A–U pair in 80% of the lupin tRNA. Similar minor differences exist between taxa in the other tRNA preparations that have been examined. There seems little or no taxonomic significance in such variations, particularly since larger variations have been found within the three forms (isoacceptors) of the tRNA for leucine in *Phaseolus* chloroplasts. The taxa sampled so far for tRNA sequencing have been highly unrepresentative of plants generally and more data are needed before a definitive judgement on the taxonomic value of sequence variation can be made.

Returning to DNA sequence comparisons in plants, which are still much more difficult to achieve than in the RNA series, there are clearly advantages in studying chloroplast rather than nuclear DNA in view of the relative simplicity of its organization and its small size (120–180 kilobase pairs) (see Table 18.2). A new approach to the comparative analysis of chloroplast DNA has recently been developed by Palmer *et al.* (1983), using restriction endonucleases. In this procedure, the chloroplast DNA of a given plant species, after separation and purification, is treated with up to ten different restriction endonucleases, enzymes that recognize and cleave the DNA duplex at specific six-base-pair sequences. For example, an endonuclease from *Proteus vulgaris* will specifically recognize and then cleave wherever there is the base sequence—CGATCG—, etc.; catalogues of such restrictive enzymes are available (e.g. Roberts, 1980). The characteristic set of fragments generated by each enzyme in turn can be resolved by agarose gel electrophoresis and visualized by ultraviolet illumination of the ethidium bromide-stained gel.

Species-specific changes in the fragment pattern produced by a given restriction enzyme are then interpretable as the consequence of either the loss or gain of a specific restriction cleavage site and are generally caused by a single base-pair substitution (i.e. a single mutation). The phylogenetic utility of this type of chloroplast DNA analysis lies in the fact that many restriction site changes are shared by two or more closely related species and thus they can be used to infer shared evolutionary histories.

This approach has been applied to species comparisons in both *Lycopersicon* (Solanaceae) (Palmer and Zamir, 1982) and in *Atriplex* (Chenopodiaceae) (Palmer *et al.*, 1983). The results in *Atriplex* generally confirm the conclusions drawn about species relationships from DNA–DNA homology experiments (see section III). For example, chloroplast genomes among C_4 species are remarkably homogeneous, with less than 1% sequence divergence. By comparison, the primitive C_3 species are distinctive and show divergence values greater than 1%. The chloroplast genome appears to provide a more sensitive guide to mutational change than the nuclear genome since the range of divergence values overall is tenfold within a given group of *Atriplex* species for chloroplast DNA and no more than twofold for the nuclear DNA.

Such an approach, using restriction enzymes on their own, does not provide a complete sequence comparison and utilization of other techniques (e.g. hybridization) is necessary to produce fully detailed results. One of the advantages of knowing complete sequences of sections of the chloroplast DNA is that comparisons can be made with the genomes of other organelles within the cell. Such a comparison in maize, *Zea mays*, has revealed a remarkable homology in a 12-kilobase segment between chloroplast and mitochondrial DNA (Stern and Lonsdale, 1982). From this result, Ellis (1982) has speculated that the chloroplast DNA must have become incorporated at some stage into the mitochondrial genome, after a duplication and transposition event. He has proposed the term 'promiscuous DNA' for that portion of the genome which has moved in this way. This discovery opens up a whole new field of possibilities of movement of genetic material between nucleus, chloroplast and mitochondrion during plant evolution. If found to occur elsewhere, this duplication of genetic information may have considerable repercussions on the utilization of DNA sequence data in taxonomic investigations. Clearly, it is a new and remarkable facet of genome organization which must be borne in mind in future comparative studies.

V. Conclusion

Compared with the proteins, nucleic acids have not so far yielded much data of consequence to the practising plant taxonomist. The relative crude comparison afforded by DNA or RNA base composition is mainly of interest with lower plants (i.e. bacteria and fungi) and such analyses have not yielded any important insights into angiosperm classification. A variety of DNA homology experiments have been conducted with DNA forms from higher plants, but considerable difficulties in both experimental technique and interpretation have meant that the procedures have not as yet been applied on any scale. The fact that DNA–DNA hybridization data can usefully complement information from other sources has been demonstrated with *Atriplex* species (Belford *et al.*, 1981) and with cereals (Flavell *et al.*, 1977).

Nucleic acid sequence analysis is yet in its infancy with higher plants so that it is difficult as yet to assess its impact on plant taxonomy. The approach is obviously one of considerable potential. However, the evolutionary significance of genomic differences is at present difficult to evaluate, although a number of models have already been proposed (Dover, 1980). The question remains whether the enormous investment in equipment and man-hours required for DNA sequence determinations will eventually pay off in new taxonomic insights. It may, in fact, be unnecessary for the taxonomist to set up a specific sequencing programme since the information may accrue as a by-product of molecular biology research. It will be interesting to see whether present trends in genetic engineering with plants will provide novel data of taxonomic utility. Clearly, they will in the case of cereal and legume crops but at present most other higher plants receive little attention from this point of view.

19

Polysaccharides

I. Introduction 461
II. Celluloses and hemicelluloses 464
III. Gum exudates 467
IV. Storage carbohydrates 470
V. Seed polysaccharides 472
VI. Conclusion 473

I. Introduction

Plant polysaccharides are a very diverse group of carbohydrate polymers, which are universally distributed throughout all tissues. A few have a relatively simple ordered structure based on a single sugar unit (e.g. the glucose polymers cellulose and amylose) but most are complex branched polymers containing as many as six or seven different sugar moieties. From the taxonomic viewpoint, they are the most tantalizing of macromolecules. Thus, on the one hand, the considerable complexity of structure in most polysaccharides is a point in their favour as taxonomic markers. On the other hand, however, the practical problems associated with structural determination and the fact that it is usually impossible to define structure other than in terms of 'repeating' units or partial sequences are significant limitations (Fig. 19.1). The most progress has been made, in fact, with microbial and algal polysaccharides from the comparative viewpoint and here significant correlations have been noted between classification and polysaccharide type (e.g. Bartnicki-Garcia, 1970; Percival and McDowell, 1967, 1981). In the case of higher plants, progress has been slower and even today it is only possible to find one or two examples where studies of polysaccharide structures have illuminated plant relationships.

The classification of plant polysaccharides is difficult and a degree of arbitrariness enters into any attempt to place the many different types into neat categories. A major dividing line can be drawn between structural and storage polysaccharide, represented in the simplest case by cellulose ($\beta 1 \rightarrow 4$ glucan) and the amylose component of starch ($\alpha 1 \rightarrow 4$ glucan) respectively. In practice, the actual mixtures of structural and storage polymers are more complex than this; also it may be difficult to distinguish them on a functional basis, particularly in the case of the polysaccharides present in seeds.

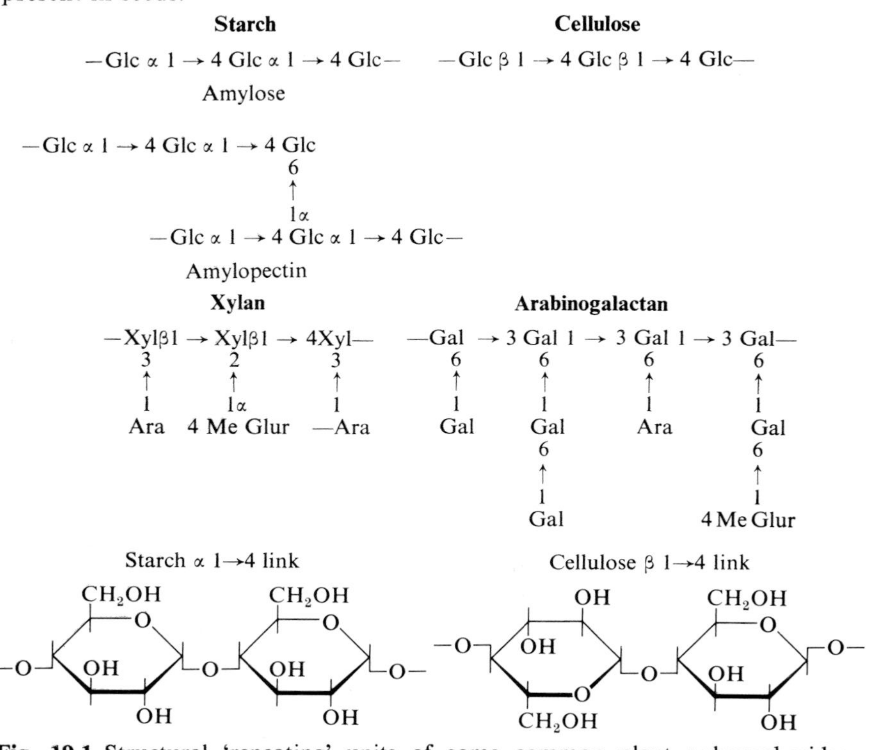

Fig. 19.1 Structural 'repeating' units of some common plant polysaccharides. *Key*: Glc, glucose; Gal, galactose; Ara, arabinose; Xyl, xylose, 4 Me Glur, 4-methylglucuronic acid.

The structural complexity of polysaccharides is due to the fact that two sugar units can be linked together through an ether linkage in a variety of ways. The reducing end of one sugar (C1) can condense with any hydroxyl residue of a second sugar (at C2, C3, C4 or C6) so that during polymerization some sugars may be substituted in two positions, giving rise to branched chain structures (see Fig. 19.1). Furthermore, the ether linkage at the reducing end can have either an α- or β-configuration, depending on the stereochemistry of the simple sugars, and both types of linkage may co-exist in the same macromolecule. Such complexity occurs in the hemicelluloses, polymers

Table 19.1 The main classes of polysaccharides in higher plants, algae and fungi

Class name[a]	Sugar unit(s)	Linkage	Distribution
Higher plants			
Cellulose	Glucose	β1→4	Universal as cell-wall material
Starch-amylose	Glucose	α1→4	Universal as storage material
Starch-amylopectin	Glucose	α1→4, α1→6	
Fructan	Fructose (some glucose)	β2→1	In artichoke, chicory etc.
Xylan	Xylose (some arabinose and uronic acid)	β1→4	Widespread e.g. in grasses
Glucomannan	Glucose, mannose	β1→4	Widespread, but especially in coniferous wood
Arabinogalactan	Arabinose, galactose	1→3, 1→6	
Pectin	Galacturonic acid (some others)	α1→4	Widespread
Galactomannan	Mannose, galactose	β1→4, α1→6,	Seed mucilages
Gum	Arabinose, rhamnose, galactose, glucuronic acid	highly branched	In *Acacia* and *Prunus* species
Algae (seaweeds)			
Laminaran	Glucose	β1→3	Phaeophyceae
Polysaccharide sulphate	Fucose (and others)	—	
Alginic acid	Mannuronic and glucuronic acids	—	
Amylopectin	Glucose	α1→4, 1→6	Rhodophyceae (red algae)
Galactan	Galactose	1→3, 1→4	
Starch	Glucose	α1→4, 1→6	Chlorophyceae (green algae)
Polysaccharide sulphate	Rhamnose, xylose, glucuronic acid	—	
Fungi			
Chitin	*N*-Acetylglucosamine	β1→4	Widespread
Chitosan	Glucosamine	β1→4	Zygomycetes
β-Glucan	Glucose	β1→3, β1→6	Widespread
α-Glucan	Glucose	α1→3, 1→4	Widespread
Mannan	Mannose	α1→2, 1→6, 1→3	Mainly Ascomycetes

[a] It is sometimes useful to distinguish between homopolysaccharides, which have only one constituent sugar (e.g. cellulose), and heteropolysaccharides, with several sugar units (e.g. glucomannan).

based on several sugars joined in a variety of linkages, which together with celluloses make up the cell wall of higher plants. Many of the gum and mucilage polysaccharides also have very complex structures. A summary of the different polysaccharide types found in the plant kingdom is given in Table 19.1.

Some comparative information on polysaccharide type can be obtained by determining the sugar composition on hydrolysis of a crude mixture of polymers, such as that in the cell wall after exhaustive extraction of leaf tissue with hot ethanol to remove low molecular-weight sugars (Andrews *et al.*, 1960). Colour tests can be used *in situ* for revealing the presence of starch or related storage carbohydrates; amyloids can be detected in seed endosperm by staining with an iodine reagent (Kooiman, 1960). Such preliminary data are liable to error and it is now generally appreciated that fractionation of the polysaccharides present is an essential preliminary to determination of monosaccharide composition.

A range of solvent extractions can be employed to separate different classes of polysaccharide from plant tissues. Ion-exchange and Sephadex chromatography can then be applied to these different fractions in order to obtain pure polysaccharides. Hydrolysis and quantitative analyses of the sugar components is simply carried by standard procedures. Determination of the sugar–sugar linkages present is more time-consuming and involves the complete methylation of the polysaccharide, then acid hydrolysis and the separation of the partially methylated sugars so released. Non-sugar components, such as inorganic ion or amino acid, may be present and need to be determined. Useful comparative information obtainable on pure polysaccharides is molecular size, viscosity and optical activity. For heteropolysaccharides, it is usually not possible to determine the complete sequence of sugars; structure can only be defined in terms of the 'repeating units' and end groups. Details of analytical procedures for polysaccharides are given in Harborne (1973b).

The chemistry and distribution of plant polysaccharides are well reviewed in the literature; see e.g. Percival (1967), Aspinall (1970), Towle and Whistler (1973), Stephen (1979), Preiss (1980) and Tanner and Loewus (1981). In this account of systematic aspects, only a few selected examples will be discussed among cell-wall constituents, gum exudates, storage carbohydrates and seed polysaccharides.

II. Celluloses and hemicelluloses

Cellulose represents a very large percentage of the combined carbon in plants and indeed is the most abundant organic compound known. It is the fibrous material of the cell wall and is responsible, with lignin, for the structural rigidity of plants. Chemically, it is a β-glucan and consists of long chains of β1 → 4 linked glucose units, the molecular weight of the chains varying from 100 000 to 200 000. Cellulose occurs in the plant cell wall as a crystalline lattice, in which long straight chains of polymer lie side by side held together by hydrogen bonding.

By its very nature and its fundamental structural role, cellulose is unlikely to vary between organisms and this is, in fact, the general case. Exceptionally there is pronounced variation in the cell walls of fungi, where cellulose is only present as such in two fungal classes (Table 19.2). Its place is taken in most other fungal groups by chitin, a structurally similar polymer which is based on the sugar *N*-acetylglucosamine instead of glucose. In the case of higher plants, the only significant variation between organisms is in the quantity present. Cotton fibres, for example, are almost pure cellulose. Again, the cellulose content of trees is 40–50%, whereas that of herbaceous plants is significantly lower. There may be some variation in the degree of polymerization of cellulose in different plants, but this has not yet been quantified on a comparative basis.

Table 19.2 The distribution of different wall polysaccharides in the fungi

Fungal group	Polymer[a]				
	Chitin	Chitosan	Cellulose	Glucan	Mannan
Phycomycetes					
Oomycetes			+	+	
Hypochytridiomycetes	+		+		
Chytridiomycetes	+			+	
Zygomycetes	+	+		(+)[b]	
Ascomycetes					
Hemiascomycetidae				+	+
Euascomycetidae	+			+	
Basidiomycetes					
Heterobasidiomycetidae	+				+
Homobasidiomycetidae	+			+	

[a]Key: chitin is based on *N*-acetylglucosamine (β1→4); chitosan is based on glucosamine (β1→4); glucan is a mixed β1→3, β1→6 glucose polymer; mannan is a mixed α1→2, α1→3 and α1→6 mannose polymer. (From Bartnicki-Garcia, 1970)
[b]Only in the spore cell wall.

Recently, attention has been drawn to the presence of related glucose polymers in cell walls of cereals (Nevins *et al.*, 1978; Stinard and Nevins, 1980). These so-called non-cellulosic β-D-glucans are associated with the cell walls of grass coleoptiles and have 30% of their sugar-sugar linkages β1 → 3, the rest being β1 → 4 as in cellulose. It is possible to survey the presence or absence of these glucans by treatment of cell wall specimens with a specific glucanase, which liberates particular oligosaccharides if mixed linkage polysaccharides are present. As a result, it has been shown that non-cellulosic β-glucans are present widely in grasses (in representatives of at least six subfamilies) but do not occur in any related monocotyledonous groups (Stinard and Nevins, 1980). These polysaccharides seem thus to be characteristic of the Gramineae, although wider surveys are needed to confirm this statement. A related, but not identical polysaccharide with mixed $\beta 1 \rightarrow 3$ and $\beta 1 \rightarrow 4$ linkages, namely lichenin occurs in the thallus of certain lichens (Clarke and Stone, 1963) so it is possible that these abnormal cellulose-like materials may have a wider distribution in Nature than appears at present.

Another distinctive feature of cell walls in grasses, particularly in unlignified tissues, that has recently been discovered is the presence of phenolic acids apparently bound by ester linkage to the cellulose matrix. Thus ferulic acid, diferulic acid and *p*-coumaric acid have been variously released on alkaline treatment of such preparations. The presence of these acids is even more easily monitored *in situ* using a fluorescence microscope, since they impart a blue ultraviolet fluorescence to the wall, the colour changing to green in the presence of ammonia vapour. The presence of phenolic acids bound to cell wall was first observed in *Lolium multiflorum* and other fodder grasses (Hartley and Jones, 1977) and a later survey of monocotyledonous families showed that some 50% of 104 species in 52 families contained this type of cell wall constituent (Harris and Hartley, 1980).

The distribution of this character is taxonomically significant in monocotyledons in that it is uniformly present in those families of the subclass Commelinidae surveyed and equally absent from families of the subclass Alismatidae. There are also scattered occurrences in the Palmae (subclass Arecidae), the Philydraceae, the Pontederiaceae and the Haemodoraceae (subclass Liliidae). An even more interesting distribution pattern for this cell wall character was found in dicotyledons, where bound phenolic acids were found to be restricted entirely to plants of the Centrospermae (Hartley and Harris, 1981). Some 251 species in 150 families were studied. The restriction to the Centrospermae (all species of both the Cactaceae and the Caryophyllaceae were positive) is of particular interest vis-a-vis the recent controversy over the anthocyanin–betalain dichotomy within this plant order. This new chemical character would seem to confirm the essential unity of this natural order of plants (see chapter 12).

Since the hemicelluloses of higher plant cell walls are highly complex compared to the celluloses, they might be expected to be of considerable comparative interest. Unfortunately, although they have been analysed in a variety of plant tissues to a varying extent, it is difficult to know at present whether they should be examined in chemosystematic programmes. Consistent differences in percentage sugar compositions have been recorded in the hemicelluloses isolated from three grass species and from three legumes by Gaillard (1965) so that they do apparently vary at the family level. Typical results are shown in Table 19.3, where Gaillard's analyses of three hemicellulose polymers from lucerne and wheat cell walls are compared. The most pronounced differences are in the sugar composition of the branched polymers. In more recent years, hemicellulose fractions have been extensively analysed in a number of higher plants and significant differences have been detected according to whether the source is gymnosperm or angiosperm or whether it is monocotyledonous or dicotyledonous. For example, in monocotyledons, the xylan fraction is usually composed of arabinoxylans, whereas in dicotyledons, it consists of chains of xyloglucans. Again, in both dicotyledons and gymnosperms, 4-*O*-methylglucuronoxylans are also characteristically present. The main features of cell wall hemicelluloses have been reviewed by Aspinall (1981) and Wilkie (1979). As yet, however, it appears that these differences in composition have not yet been exploited to any extent in taxonomic work.

Table 19.3 Sugar analyses of hemicellulose polymers from lucerne and wheat

Plant	Polymer	Composition of sugars after acid hydrolysis[a] (%) Uronic acid	Gal	Ara	Xyl	Glc
Medicago	Linear A	6.6	—	—	93.4	—
sativa	Linear B	1.5	—	8.3	90.1	(11.4)[b]
(Leguminosae)	Branched B	22.3	31.1	34.2	3.1	9.3
Triticum	Linear A	2.1	—	5.7	92.2	—
vulgare	Linear B	0.3	—	11.1	88.6	(12.0)[b]
(Gramineae)	Branched B	7.9	9.8	26.5	55.8	—

[a]Key: uronic acid, glucuronic + galacturonic acids; Gal, galactose; Ara, arabinose; Xyl, xylose; Glc, glucose.
[b]Due to contamination with a β-glucan, other sugar percentages are calculated after allowing for this impurity.

III. Gum exudates

Gum exudation is a distinctive feature of a range of higher plants, the gum being exuded from stem, root, leaf or fruit. For example, *Acacia* trees are well known to exude such gums from the trunk and other parts; this gum is collected and used commercially as gum arabic. Chemical analyses have revealed that these gums are almost completely pure polysaccharide. Gums are thus ideal starting materials for polysaccharide determinations in chemotaxonomic programmes. The main problem is their very considerable structural complexity; for full analysis, 100 g is needed and several complex analytical procedures are involved. Although commonly arabinogalactans, they also contain glucose and rhamnose, together with one or more rarer monosaccharide components (Fig. 19.2).

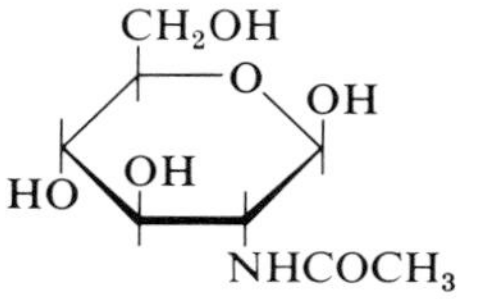

N-Acetylglucosamine
(as chitin in fungal cell walls)

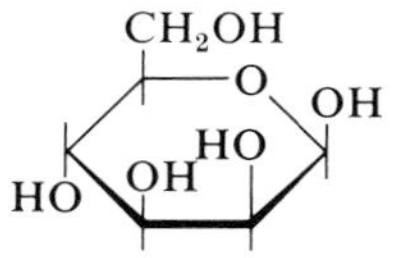

Mannose
(in endosperm polysaccharide of the Palmae)

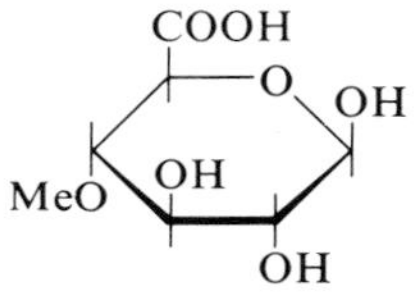

4-*O*-Methylglucuronic acid
(in *Acacia* gum polysaccharides)

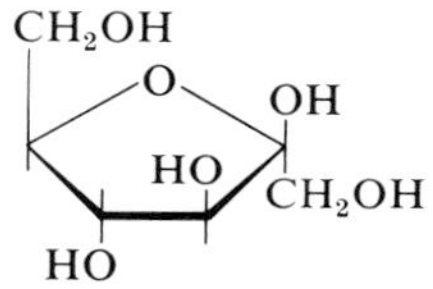

D-Fructose
(in storage polysaccharides of the Compositae)

Fig. 19.2 Some polysaccharide sugar components of special taxonomic interest.

The chemistry of plant gums has now been studied in about 40 genera representing 30 families and 20 orders of flowering plants (Stephen, 1979). Chemotaxonomically useful information has only been produced in a very few cases, notably in the genus *Acacia* and to a much lesser extent in *Combretum* (Combretaceae) (Anderson and Bell, 1977). Our knowledge in the case of *Acacia* is due to the research efforts of Anderson and his co-workers (see Anderson and Dea, 1969; Anderson, 1978) who have determined the gum composition of over 90 species.

The most interesting variables in these gums are the percentage compositions of the different monosaccharides released by acid hydrolysis. Useful comparative information has also been obtained by measuring optical rotations and molecular weights on the original gums. Other parameters examined include nitrogen content (indicating some amino acid content) and intrinsic viscosity. From such analyses, it is apparent that the gum produced by an *Acacia* plant contains in its polysaccharide composition a characteristic fingerprint for that species. Furthermore, gums of related species are broadly similar to each other although recognizably separable on detailed analysis) but differ from those of unrelated species.

Typical results indicating correlations between polysaccharide analyses and series classification in *Acacia* are shown in Table 19.4. Besides noticeable differences in the amounts of constituent sugars, there are variations in the optical rotations of the gums at the series level. In the series *Botryocephalae*, it is interesting that two distinct groups of species exist based on gum analysis, types A and B. Furthermore, type B species show distinct resemblances to the gums of *Phyllodineae* taxa, suggesting that there must be a chemical link between the species of these otherwise morphologically different series.

Table 19.4 Variation in gum composition with series classification in the genus *Acacia*

Acacia series	Average percentage of monosaccharide components[a] 4-MeGlur	Glur	Gal	Ara	Rha	Specific rotations of polysaccharides
Phyllodineae	2.9	4.4	78	14	1	Slightly positive
Juliflorae	9.9	21	54	13	3	Most positive
Botryocephalae, type A	5.0	8.4	43	38	6	Strongly negative
Botryocephalae, type B	4.4	2.8	78	13	2	Slightly positive or weakly negative
Gummiferae	3.6	5.6	38	51	2	Strongly positive
Vulgares	2.3	11	47	28	11	Negative

[a]Average of four species of each series, except in the case of *vulgares*, where only three species were examined (Anderson, 1978). Sugar key: 4-MeGlur, 4-*O*-methylglucuronic acid; Glur, glucuronic acid; Gal, galactose; Ara, arabinose; Rha, rhamnose.

Taxonomically, the main use that has been made of these gum data has been at the species or subspecies level. Here, as Anderson puts it:

> even allowing for the now established limits of seasonal and natural variation in the composition of the gum from any one species, the analytical parameters, taken overall, establish a form of fingerprint which serves to characterise that particular *Acacia* species.

One example of such use is with a morphologically atypical specimen of *A. senegal* which had thorns in pairs instead of the usual triads. Gum analysis clearly showed it to be chemically a true *A. senegal* (Anderson, 1978). In another case, gum analysis showed that *A. raddiana* could acceptably be reallocated, as a result of taxonomic revision, to become a subspecies of *A. tortilis* (Anderson and Brenan, 1975).

Surely, as more species are examined for gum composition, the results will considerably aid in the placing of species into series and sections in this large and complex genus. One incentive for further polysaccharide analysis may be the fact that considerable data from other types of chemical analyses, notably of non-protein amino acids (see chapter 6) and flavonoids (see chapter 7) are also now available for these economically important plants.

IV. Storage carbohydrates

In contrast to the situation in the algae where storage carbohydrates vary according to algal class (Percival and McDowell, 1967, 1981; Craigie, 1974), in higher plants there is only one common storage material, namely starch. Such variation there is in starch is essentially physical, i.e. in the way that the starch granules are laid down within the cell, and is the subject of anatomical (or microscopic) rather than chemical investigation. It is true that starch can be separated into two chemically distinctive components: amylose, a straight chain $\alpha 1 \rightarrow 4$ linked glucose polymer; and amylopectin, a branched polymer containing both $\alpha 1 \rightarrow 4$ and $\alpha 1 \rightarrow 6$ linked glucose units. Even here, however, the relative proportions of the two components are extremely regular and do not vary significantly from the standard ratio of one amylose to four amylopectins. Genetic mutants are known in wrinkled pea *Pisum sativum* where the ratio is 3 : 2, and in waxy maize *Zea mays* where amylopectin is the only component. However, such extremes have not been recorded in wild type materials.

The only significant difference in storage carbohydrate among the flowering plants is where the glucose-based starch is replaced by fructose-containing polymers. This happens in two groups of plants, in the Campanulaceae, Compositae and related families of the dicotyledons and in the Gramineae and some members of the Liliales of the monocotyledons (Table 19.5). The type of storage fructan found in the two groups is structurally different (Meier and Reid, 1982).

In the Compositae, the fructans are known as inulins and consist of some 30–35 units of β-fructofuranose, joined through $2 \rightarrow 1$ linkages and terminated by a sucrose moiety. Inulins differ from starches in their relatively low molecular weight, their ready solubility in water and their inability to stain with iodine so that their presence or absence can be determined fairly readily. Although inulins occur in particularly high concentration in perennial members of the family, e.g. in storage organs of chicory, *Cichorium intybus*, they have been detected widely throughout the Compositae. The only negative reports are from a few annual species (Hegnauer, 1978).

Inulins are also widely present in members of the Campanulaceae; these fructans thus represent a significant link between the two families. The Campanulaceae and Compositae are also connected chemically by the fact that polyacetylenes are well represented in both groups (see chapter 8). There is, however, only limited biological support for a particularly close alliance between these two families (Cronquist, 1977a); nevertheless, they are usually placed near to each other in most systems of classification.

Table 19.5 Distribution of different fructosides in flowering plant families[a]

Plant families	Number of species
With inulins (and isokestose)	
Compositae	many
Menyanthaceae	3
Boraginaceae	19
Goodeniaceae	5
Campanulaceae	26
Stylidaceae	5
Brunoniaceae	1
Calyceraceae	2
With both inulins and levans	
Polemoniaceae	10 of 13
Clethraceae	3
Monotropaceae	2
With levans	
Gramineae	21 spp. of Festucoideae
Cyperaceae	5
Amaryllidaceae	15
Alstromeriaceae	1
Agavaceae	10
Cyanastraceae	1
Haemodoraceae	1
Hypoxidaceae	2
Iridaceae	23
Liliaceae	31
Ruscaceae	2
Xanthorrhoeaceae	1
Zosteraceae	1

[a]Data from Pollard and Amuti (1981) and Pollard (1982)

In their natural occurrence, the inulins may be accompanied by a structurally related trisaccharide isokestose (a glucosyldifructoside) and stems of some 550 species of dicotyledons have recently been surveyed for both these carbohydrates (Pollard and Amuti, 1981). Occurrences in a further six families, and notably in the Boraginaceae and Menyanthaceae, have been confirmed as a result (Table 19.5). Additionally, the presence of inulins with levans (the characteristic fructosides of monocotyledons, see below) has been recorded in the Polemoniaceae (but not in all species), the Clethraceae and the Monotropaceae. These results are of some taxonomic importance in relation to the uncertainties surrounding the classification of several of these families. For example, this chemical link between the Polemoniaceae and the

Menyanthaceae is of significance since, although the Menyanthaceae is traditionally placed with the Gentianaceae in the Gentianales, there is good anatomical evidence indicating an association with the polemoniads.

The other well known source of fructan storage material in plants is the family Gramineae. Here, the polysaccharides are called levans and are known to consist of 20–50 units of β-fructofuranose joined by 2 → 6 links, and with a terminal sucrose unit. Branching through 2 → 1 links may or may not be present as an additional feature. Slightly different levans have been found to occur consistently within certain generic groupings within the tribes Triticeae and Festuceae (Bacon, 1960). Branched fructans have also been recorded in several other monocotyledons, *e.g.* in *Agave vera-cruz* (Agavaceae) and *Cordyline terminalis* (Liliaceae) and a recent survey of the Monocotyledoneae showed them to be present in some 10 families within the lilialian alliance (Table 19.5) (Pollard, 1982). Interestingly, these fructosides could not be detected in eight other monocotyledon families, whose placement in the Liliales is controversial or dubious. This fructan character thus offers support to those taxonomists who prefer a narrow circumscription for this order of higher plants.

V. Seed polysaccharides

One of the first to study chemical variation in seed endosperm polysaccharides was Kooiman (1960), who tested some 2500 species for the presence or absence of amyloids. Amyloids are so called because they can be detected through the blue colour they yield in solution with an iodine–potassium iodide–sodium sulphate reagent. They are thought to be β1 → 4 linked polysaccharides containing glucose, galactose and xylose. Amyloid-positive species were found by Kooiman (1960) in 16 of 208 dicotyledon families, but were uniformly absent from the monocotyledon (25 families tested). Within the dicotyledons, there were some interesting distribution patterns. For example, in the Leguminosae, only members of the subfamily Caesalpinioideae were positive and furthermore only members of the three tribes Cynometreae, Sclerobieae and Amherstieae. Again, in the Acanthaceae, all ten positive-reacting species were from the same tribe, the Justicieae. Yet again, all *Paeonia* species examined were positive, but other Ranunculaceae (30 spp. tested) were uniformly negative.

In summary, the surveys of Kooiman established the potentiality of seed mucilages for chemotaxonomic comparison. Subsequent chemical investigations, largely on non-amyloid containing seeds, have confirmed this and revealed the presence of several distinctive patterns in seed carbohydrates. Undoubtedly the most common type of polymer in seed endosperm is galactomannan. These are widespread in legume seeds, where they are of some taxonomic interest due to variations in galactose/mannose ratios and in the amounts present in different species (Bailey, 1971). Galactomannans are also recorded in seeds of Annonaceae, Convolvulaceae, Rubiaceae and Palmae. Structurally, they contain β1 → 4 mannose chains, which have galactose units linked α1 → 6. There are considerable variations in the frequency of the galactose units and in the mode of their attachment. Where detailed studies of seed galactomannans have been conducted, it is clear that some of the finer details of polysaccharide structure are characteristic at the species level.

Other sugars besides galactose and mannose may be found in seed mucilages. Thus, a galactoglucomannan has been reported specifically to occur in seven species of Liliaceae and two of Iridaceae (Jakimow-Barras, 1973). In this work, the polysaccharide of one of the members of the Liliaceae, *Asparagus officinalis*, was shown to have glucose, mannose and galactose in the proportion 43 : 49 : 7. The mannose and glucose were present in a linear β1 → 4 linked chain, with the galactose present as α1 → 6 side chains.

In the seed endosperm of two Palmae, *Phytelephas macrocarpa* and *Phoenix dactylifera*, two β1 → 4 mannans have been characterized instead of the more usual galactomannans. The same unusual type has also been reported in the unrelated *Carum carvi* (Umbelliferae) (Hopf and Kandler, 1977). Finally, mention may be made of acidic polysaccharides, based on galacturonic acid and rhamnose, in several seed mucilages. Such materials have been reported from *Brassica* and *Lepidium* (Cruciferae), *Althaea* and *Hibiscus* (Malvaceae), *Plantago* (Plantaginaceae) and *Linum* (Linaceae) (see Stephen, 1979).

VI. Conclusion

In the fungi and the algae, chemical analyses of both cell wall and storage polysaccharides have been developed to the point where they have yielded valuable new data for using in both phenetic and phylogenetic classification. Crucial factors in such studies have been the inherent structural variability and the relative ease of isolating pure polysaccharide components. In the case of the fungi, there has been the additional point that few other chemical approaches are available for comparative purposes. With the algae, the commercial exploitation of seaweed polysaccharides has been a significant incentive for detailed chemical investigations.

In the case of the higher plants, much less progress has been made in the comparative biochemistry of the polysaccharides. This must be due in part to the considerable experimental difficulties of separating and purifying these polymers prior to analysis. Another disincentive has been the apparent uniformity in both cell wall and storage polymers. This uniformity may be more apparent than real and it is possible, at least with the hemicelluloses of cell walls, that taxonomically distinctive fractions may be present. The one place where most progress has been made is with the gum exudates, materials that are attractive for analysis because they are essentially pure polysaccharide to start with. Gum exudation is, however, a character well represented in certain families (e.g. Leguminosae, Rosaceae) but in general of limited distribution in the plant kingdom as a whole.

Perhaps the most promising area for more widespread exploitation are the seed endosperm polysaccharides. Clearly, they are very widely distributed and of many diverse structures. It may be suggested that they represent an interesting macromolecular approach to higher plant systematics worthy of further exploration.

20
Palaeobiochemistry

Like comparative systematics at the level of the organism, any comprehensive treatment of chemosystematics would appear incomplete without an account of the fossil record of organic compounds. This field has been termed organic geochemistry by many workers and, beginning about 1963, several texts on the subject appeared (Breger, 1963; Degens, 1965; Eglinton and Murphy, 1969). Nearly all of these were concerned, primarily, with 'free organic molecules' in the sense that they are found freed from the bodies in which they were produced, occurring mostly in heterogeneous mixtures of hydrocarbons such as petroleum, shales and coals.

Much of the initial interest in organic geochemistry was undoubtedly due to the possible use of fossil chemicals as substrate markers in oil exploration, but it soon became apparent that most organic shales and petroleum throughout the world are quite similar in the kinds of molecules represented (Van Hoeven *et al.*, 1966). After this initial flurry of textbook writing, the more dedicated workers in the field began to sustain themselves upon the intellectual application of their work, namely evolutionary biology.

Considering the structural nature of organic molecules, the number of fossil compounds that exist in the earth's sediments are likely to be considerable. Thus Blumer (1973) notes that, whereas something like 90 000 species of fossil animals are known at present, 'the almost limitless possibility of structural variation in organic chemistry and the complexity of the processes in the subsurface has produced in a single Triassic sediment more than 2×10^6 distinct derivatives of chlorophyll *a* alone'.

Because of the widespread interest in biochemical evolution, especially as related to the origin of life and metabolic processes relating to this, much effort has been made to relate fossil compounds of subsurface strata with the fossil organisms that occupy these strata (Gelpi *et al.*, 1970; Schopf, 1970; Smith *et al.*, 1970). Nothing really unexpected, evolutionarily speaking, has come out of these studies except for the recognition that such a wide assortment of organic substances might actually persist as complex structural units for so long. After initial deposition, of course, structural changes occur in the component compounds themselves, which accounts for the unusually large array of compounds detected, such as noted for chlorophyll *a*, above. Indeed, quite simple compounds such as the normal L-amino acids racemize with time, so that the ratio between L- and D-forms of a given amino acid can be used to date a given fossil horizon (Kvenvolden *et al.*, 1970).

Table 20.1 Organic acids from lignite coal identified as their methyl esters by GC–MS and high resolution MS. (From Hayatsu *et al.*, 1978.)

Peak no.	Compound
1	Succinic acid
2	Glutaric acid
3	Methylfuranmonocarboxylic acid
4	Mixture of 3-methyl, 2,3-dimethyl and 3,3′-6,6′-tetramethyl dibasic acids
5	Benzoic acid
6	Methylbenzoic acid
7	Dimethylbenzoic acid
8	Methoxybenzoic acid (*m*- or *p*-)
9	*o*-Methoxybenzoic acid
10	Methoxymethylbenzoic acid
11	*o*-Methoxymethylbenzoic acid
12	1,2-Benzenedicarboxylic acid
13	1,4-Benzenedicarboxylic acid
14	1,3-Benzenedicarboxylic acid
15	Dimethoxydimethylbenzofuran (T)
16	Dimethylfurandicarboxylic acid
17	*o*-Methylbenzenedicarboxylic acid
18	Methylbenzenedicarboxylic acid
19	Dimethylbenzenedicarboxylic acid
20	*o*-Methoxybenzenedicarboxylic acid
21	Methoxybenzenedicarboxylic acid
22	Methoxymethylbenzenedicarboxylic acid
23	1,2,4-Benzenetricarboxylic acid
24	1,2,3-Benzenetricarboxylic acid
25	1,3,5-Benzenetricarboxylic acid
26	Methylbenzenetricarboxylic acid
27	Methoxyphenyldihydrobenzofuran (T)
28	Mixture of compounds including methoxyphenyldihydrobenzofuran (T)
29	Dimethylbenzenetricarboxylic acid
30	*o*-Methoxybenzenetricarboxylic acid
31	Methoxybenzenetricarboxylic acid
32	Homologue of dehydroabietic acid (see text, T)
33	Methoxybenzyldihydrobenzofuran (T)
34	Methoxymethylphenyldihydrobenzofuran (T)
35	Dehydroabietic acid
36	Methoxymethyldibenzofuranmonocarboxylic acid (T)

T means that identification is tentative.

Although a wide array of organic geochemicals exists, the most commonly studied classes of fossil chemicals are those listed below (Eglinton, 1973): alkanes, alkenes, aromatics, alcohols, ketones, fatty acids, quinones, porphyrins, chlorins, carotenoids, carbohydrates, amino acids, biopolymers and kerogen.

The most thoroughly studied, or at least most significant from an evolutionary viewpoint, are the porphyrins (Blumer, 1973); these, along with n-alkanes and isoprenoids from early precambrian sediments, suggest that 'organisms exhibiting nucleic acid-directed enzyme synthesis and complex biosynthetic pathways were extant more than 3 billion years ago' (Schopf, 1970). Even so, the possible utility of biochemical fossils in kerogen from pre-Phanerozoic sediments seems quite limited since 'the least metamorphosed of these ancient sediments have evolved far toward amorphous carbon or graphite and do not yield biochemical fossils' (Leventhal *et al.*, 1975).

Hayatsu *et al.* (1978) examined the types of organic acids extractable from lignite and bituminous fossil coals from Wyoming. They identified 36 or more compounds (Table 20.1). Two diterpenoid acids in the lignite extract, dihydroabietic acid and its homologue, were believed to be derived from abietic acid, which is commonly present in conifer resins. In short, along with other data presented by these authors, it might be reasonably concluded that the Wyoming lignites were largely derived from forests dominated by conifers.

Fossil chemicals removed from the organisms in which they occur, while open to broad evolutionary interpretations, are not especially instructive taxonomically. More helpful would be chemical data tied to a particular fossil organism. Such studies are better developed for animals than plants (Voss-Foucard and Gregoire, 1971; Matter and Miller, 1972; Wyckoff, 1972; Halstead and Wood, 1973).

Two of the more promising studies have been those of King and Hare (1972) and Niklas and Gensel (1977). The former authors analysed the amino acids from calcified proteins of both extant and fossilized shells of planktonic Foraminifera (amoeboid calcareous protozoans) and found that the shells of each of 16 living species had unique patterns that differed 'to a greater or lesser degree' from the composition patterns of all the other species examined (Table 20.2). Analyses of two early Miocene species of this group (Fig. 20.1) suggest that characteristic differences in composition are sufficiently preserved over geological time (at least up to 18 million years) to determine phylogenetic affinities among taxa of Foraminifera generally. Although discrimination among these taxa is based largely on quantitative differences in amino acid composition, the authors contend that the approach will be useful in tracing evolutionary lineages using palaeontological materials and that this may prove useful in discriminating between species and subspecies at a given horizon in time. Niklas and Gensel (1977) examined 27 Palaeozoic plant taxa and, with the use of principal component analysis, concluded that, in spite of considerable thermolytic changes in the fossil organic material, that a clustering was obtained 'indicating degree of similarity, similar to that based on morphological-anatomical methods'.

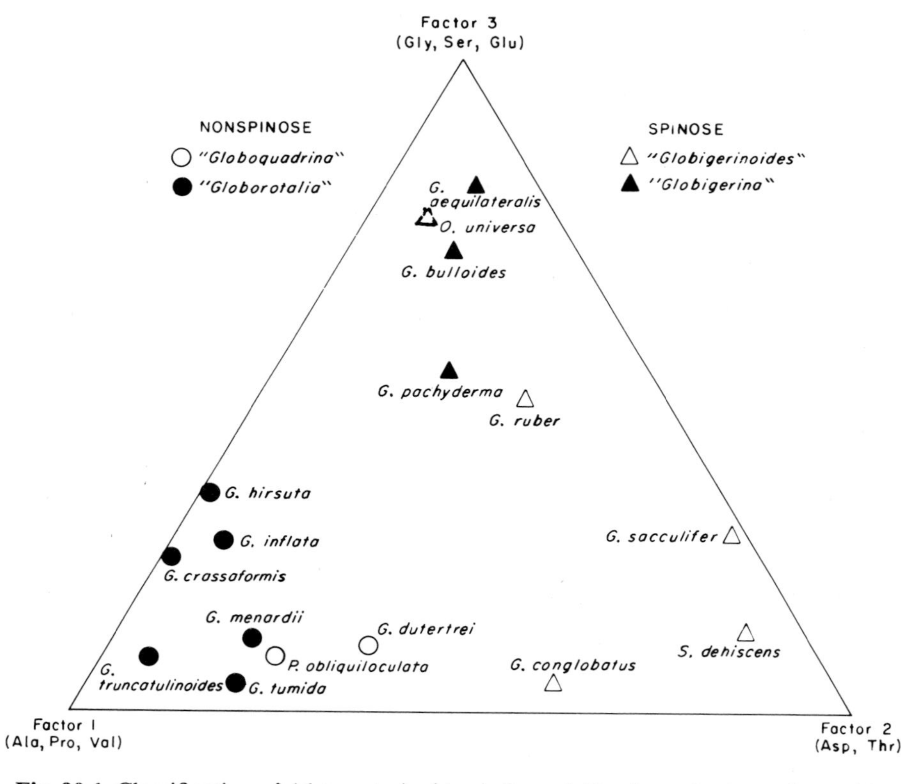

Fig. 20.1 Classification of 16 recent planktonic foraminiferal species by amino acid composition by means of Q-mode factor analysis. The symbols represent groups of morphologically similar species in a current taxonomic scheme. (From King and Hare, 1972.)

Relatively few studies have been reported for plants generally, presumably because in the fossilization process, plants become more mineralized than does the matrix of animal remains such as turtle shells, teeth and bone. Some of the more interesting early studies of fossil chemicals from plants have been those of Knocke and Ourisson (1967) and Ourisson (1974) who report a striking correlation with respect to hydrocarbon constituents between the extant *Equisetum sylvaticum* and the fossil *E. brongniarti*. Bonnett *et al.* (1972)

note that the relatively stable compound, equisetolic acid, which occurs in the spores and cones of selected extant species of *Equisetum* but not in others, might prove useful as a specific geochemical marker, but the compound was not found in the two fossil compressions of this genus which they examined.

Table 20.2 Comparison of amino acid concentrations in two extinct Early Miocene species (approximately 18 million years ago) and a Recent species; all three species are of the same genus, *Globoquadrina*. (From King and Hare, 1972.)

	Concentration nmol (g of shell material)$^{-1}$		
Amino acid	Recent	Early Miocene	
	G. dutertrei	*G. altispira*	*G. dehiscens*
Asp	467	398	273
Thr	172	11	11
Ser	166	15	18
Glu	252	193	168
Pro	139	118	115
Gly	273	164	155
Ala	215	265	230
Cys	25		
Val	136	129	99
Met	20		
Allo-Ile		56	46
Ile	81	43	35
Leu	95	65	49
Tyr	58	28	22
Phe	87	32	31
His	40	19	19
Lys	54	17	17
Arg	41		
Total	2321	1553	1288
Allo-Ile/Ile	0	1.30	1.31

Plant remains are likely to contain considerable fossilized cutin, the latter being characteristic of most vascular plants and, according to Hunneman and Eglinton (1972), can be used to distinguish among the extant major plant divisions. Thus, the cutin of Gnetophyta and angiosperms are characterized by 10,16-dihydroxyhexadecanoid acid, that of gymnosperms by 9-hydroxylated compounds, whereas the cutin of ferns and lycopods are dominated by 16-hydroxyhexadecanoic acid. In fact, these workers tentatively propose a progression in cutin acid composition which is paralleled by that of the evolutionary advancement of the taxa concerned: 'the higher the plant on the evolutionary scale the higher the degree of oxidation of the cutin acids', discovering again the earlier, often discredited, views of McNair (1935, 1965).

Unfortunately, no extensive attempt has been made to ascertain the 'survival' of such marker-molecules among fossil vascular plants so as to determine their diagnostic utility.

Niklas (1976a), presumably using somewhat more refined techniques than previous investigators, reports the successful use of palaeochemistry in positioning several non-vascular Palaeozoic plants. Organic constituents from compression fossils of four problematical fossil genera were compared with those of living taxa and, as a result of numerical cluster analysis of the isoprenoids, branched hydrocarbons and steroids identified, it was suggested that these taxa have algal affinities. Similarly, the controversial genus *Parka*, variously suggested as belonging to the animal or plant kingdoms by its hydrocarbon, amino acid and carbohydrate fractions, is seemingly best referred to the Chlorophyta (Niklas, 1976a). Chemical evidence of terrestrial adaptation (or at least to periodic desiccation) for algal-like fossils has also been reported by Niklas (1976c) but the data are sparse and open to conflicting interpretations, in that, as noted by Niklas for the *Prototaxities*, such fossils appear to be morphologically ill-adapted to terrestrial habitats.

In similar studies, Niklas and Pratt (1980) reported upon the palaeobiochemistry of some of the oldest vascular-like plant remains from Silurian beds of the eastern U.S.A. Following pyrolysis of fossil material they detected various phenolic aldehydes and aromatic compounds. This suggests that the fossilized tissue was composed of lignin or lignin-like substances. Taken together with morphological observations, they conclude that the tissue concerned 'might have functioned as water-conducting cells'.

One of the more remarkable studies of fossil chemicals reported to date has been that of Niklas and Giannasi (1977a) on 'fossil flavonoids' of the Ulmaceae from the Miocene (Succor Creek Flora) of Oregon. They report that organic solvent extractions from fossil leaves of the genus *Zelkovia* revealed the chemical preservation of kaempferol and dihydrokaempferol for something like 25–36 million years. Interestingly, both compounds may be found in extant species of *Zelkovia*. They believe, however, that unusual preservational features account for the consortium of organic compounds detected. Further, from simulated thermolytic experiments (Niklas and Giannasi, 1977b), they conclude that the flavonoids concerned did not experience postdepositional temperatures in excess of 80°C or extreme pH shifts.

In a series of similar extracts from fossil leaves of both *Celtis* and *Ulmus* from the same geological beds, Giannasi and Niklas (1977) report the isolation of quercetin 3-*O*-glycosides, apigenin and luteolin carbon glycosides. The fossil flavonoids were compared with extant flavonoids from the two genera concerned and a 'high level of phytochemical fidelity' is noted (Table 20.3). Niklas and Giannasi (1978) also analysed fossil leaves of the genera *Acer* and *Quercus* from the Succor Creek Flora and conclude that, in combination with the data from *Celtis*, *Ulmus* and *Zelkovia*, little flavonoid evolution has occurred in these taxa since post-Miocene time. Their data also suggest that the present day species of *Quercus* and *Zelkovia* from Asia show a greater affinity to the fossil assemblages from North America than they do to the living species in this latter region.

Table 20.3 Distribution of flavonoid glycosides in fossil and extant species of *Celtis* and *Ulmus*. The compounds are identified tentatively and remain to be compared with authentic samples. (From Giannasi and Niklas, 1977.)

Compound		Distribution		
No.	Identification	*Celtis* sp. (fossil)	*C. occidentalis* (extant)	*Ulmus* sp. (fossil)
1	Luteolin-8-*C*-glycoside (orientin)	+	+	
2	Apigenin-8-*C*-glycoside (vitexin)	+		
3	Luteolin-6-*C*-glycoside (isoorientin)	+	+	
4	Apigenin-6-*C*-glycoside (isovitexin)	+	+	
5	Luteolin-*C*-glycoside	+		
7	Luteolin-8-*O*-glycosyl-*C*-glycoside	+	+	
8	Apigenin-8-*O*-glycosyl-*C*-glycoside	+	+	
9	Chrysoeriol-*C*-glycoside		+	
10	Apigenin-*C*-glycoside		+	
10a	Apigenin-*C*-glycoside		+	
14	Quercetin-3-*O*-glycoside			+
15	Quercetin-3-*O*-glycoside			+
16	Quercetin-3-*O*-glycoside			+

Niklas and Brown (1981) call attention to the remarkable preservation of cellular organelles in leaf tissue of Miocene age. The genera involved are *Betula*, *Castanea*, *Hydrangea*, *Persea*, *Platanus* and *Quercus*. According to the authors, 'ultrastructural detail in preservation . . . correlate with their palaeobiochemical profiles'. They conclude that, during fossilization, the cellulosic cell wall and chloroplasts are the most stable cellular fractions, and that the endoplasmic reticulum, nuclei and mitochondria are more labile.

Finally, Giannasi and Niklas (1981), report upon the palaeobiochemistry of a newly discovered *Fagus*-like fossil (*Pseudofagus idahoensis*) from the Clarkia Lake Miocene beds of Idaho. Examination of its chemistry and comparison with that obtained from extant species of *Fagus* and related genera, indicate that (1) the steroid chemistry suggests a position in the Fagaceae, (2) like all extant taxa of *Fagus*, the fossil material lacked ellagic acid, which separates it from other extant fagoid genera and (3) flavonoid pigments position it closest to the extant North American *Fagus grandiflora* or the Eurasian *F. sylvatica*. However, the exclusive presence of several other chemicals (e.g., isorhamnetin, onocerane and 5-α-cholestane) mark *Pseudofagus* as distinct.

The remarkable preservation of such leaves along with at least some of the secondary constituents suggests that palaeobiochemistry will become increasingly important in chemosystematics, a development that would have been difficult to envision some 20 years ago.

21

Epilogue and Future Perspectives

I.	The past two decades	483
II.	Micromolecular perspectives	485
III.	Macromolecular perspectives	486

I. The past two decades

Over the past two decades there has been an exponential increase in the publication of research having to do with chemosystematics or molecular evolution. This is especially true in the field of macromolecular systematics where automated instruments have permitted rapid sequential analyses of both proteins and DNA. What is most remarkable is the enormous range of chemical characters that are now available for systematic purposes. Some idea of the variety of characters that have been employed in recent years within the plant kingdom is indicated in Table 21.1; most of these have been discussed *inter alia* in earlier chapters. A considerable range of such characters have likewise been employed in animal taxonomy. While most emphasis in animals has been at the macromolecular level and especially with proteins (see Wright, 1974; Ferguson, 1980), it may be noted that micromolecular data are also becoming of interest, particularly with regard to pheromonal and defence secretions in insects (Howse and Bradshaw, 1980).

In 1958 when the junior author first began his interest in this field with the late R. E. Alston he truly had no appreciation of the kind of developments that might be possible in the area. Indeed, some of his more forward-looking colleagues suggested that he become immersed in the chemistry of enzymes and DNA, that the 'ultimates' lay in this approach to systematics. Maybe so, I said, but that is too far in the future. Now, bewildered by developments in the field, I realize that their advice was sound. But the truth is the junior author is first and foremost a plant systematist interested in the morphology and natural history of organisms, while the senior author is first and foremost a comparative phytochemist interested in the distribution and role of micromolecular compounds in organisms generally. And interests pretty much predicate directions in research.

Table 21.1 The range of chemical characters available for systematic purposes in the plant kingdom

Phylum	Representative chemical characters
Bacteria	Pyrolysis GC–MS for 'fingerprinting' strains Degradative pathways of aromatic compounds DNA base ratios and DNA–DNA hybridization Ribosomal RNA base sequences
Fungi	Cell wall composition Pathways of lysine biosynthesis Tryptophan synthesizing enzymes
Algae	Storage and structural polysaccharides Carotenoid and chlorophyll pigments, biliproteins Pathways of fatty acid biosynthesis
Lichens	Depsides, depsidones and anthraquinones
Bryophyta	Conjugation of D-methionine Sesquiterpenoids Flavonoids
Ferns	Acylphloroglucinol derivatives Triterpenoid hydrocarbons Mannans DNA–DNA hybridization
Gymnosperms	Essential oils Biflavonyls Polyols
Angiosperms	Pigments, bitter principles, volatiles, hidden metabolites, etc. Isozymes of the shikimate pathway Phytoalexin induction Protein sequences and band patterns Repetitive DNA sequences

But even so, it is unlikely that any person can ever hope to be 'master' of the entire field of molecular evolution, much less systematics in the morphological sense. Specialization in both concept and method will dominate the deeper chemosystematic probes, whereas the broadly trained classical systematist dealing with morphological data will necessarily have to draw upon his more molecular-oriented colleagues for informed application. And *vice versa*, of course, but it is likely that most of the 'leaning' will be in the former direction. The production of this text has been a joint effort between systematist and biochemist and an important lesson learnt over the last two decades is that major developments in chemosystematics can only take place with proper research collaboration between the two disciplines.

II. Micromolecular perspectives

As indicated several times in this text, the use of flavonoids, terpenoids and other chemical markers among plant groups generally, especially at the specific and generic levels, will require more extensive sampling techniques in order to be maximally convincing in their application to systematic problems. This will require both rigorous qualitative and quantitative analysis of a wide array of compounds so that these might be treated numerically (for populational purposes), much as Adams (1983) has approached infraspecific variation in *Juniperus*. Numerical methods for phylogenetical purposes are also possible, for example, Seaman and Funk's (1983) cladistical analysis of sesquiterpene lactones, or Otto Gottlieb's (1982) account, or yet others for supraspecific relationships. Cladistic treatments (e.g. Platnick and Funk, 1983) with their emphasis on synapomorphies and apomorphies ('ancestral' versus derived states) are ideal for chemical characters in that these can be readily polarized and branching patterns inferred from them. The ready availability of appropriate algorithms for computer-generated phyletic diagrams, with parsimony as the conceptual fulcrum, makes the appearance of such studies almost passé.

No doubt massive data banks for the storage of micromolecular information will be needed in due course, along with clear and unequivocal documentation of the taxa from which such compounds were isolated. The reason data banks for chemical information is more critical than data banks for morphological data is, obviously, that the chemistry is not stored in museum cases in visible form. It would be convenient in the extreme to have access to a central storage centre in which such data are assembled and made available to the monographer, co-evolutionary biologist, or whatever.

A start has been made in this direction by F. A. Bisby of the University of Southampton, who has prepared a data bank for chemical characters present in members of the tribe Vicieae of the Leguminosae. Nomenclatural, morphological and geographical data are also included. A major problem remains, however, of the relative scarcity of comparative chemical data on any large group of plants, at the family level or above. Even where much chemical effort has already been expended (e.g. on the pigments present in families of the Centrospermae), the overall coverage species by species is still quite low. Concerted surveys should be made in such cases to fill in the gaps with particularly interesting plant groups.

Ultimately, as Crawford (1978) rightly notes, the chemosystematist dealing with micromolecular compounds must come to grips with the homology of biochemical pathways leading to their products. This is especially true if one is comparing such compounds from distantly related species, say among different orders or divisions. Thus, one is soon faced with comparative enzymology and it is likely that this field, combined with an ever-increasing awareness of the genetic control of metabolic pathways, will expand greatly in the not too distant future so as to provide considerable insight into phylogenetic arrangements. In this sense, the role of the micromolecular systematist interfaces with that of the macromolecular systematist, which will be discussed briefly below.

III. Macromolecular perspectives

Because of the extraordinary financial support of biomedical research over the past two decades, with emphasis upon an understanding of the mechanisms which control the onset of various diseases in man, an exceptionally large amount of macromolecular data immediately relevant to systematics has become available. So much so that some of the more curious minds in the field have turned their attention to the evolutionary study of entire metabolic pathways (Yeh and Ornston, 1980; Yeh *et al.*, 1982), searching for the 'primitive gene' or genes which gave rise to such systems (Bloch *et al.*, 1982) if not for the beginnings of the primordial reproductive biochemical feed-back system itself which is referred to, simplistically, as 'life'.

The potential of such macromolecular studies for resolving the major phyletic problems among living organisms is considerable. Hardly a year passes but what one or more major symposia are published (e.g. Goodman, 1982) and methodological and conceptual 'breakthroughs', especially in DNA research, is seemingly almost routine.

The complexity and exceptional accumulation of sequential data for this or that DNA, or derived peptide chain, has created the need, almost immediately, for a centralized data bank which the large number of workers engaged in such activity might consult for comparative purposes. Although most of these data are currently those derived from human macromolecules, in due course similar information for yet other organisms will be stored there and through such visionary efforts a sound approximation of the phyletic arrangements of the *major* lineages of the earth will have been achieved.

Because of sampling problems especially at the infra- and interpopulational levels, mainly a consequence of the considerable macromolecular variability among both enzymes and genes (DNA) that code them (e.g. Dowsett and Young, 1982), classification at the familial level or lower is likely to be more intractable. We tend to believe, therefore, that the classical morphological monographer of the future will still be hard at his chosen interest: working with groups of closely related organisms (usually at the genus level or lower), attempting to make order out of artificial groupings, or make greater order out of lesser order, etc. To do this we believe the competent monographer cannot afford to be ignorant of, or ignore, comparative chemistry. This is especially true if a phylogenetic treatment of the organisms under study is sought.

We end this book with a series of questions posed by Alston and Turner (1963a) at the end of their seminal text, *Biochemical Systematics*:

> Are we really at the beginning of a new period of taxonomic history? Will taxonomically oriented biochemical investigations yield data that make possible a better phylogenetic scheme? Will they give answers to taxonomic questions that previous methods did not permit? Will chemotaxonomy become as significant in the next half-century as cytotaxonomy has during the last? Is the time at hand for this molecular approach?

The answers to all now seem so obvious: yes, ah, yes.

Bibliography

Abbott, H. C. (1886). Certain chemical constituents of plants considered in relation to their morphology and evolution. *Bot. Gaz.* **11**, 270–272.

Abbott, H. C. (1887). Comparative chemistry of higher and lower plants. *Amer. Nat.* **21**, 719–753.

Abrahamson, W. G. and Solbrig, O. T. (1970). Soil preference and variation in flavonoid pigments in species of *Aster*. *Rhodora* **72**, 251–263.

Adams, R. P. (1970a). Seasonal variation of terpenoid constituents in natural populations of *Juniperus pinchotii* Sudw. *Phytochemistry* **9**, 397–402.

Adams, R. P. (1970b). Contour mapping and differential systematics of geographical variation. *System. Zool.* **19**, 385–390.

Adams, R. P. (1972). Numerical analyses of some common errors in chemosystematics. *Brittonia* **24**, 9–21.

Adams, R. P. (1974). On 'numerical chemotaxonomy' revised. *Taxon* **23**, 336–338.

Adams, R. P. (1975). Gene flow versus selection pressure and ancestral differentiation in the composition of species: analysis of populational variation of *Juniperus ashei* Buch. using terpenoid data. *J. Mol. Evol.* **5**, 177–185.

Adams, R. P. (1977). Chemosystematics—Analysis of populational differentiation and variability of ancestral and recent populations of *Juniperus ashei*. *Ann. Missouri Bot. Gard.* **64**, 184–209.

Adams, R. P. (1982). A comparison of multivariate methods for the detection of hybridisation. *Taxon* **31**, 646–661.

Adams, R. P. (1983). Infraspecific terpenoid variation in *Juniperus scopulorum*: evidence for pleistocene refugia and recolonisation in western North America. *Taxon* **32**, 30–46.

Adams, R. P. and Hagerman, A. (1976). A comparison of the volatile oils of mature versus young leaves of *Juniperus scopulorum*: chemosystematic significance. *Biochem. Syst. Ecol.* **4**, 75–79.

Adams, R. P. and Hagerman, A. (1977). Diurnal variation in the volatile terpenoids of *Juniperus scopulorum* (Cupressaceae). *Amer. J. Bot.* **64**, 278–285.

Adams, R. P. and Powell, R. A. (1976). Seasonal variation of sexual differences in the volatile oil of *Juniperus scopulorum*. *Phytochemistry* **15**, 509–510.

Adams, R. P. and Turner, B. L. (1970). Chemosystematic and numerical studies of natural populations of *Juniperus ashei* Buch. *Taxon* **19**, 728–751.

Adams, R. P., von Rudloff, E., Zanoni, T. A. and Hogge, L. (1980). The terpenoids of an ancestral/advanced species pair of *Juniperus*. *Biochem. Syst. Ecol.* **8**, 35–37.

Adams, R. P., Palma, M. M. and Moore, W. S. (1981). Volatile oils of mature and juvenile leaves of *Juniperus horizontalis*: chemosystematic significance. *Phytochemistry* **20**, 2501–2502.

Adams, R. P., Zanoni, T. A., von Rudloff, E. and Hogge, L. (1981). The southwestern USA and northern Mexico one-seeded junipers: their volatile oils and evolution. *Biochem. Syst. Ecol.* **9**, 93–96.

Adcock, J. W. and Betts, T. J. (1974). A chemotaxonomic survey of essential oil constituents in the tribe Laserpitieae (Umbelliferae). *Planta Medica* **26**, 52–64.

Adesida, G. A., Adesogan, E. K., Okorie, D. A., Taylor, D. A. H. and Styles, B. T. (1971). The limonoid chemistry of the genus *Khaya* (Meliaceae). *Phytochemistry* **10**, 1845–1853.

Adzet, T., Granger, R., Passet, J. and San Martin, R. (1977). Le Polymorphisme Chimique dans le Genre *Thymus*: sa Signification Taxonomique. *Biochem. Syst. Ecol.* **5**, 269–272.

Akabori, Y. (1978). Flavonoid pattern in the Pteridaceae IV. Seasonal variation of the flavonoids in the fronds of *Adiantum monochlemys. Bot. Mag. Tokyo* **91**, 137–139.

Akahori, A. (1965). Steroidal sapogenins contained in Japanese *Dioscorea* spp. *Phytochemistry* **4**, 97–106.

Albach, R. F. and Redman, G. H. (1969). Composition and inheritance of flavanones in citrus fruit. *Phytochemistry* **8**, 127–143.

Ali, S. I. and Qaiser, M. (1980). Hybridisation in *Acacia nilotica* (Mimosoideae). *Bot. J. Linn. Soc.* **80**, 69–77.

Allan, R. D., Wells, R. J., Correll, R. L. and Macleod, J. K. (1978). The presence of quinones in the genus *Cyperus* as an aid to classification. *Phytochemistry* **17**, 263–266.

Allison, M. J. (1973). Genetic studies on the α-amylase isozymes of barley malt. *Genetica* **44**, 1–15.

Alston, R. E. (1965). Flavonoid chemistry of *Baptisia*: a current evaluation of chemical methods in the analysis of interspecific hybridisation. *Taxon* **14**, 268–274.

Alston, R. E. (1967). Biochemical systematics. *In* 'Evolutionary Biology' (Dobzhansky, T., Hecht, M. K. and Steere, W. C., eds.), Vol. 1, pp. 197–305. Appleton-Century-Crofts, New York.

Alston, R. E. and Hempel, K. (1964). Chemical documentation of interspecific hybridization. *J. Hered.* **55**, 267–269.

Alston, R. E. and Simmons, J. (1962). A specific and predictable biochemical anomaly in interspecific hybrids of *Baptisia viridis* × *B. leucantha. Nature (Lond.)* **195**, 825.

Alston, R. E. and Turner, B. L. (1962). New techniques in analysis of complex natural hybridization. *Proc. Natl. Acad. Sci. U.S.A.* **48**, 130–137.

Alston, R. E. and Turner, B. L. (1963a). 'Biochemical Systematics'. Prentice-Hall, Englewood Cliffs, New Jersey.

Alston, R. E. and Turner, B. L. (1963b). Natural hybridization among four species of *Baptisia* (Leguminosae). *Amer. J. Bot.* **50**, 159–173.

Alston, R. E., Turner, B. L., Lester, R. N. and Horne, D. (1962). Chromatographic validation of two morphologically similar hybrids of different origin. *Science* **137**, 1048–1050.

Alston, R. E., Rosler, H., Naifeh, K. and Mabry, T. J. (1965). Hybrid compounds in natural interspecific hybrids. *Proc. Natl. Acad. Sci. U.S.A.* **54**, 1458–1465.

Altosaar, I., Bohm, B. A. and Ornduff, R. (1974). Disc electrophoresis of albumin and globulin fractions from dormant achenes of *Lasthenia. Biochem. Syst. Ecol.* **2**, 67–72.

Anderson, D. M. W. (1978). Chemotaxonomic aspects of the chemistry of *Acacia* gum exudates. *Kew Bull.* **32**, 529–536.

Anderson, D. M. W. and Bell, P. C. (1977). The composition of the gum exudates from some *Combretum* species. *Carbohydrate Research* **57**, 215–221.
Anderson, D. M. W. and Brenan, J. P. M. (1975). Chemotaxonomic aspects of the gum exudates from some subspecies of *Acacia tortilis*. *Boissiera* **24**, 307–309.
Anderson, D. M. W. and Dea, I. C. M. (1969). Chemotaxonomic aspects of *Acacia* exudates. *Phytochemistry* **8**, 167–176.
Anderson, E. (1949). 'Introgressive Hybridization.' John Wiley and Sons, New York.
Anderson, E. (1953). Introgressive hybridisation. *Biol. Rev.* **28**, 280–307.
Anderson, E. (1957). An experimental investigation of judgements concerning genera and species. *Evolution* **11**, 260–262.
Andrews, P. G., Hough, L. and Stacey, B. (1960). Polysaccharide composition of leaves. *Nature (Lond.)* **185**, 166–167.
Aplin, R. T., Cambie, R. C. and Rutledge, P. S. (1963). The taxonomic distribution of some diterpene hydrocarbons. *Phytochemistry* **2**, 205–214.
Appelqvist, L. A. (1976). Lipids in the Cruciferae. *In* 'The Biology and Chemistry of the Cruciferae' (Vaughan, J. G., Macleod, A. J. and Jones, B. M. G., eds.), pp. 221–278. Academic Press, London and New York.
Asker, S. and Fröst, S. (1972a). Plant age and chromatographic pattern in *Potentilla*. *Hereditas* **72**, 149–152.
Asker, S. and Fröst, S. (1972b). Chromatographic studies of phenolic compounds in apomictic *Potentilla* L. *Hereditas* **65**, 241–250.
Aspinall, G. O. (1970). 'Polysaccharides.' Pergamon Press, Oxford.
Aspinall, G. O. (1981). Constitution of plant cell wall polysaccharides. *In* 'Extracellular Carbohydrates' (Tanner, W. and Loewus, F. A., eds.), pp. 1–8. Springer Verlag, Berlin.
Averett, J. E. (1970). New combinations in the Solaneae (Solanaceae) and comments regarding the taxonomic status of *Leucophysalis*. *Ann. Missouri Bot. Gard.* **57**, 380–382.
Axelrod, D. I. (1950). 'Evolution of Desert Vegetation in Western North America.' Publication of the Carnegie Institute, Washington, No. 590.
Ayala, F. J. (ed.) (1976). 'Molecular Evolution.' Sinauer Associates. Sunderland, Mass.
Ayala, F. J. (1982). Genetic variation in natural populations: Problem of electrophoretically cryptic alleles. *Proc. Natl. Acad. Sci. U.S.A.* **79**, 550–554.
Ayala, F. J., Powell, J. R., Tracey, M. L., Mourao, C. A. and Perez-Salas, S. (1972). Enzyme variability in the *Drosophila willistoni* group. IV. *Genetics* **70**, 113–139.
Aynilian, G. H., Farnsworth, N. R. and Trojanek, J. (1974). The use of alkaloids in determining the taxonomic position of *Vinca libanotica* (Apocynaceae). *In* 'Chemistry in Botanical Classification' (Bendz, G. and Santesson, J., eds.), pp. 189–204. Academic Press, New York.
Baagøe, J. (1977). Microcharacters in the ligules of the Compositae. *In* 'The Biology and Chemistry of the Compositae' (Heywood, V. H., Harborne, J. B. and Turner, B. L., eds.), pp. 119–140. Academic Press, London.
Bacon, J. D. (1978). Taxonomy of *Nerisyrenia* (Cruciferae). *Rhodora* **80**, 159–227.
Bacon, J. S. D. (1960). The oligofructosides. *Bull. Soc. Chim. Biol.* **42**, 1441–1449.
Baetcke, K. P. and Alston, R. E. (1968). The composition of a hybridizing population of *Baptisia sphaerocarpa* and *Baptisia leucophaea*. *Evolution* **22**, 157–165.
Bailey, D. K., Snajberk, K. and Zavarin, E. (1982). On the question of natural hybridization between *Pinus discolor* and *P. cembroides*. *Biochem. Syst. Ecol.* **10**, 111–119.

Bailey, I. W. (1944). The development of vessels in angiosperms and its significance in morphological research. *Amer. J. Bot.* **31**, 421–428.

Bailey, J. A. and Mansfield, J. W. (1982). 'Phytoalexins ' Blackie, Glasgow and London.

Bailey, J. A., Vincent, G. G. and Burden R. S. (1974). Diterpenes from *Nicotiana glutinosa* and their effect on fungal growth. *J. gen. Microbiol.* **85**, 57–64.

Bailey, R. W. (1965). 'Oligosaccharides'. Macmillan Co., New York.

Bailey, R. W. (1971). Polysaccharides in the Leguminosae. *In* 'Chemotaxonomy of the Leguminosae' (Harborne, J. B., Boulter, D. and Turner, B. L., eds.), pp. 503–541. Academic Press, London.

Baker, H. G. and Baker, I. (1976). Analyses of amino acids in flower nectars of hybrids and their parents, with phylogenetic implications. *New Phytol.* **76**, 87–98.

Baker, H. G., Opler, P. A. and Baker, I. (1978). A comparison of the amino acid complements of floral and extrafloral nectars. *Bot. Gaz.* **139**, 322–332.

Baker, R. T. and Smith, H. G. (1902). Research on the eucalypts especially in regard to their essential oils. Sydney Technological Museum, New South Wales Technical Educational Series No. 13 (2nd edn. published 1920).

Baldwin, B. C. (1976). Xenobiotic metabolism in plants. *In* 'Drug Metabolism—from Microbe to Man' (Parke, D. V. and Smith, R. L., eds), pp. 191–217. Taylor and Francis, London.

Ball, G. A., Jr., Beal, E. O. and Flecker, E. A. (1967). Variation of chromatographic spot patterns of two species of clonal plants grown under controlled environmental conditions. *Brittonia* **19**, 273–279.

Ballard, R. E., McClure, J. W., Eshbaugh, W. H. and Wilson, K. G. (1970). A chemosystematic study of *Capsicum*. *Amer. J. Bot.* **57**, 225–233.

Bandaranayake, W., Karunanayake, S., Sotheeswaran, S., Sultanbawa, M. U. S. and Balasubramaniam, S. (1977). Triterpenoid taxonomic markers for *Stemonoporus* and other genera of the Dipterocarpaceae. *Phytochemistry* **16**, 699–701.

Bandoni, A. L., Stermitz, F. R., Rondina, R. V. D. and Coussio, J. D. (1975). Alkaloidal content of Argentine *Argemone*. *Phytochemistry* **14**, 1785–1788.

Bannister, M. H., Brewerton, H. V. and McDonald, I. R. C. (1959). Vapour phase chromatography in a study of hybrids of *Pinus*. *Svensk Papparstidning* **62**, 567–573.

Banthorpe, D. V. and Wirz-Justice, A. (1969). Terpene biosynthesis. Part I. Preliminary tracer studies on terpenoids and chlorophyll of *Tanacetum vulgare* L. *J. Chem. Soc.* (*London*) 1969, 541–549.

Barker, R. E. and Hovin, A. W. (1974). Inheritance of indole alkaloids in reed canary grass (*Phalaris arundinacea* L.). I. Heritability estimates for alkaloid concentration. *Crop Sci.* **14**, 50–53.

Bartnicki-Garcia, S. (1970). Cell wall composition and other biochemical markers in fungal phylogeny. *In* 'Phytochemical Phylogeny' (Harborne, J. B., ed.), pp. 81–104. Academic Press, London.

Baskin, S. I. and Bliss, C. A. (1969). Sugar occurring in the extrafloral exudates of the Orchidaceae. *Phytochemistry* **8**, 1139–1145.

Bate-Smith, E. C. (1962). The phenolic constituents of plants and their taxonomic significance. I. Dicotyledons. *J. Linn. Soc.* (*Bot.*) **58**, 95–173.

Bate-Smith, E. C. (1964). Chemistry and Taxonomy of *Fouquieria splendens*: a new member of the asperuloside group. *Phytochemistry* **3**, 623–625.

Bate-Smith, E. C. (1965). The chemical taxonomy of some phenolic constituents of plants. *Mem. Soc. Bot. Fr.* 16–28.

Bate-Smith, E. C. (1973). Chemotaxonomy of *Geranium*. *Bot. J. Linn. Soc.* **67**, 347–359.

Bate-Smith, E. C. and Richens, R. H. (1973). Flavonoid chemistry and taxonomy in *Ulmus*. *Biochem. Syst.* **1**, 141–146.

Bate-Smith, E. C. and Swain, T. (1966). The Asperulosides and the Aucubins. *In* 'Comparative Phytochemistry' (Swain, T. ed.), pp. 159–174. Academic Press, London.

Bate-Smith, E. C., Sell, P. D. and West, C. (1968). Chemistry and taxonomy of *Hieracium* L. and *Pilosella* Hill. *Phytochemistry* **7**, 1165–1169.

Bate-Smith, E. C., Ferguson, I. K., Hutson, K., Jensen, S. R., Nielsen, B. J. and Swain, T. (1975). Phytochemical interrelationships in the Cornaceae. *Biochem. Syst. Ecol.* **3**, 79–89.

Belford, H. S. and Thompson, W. F. (1981). Single copy DNA homologies in *Atriplex*. II. Hybrid thermal stabilities and molecular phylogeny. *Heredity* **46**, 109–122.

Belford, H. S., Thompson, W. F. and Stein, D. B. (1981). DNA Hybridisation techniques for the study of plant evolution. *In* 'Phytochemistry and Angiosperm Phylogeny' (Young, D. A. and Seigler, D. S., eds.), pp. 1–18. Praeger, New York.

Bell, E. A. (1958). Canavanine and related compounds in Leguminosae. *Biochem. J.* **70**, 617–619.

Bell, E. A. (1966). Amino acids and related compounds. *In* 'Comparative Phytochemistry' (Swain, T., ed.), pp. 195–209. Academic Press, London.

Bell, E. A. (1971). Comparative biochemistry of non-protein amino acids. *In* 'Chemotaxonomy of the Leguminosae' (Harborne, J. B., Boulter, D. and Turner, B. L., eds.) pp. 179–206. Academic Press, London.

Bell, E. A. (1972). Toxic amino acids in the Leguminosae. *In* 'Phytochemical Ecology' (Harborne, J. B., ed.), pp. 163–178. Academic Press, London.

Bell, E. A. (1978). Toxins in seeds. *In* 'Biochemical Aspects of Plant and Animal Coevolution' (Harborne, J. B., ed.), pp. 143–162. Academic Press, London.

Bell, E. A. (1980). Non-protein amino acids. *In* 'Encyclopedia of Plant Physiology' (Bell, E. A. and Charlwood B. V., eds.), New Series, Vol. 8, pp. 403–432. Springer-Verlag, Berlin.

Bell, E. A. and Evans, C. S. (1978). Biochemical evidence of a former link between Australia and the Mascarene Islands. *Nature* (*Lond.*) **273**, 295–296.

Bell, E. A. and Janzen, D. L. (1971). Medical and ecological considerations of L-Dopa and 5-HTP in seeds. *Nature* (*Lond.*) **229**, 136–137.

Bell, E. A., Lackey, J. A. and Polhill, R. M. (1978). Systematic significance of canavanine in the Papilionoideae. *Biochem. Syst. Ecol.* **6**, 201–212.

Belzer, N. F. and Ownbey, M. (1971). Chromatographic comparison of *Tragopogon* species and hybrids. *Amer. J. Bot.* **58**, 697–790.

Bendich, A. J. and Bolton, E. T. (1967). Relatedness among plants as measured by the DNA-agar technique. *Plant Physiol.* **42**, 959–967.

Bendz, G. and Santesson, J. (eds.) (1974). Chemistry in Botanical Classification. Nobel Symposium 25. Academic Press, London.

Bennett, M. D. and Smith, J. B. (1976). Nuclear DNA amounts in angiosperms. *Phil. Trans. R. Soc. Lond.* **274B**, 227–273.

Bennett, R. D., Ko, S. and Heftmann, E. (1966). Estrone and cholesterol from the date palm, *Phoenix dactylifera*. *Phytochemistry* **5**, 231–235.

Bentham, G. (1867). Umbelliferae. *In* 'Genera Plantarum' (Bentham, G. and Hooker, J. D., eds.), Vol. 1, pp. 859–931. Spottiswood, London.

Bentham, G. (1873). Notes on the classification, history and geographical distribution of Compositae. *J. Linn. Soc. Lond. Bot.* **13**, 335–577.

Bentley, R. K. (1975). Biosynthesis of quinones. *In* 'Biosynthesis' (Geissman, T. A., ed.), Vol. 3, pp. 181–246. The Chemical Society of London.

Bentley, R. K., Jenkins, J. K., Jones, E. R. H. and Thaller, V. (1969). Polyacetylenes from the Campanulaceae: tetrahydropyranyl polyacetylenic alcohols from the clustered bellflower, *Campanula glomerata. J. Chem. Soc. C*, 830–832.

Bergman, F. (1973). Genetische Untersuchungen bei *Picea abies* mit Hilfe der Isoenzym-Identifizierung. *Silvae Genetica* **22**, 63–66.

Bergström, G. (1978). Role of volatile chemicals in *Ophrys*-pollinator interactions: *In* 'Biochemical Aspects of Plant and Animal Coevolution' (Harborne, J. B., ed.), pp. 207–232. Academic Press, London and New York.

Bernard, N. (1911). Sur la fonction fungicide des bulbes d'Ophrydées. *Ann. Sci. Nat. Bot.* **14**, 221–234.

Bernhard, R. A. (1970). Chemotaxonomy: distribution studies of sulphur compounds in *Allium. Phytochemistry* **9**, 2019–2027.

Bieleski, R. L. and Johnson, P. N. (1972). The external location of phosphatase activity in phosphorus-deficient *Spirodela oligorhiza. Aust. J. Biol. Sci.* **25**, 707–720.

Bierner, M. W. (1973a). Sesquiterpene lactones and the systematics of *Helenium quadridentatum* and *H. elegans. Biochem. System.* **1**, 95–96.

Bierner, M. W. (1973b). Chemosystematic aspects of flavonoid distribution in twenty-two taxa of *Helenium. Biochem. System.* **1**, 55–57.

Binns, W. W. and Blunden, G. (1969). Effects of hybridization on leaf constituents in the genus *Salix. Phytochemistry* **8**, 1235–1239.

Bisset, N. G., Diaz, M. A., Ehret, C., Ourisson, G., Palmade, M., Patil, F., Pesnelle, P. and Streith, J. (1966). Chemotaxonomic studies in the Dipterocarpaceae. II. Constituents of the genus *Dipterocarpus*. Phytochemistry **5**, 865–880.

Biswas, S. B. and Sarkar, A. K. (1970). Deoxyribonucleic acid base composition of some angiosperms and its taxonomic significance. *Phytochemistry* **9**, 2425–2430.

Bjeldanes, L. F. and Geissman, T. A. (1969). Euparinoid constituents of *Eucelia californica. Phytochemistry* **8**, 1293–1296.

Bjeldanes, L. F. and Geissman, T. A. (1971). Sesquiterpene lactones: constituents of an F_1 hybrid *Encelia farinosa* × *E. californica. Phytochemistry* **10**, 1079–1081.

Block, D., McArthur, B., Widdowson, R., Spector, D., Guimares, R. and Smith, J. (1982). Sequence homologies between t- and r-DNAs and their relation to the evolutionary origin of the RNAs. (Abs.) *J. Cell. Biol.* **95**, 468.

Blumer, M. (1973). Chemical fossils: trends in organic geochemistry. *J. Pure Appl. Chem.* **34**, 591–610.

Bohannon, M. B., Hagemann, J. W. and Earle, F. R. (1974). Screening seed of *Trigonella* and three related genera for diosgenin. *Phytochemistry* **13**, 1513–1514.

Bohlmann, F., Burkhardt, T. and Zdero, C. (1973). 'Naturally Occurring Acetylenes.' Academic Press, London and New York.

Bohm, B. A. (1975). Chalcones, aurones and dihydrochalcones. *In* 'The Flavonoids' (Harborne, J. B., Mabry, T. J. and Mabry, H., eds.), pp. 442–504. Chapman and Hall, London.

Bohm, B. A. (1977). Heliantheae-chemical review. *In* 'The Biology and Chemistry of the Compositae' (Heywood, V. H., Harborne, J. B. and Turner, B. L., eds.), pp. 739–759. Academic Press, London.

Bonnett, R., Middlemiss, F. A. and Noro, T. (1972). Distribution of equisetolic acid in the Equisetales. *Phytochemistry* **11**, 2801–2802.

Booth, T. A. and Richards, A. J. (1978). The *Hordeum murinum* aggregate: disc electrophoresis of seed proteins. *Bot. J. Linn. Soc.* **76**, 115–125.

Boudet, A. M., Boudet, A. and Bouyson, H. (1977). Taxonomic distribution of isoenzymes of dehydroquinate hydrolyase in the angiosperms. *Phytochemistry* **16**, 919–922.

Boulter, D. (1980). The use of amino acid sequence data in phylogenetic studies with special reference to plant proteins. *In* 'Chemosystematics, Principles and Practice' (Bisby, F. A., Vaughan, J. G. and Wright, C. A., eds.), pp. 235–240. Academic Press, London.

Boulter, D., Derbyshire, E., Frahm-Lelineld, J. A. and Polhill, R. M. (1970). Observation on the cytology and seed-proteins of various African species of *Crotalaria* L. (Leguminosae). *New Phytol.* **69**, 117–131.

Boulter, D. and Parthier, B. (eds.) (1982). Structure, Biochemistry and Physiology of Proteins. 'Encyclopedia of Plant Physiology,' New Series, Vol. 14A. Springer-Verlag, Berlin.

Boulter, D. and Ramshaw, J. A. M. (1975). Amino acid sequence Analysis of Proteins. *In* 'The Chemistry and Biochemistry of Plant Proteins' (Harborne, J. B. and Van Sumere, C. F., eds.), pp. 1–30. Academic Press, London.

Boulter, D. and Thurman, D. A. (1968). Acrylamide gel electrophoresis of proteins in plant systematics. *In* 'Chemotaxonomy and Serotaxonomy' (Hawkes, J. G. ed.), pp. 39–48. Academic Press, New York.

Boulter, D., Thurman, D. A. and Turner, B. L. (1966). The use of disc electrophoresis or plant proteins in systematics. *Taxon* **15**, 135–143.

Boulter, D., Thurman, D. A. and Derbyshire, E. (1967). A disc electrophoretic study of globular proteins of legume seeds with reference to their systematics. *New Phytol.* **66**, 27–36.

Boulter, D., Ramshaw, J. A. M., Thompson, E. W., Richardson, M. and Brown, R. H. (1972). A phylogeny of higher plants based on the amino acid sequences of cytochrome *c* and its biological implications. *Proc. R. Soc. Lond.* **B181**, 441–455.

Boulter, D., Haslett, B. G., Peacock, D., Ramshaw, J. A. M. and Scawen, M. D. (1977). Chemistry, Function and Evolution of Plastocyanin. *In* 'International Review of Biochemistry' (Northcote, D. H., ed.), pp. 1–40. University Park Press, Baltimore.

Boulter, D., Peacock, D., Guise, A., Gleaves, J. T. and Estabrook, G. (1979). Relationships between the partial amino acid sequences of plastocyanin from members of ten families of flowering plants. *Phytochemistry* **18**, 603–608.

Boyden, A. (1954). The measurement and significance of serological correspondence among proteins. *In* 'Serological Approaches to Studies of Protein Structure and Metabolism' (Cole, W. H., ed.), pp. 74–97. Rutgers University Press, New Brunswick.

Braga de Oliviera, A., Gottlieb, O. R., Ollis, W. D. and Rizzini, C. T. (1971). A phylogenetic correlation of the genera *Dalbergia* and *Machaerium*. *Phytochemistry* **10**, 1863–1876

Bragg, L. H., Bacon, J. D., McMillan, C. and Mabry, T. J. (1978). Flavonoid patterns in the *Prosopis juliflora* complex. *Biochem. Syst. Ecol.* **6**, 113–116.

Breger, I. A. (ed.) (1963). 'Organic Geochemistry.' Macmillan, New York.

Brehm, B. G. (1966). Taxonomic implications of variation in chromatographic pattern components. *Brittonia* **18**, 194–202.

Brehm, B. G. (1969). Molecular data in plant systematics. *Natl. Acad. Sci. Publ.* 1692, 312–318.

Brehm, B. G. and Alston, R. E. (1964). A chemotaxonomic study of *Baptisia leucophaea* var. *laevicaulis* (Leguminosae). *Amer. J. Bot.* **51**, 644–650.

Brehm, B. G. and Ownbey, M. (1965). Variation in chromatographic patterns in the *Tragopogon dubius–pratensis–porrifolius* complex (Compositae). *Amer. J. Bot.* **52**, 811–818.

Brewer, G. J. (1970). 'An Introduction to Isozyme Techniques.' Academic Press, New York.

Britten, R. J. and Davidson, E. H. (1971). Repetitive and non-repetitive DNA sequences and a speculation on the origins of evolutionary novelty. *Quart. Rev. Biol.* **46**, 111–138.

Britten, R. J., Graham, D. E. and Neufield, B. R. (1974). Analysis of repeating DNA sequences by reassociation. *Methods in Enzymology* **29**, 363–418.

Britton, D. M. and Widen, C.-J. (1974). Chemotaxonomic studies on *Dryopteris* from Quebec and eastern North America. *Canad. J. Bot.* **52**, 627–638.

Brody, L. and Mendlinger, S. (1980). Species relationships and genetic variation in the diploid wheats as revealed by starch gel electrophoresis. *Pl. Syst. Evol.* **136**, 247–258.

Brown, B. V. and Smith, B. N. (1975). The genus *Dichanthelium*, Gramineae. *Bull. Torrey Bot. Club* **102**, 10–13.

Brown, R. (1833). On the organs and mode of fecundation in Orchideae and Asclepedieae. *Trans. Linn. Soc.* **16**, 685–738.

Bruun, H. G. (1932). Cytological studies in *Primula. Symbolae Botan. Upsalienses* **1**, 1–239.

Bui, A. M., Debray, M. M., Boiteau, P. and Potier, M. (1977). Etude chimiotaxonomique de quelques espèces de *Hazunta. Phytochemistry* **16**, 703–706.

Burnett, W. C., Jones, S. B. and Mabry, T. J. (1978). The role of sesquiterpene lactones in plant-animal coevolution. *In* 'Biochemical Aspects of Plant and Animal Coevolution' (Harborne, J. B., ed.), pp. 233–257. Academic Press, London.

Burtt, B. L. (1962). Studies on the Gesneriaceae of the Old World. XXIV. Tentative keys to the tribes and genera. *Notes Roy. Botan. Garden Edinb.* **24**, 205–220.

Buttery, B. R. and Buzzell, R. I. (1971). Properties of inheritance of urease isoenzymes in soybean seeds. *Canad. J. Bot.* **49**, 1101–1105.

Buzzati-Traverso, A. A. (1953). Paper chromatographic patterns of genetically different tissues: a contribution to the biochemical study of individuality. *Proc. Nat. Acad. Sci.* U.S.A. **39** 376–391.

Byng, G., Whitaker, R., Flick, C. and Jensen, R. A. (1981). Enzymology of L-tyrosine biosynthesis in *Zea mays. Phytochemistry* **20**, 1289–1292.

Cagnin, M. A. H., Gomes, C. M. R., Gottlieb, O. R., Marx, M. C., Rocha, A. I., Silva, G. F. and Temperini, J. A. (1977). Biochemical systematics: methods and principles. *Plant Syst. Evol., Suppl.* **1**, 53–76.

Camm, E. L., Towers, G. H. N. and Mitchell, J. C. (1975). UV-mediated antibiotic activity of some Compositae species. *Phytochemistry* **14**, 2007–2011.

Camp, W. H. (1949). Cinchona at high altitudes in Ecuador. *Brittonia* **6**, 394–430.

Camp, W. H. and Gilly, C. L. (1943). The structure and origin of species. *Brittonia* **4**, 323–335.

Carpenter, I., Locksley, H. D. and Scheinmann, F. (1969). Xanthones in higher plants: biogenetic proposals and a chemotaxonomic survey. *Phytochemistry* **8**, 2013–2026.

Carter, L. C. and Brehm, B. G. (1969). Chemical and morphological analysis of introgressive hybridization between *Iris tenax* and *I. chrysophylla*. *Brittonia* **21**, 44–54.

Challice, J. S. (1972). Phenolics of *Pyrus* interspecific hybrids. *Phytochemistry* **1**, 3015–3018.

Challice, J. S. (1973). Phenolic compounds of the subfamily Pomoideae: a chemotaxonomic survey. *Phytochemistry* **12**, 1095–1101.

Challice, J. S. (1974). Rosaceae chemotaxonomy and the origins of the Pomoideae. *Bot. J. Linn. Soc.* **69**, 239–259.

Challice, J. S. and Westwood, M. N. (1973). Numerical taxonomic studies of the genus *Pyrus* using both chemical and botanical characters. *Bot. J. Linn. Soc.* **67**, 121–148.

Challice, J. S. and Williams, A. H. (1970). Phenolic compounds of the genus *Pyrus*: a chemotaxonomic study of further Oregon specimens. *Phytochemistry* **9**, 1271–1276.

Chan, P. H., Sakane, K., Singh, S. and Wildman, S. G. (1972). Crystalline fraction I protein: preparation in large yield. *Science* **176**, 1145–1146.

Chandler, R. F. and Hooper, S. N. (1979). Friedelin and associated triterpenoids. *Phytochemistry* **18**, 711–724.

Chaney, R. W. (1938). Palaoecological interpretations in Cenozoic plants in western North America. *Bot. Rev.* **4**, 371–396.

Chang, C. P. and Mabry, T. J. (1973). The constitution of the Order Centrospermae: rRNA–DNA hybridization studies among betalain- and anthocyanin-producing families. *Biochem. Syst.* **1**, 185–190.

Charig, A. (1981). Cladistics: a different point of view. *Biologist* **28**, 19–20.

Charlwood, B. V. and Bell, E. A. (1980). Automatic amino acid analysis data handling. *In* 'Chemosystematics, Principles and Practice' (Bisby, F. A., Vaughan, J. G. and Wright, C. A., eds.), pp. 91–102. Academic Press, London.

Chen, K., Gray, J. C. and Wildman, S. G. (1975). Fraction I protein and the origin of polyploid wheats. *Science* **190**, 1304–1306.

Chen, K. and Wildman, S. G. (1981). Differentiation of fraction I protein in relation to age and distribution of angiosperm groups. *Pl. Syst. Evol.* **138**, 89–113.

Chen, J. and Meeuse, B. J. D. (1971). Production of free indole by some Aroids. *Acta Botan. Neerl.* **20**, 627–635.

Chen, S.-L., Towill, L. R. and Loewenberg, J. R. (1970). Isoenzyme patterns in developing *Xanthium* leaves. *Physiol. Plant.* **23**, 434–443.

Cherry, J. P. and Katterman, F. R. H. (1971). Non-specific esterase isozyme polymorphism in natural populations of *Gossypium thurberi*. *Phytochemistry* **10**, 141–145.

Cherry, J. P., Katterman, F. R. H. and Endrizzi, J. E. (1971). A comparative study of seed proteins of allopolyploids of *Gossypium* by gel electrophoresis. *Canad. J. Genet. Cytol.* **13**, 155–158.

Chrispeels, M. J. and Baumgartner, B. (1978). Serological evidence confirming the assignment of *Phaseolus aureus* and *P. mungo* to the genus *Vigna*. *Phytochemistry* **17**, 125–126.

Clarke, A. E. and Stone, B. A. (1963). Chemistry and biochemistry of β-1,3-glucans. *Rev. Pure Appl. Chem.* **13**, 134–156.

Clarkson, R. B. and Fairbrother, D. E. (1970). A serological and electrophoretic investigation of eastern North American *Abies* (Pinaceae). *Taxon* **19**, 720–727.

Clausen, J. (1951). 'Stages in the Evolution of Plant Species'. Cornell University Press, Ithaca.

Clauss, von E. (1970). Die frien Aminosauren im Endosperm der Fruhen Samonentwicklung bei verschieden valenzstufen. *Biol. Zbl.* **89**, 577–590.

Clegg, M. T. and Allard, R. W. (1972) Patterns of genetic differentiation in the slender wild oat species *Avena barbata*. *Proc. Natl. Acad. Sci. U.S.A.* **69**, 1820–1824.

Clifford, H. T. and Harborne, J. B. (1969). Flavonoid pigmentation in the sedges, Cyperaceae. *Phytochemistry* **8**, 123–126.

Cobb, F. W. Jr., Zavarin, E. and Bergot, J. (1972). Effect of air pollution on the volatile oil from leaves of *Pinus ponderosa*. *Phytochemistry* **11**, 1815–1818.

Comas, C. I., Hunziker, J. H. and Crisci, J. V. (1979). Species relationships in *Bulnesia* as shown by seed protein electrophoresis. *Biochem. Syst. Ecol.* **7**, 303–308.

Conn, E. E. (1978). Cyanogenesis, the production of hydrogen cyanide by plants. *In* 'Effects of Poisonous Plants on Livestock' (Keeler, R. F., Van Kempen, K. R. and James, L. F., eds.), pp. 301–310. Academic Press, New York.

Conner, H. E. and Purdie, A. W. (1976). Inheritance of triterpene methyl ethers in *Cortaderia* (Graminae). *Phytochemistry* **15**, 1937–1939.

Connolly, J. D., Overton, K. H. and Polonsky, J. (1970). The chemistry and biochemistry of the limonoids and quassinoids. *Prog. Phytochem.* **2**, 385–456.

Cooper-Driver, G. A. and Swain, T. (1976). Cyanogenic polymorphism in bracken in relation to herbivore predation. *Nature* (*Lond.*) **260**, 604.

Constance, L. (1955). The systematics of the angiosperms. *In* 'A Century of Progress in the Natural Sciences, 1853–1953,' pp. 405–503. Californian Academy of Science, San Francisco.

Coradin, L. and Ginnasi, D. E. (1980). The effects of chemical preservations on plant collections to be used in chemotaxonomic surveys. *Taxon* **29**, 33–40.

Cordell, G. A. (1981). 'Introduction to Alkaloids, a Biogenetic Approach.' John Wiley and Sons, New York.

Corner, E. J. H. (1981). Angiosperm classification and phylogeny: a criticism. *Bot. J. Linn. Soc.* **82**, 81–87.

Corrigan, D., Timoney, R. F. and Donnelly, D. M. X. (1978). n-Alkanes and ω-hydroxyalkanoic acids from the needles of twenty-eight *Picea* species. *Phytochemistry* **17**, 907–910.

Courtois, J. E. and Percheron, F. (1971). Distribution des Monosaccharides, Oligosaccharides et Polyols. *In* 'Chemotaxonomy of the Leguminosae' (Harborne, J. B., Boulter, D. and Turner, B. L., eds.), pp. 207–230. Academic Press, London.

Craig, I. L., Murray, B. E. and Rajhathy, T. (1972). Leaf esterase isozymes in *Avena* and their relationship to the genomes. *Canad. J. Genet. Cytol.* **14**, 581–589.

Craigie, J. S. (1974). Storage Products. *In* 'Algal Physiology and Biochemistry' (Stewart, W. D. P., ed.), pp. 206–235. Blackwell Scientific, Oxford.

Crawford, D. J. (1970). Systematic studies on Mexican *Coreopsis* (Compositae). *Coreopsis mutica*: flavonoid chemistry, chromosome numbers, morphology and hybridization. *Brittonia* **22**, 93–111.

Crawford, D. J. (1972). The morphology and flavonoid chemistry of synthetic infra-specific hybrids in *Coreopsis mutica* (Compositae) *Taxon* **21**, 27–38.

Crawford, D. J. (1974). A morphological and chemical study of *Populus acuminata*. *Brittonia* **26**, 74–89.

Crawford, D. J. (1978). Okanin 4′-diglucoside from *Coreopsis petrophiloides* and comments on anthochlors and evolution in *Coreopsis*. *Phytochemistry* **17**, 1680–1681.

Crawford, D. J. (1979). Flavonoid chemistry and angiosperm evolution. *Bot. Rev.* **44**, 431–456.

Crawford, D. J. and Julian, E. A. (1976). Seed protein profiles in the narrow-leaved species of *Chenopodium* of the Western USA: taxonomic value and comparison with flavonoids. *Amer. J. Bot.* **63**, 302–308.

Crawford, D. J. and Levy, M. (1978). Flavonoid profile affinities and genetic similarity. *System. Bot.* **3**, 369–373.

Crawford, D. J. and Mabry, T. J. (1978). Flavonoid chemistry of *Chenopodium fremontii*. Infra-specific variation and systematic implications at the interspecific level. *Biochem. Syst. Ecol.* **6**, 189–192.

Crawford, D. J. and Smith, E. B. (1980). Flavonoid chemistry of *Coreopsis grandiflora* (Compositae). *Brittonia* **32**, 154–159.

Crawford, D. J. and Stuessy, T. F. (1981). The taxonomic significance of anthochlors in the subtribe Coreopsidinae (Compositae, Heliantheae). *Amer. J. Bot.* **68**, 107–117.

Crawford, D. J., Smith, E. B. and Mueller, A. M. (1980). Leaf flavonoid chemistry of *Coreopsis* (Compositae) Section Palmatae. *Brittonia* **32**, 452–463.

Crews, P., Ng, P., Kho-Wiseman, E. and Pace, C. (1976). Halogenated monoterpenes of the red alga *Microcladia*. *Phytochemistry* **15**, 1707–1711.

Cristofolini, G. (1980). Interpretation and analysis of serological data. *In* 'Chemosystematics, Principles and Practice' (Bisby, F. A., Vaughan, J. G. and Wright C. A., eds.), pp. 269–288. Academic Press, London.

Cristofolini, G. (1981). Serological systematics of the Leguminosae. *In* 'Advances in Legume Systematics (Polhill, R. M. and Raven, P. H., eds.), pp. 513–532. Royal Botanic Gardens, Kew.

Critchfield, W. B. (1967). Crossability and relationships of the closed-cone pines. *Silvae Genetica* **16**, 89–97.

Croft, L. R. (1980). 'Introduction to Protein Sequence Analysis.' John Wiley and Sons, Chichester.

Cromwell, B. T. (1955). Plant alkaloids. *In* 'Modern Methods of Plant Analysis' (Paech, K. and Tracey, M. V., eds.), Vol. IV, pp. 367–516. Springer Verlag, Berlin.

Cromwell, B. T. and Richardson, M. (1966). Trimethylamine in *Chenopodium vulvaria*. *Phytochemistry* **5**, 735–746.

Cronquist, A. (1955). Phylogeny and taxonomy of the Compositae. *Amer. Midland Nat.* **53**, 478–511.

Cronquist, A. (1957). Outline of a new system of families and orders of dicotyledons. *Bull. Jard. Bot. Bruxelles* **27**, 13–40.

Cronquist, A. (1968). 'The Evolution and Classification of Flowering Plants.' 396 pp. T. Nelson and Sons, London.

Cronquist, A. (1976). The taxonomic significance of the structure of plant proteins: a classical taxonomist's view. *Brittonia* **28**, 1–27.

Cronquist, A. (1977a). The Compositae revisited. *Brittonia* **29**, 137–153.

Cronquist, A. (1977b). On the taxonomic significance of secondary metabolites in angiosperms. *Plant. Syst. Evol., Suppl.* **1**, 179–189.

Cronquist, A. (1979). Once again, what is a species? *Agricultural Res.* **2**, 3–20.

Cronquist, A. (1981). 'An Integrated System of Classification of Flowering Plants.' Colombia University Press, New York.

Crow, W. B. (1926). Phylogeny and the natural system. *Genetics* **17**, 6–155.

Crowden, R. K., Harborne, J. B. and Heywood, V. H. (1969). Chemosystematics of the Umbelliferae—a general review. *Phytochemistry* **8**, 1963–1984.

Cruickshank, I. A. M. and Perrin, D. R. (1960). Isolation of a phytoalexin from *Pisum sativum*. *Nature* (*Lond.*) **187**, 799–800.

Culberson, C. F. (1969). 'Chemical and Botanical Guide to Lichen Products.' University of North Carolina Press, Chapel Hill.

Culberson, C. F. and Culberson, W. L. (1976). Chemosyndromic variation in lichens. *Syst. Bot.* **1**, 325–339.

Culberson, C. F. and Hale, M. E., Jr. (1973). Chemical and morphological evolution in *Parmelia* sect. *hypotrachyna*: product of ancient hybridization. *Brittonia* **25**, 162–173.

Culberson, C. F., Culberson, W. L. and Esslinger, T. L. (1977). Chemosyndromic variation in the *Parmelia pulla* group. *The Bryologist* **80**, 125–135.

Culberson, W. L. (1969). The behaviour of the species of the *Ramalina siliquosa* group in Portugal. *Osterr. Bot. Z.* **116**, 85–94.

Culberson, W. L. and Culberson, C. F. (1967). Habitat selection by chemically different races of lichens. *Science* **158**, 1195–1197.

Culberson, W. L., Culberson, C. F. and Johnson, A. (1977). Correlations between secondary product chemistry and ecogeography in the *Ramalina siliquosa* group (lichens). *Plant Syst. Evol.* **127**, 191–200.

Culvenor, C. C. J. (1978). Pyrrolizidine alkaloids—occurrence and systematic importance in angiosperms. *Bot. Notiser* **131**, 473–486.

Curtis, P. J. and Meade, P. M. (1971). Cucurbitacins from the Cruciferae. *Phytochemistry* **10**, 3081–3083.

Cutler, D. F. (1969). Restionaceae. *In* Anatomy of the Monocotyledons, Vol. IV Juncales (Metcalfe, C. R., ed.), pp. 116–331. Clarendon Press, Oxford.

Dahlgren, R. (1975). A system of classification of the angiosperms to be used to demonstrate the distribution of characters. *Bot. Notiser* **128**, 119–147.

Dahlgren, R. (1980). A revised system of classification of the angiosperms. *Bot. J. Linn. Soc.* **80**, 91–124.

Dahlgren, R., Jensen, S. R. and Nielsen, B. J. (1976). Iridoids in Fouquieriaceae and notes on its possible affinities. *Bot. Notiser* **129**, 207–212.

Darlington, C. D. (1956). 'Chromosome Botany.' George Allen and Unwin, London.

Dass, H. C. (1972). Phylogenetic affinities in Triticinae studied by thin-layer chromatography. *Canad. J. Genet. Cytol.* **14**, 703–712.

Dass, H. C. and Nybom, N. (1967). The relationships between *Brassica nigra*, *B. campestris*, *B. oleracea*, and their aphiploid hybrids studied by means of numerical chemotaxonomy. *Canad. J. Genet. Cytol.* **9**, 880–890.

Dass, H. C., Randhawa, G. S. and Kaur, M. (1975). Variation in flavonoid patterns and species relationships in the genus *Annona*. *Ind. J. Hort.* **32**, 111–116.

Daubenmire, R. (1972). On the relation between *Picea pungens* and *P. engelmannii* in the Rocky Mountains. *Canad. J. Bot.* **50**, 733–742.

Davies, B. H. (1976). Carotenoids. *In* 'Chemistry and biochemistry of plant Pigments' (Goodwin, T. W. ed.), 2nd edn., Vol. 2, pp. 38–165. Academic Press, London.

Davis, P. H. and Heywood, V. H. (1963). 'Principles of Angiosperm Taxonomy.' D. van Nostrand Co., Princeton.

Daussant, J. (1975). Immunochemical investigations of plant proteins. *In* 'The Chemistry and Biochemistry of Plant Proteins' (Harborne, J. B. and Van Sumere, C. F., eds.), pp. 31–69. Academic Press, London.

Dayhoff, M. O. (1972). 'Atlas of Protein Sequence and Structure.' Vol. 5. National Biomedical Research Foundation, Silver Spring, Maryland.

Dayhoff, M. O. and Eck, R. V. (1966). 'Atlas of Protein Sequence and Structure,' Volume 2. National Biomedical Research Foundation, Maryland.

Dean, P. D. G., Exley, D. and Goodwin, T. W. (1971). Re-estimation of oestrone in pomegranate seeds. *Phytochemistry* **10**, 2215–2216.

De Candolle, A. P. (1804). "Essai sur les Proprieties Medicales des Plantes . . . etc." Ed. 1.

Degens, E. T. (1965). 'Geochemistry of Sediments,' 342 pp. Prentice-Hall, New York.

De Jong, D. W. (1972). Detergent extraction of enzymes from tobacco leaves varying in maturity. *Plant Physiol.* **50**, 733–737.

De Jong, D. W. (1973). Effect of temperature and daylength on peroxidase and malate (NAD) dehydrogenase isozyme composition in tobacco leaf extracts. *Amer. J. Bot.* **60**, 846–852.

Dement, W. A. and Mabry, T. J. (1972). Flavonoids of North American species of *Thermopsis*. *Phytochemistry* **11**, 1089–1093.

Dement, W. A. and Mabry, T. J. (1975). Biological implications of flavonoid chemistry in *Baptisia* and *Thermopsis*. *Biochem. Syst. Ecol.* **3**, 91–94.

Dement, W. A. and Raven, P. H. (1973). Distribution of the chalcone isosalipurposide in the Onagraceae. *Phytochemistry* **12**, 807–808.

Derbyshire, E., Wright, D. J. and Boulter, D. (1976). Legumin and vicilin, storage proteins of legume seeds. *Phytochemistry* **15**, 3–24.

Desborough, S. and Peloquin, S. J. (1966). Disc electrophoresis of tuber proteins from *Solanum* species and interspecific hybrids. *Phytochemistry* **5**, 727–733.

de Ley, J. (1968). Molecular biology and bacterial phylogeny. *Evolutionary Biology* **2**, 104–156.

de Wet, J. M. J. and Scott, B. D. (1965). Essential oils as taxonomic criteria in *Bothriochloa*. *Bot. Gaz.* **126**, 209–214.

de Wet, J. M. J., Timothy, D. H., Hilu, K. W. and Fletcher, G. B. (1981). Systematics of South American *Tripsacum* (Gramineae). *Amer. J. Bot.* **68**, 269–276.

Dickinson, W. J. (1980). Tissue specificity of enzyme expression regulated by diffusible factors: evidence in *Drosophila* hybrids. *Science* **207**, 995–997.

Dittrich, P., Gietl, M. and Kandler, O. (1971). 1-*O*-Methyl-*muco*-inositol in higher plants. *Phytochemistry* **11**, 245–250.

Dodson, C. H. (1975). Coevolution of orchids and bees. *In* 'Coevolution of Animals and Plants' (Gilbert, L. E. and Raven, P. H., eds.), pp. 91–99. University of Texas Press, Austin and London.

Dodson, C. H. and Hills, H. G. (1966). Gas chromatography of orchid fragrances. *Amer. Orchid Soc. Bull.* **35**, 720–725.

Dodson, C. H., Dressler, R. L., Hills, H. G., Adams, R. M. and Williams, N. H. (1969). Biologically active compounds in orchid fragrances. *Science* **164**, 1243–1249.

Döpp, H. and Musso, H. (1973). Colouring matters from fly agaric. II. Isolation of pigments from *Amanita muscaria* and their chromophores. *Chem. Ber.* **106**, 3473–3482

Doolittle, R. F. (1981). Similar amino acid sequences: chance or common ancestry? *Science* **214**, 149–159.

Dossaji, S. F., Mabry, T. J. and Bell, E. A. (1975). Biflavanoids of the Cycadales. *Biochem. Syst. Ecol.* **2**, 171–175.

Doty, P., Marmur, J., Eigner, J. and Schildkrant, C. (1960). Strand separation and specific recombination in DNAs: physical chemical studies. *Proc. Natl. Acad. Sci.* **46**, 461–476.

Douglas, A. G. and Eglinton, G. (1967). The distribution of alkanes. *In* 'Comparative Phytochemistry' (Swain, T., ed.), pp. 57–78. Academic Press, London.

Dover, G. A. (1980). Problems in the use of DNA for the study of species relationships and the evolutionary significance of genomic differences. *In* 'Chemosystematics, Principles and Practice' (Bisby, F. A., Vaughan, J. G. and Wright, C. A., eds.), pp. 241–268. Academic Press, London.

Dowsett, A. P. and Young, M. W. (1982). Differing levels of dispersed repetitive DNA among closely related species of *Drosophila. Proc. Natl. Acad. Sci. U.S.A.* **79**, 4570–4574.

Dreyer, D. L. (1966). Citrus bitter principles V. Botanical distribution and chemotaxonomy in the Rutaceae. *Phytochemistry* **5**, 367–378.

Dreyer, D. L. and Trousdale, E. K. (1978). Cucurbitacins in *Purshia tridentata. Phytochemistry* **17**, 325–326.

Dreyer, D. L., Pickering, M. V. and Cohan, P. (1972). Distribution of limonoids in the Rutaceae. *Phytochemistry* **11**, 705–713.

Drozdz, B. and Bloszyk, E. (1978). Sesquiterpene lactones of Compositae. 21. Selective detection by thin-layer chromatography. *Planta Medica* **33**, 379–384.

Drude, O. (1897–8). Umbelliferae. *In* 'Die Naturlichen Pflanzenfamilien' (Engler, A. and Prantl, K., eds.), Vol. 3, part 8, pp. 63–75. W. Engelmann, Leipzig.

Dubois, J. A. and Harborne, J. B. (1975). Anthocyanin inheritance in petals of flax, *Linum usitatissimum. Phytochemistry* **14**, 2491–2494.

Dunn, C. and Everitt, B. S. (1982). 'An Introduction to Mathematical Taxonomy.' Cambridge Studies Math. Biol. 5. Cambridge University Press, Cambridge.

Dunnhill, P. M. and Fowden, L. (1965). The amino acids of seeds of the Cucurbitaceae. *Phytochemistry* **4**, 933–944.

Earle, F. R., Glass, G. A., Gessinger, G. C., Wolff, I. A., Bagby, M. O. and Jones, Q. (1960). Search for new industrial oils. *J. Amer. Oil Chem. Soc.* **37**, 440–447.

Eckenwalder, J. E. (1982). *Populus* × *inopina* hybr. nov. (Salicaceae), a natural hybrid between *P. fremontii* Wats. and *P. nigra* L. *Madroño* **29**, 67–78.

Edmonds, J. M. and Glidewell, S. M. (1977). Acrylamide gel electrophoresis of seed proteins from some *Solanum* species. *Pl. Syst. Evol.* **127**, 277–291.

Efron, Y., Peleq, M. and Ashri, A. (1973). Alcohol dehydrogenase allozymes in the safflower genus *Carthamus* L. *Biochem. Genet.* **9**, 299–308.

Eglinton, G. (1973). Chemical fossils: a combined organic geochemical and environmental approach. *J. Pure Appl. Chem.* **34**, 611–632.

Eglinton, G., Gonzales, A. G., Hamilton, R. J. and Raphael, R. A. (1962). Hydrocarbon constituents of the wax coatings of plant leaves; a taxonomic survey. *Phytochemistry* **1**, 89–102.

Eglinton, G. and Hamilton, R. J. (1963). The distribution of alkanes. *In* 'Chemical Plant Taxonomy' (Swain, T., ed.), pp. 187–218. Academic Press, London.

Eglinton, G. and Murphy, M. T. J. (1969). 'Organic Geochemistry'. Springer-Verlag, New York.

Ehrendorfer, F. (1968). Geographical and ecological aspects of infraspecific differentiation. *In* 'Modern Methods in Plant Taxonomy' (Heywood, V. H., ed.), pp. 261–296. Academic Press, London.

Ehrlich, P. R. and Raven, P. H. (1965). Butterflies and plants: a study in coevolution. *Evolution* **18**, 586–608.

Eisner, T. (1964). Catnip: its raison d'etre. *Science* **146**, 1318–1320.

Ellis, B. E. (1974). Degradation of aromatic compounds in plants. *Lloydia* **37**, 168–184.

Ellis, J. R. S. (1979). The application of starch gel electrophoresis of gliadin proteins to the identification of wheat varieties. *In* 'Recent Advances in the Biochemistry of Cereals' (Laidman, D. L., and Wyn Jones, R. G., eds.), pp. 349–354. Academic Press, London.

Ellis, J. R. S. (1982). Promiscuous DNA-chloroplast genes inside plant mitochondria. *Nature* (*Lond.*) **299**, 678–679.

Ellis, J. R. S. and Beminster, C. H. (1977). Identification of British wheat varieties by means of starch gel electrophoresis. *J. Nat. Inst. Agr. Bot.* **14**, 221–231.

Ellison, W. (1964). A systematic study of the genus *Bahia* (Compositae). *Rhodora* **66**, 67–68.

Ellison, W. L., Alston, R. E. and Turner, B. L. (1962). Methods of presentation of crude biochemical data for systematic purposes, with particular reference to the genus *Bahia*. *Amer. J. Bot.* **49**, 599–604.

Emboden, W. A. and Lewis, H. (1967). Terpenes as taxonomic characters in *Salvia* section *Audibertia*. *Brittonia* **19**, 152–160.

Endler, J. A. (1977). Geographic Variation, Speciation and Clines. *Monographs Pop. Biol.* **10**, 1–246.

Endlicher, S. L. (1836). Genera Plantarum Secundum Ordines Naturales Disposita. Vindobonae.

Engler, A. and Diels, L. (eds.) (1936). 'Syllabus der Pflanzenfamilien', 11th edn. Boerntrager, Berlin.

Erdtman, H. (1963). Some aspects of chemotaxonomy. *In* 'Chemical Plant Taxonomy' (Swain, T., ed.), pp. 89–125. Academic Press, London.

Erdtman, H. (1973). Molecular taxonomy. *In* 'Phytochemistry III' (Miller, L. P., ed.), Van Nostrand Reinhold, New York.

Erdtman, H. and Norin, T. (1966). The chemistry of the order Cupressales. *Fortschr. Chem. Org. Naturst.* **24**, 206–287.

Erdtman, H., Kimland, B. and Norin, T. (1966). Wood constituents of *Ducampopinus krempfii* (*Pinus krempfii*). *Phytochemistry* **5**, 927–931.

Ergle, D., Katterman, F. R. and Richmond, W. (1964). Aspects of nucleic acid composition in *Gossypium*. *Plant Physiol.* **39**, 145–150.

Erickson, J. M. and Feeney, P. (1974). Sinigrin: a chemical barrier to larvae of the black swallowtail butterfly, *Papilio polyxenes*. *Ecology* **55**, 103–111.

Ettlinger, M. G. and Kjaer, A. (1968). Sulphur compounds in plants. *Recent Adv. Phytochem.* **1**, 59–144.

Eugster, C. H. (1980). Terpenoid, especially diterpenoid pigments. *In* 'Pigments in Plants' (Czygan, F. C., ed.), pp. 149–186. Gustav Fischer, Stuttgart.

Evans, C. S. and Bell, E. A. (1978). Uncommon amino acids in the seeds of 64 species of Caesalpinieae. *Phytochemistry* **17**, 1127–1129.

Evans, C. S., Qureshi, M. Y. and Bell, E. A. (1977). Free amino acids in the seeds of *Acacia* species. *Phytochemistry* **16**, 565–570.

Evans, D., Knights, B. A., Math, V. B. and Ritchie, A. L. (1957a). α-Diketones in *Rhododendron* waxes. *Phytochemistry* **14**, 2447–2451.

Evans, D., Knights, B. A. and Math, V. B. (1975b). Steryl acetates in *Rhododendron* waxes. *Phytochemistry* **14**, 2453–2454.

Fahselt, D. (1971). Flavonoid components of *Dicentra canadensis* (Fumariaceae). *Canad. J. Bot.* **49**, 1559–1563.

Fahselt, D. (1972). The use of flavonoid components in the characterisation of the genus *Corydalis* (Fumariaceae). *Canad. J. Bot.* **50**, 1605–1610.

Fahselt, D. and Ownbey, M. (1968). Chromatographic comparison of *Dicentra* species and hybrids. *Amer. J. Bot.* **55**, 334–345.

Fairbairn, J. W. and El-Muhtadi, F. J. (1972). Chemotaxonomy of anthraquinones in *Rumex*. *Phytochemistry* **11**, 263–268.

Fairbairn, J. W. and Wassel, G. (1964). The alkaloids of *Papaver somniferum* L. I. Evidence for a rapid turnover of the major alkaloids. *Phytochemistry* **3**, 253–258.

Fairbrother, D. E., Mabry, T. J., Scogin, R. L. and Turner, B. L. (1975). The bases of angiosperm phylogeny: Chemotaxonomy. *Ann. Missouri Bot. Gard.* **62**, 765–800.

Farnsworth, N. R. and Bingel, A. S. (1977). Problems and prospects of discovering new drugs from higher plants by pharmacological screening. *In* 'New Natural Products and Plant Drugs with Pharmacological, Biological or Therapeutical Activity' (Wagner, H. and Wolff, P., eds.), pp. 1–22. Springer-Verlag, Berlin.

Fassett, N. C. (1944). Hybrid swarms of *Juniperus virginiana* and *J. scopulorum*. *Bull. Torrey Bot. Club* **71**, 475–483.

Feret, P. P. (1972). Peroxidase isoenzyme variation in interspecific elm hybrids. *Canad. J. Forest Res.* **2**, 264–270.

Ferguson, A. (1980). 'Biochemical Systematics and Evolution.' Blackie, Glasgow.

Fiasson, J. L., Lebreton, P. and Arpin, N. (1968). Les carotenoides des champignons. *Bull. Soc. Naturalistes et Archeologues de l'Ain* **82**, 47–67.

Fieldes, M. A. and Tyson, H. (1972). Activity and relative mobility of peroxidase isoenzymes in genotrophs and genotypes of flax (*Linum usitatissimum* L.) *Canad. J. Genet. Cytol.* **14**, 625–636.

Fikenscher, L. H. and Hegnauer, R. (1977). Uber die cyanogenen Verbindungen bei einigen Compositae, bei den Oliniaceae und in der Rutaceen-Gattung *Zieria*. *Pharm. Weekblad* **112**, 11–20.

Fischer, M. and Kandler, O. (1975). Identifizierung von selaginose und deren verbreitung in der gattung *Selaginella*. *Phytochemistry* **14**, 2629–2633.

Fischer, N. H., Oliver, E. J. and Fischer, H. D. (1979). The biogenesis and chemistry of sesquiterpene lactones. *Fortsch. Chem. Natur.* **38**, 1–390.

Fitch, W. M. (1976). Molecular evolutionary clocks. *In* 'Molecular Evolution' (Ayala, J., ed.), pp. 160–178. Sinauer Associates Inc., Sunderland, Mass.

Fitch, W. M. and Margoliash, E. (1967). Construction of phylogenetic trees. *Science* **155**, 279–284.

Fitch, W. M. and Margoliash, E. (1968). The construction of phylogenetic trees. II. How well do they reflect past history? *Brookhaven Symposia Biol.* **21**, 217–242.

Fitch, W. M. and Yasunobu, K. T. (1975). Phylogenies from amino acid sequences aligned with gaps: the problem of gap weighting. *J. Mol. Evol.* **5**, 1–24.

Flake, R. H. and Turner, B. L. (1968). Numerical classification for taxonomic problems. *J. Theoret. Biol.* **20**, 260–270.

Flake, R. H. and Turner, B. L. (1973). Volatile constituents, especially terpenes, and their utility and potential as taxonomic characters in populational studies. *Nobel Symposium* **25**, 123–128.

Flake, R. H., Rudloff, E. von and Turner, B. L. (1969). Quantitative study of clinal variation in *Juniperus virginiana* using terpenoid data. *Proc. Natl. Acad. Sci. U.S.A.* **64**, 487–494.

Flake, R. H., Rudloff, E. von and Turner, B. L. (1973). Configuration of a clinal pattern of chemical differentiation in *Juniperus virginiana* from terpenoid data obtained in successive years. *In* 'Terpenoids: Structure, Biogenesis, and Distribution' (Runeckles, V. C. and Mabry, T. J. eds.), pp. 215–228. Academic Press, New York.

Flake, R. H., Urbatsch, L. and Turner, B. L. (1978). Chemical documentation of allopatric introgression in *Juniperus*. *Syst. Bot.* **3**, 129–144.

Flashman, S. M. and Levings, C. S. (1981). Enzymatic cleavage of DNA: biological role and application to sequence analysis. *In* 'Proteins and Nucleic Acids' (Marcus, A., ed.), pp. 83–109. Academic Press, New York.

Flavell, R. B., O'Dell, M. and Smith, D. B. (1979). Repeated sequence DNA comparisons between *Triticum* and *Aegilops* species. *Heredity* **42**, 309–322.

Flavell, R. B., Rimpau, J. and Smith, D. B. (1977). Repeated sequence DNA relationship in four cereal genomes. *Chromosoma* **63**, 205–222.

Florkin, M. and Stotz, E. H. (eds.) (1974). Comparative Biochemistry, Molecular Evolution. Vol. 29. Comprehensive Biochemistry. Elsevier, Amsterdam.

Forde, M. B. (1964). Inheritance of turpentine composition in *Pinus attenuata* × *radiata* hybrids. *N.Z. J. Bot.* **2**, 53–59.

Forde, M. B. and Blight, M (1964). Geographical variation in the turpentine of Bishop pine. *N.Z. J. Bot.* **2**, 44–52.

Forsyth, W. J. C. and Simmonds, N. W. (1957). Anthocyanidins of *Lochnera rosea*. *Nature* (*Lond.*) **180**, 247.

Fowden, L. (1965). The chemical approach to plants. *Sci. Prog.* **53**, 583–599.

Fowden, L. (1970). The non-protein amino acids of plants. *Progr. Phytochem.* **2**, 203–266.

Fowden, L. (1972). Amino acid complement of plants. *Phytochemistry* **11**, 2271–2276.

Fowden, L. (1973a). Phytochemistry: retrospect and prospect. *In* 'Phytochemistry, Vol. III' (Miller, C. P., ed.), pp. 400–406. Van Nostrand-Reinhold, New York.

Fowden, L. (1973b). Amino acids. *In* 'Phytochemistry, Vol. II Organic Metabolites' (Miller, L. P., ed.), pp. 1–29. Van Nostrand-Reinhold, New York.

Fowden, L. and Pratt, H. (1973). Cyclopropylamino acids of the genus *Acer*: Distribution and biosynthesis. *Phytochemistry* **12**, 1677–1681.

Fowden, L. and Steward, F. C. (1957). Nitrogenous compounds and nitrogen metabolism in the Liliaceae. *Ann. Bot.* **21**, 53–67.

Fox, G. E. (plus 17 other authors) (1980). The phylogeny of prokaryotes. *Science* **209**, 457–463.

Franklin, E. C. and Snyder, E. B. (1971). Variation and inheritance of monoterpene composition in longleaf pine. *Forest Sci.* **17**, 178–179.

Fröst, S. and Bose, S. (1966). An investigation of the phenolic compounds in two species of jute (*Corchorous olitorius* and *C. capsularis*) and their supposed hybrid, using the thin layer chromatographic technique. *Hereditas* **55**, 183–187.

Fröst, S. and Holm, G. (1971). Thin-layer chromatographic studies of phenolic compounds in twenty varieties of barley. *Hereditas* **69**, 25–34.

Funk, V. A. and Brooks, D. R. (1981). 'Advances in Cladistics.' New York Botanical Garden, Bronx, New York.

Gaillard, B. D. E. (1965). Comparison of the hemicelluloses from plants belonging to two different plant families. *Phytochemistry* **4**, 631–634.

Gaines, C. G., Byng, G. S., Whitaker, R. J. and Jensen, R. A. (1982). L-Tyrosine regulation and biosynthesis via arogenate dehydrogenase in suspension-cultured cells of *Nicotiana sylvestris*. *Planta* **156**, 233–240.

Gallez, G. P. and Gottlieb, L. D. (1982). Genetic evidence for the hybrid origin of the diploid plant *Stephanomeria diegensis*. *Evolution* **36**, 1158–1167.

Galliard, T. and Mercer, E. I. (eds.) (1975). 'Recent Advances in the Chemistry and Biochemistry of Plant Lipids.' Academic Press, London and New York.

Garber, E. (1974). Enzymes as taxonomic and genetic tools in *Phaseolus* and *Aspergillus*. *Israel J. Med. Sci.* **10**, 268–277.

Gardner, R. C. (1974). Systematics of *Cirsium* (Compositae) in Wyoming. *Madroño* **22**, 239–249.

Gaskin, P., MacMillan, J., Firn, R. D. and Price, J. R. (1971). Parafilm: a convenient source of *n*-alkane standards for the determination of gas chromatographic retention indices. *Phytochemistry* **10**, 1155–1157.

Gauss, D. H. and Sprinzl, M. (1981). Compilation of tRNA sequences. *Nucl. Acids Res.* **9**, rl–r23.

Geissman, T. A. (1941). Anthochlor pigments. The pigment of *Coreopsis douglasii*. *J. Amer. Chem. Soc.* **63**, 656–658.

Geissman, T. A. and Crout, D. H. G. (1969). 'Organic Chemistry of Secondary Plant Metabolism.' Freeman, Cooper and Co., San Francisco.

Gell, P. G., Hawkes, J. G. and Wright, S. T. C. (1960). The application of immunological methods to the taxonomy of species within the genus *Solanum*. *Proc. Roy. Soc. London* **151B**, 364–383.

Gelpi, E., Schneider, H., Mann, J. and Oro, J. (1970). Hydrocarbons of geochemical significance in microscopic algae. *Phytochemistry* **9**, 603–612.

Gerhold, H. D. and Plank, G. H. (1970). Monoterpene variations in vapours from white pines and hybrids. *Phytochemistry* **9**, 1393–1398.

Gershenzon, J. and Mabry, T. J. (1983). Secondary metabolites and the higher classification of angiosperms. *Nord. J. Bot.* **3**, 5–34.

Gershenzon, J., Lincoln, D. E. and Langenheim, J. H. (1978). The effect of moisture stress on monoterpenoid yield and composition in *Satureja douglasii*. *Biochem. Syst. Evol.* **6**, 33–43.

Gershenzon, J., Pfeil, R. M., Lin, Y. L., Mabry, T. J. and Turner, B. L. (1984). Sesquiterpene lactones from two newly-described species, *Vernonia jonesii* and *V. pooleae*. *Phytochemistry* **23**, in press.

Gertz, O. (1938). The distribution of anthochlors in the Compositae. *Kungl. Fysiog. Sallsk. Forh.* **8**, 62–70.

Gewitz, H. S., Lorimer, G. H., Solomonson, L. P. and Vennesland, B. (1974). Presence of HCN in *Chlorella vulgaris* and its possible role in controlling the reduction of nitrate. *Nature* (*Lond.*) **249**, 79–81.

Giannasi, D. E. (1978). Generic relationships in the Ulmaceae based on flavonoid chemistry. *Taxon* **27**, 331–344.

Giannasi, D. E. and Niklas, K. J. (1977). Flavonoid and other chemical constituents of fossil Miocene *Celtis* and *Ulmus* (Succor Creek Flora). *Science* **197**, 765–767.

Giannasi, D. E. and Niklas, K. J. (1981). Comparative paleobiochemistry of some fossil and extant Fagaceae. *Amer. J. Bot.* **68**, 762–770.

Gibbons, G. F., Goad, L. J. and Goodwin, T. W. (1967). The sterols of some marine red algae. *Phytochemistry* **6**, 677–683.

Gibbs, R. D. (1958). Chemical evolution in plants. *J. Linn. Soc.* (*Bot.*) **56**, 49–57.
Gibbs, R. D. (1963). History of chemical taxonomy. *In* 'Chemical Plant Taxonomy' (Swain, T. ed.), pp. 41–88. Academic Press, London and New York.
Gibbs, R. D. (1974). 'Chemotaxonomy of Flowering Plants.' McGill-Queen's University Press, Montreal.
Gibby, M., Widen, C. and Widen, H. K. (1978). Cytogenetic and phytochemical investigations in hybrids of Macaronesian *Dryopteris* (Pteridophyta: Aspidiaceae). *Pl. Syst. Evol.* **130**, 235–252.
Gill, L. S., Lawrence, B. M. and Morton, J. K. (1973). Variation in *Mentha arvensis* L. (Labiatae). I. The North American populations. *Bot. J. Linn. Soc.* **67**, 213–232.
Glasby, J. S. (1975–1977). 'Encyclopedia of the Alkaloids,' Vols. 1–3. Plenum Publishing Corp., New York.
Glennie, C. W., Harborne, J. B., Rowley, G. D. and Marchant, C. (1971). Correlations between flavonoid chemistry and plant geography in the *Senecio radicans* complex. *Phytochemistry* **10**, 2413–2417.
Glynn, A. N. and Reid, J. (1969). Electrophoretic patterns of soluble fungal proteins and their possible uses as taxonomic criteria in the genus *Fusarum. Canad. J. Bot.* **47**, 1823–1831.
Goldblatt, P., Nowicke, J. W., Mabry, T. J. and Behnke, H. D. (1976). Gyrostemonaceae: Status and affinity. *Bot. Notiser* **129**, 201–206.
Gomes, C. M. R. and Gottlieb, O. R. (1980). Alkaloid evolution and angiosperm systematics. *Biochem. Syst. Ecol.* **8**, 81–87.
Gomes, C. M. R., Gottlieb, O. R., Bettolo, G. B. M., Monaches, F. D. and Polhill, R. M. (1981). Systematic significance of flavonoids in *Derris* and *Lonchocarpus. Biochem. Syst. Ecol.* **9**, 129–147.
Gonnet, J. F. (1980). La Systematique infraspecifique de l'*Anthyllis vulneraria* (Leguminosae). Vue a la Lumiere de la Biochemie Flavonique. *Biochem. Syst. Ecol.* **8**, 55–63.
Goodman, M. (Ed.) (1982). 'Macromolecular Sequences in Systematic and Evolutionary Biology.' Plenum Press, New York.
Goodwin, J. F. (1975). Spontaneous hybrids in *Dodecatheon* (Primulaceae). *Madroño* **23**, 81–87.
Goodwin, T. W. (ed.) (1976a). 'Chemistry and Biochemistry of Plant Pigments.' 2 Vols, 2nd edn. Academic Press, London.
Goodwin, T. W. (ed.) (1976b). Distribution of carotenoids. *In* 'Chemistry and Biochemistry of Plant Pigments,' 2nd edn., Vol. 1, pp. 225–261. Academic Press, London.
Goodwin, T. W. (1980). 'The Biochemistry of the Carotenoids.' Volume 1: Plants. Chapman and Hall, London.
Goodwin, T. W. and Goad, L. J. (1970). Carotenoids and triterpenoids. *In* 'The Biochemistry of Fruits and Their Products' (Hulme, A. C., ed.), Vol. 1, pp. 305–368. Academic Press, London.
Goodwin, T. W. and Mercer, E. I. (1982). 'Introduction to Plant Biochemistry,' 2nd edn. Pergamon Press, Oxford.
Gorham, J. (1977). Lunularic acid and related compounds in liverworts, algae and *Hydrangea. Phytochemistry* **16**, 249–253.
Gornall, R. J., Bohm, B. A. and Dahlgren, R. (1979). The distribution of flavonoids in the angiosperms. *Bot. Notiser* **132**, 1–30.
Gottlieb, L. D. (1973a). Genetic differentiation, sympatric speciation and the origin of a diploid species of *Stephanomeria. Amer. J. Bot.* **60**, 545–554.

Gottlieb, L. D. (1973b). Enzyme differentiation and phylogeny in *Clarkia franciscana*, *C. rubicunda* and *C. amoena*. *Evolution* **27**, 205–214.

Gottlieb, L. D. (1973c). Genetic control of glutamate oxaloacetate transaminase isozymes in the diploid plant *Stephanomeria exigua* and its allotetraploid derivative. *Biochem. Genet.* **9**, 97–107.

Gottlieb, L. D. (1974). Genetic confirmation of the origin of *Clarkia lingulata*. *Evolution* **28**, 244–250.

Gottlieb, L. D. (1981). Electrophoretic evidence and plant populations. *Prog. Phytochem.* **7**, 1–46.

Gottlieb, L. D. (1982). Conservation and duplication of isozymes in plants. *Science* **216**, 373–380.

Gottlieb, O. R. (1972). Chemosystematics of the Lauraceae. *Phytochemistry* **11**, 1537–1570.

Gottlieb, O. R. (1982). 'Micromolecular Evolution, Systematics and Ecology.' Springer-Verlag, Berlin.

Gough, L. J. (1966). A comparative study of some resins from the family Cupressaceae. *4th Intern. Symp. IUPAC, The Chemistry of Natural Products, Abstracts*, pp. 175–176.

Gough, L. J. and Welch, H. J. (1978). Nomenclatural transfer of *Chamaecyparis obtusa* 'Sanderi' to *Thuja orientalis* on the basis of phytochemical evidence. *Bot. J. Linn. Soc.* **77**, 217-221.

Grabar, P. and Williams, C. A. (1953). Methods permettant l'etude conjuguee des proprietes electrophoretiques et immunochimiques d'un melange de proteines. *Biochim. Biophys. Acta* **10**, 193–194.

Granger, R. and Passet, J. (1973). *Thymus vulgaris* spontane de France: Races Chimiques et chemotaxonomie. *Phytochemistry* **12**, 1683–1691.

Grant, V. (1959). 'History of the Phlox Family,' Vol. 1. Martinus Nyhoff, The Hague.

Grant, V. (1977). 'Organismic Evolution'. Freeman and Co., San Francisco.

Grant, V. (1981). 'Plant Speciation,' 2nd edn. Columbia University Press, New York.

Grant, W. F. (1960). The categories of classical and experimental taxonomy and the species concept. *Revue Canadienne de Biologie* **19**, 241–262.

Grant, W. F. and Sidhu, B. B. (1967). Basic chromosome number, cyanogenetic glucoside variation and geographical distribution of *Lotus* species. *Canad. J. Bot.* **45**, 639-647.

Gray, A. I. and Waterman, P. G. (1978). Coumarins in the Rutaceae. *Phytochemistry* **17**, 845–864.

Gray, J. C. (1980). Fraction I protein and plant phylogeny. *In* 'Chemosystematics, Principles and Practice (Bisby, F. A., Vaughan, J. G. and Wright, C. A., eds.), pp. 167–194. Academic Press, London.

Gray, J. C., Kung, S. D., Wildman, S. G. and Sheen, S. J. (1974). Origin of *Nicotiana tabacum* detected by polypeptide composition of fraction I protein. *Nature (Lond.)* **252**, 226–227.

Grayer-Barkmeijer, R. J. (1973). A chemosystematic study of *Veronica:* iridoid glucosides. *Biochem. Syst. Ecol.* **1**, 101–110.

Green, B. R. (1971). Isolation and base composition of DNAs of primitive land plants. I. Ferns and fern-allies. *Biochim. Biophys. Acta* **254**, 402–406.

Green, B. R. and Dick, M. W. (1972). DNA base composition and the taxonomy of the oomycetes. *Can. J. Microbiol.* **18**, 963–968.

Greene, E. L. (1909), 'Landmarks of Botanical History.' Part I. Smithsonian Institute, Washington, Misc. Coll. 54, New York.

Greenhalgh, J. R. and Mitchell, N. D. (1976). Involvement of flavour volatiles in the resistance to downy mildew of wild and cultivated forms of *Brassica oleracea*. *New Phytol.* **77**, 391–398.

Greger, H. (1977). Anthemideae-chemical review. *In* 'The Biology and Chemistry of the Compositae' (Heywood, V. H., Harborne, J. B. and Turner, B. L., eds.), Vol. II, pp. 899–942. Academic Press, London.

Greger, H. (1978). Comparative phytochemistry and systematics of *Anacyclus*. *Biochem. Syst. Ecol.* **6**, 11–17.

Greger, H. (1979). Aromatic acetylenes and dehydrofalcarinone derivatives within the *Artemisia dracunculus* complex. *Phytochemistry* **18**, 1319–1322.

Greger, H. and Ernet, D. (1973). Flavonoid-Muster, Systematik und Evolution bei *Valerianella*. *Phytochemistry* **12**, 1693–1699.

Gregory, W. C. (1941). Phylogenetic and cytological studies in the Ranunculaceae. *Trans. Am. Phil. Soc.* **NS31**, 443–521.

Grierson, D. (1977). The nucleus and the organisation and transcription of nuclear DNA. *In* 'The Molecular Biology of Plant Cells' (Smith, H., ed.), pp. 213–255. Blackwell Scientific, Oxford.

Grieve, C. M. and Scora, R. W. (1980). Flavonoid distribution in the Aurantioideae (Rutaceae). *System. Bot.* **5**, 39–53.

Grund, C., Gilroy, J., Gleaves, T., Jensen, U. and Boulter, D. (1981). Systematic relationships of the Ranunculaceae based on amino acid sequence data. *Phytochemistry* **20**, 1559–1566.

Gunstone, F. D. (1967). 'An introduction to the Chemistry and Biochemistry of Fatty Acids and their Glycerides', 2nd edn. Chapman and Hall, London.

Guppy, G. A. and Bohm, B. A. (1976). Flavonoids of five *Hieracium* species of British Columbia. *Biochem. Syst. Ecol.* **4**, 231–234.

Gurni, A. A. and Kubitzki, K. (1981). Flavonoid chemistry and systematics of the Dilleniaceae. *Biochem. Syst. Ecol.* **9**, 109–114.

Gussin, A. F. S. (1972). Does trehalose occur in Angiospermae? *Phytochemistry* **11** 1827–1828.

Habeck, J. R. and Weaver, T. W. (1969). A chemosystematic analysis of some hybrid spruce (*Picea*) populations in Montana. *Can. J. Bot.* **47**, 1565–1570.

Hale, M. E. (1974). 'The Biology of Lichens.' 181 pp. 2nd edn. Edward Arnold, London.

Halim, A. F. and Collins, R. P. (1973). Essential oil analysis of the Myricaceae of the Eastern United States. *Phytochemistry* **12**, 1077–1083.

Hall, O. (1959). Immuno-electrophoretic analyses of allopolyploid ryewheat and its parental species. *Hereditas* **45**, 495–504.

Hall, O. and Johnson, B. L. (1963). Electrophoretic analysis of the amphiploid *Stipa viridula* × *Oryzopsis hymenoides* and its parental species. *Hereditas* **48**, 530–535.

Hall, T. C. and Davies, J. W. (eds.) (1979). 'Nucleic Acids in Plants,' 2 vols. CRC Press Inc., Boca Raton, Florida.

Hall, H. M. and Clements, F. E. (1923). The genus *Atriplex*. *Carnegie Institute Wash. Publ.* **326**, 253–324.

Halstead, L. B. and Wood, L. (1973). Amino acid components in fossil tortoiseshell from the Oligocene of the Isle of Wight. *Nature* (*Lond.*) **244**, L82.

Hamrick, J. L. and Allard, R. W. (1972). Microgeographical variation in allozyme frequencies in *Avena barbata*. *Proc. Nat. Acad. Sci. U.S.A.* **69**, 2100–2104.

Hancock, J. F. (1982). Alcohol dehydrogenase isozymes in *Gossypium hirsutum* and its putative diploid progenitors: the biochemical consequences of enzyme multiplicity. *Pl. System. Evol.* **140**, 141–150.

Hanover, J. W. (1966). Environmental variation in the monoterpenes of *Pinus monticola* Dougl. *Phytochemistry* **5**, 713–717.
Hanover, J. W. and Wilkinson, R. C. (1969). A new hybrid between blue spruce and white spruce. *Canad. J. Bot.* **47**, 1693–1700.
Hanover, J. W. and Wilkinson, R. C. (1970). Chemical evidence for introgressive hybridization in *Picea*. *Silvae Genetica* **19**, 17–22.
Hanson, J. R. (1968). Recent advances in the chemistry of the tetracyclic diterpenes. *Prog. Phytochem.* **1**, 161–189.
Harborne, J. B. (1963a). Distribution of anthocyanins in higher plants. *In* 'Chemical Plant Taxonomy' (Swain, T., ed.), pp. 359–388. Academic Press, London.
Harborne, J. B. (1963b). Coumarin metabolism in Higher Plants. *Ann. Rep. John Innes Inst.* **54**, 44–45.
Harborne, J. B. (1966). Flavonols as yellow flower pigments. *Phytochemistry* **4**, 647–657.
Harborne, J. B. (1967a). 'Comparative Biochemistry of the Flavonoids.' 384 pp. Academic Press, London.
Harborne, J. B. (1967b). Correlations between chemistry, pollen morphology and Systematics in the Family Plumbaginaceae. *Phytochemistry* **6**, 1415–1428.
Harborne, J. B. (1967c). Flavonoid patterns in the Bignoniaceae and Gesneriaceae. *Phytochemistry* **6**, 1643–1651.
Harborne, J. B. (1968). Correlations between flavonoid pigmentation and systematics in the family Primulaceae. *Phytochemistry* **7**, 1215–1230.
Harborne, J. B. (1969a). Chemosystematics of the Leguminosae. Flavonoid and isoflavonoid patterns in the tribe Genisteae. *Phytochemistry* **8**, 1449–1456.
Harborne, J. B. (1969b). Hirsutin and gossypetin in *Dionysia* (Primulaceae). *Phytochemistry* **10**, 472–475.
Harborne, J. B. (ed.) (1970). 'Phytochemical Phylogeny.' Academic Press, New York.
Harborne, J. B. (1971). Flavonoid and phenylpropanoid patterns in the Umbelliferae. *In* 'Chemistry and Biochemistry of the Umbelliferae' (Harborne, J. B., ed.), pp. 293–314. Academic Press, London.
Harborne, J. B. (1971). Distribution of flavonoids in the Leguminosae. *In* 'Chemotaxonomy of the Leguminosae' (Harborne, J. B., Boulter, D. and Turner, B. L., eds.), pp. 257–283. Academic Press, London.
Harborne, J. B. (1973a). Flavonoids. *In* 'Phytochemistry II' (Miller, L. P., ed.), pp. 344–380. Van Nostrand-Reinhold, New York.
Harborne, J. B. (1973b). 'Phytochemical Methods.' Chapman and Hall, London.
Harborne, J. B. (1975a). Biochemical systematics of flavonoids. *In* 'The Flavonoids' (Harborne, J. B., Mabry, T. J. and Mabry, H., eds.), pp. 1056–1095. Chapman and Hall, London.
Harborne, J. B. (1975b). Flavonoid bisulphates and their co-occurrences with ellagic acid in the Bixaceae, Frankeniaceae and related families. *Phytochemistry* **14**, 1331–1337.
Harborne, J. B. (1976). A unique pattern of anthocyanins in *Daucus carota* and other Umbelliferae. *Biochem. Syst. Ecol.* **4**, 31–35.
Harborne, J. B. (1977a). Flavonoid profiles in the Compositae. *In* 'The Biology and Chemistry of the Compositae' (Heywood, V. H., Harborne, J. B. and Turner, B. L., eds.), pp. 359–384. Academic Press, London.
Harborne, J. B. (1977b). Variations in pigment patterns in *Pyrrhopappus* and related taxa of the Cichorieae. *Phytochemistry* **16**, 927–928.
Harborne, J. B. (1977c). Flavonoid sulphates—a new class of natural product of ecological significance in plants. *Progr. Phytochem.* **4**, 189–208.

Harborne, J. B. (1978). The rare flavone isoetin as a yellow flower pigment in *Heywoodiella oligocephala* and in other Cichorieae. *Phytochemistry* **17**, 915–917.
Harborne, J. B. (1979a). Correlations between flavonoid chemistry, anatomy and geography in the Restionaceae. *Phytochemistry* **18**, 1323–1327.
Harborne, J. B. (1979b). Plant Phenolics. *In* 'Secondary Plant Products, Encyclopedia of Plant Physiology' (Bell, E. A. and Charlwood, B. V., eds.), New series, Vol. 8, pp. 329–402. Springer-Verlag, Berlin.
Harborne, J. B. (1980). Phenolic compounds derived from shikimate. *In* "Biosynthesis" (Bu-Lock, J., ed.), pp. 40–75. The Chemical Society, London.
Harborne, J. B. (1982). 'Introduction to Ecological Biochemistry,' 2nd edn. Academic Press, London and New York.
Harborne, J. B. and Green, P. S. (1980). A Chemotaxonomic survey of flavonoids in leaves of the Oleaceae. *Bot. J. Linn. Soc.* **81**, 155–167.
Harborne, J. B. and Ingham, J. L. (1978). Biochemical aspects of the coevolution of higher plants with their fungal parasites. *In* 'Biochemical Aspects of Plant and Animal Coevolution (Harborne, J. B. ed.), pp. 343–405. Academic Press, London.
Harborne, J. B. and Mabry, T. J. (eds.) (1982). 'The Flavonoids, Advances in Research.' Chapman and Hall, London.
Harborne, J. B. and Smith, D. M. (1978a). Correlations between anthocyanin chemistry and pollination ecology in the Polemoniaceae. *Biochem. Syst. Ecol.* **6**, 127–130.
Harborne, J. B. and Smith, D. M. (1978b). Anthochlors and other flavonoids as honey guides in the Compositae. *Biochem. Syst. Ecol.* **6**, 287–291.
Harborne, J. B. and Swain, T. (1969). 'Perspectives in Phytochemistry.' Academic Press, New York.
Harborne, J. B. and Williams, C. A. (1973a). A chemotaxonomic survey of flavonoids and simple phenols in leaves of the Ericaceae. *Bot. J. Linn. Soc.* **66**, 37–54.
Harborne, J. B. and Williams, C. A. (1973b). Species-specific kaempferol derivatives in ferns of the Appalachian *Asplenium* complex. *Biochem. Syst.* **1**, 51–54.
Harborne, J. B. and Williams, C. A. (1977). Vernonieae-chemical review. *In* 'The Biology and Chemistry of the Compositae' (Heywood, V. H., Harborne, J. B. and Turner, B. L., eds.), pp. 523–538. Academic Press, London.
Harborne, J. B., Heywood, V. H. and Williams, C. A. (1969). Distribution of myristicin in seeds of the Umbelliferae. *Phytochemistry* **8**, 1729–1732.
Harborne, J. B., Mabry, T. J. and Mabry, H. (1975). 'The Flavonoids.' Chapman and Hall, London.
Harborne, J. B., Ingham, J. L., King, L. and Payne, M. (1976). The isopentenyl isoflavone luteone as a preinfectional antifungal agent in the genus *Lupinus*. *Phytochemistry* **15**, 1485–1487.
Harborne, J. B., Williams, C. A. and Wilson, K. L. (1982). Flavonoids in leaves and inflorescences of Australian *Cyperus* species. *Phytochemistry* **21**, 2491–2507.
Hardman, R. and Benjamin, T. V. (1976). The co-occurrence of ecdysones with bufadienolides and steroidal saponins in the genus *Helleborus*. *Phytochemistry* **15**, 1515–1516.
Hargreaves, J. A., Mansfield, J. W. and Coxon, D. T. (1976). Identification of medicarpin as a phytoalexin in the broad bean plant. *Nature* (*Lond.*) **262**, 318–319.
Harley, R. N. and Bell, M. G. (1967). Taxonomic analysis of herbarium material by gas chromatography. *Nature* (*Lond.*) **213**, 1241–1242.
Harney, P. M. (1966). A chromatographic study of species presumed ancestral to *Pelargonium* × *hortorum* Bailey. *Canad. J. Genet. Cytol.* **8**, 780–787.

Harney, P. M. and Grant, W. F. (1963). Biochemical anomaly of interspecific hybrids between *Lotus* species. *Science* **142**, 1061.

Harney, P. M. and Grant, W. F. (1964a). A polygonal presentation of chromatographical investigations on the phenolic content of certain species of *Lotus*. *Canad. J. Genet. Cytol.* **7**, 40–51.

Harney, P. M. and Grant, W. F. (1964b). The cytogenetics of *Lotus* (Leguminosae). VII. Segregation and recombination of chemical constituents in interspecific hybrids between species closely related to *L. corniculatus* L. *Canad. J. Genet. Cytol.* **6**, 140–146.

Harney, P. M. and Grant, W. F. (1964c). A chromatographic study of the phenolics of species of Lotus closely related to *L. corniculatus* and their taxonomic significance. *Amer. J. Bot.* **51**, 621–627.

Harper, R., Bate-Smith, E. C. and Land, D. G. (1968). 'Odour Description and Odour Classification,' 191 pp. Churchill, London.

Harris, H. (1971). Protein polymorphism in man. *Canad. J. Genet. Cytol.* **13**, 381–396.

Harris, P. J. and Hartley, R. D. (1980). Phenolic constituents of the cell walls of monocotyledons. *Biochem. Syst. Ecol.* **8**, 153–160.

Hartley, R. D. and Harris, P. J. (1981). Phenolic constituents of the cell walls of dicotyledons. *Biochem. Syst. Ecol.* **9**, 189–203.

Hartley, R. D. and Jones, E. C. (1977). Phenolic components and degradability of cell walls of grass and legume species. *Phytochemistry* **16**, 1531–1534.

Hartley, T. G., Dunstone, E. A., Fitzgerald, J. S., Johns, S. R. and Lamberton, J. A. (1973). A survey of New Guinea plants for alkaloids. *Lloydia* **36**, 217–319.

Hartmann, T., Ilert, H. I. and Steiner, M. (1972). Aldehydramierung, der bevorzugte Biosyntheseweg fur primare, aliphatische Monoamine in Blutenpflanzen. *Z. Pflanzenphysiol.* **68**, 11–18.

Haufler, C. H. and Giannasi, D. E. (1982). A chemosystematic survey of the fern genus *Bommeria*. *Biochem. Syst. Evol.* **10**, 107–110.

Hausen, B. M. (1978). Occurrence of the contact allergen primin and other quinonoids in species of the Primulaceae. *Arch. Dermatol. Res.* **261**, 311–321.

Hawkes, J. G. (ed.) (1968). 'Chemotaxonomy and Serotaxonomy.' Academic Press, New York.

Hayashi, K. and Abe, Y. (1956). Papierchromatographische Ubersicht der Anthocyane im Pflanzen reich. *Bot. Mag. Tokyo* **69**, 227–235.

Hayashi, N. and Komae, H. (1974). Geographical variation in terpenes from *Lindera umbellata* and *L. sericea*. *Phytochemistry* **13**, 2171–2174.

Hayatsu, R., Winans, R., Scott, R. G., Moore, L. P. and Studier, M. H. (1978). Characterisation of organic acids trapped in coals. *Nature* (*Lond.*) **275**, 116–118.

Hebert, P. D. N. and Crease, T. J. (1980). Clonal coexistence in *Daphnia pulex* (Leydig): another planktonic paradox. *Science* **207**, 1363–1365.

Heckard, L. R., Morris, M. I. and Chuang, T. I. (1980). Origin and taxonomy of *Castilleja montigena* (Scrophulariaceae). *System. Bot.* **5**, 71–85.

Hecht, A. and Tandon, S. L. (1953). Chromosomal interchanges as a basis for the determination of species in *Oenothera*. *Science* **118**, 557–558.

Hedberg, O. (ed.) (1958). 'Systematics of Today.' Almquist and Wiksells, Uppsala.

Hefendehl, F. W. (1977). Monoterpene composition of a carvone containing polyploid strain of *Meuthon longifolia*. *Herba Hungarica* **16**, 39–42.

Hefendehl, F. W. and Murray, M. J. (1972). Changes in monoterpene composition in *Mentha aquatica* produced by gene substitution. *Phytochemistry* **11**, 189–195.

Hegnauer, R. (1960). Die systematische Bedeutung des Blausaure Markmals. *Pharm. Zentrahalle* **99**, 322–329.

Hegnauer, R. (1962–1973). 'Chemotaxonomie der Pflanzen.' Vols. 1–6. Birkhauser-Verlag, Basle.

Hegnauer, R. (1963). The taxonomic significance of alkaloids. *In* 'Chemical Plant Taxonomy', (Swain, T., ed.), pp. 389–428. Academic Press, London.

Hegnauer, R. (1964). Die Pseudoindikane der Dikotyledonen und einige ihnen nahe stehende Verbindungen. 'Chemotaxonomie der Pflanzen', Vol. 3, pp. 29–34. Birkhauser-Verlag, Basle.

Hegnauer, R. (1965). Chemotaxonomy—past and present. *Lloydia* **28**, 267–278.

Hegnauer, R. (1966). Comparative phytochemistry of alkaloids. *In* 'Comparative Phytochemistry' (Swain, T., ed.), pp. 211–230. Academic Press, London.

Hegnauer, R. (1967). Chemical characters in plant taxonomy: some possibilities and limitations. *Intern. Union Pure Appl. Chem.* **4**, 173–187.

Hegnauer, R. (1969). Chemical evidence for the classification of some plant taxa. *In* 'Perspectives in Phytochemistry' (Harborne, J. B. and Swain, T., eds.), pp. 121–138. Academic Press, London.

Hegnauer, R. (1971). Chemical patterns and relationships of Umbelliferae. *In* 'The Biology and Chemistry of the Umbelliferae' (Heywood, V. H., ed.), pp. 267–278. Academic Press, London.

Hegnauer, R. (1973). Zur systematischen Bedeutung des Markmals der Cyanogenese. *Biochem. System.* **1**, 191–197.

Hegnauer, R. (1976). Accumulation of secondary products and its significance for biological systematics. *Nova Acta Leopoldina Suppl.* **7**, 45–76.

Hegnauer, R. (1977). Cyanogenic glycosides as systematic markers in Tracheophyta. *Plant Syst. Evol. Suppl.* **1**, 191–210.

Hegnauer, R. (1978). The chemistry of the Compositae. *In* 'The Biology and Chemistry of the Compositae' (Heywood, V. H., Harborne, J. B. and Turner, B. L., eds), pp. 283–336. Academic Press, London.

Hegnauer, R. and Kooiman, P. (1978). Die Systematische Bedeutung von iridoiden Inhaltsstoffen im Rahmen von Wettstein's Tubiflorae. *Planta Medica* **33**, 1–33.

Heiser, C. B. (1973). Introgression Re-examined. *Bot. Rev.* **39**, 347–366.

Hendrie, M. S. and Shewan, J. M. (1979). The identification of pseudomonads. *In* 'Identification Methods for Microbiologists' (Skinner, F. A. and Lovelock, D. W., eds), 2nd edn., pp. 1–14. Academic Press, London.

Herbin, G. A. and Robins, P. A. (1968). The chemotaxonomy of alkanes and alkenes in the genus *Aloe* (Liliaceae). *Phytochemistry* **7**, 239–255.

Herbin, G. A. and Robins, P. A. (1969). Patterns of variation and development in leaf wax alkanes. *Phytochemistry* **8**, 1985–1998.

Herold, A. and Lewis, D. H. (1977). Mannose and green plants: occurrence, physiology and metabolism. *New Phytol.* **79**, 1–40.

Herz, W. (1977). Sesquiterpene lactones of the Compositae. *In* 'The Biology and Chemistry of the Compositae' (Heywood, V. H., Harborne, J. B. and Turner, B. L., eds), pp. 337–358. Academic Press, London.

Herz, W., Murari, R. and Govindan, S. V. (1979). Sesquiterpene lactones of *Eupatorium anomalum* and *E. mohrii*. *Phytochemistry* **18**, 1337–1341.

Heslop-Harrison, J. (1953). 'New Concepts in Flowering Plant Taxonomy.' Heinemann, London.

Heslop-Harrison, J. (1963). Species concepts: theoretical and practical aspects. *In* 'Chemical Plant Taxonomy' (T. Swain, ed.), pp. 17–40. Academic Press, New York.

Heslop-Harrison, J. (1968). Chairman's summing-up. *In* 'Chemotaxonomy and Serotaxonomy' (Hawkes, J. G. ed.), pp. 279–284. Academic Press, New York.

Hess, von Dieter (1971). Chemogenetische Untersuchungen zur Synthese der Blutenfarbstoffe in der Gattung *Torenia* spec. (Scrophulariaceae). *Biochem. Physiol. Pflanzen* **162**, 386–389.

Heywood, V. H. (1966). Phytochemistry and Taxonomy. *In* 'Comparative Phytochemistry' (Swain, T. ed.), pp. 1–20. Academic Press, New York.

Heywood, V. H. (1973). The role of chemistry in plant systematics. *J. Pure Appl. Chem.* **34**, 355–375.

Heywood, V. H., Harborne, J. B. and Turner, B. L. (eds.) (1977). 'The Biology and Chemistry of the Compositae.' 2 Vols. Academic Press, London and New York.

Hickok, L. G. and Anway, J. C. (1972). A morphological and chemical analysis of geographical variation in *Tilia* L. of eastern North America. *Brittonia* **24**, 2–8.

Higgins, V. J. and Miller, R. L. (1968). Phytoalexin production by alfalfa in response to infection by *Colletotrichum phomoides*, *Helminthosporium turcicum*, *Stemphylium loti* and *S. botryosum*. *Phytopathology* **58**, 1377–1383.

Hilditch, T. P. and Lovern, J. A. (1936). The evolution of natural fats, a general survey. *Nature* (*Lond.*) **137**, 478–481.

Hill, R. J. (1977). Variability of soluble seed proteins in populations of *Mentzelia* L. (Loasaceae) from Wyoming and adjacent states. *Bull. Torrey Bot. Club* **104**, 93–101.

Hillebrand, G. R. and Fairbrother, D. E. (1970). Serological investigation of the Caprifoliaceae. I. Correspondence with selected Rubiaceae and Cornaceae. *Amer. J. Bot.* **57**, 810–815.

Hillis, W. E. (1966). Variation in polyphenol composition within species of *Eucalyptus*. *Phytochemistry* **5**, 541–546.

Hinton, F. W. (1970). The taxonomic status of *Physalis lanceolata* (Solanaceae) in the Carolina Sandhills. *Brittonia* **22**, 14–19.

Hinton, W. F. (1975). Natural hybridization and extinction of a population of *Physalis virginiana* (Solanaceae). *Amer. J. Bot.* **62**, 198–202.

Hiraoka, A. (1978). Flavonoid patterns in Athyriaceae and Dryopteridaceae. *Biochem. Syst. Ecol.* **6**, 171–175.

Hitchcock, C. and Nichols, B. W. (1971). 'Plant Lipid Biochemistry.' Academic Press, London and New York.

Hohn, M. and Meinschein, W. G. (1976). Seed oil fatty acids: evolutionary significance in the Nyssaceae and Cornaceae. *Biochem. Syst. Ecol.* **4**, 193–199.

Hoisington, A. M. R. and Hancock, J. F. (1981). Effect of allopolyploidy on the activity of selected enzymes in *Hibiscus*. *Pl. Syst. Evol.* **138**, 189–198.

Holley, R., Apgar, J., Everett, G. A., Madison, J. T., Marquisee, M., Merrill, S. H., Penswick, J. and Zamir, A. (1965). Structure of a ribonucleic acid. *Science* **147**, 1462–1465.

Holligan, P. M. and Drew, E. A. (1971). Quantitative analysis of standard carbohydrates and the estimation of soluble sugars and polyols from plant tissues. *New Phytol.* **70**, 271–297.

Holmquist, R. and Pearl, D. (1980). Theoretical foundations for quantitative paleogenetics. *J. Mol. Evol.* **16**, 211–267.

Hopf, H. and Kandler, O. (1974). Biosynthesis of Umbelliferose in *Aegopodium podagraria*. *Plant Physiol.* **54**, 13–14.

Hopf, H. and Kandler, O. (1977). Characterisation of the reserve cellulose of the endosperm of *Carum carvi* as a β(1-4)-mannan. *Phytochemistry* **16**, 1715–1718.

Hopkins, C. Y., Chisholm, M. J. and Cody, W. J. (1969). Fatty acid components of some Santalaceae seed oils. *Phytochemistry* **8**, 161–165.

Horn, D. H. S., Kranz, Z. H. and Lamberton, J. A. (1964). The composition of *Eucalyptus* and some other leaf waxes. *Aust. J. Chem.* **17**, 464–476.

Horne, D. B. (1965). The distribution of flavonoids in *Baptisia nuttalliana* and *B. lanceolata* and their taxonomic implications. Doctoral thesis, the University of Texas, Austin.

Hostettmann, K. and Wagner, H. (1977). Xanthone glycosides. *Phytochemistry* **16**, 821–830.

Houts, K. P. and Hillebrand, G. R. (1976). An electrophoretic and serological investigation of seed proteins in *Galeopsis tetrahit* L. (Labiatae) and its putative parental species. *Amer. J. Bot.* **63**, 156–165.

Howse, P. F. and Bradshaw, J. W. S. (1980). Chemical systematics of social insects with particular reference to ants and termites. *In* 'Chemosystematics, Principles and Practice' (Bisby, F. A., Vaughan, J. G. and Wright, C. A., eds.), pp. 71–90. Academic Press, London.

Hrazdina, G. (1982). Anthocyanins. *In* 'The Flavonoids, Advances in Research' (Harborne, J. B. and Mabry, T. J., eds.), pp. 135–188. Chapman and Hall, London.

Hsiao, J. and Li, H. (1973). Chromatographic studies on the red horse-chestnut (*Aesculus* × *carnea*) and its putative parent species. *Brittonia* **25**, 57–63.

Huang, F. H., Cech, F. C. and Clarkson, R. B. (1975). Comparative investigation of soluble protein polymorphism in *Robinia pseudoacacia* roots by polyacrylamide gel electrophoresis. *Biochem. Syst. Ecol.* **3**, 143–147.

Huber, H. (1977). The treatment of the Monocotyledons in an evolutionary system of classification. *Plant Syst. Evol. Suppl.* **1**, 285–298.

Huizing, H. J., Gadella, T. W. J. and Kliphuis, E. (1982). Chemotaxonomical investigations of the *Symphytum officinale* polyploid complex and *S. asperum* (Boraginaceae): The pyrrolizidine alkaloids. *Pl. Syst. Evol.* **140**, 279–292.

Hull, P. (1974). Differences in esterase distribution detected by electrophoresis as evidence of continuing interspecific hybridization in the genus *Senecio*. *Ann. Bot.* **38**, 697–700.

Hultin, E. and Torssel, K. (1965). Alkaloid screening of Swedish plants. *Phytochemistry* **4**, 425–433.

Hunneman, D. H. and Eglinton, G. (1972). The constituent acids of gymnosperm cutins. *Phytochemistry* **11**, 1989–2001.

Hunt, R. S. and von Rudloff, E. (1974). Chemosystematic studies in the genus *Abies*. I. Leaf and twig oil analysis of alpine and balsam firs. *Can. J. Bot.* **52**, 477–487.

Hunt, R. S. and von Rudloff, E. (1979). Chemosystematic studies in the genus *Abies*. IV. Introgression in *Abies lasiocarpa* and *Abies bifolia*. *Taxon* **28**, 297–305.

Hunter, G. E. (1967). Chromatographic documentation of interspecific hybridisation in *Vernonia*: Compositae. *Amer. J. Bot.* **54**, 437–477.

Hunter, R. B. and Kannenberg, L. W. (1971). Isozyme characterization of corn (*Zea mays*) inbreds and its relationship to single cross hybrid performance. *Can. J. Genet. Cytol.* **13**, 649–655.

Hunter, R. L. (1957). Histochemical demonstration of enzymes separated by electrophoresis in starch gels. *Science* **125**, 1294–1295.

Hunziker, J. H. (1967). Chromosome and protein differentiation in the *Agropyron scabriglume* complex. *Taxon* **16**, 259–266.

Hunziker, J. H. (1969). Molecular data in plant systematics. Systematic Biology. *Proc. Int. Conf., Natl. Acad. Sci. US*, Washington DC.

Hunziker, J. H. (1971). El uso simultaneo de datos citogeneticos y moleculares en taxonomia experimental. *In* 'Recientes Adelantos en Biologia' (Mejia, R. H. and Moguilevsky, J. A., eds.), pp. 129–137. Buenos Aires.

Hunziker, J. H., Palacios, R. A., de Valesi, A. G. and Poggio, L. (1972). Species disjunctions in *Larrea*: evidence from morphology, cytogenetics, phenolics and seed albumins. *Ann. Missouri Bot. Gard.* **59**, 224–233.

Hurka, H. (1980). Enzymes as a taxonomic tool: a botanist's approach. *In* 'Chemosystematics, Principles and Practice' (Bisby, F. A., Vaughan, J. G. and Wright, C. A., eds.), pp. 103–122. Academic Press, London.

Hutchinson, J. (1959). 'The Families of Flowering Plants.' Vol. 1, Dicotyledons. University Press, Oxford.

Huxley, J. (ed.) (1942). 'Evolution: The Modern Synthesis'. Harper and Brothers, New York.

Ikenaga, T., Abe, M., Itakura, A. and Ohashi, H. (1979). Studies on the production of *Dubosia* in Japan. Alkaloid content in leaves of artificial interspecific hybrids between *Dubosia myoporioides* and *D. reichardtii*. *Planta Medica* **35**, 51–55.

Ilert, H. I. and Hartmann, T. (1972). Dunnschichtchromatographie der 2,4-dinitrophenylderivate wasserdampf-fluchtiger amine und ihre Anwendung auf die Trennung pflanzlicher amine. *J. Chromatog.* **71**, 119–125.

Ingham, J. L. (1972). Phytoalexins and other natural products as factors in plant disease resistance. *Botan. Rev.* **38**, 343-424.

Ingham, J. L. (1976). Induced and constitutive isoflavonoids from stems of chick-peas inoculated with spores of *Helminthosporium carbonum*. *Phytopath. Z.* **87**, 353.

Ingham, J. L. (1978). Isoflavonoid and stilbene phytoalexins of the genus *Trifolium*. *Biochem. Syst. Ecol.* **6**, 217–223.

Ingham, J. L. (1981). Phytoalexin induction and its taxonomic significance in the Leguminosae. *In* 'Advances in Legume Systematics' (Polhill, R. M. and Raven, P. H. eds.), pp. 599–626. Her Majesty's Stationery Office, London.

Ingham, J. L. and Harborne, J. B. (1976). Phytoalexin induction as a new dynamic approach to the study of systematic relationships among higher plants. *Nature (Lond.)* **260**, 241–243.

Irving, R. S. (1980). The systematics of *Hedeoma* (Labiatae). *Sida* **8**, 218–295.

Irving, R. S. and Adams, R. P. (1973). Genetic and biosynthetic relationships of monoterpenes. *Recent Adv. Phytochem.* **6**, 187–214.

Ishikura, N. (1975). Anthocyanin of *Acanthopanax divaricatus*. *Phytochemistry* **14**, 1439.

Ivanov, S. L. (1926). The fundamental biochemical law of the evolution of substances in organisms. *Bull. Appl. Bot., Genet. Pl. Breed.* **16**, 89–122.

Isler, O. (ed.) (1971). 'Carotenoids.' Birkhauser Verlag, Basle.

Jaaska, V. (1969). Electrophoretic studies of seedling phosphatases, esterases and peroxidases in the genus *Triticum* L. *Eesti NSV Tead. Akad. Toim., Biol.* **18**, 170–183.

Jackson, P., Milton, J. M. and Boulter, D. (1967). Fingerprint patterns of the globulin fraction obtained from seeds of various species of the Fabaceae'. *New Phytol.* **66**, 47–56.

Jakimow-Barras, N. (1973). Les polysaccharides des graines de quelques Liliacées et Iridacées. *Phytochemistry* **12**, 1331–1339.

Jamieson, G. R. and Reid, E. H. (1971). The leaf lipids of some conifer species. *Phytochemistry* **11**, 269–275.

Jana, S. and Seyffert, W. (1971). Simulation of quantitative characters by genes with biochemically definable action. III. *Theor. Appl. Genet.* **41**, 329–337.

Jaretzsky, R. (1926). Beitrage zur Systematik der Polygonaceae unter Berucksichtigung des Oxymethyl-anthrachinon-vorkommens. *Fedde Rep. Spec. Nov.* **22**, 49.

Jaworska, H. and Nybom, N. (1966). A thin layer chromatographic study of *Saxifraga caesia*, *S. aizoides* and their putative hybrid. *Hereditas* **57**, 159–175.

Jay, M. (1969). Chemotaxonomic researches on vascular plants. XIX. Flavonoid distribution in the Pittosporaceae. *Bot. J. Linn. Soc.* **62**, 423–429.

Jensen, R. A. and Pierson, D. L. (1975). Evolutionary implications of different types of microbial enzymology for L-tyrosine biosynthesis. *Nature (Lond.)* **254**, 667–672.

Jensen, S. R., Kjaer, A. and Nielsen, B. J. (1975a). The genus *Cornus*: non-flavonoid glucosides as taxonomic markers. *Biochem. Syst. Ecol.* **3**, 75–78.

Jensen, S. R., Nielsen, B. J. and Dahlgren, R. (1975b). Iridoid compounds, their occurrence and systematic importance in the Angiosperms. *Bot. Notiser* **128**, 148–180.

Jensen, U. (1965). Serologische Untersuchung zur Frage der systematischen Einordnung der Didiereaceae. *Bot. Jahbr. Syst.* **84**, 233–253.

Jensen, U. (1968). Serologische Beitrage zur Systematik der Ranunculaceae. *Bot. Jahrb. Syst.* **88**, 204–268.

Jensen, U. and Buttner, C (1981). The distribution of storage proteins in Magnoliophytina and their serological similarities. *Taxon* **30**, 404–419.

Jensen, U. and Penner, R. (1980). Investigation of serological determinants from single storage plant proteins. *Biochem. Syst. Ecol.* **8**, 161–170.

Jeremias, K. (1962). Der Einfluss der Temperatur auf die Speicherung der Raffinosezucker. *Ber. deut. bot. Ges.* **75**, 313–322.

John, J. (1978). Serological contribution to the taxonomy of the Primulaceae. *Biochem. Syst. Ecol.* **6**, 323–327.

Johnson, A. E., Nursten, H. E. and Williams, A. A. (1971). Vegetable volatiles: a survey of components identified. *Chem. Ind.*, 556–565, 1212–1224.

Johnson, B. L. (1967). Tetraploid wheats: seed protein electrophoretic patterns of the Emmer and Timopheeri groups. *Science* **158**, 131–132.

Johnson, B. L. (1972). Seed protein profiles and the origin of the hexaploid wheats. *Amer. J. Bot.* **59**, 952–960.

Johnson, B. L. (1975). Identification of the apparent B-genome donor of wheat. *Canad. J. Genet. Cytol.* **17**, 21–39.

Johnson, B. L. and Hall, O. (1965). Analysis of phylogenetic affinities in the Triticinae by protein electrophoresis. *Amer. J. Bot.* **52**, 506–513.

Johnson, R. G. and Fairbrother, D. E. (1975). A comparative disc electrophoretic study of pollen proteins of *Betula populifolia*. *Biochem. Syst. Ecol.* **3**, 205–208.

Jones, A. G. and Seigler, D. S. (1975). Flavonoid data and populational observations in support of hybrid status for *Populus acuminata*. *Biochem. Syst. Ecol.* **2**, 201–206.

Jones, D. A. (1972). Cyanogenic glucosides and their function. *In* 'Phytochemical Ecology' (Harborne, J. B. ed.), pp. 103–124. Academic Press, London.

Jones, D. A., Keymer, R. J. and Ellis, W. M. (1978). Cyanogenesis in plants and animal feeding. *In* 'Biochemical Aspects of Plant and Animal Coevolution' (Harborne, J. B., ed.)., pp. 21–34. Academic Press, London.

Jones, G. N. (1968). 'Taxonomy of American species of Linden (*Tilia*).' Illinois Press, Urbana.

Jones, S. B., Jr. (1972). Hybridization of *Vernonia acaulis* and *V. noveboracensis* (Compositae) in the piedmont of North Carolina. *Castanea* **37**, 244–253.

Jones, S. B. (1977). Vernonieae—systematic review. *In* 'The Biology and Chemistry of the Compositae' (Heywood, V. H., Harborne, J. B. and Turner, B. L., eds.), pp. 503–522. Academic Press, London.

Jones, S. J. (1973). Ecological genetics and natural selection in Molluscs. *Science* **182**, 546–552.

Jukes, T. H. (1966). 'Molecules and Evolution.' Columbia University Press, New York.

Jukes, T. H. (1980). Neutral changes revisited. *In* 'The Evolution of Protein Structure and Function' (Sigman, D. S. and Brazier, M. A. B., eds.), pp. 204–220. Academic Press, New York.

Just, T. (1948). Gymnosperms and the origin of angiosperms. *Bot. Gaz.* **110**, 91–103.

Juvonen, S. (1966). Uber die Terpensynthese beeinflussenden Faktoren in *Pinus silvestris* L. *Acta Bot. Fenn*, **71**, 1–92.

Kagawa, T., McGregor, D. I. and Beevers, H. (1973). Development of enzymes in the cotyledons of watermelon seedlings. *Plant Physiol.* **51**, 66–71.

Kahn, A. A., Gaspar, T., Roe, C. H., Bouchet, M. and Dubucq, M. (1972). Synthesis of isoperoxidases in lentil embryonic axis. *Phytochemistry* **11**, 2963–2969.

Kallunki, J. A. (1976). Population studies in *Goodyera* (Orchidaceae) with emphasis on the hybrid origin of *G. tesselta. Brittonia* **28**, 53–75.

Kaltsikes, P. J. and Dedio, W. (1970). A thin-layer chromatographic study of the phenolics of the genus *Aegilops*. II. *Canad. J. Bot.* **48**, 1781–1786.

Kanamori, T., Konishi, S. and Takahashi, E. (1972). Inducible formation of glutamate dehydrogenase in rice plant roots by the addition of ammonia to the media. *Physiol. Plant.* **26**, 1–6.

Kandler, O. (1964). Moglichkeiten zur Verwendung von C^{14} für chemotaxonomische Untersuchungen. *Ber. Deutsch. Bot. Gesell.* **78**, 62–73.

Kandler, O. and Schleifer, K. H. (1980). Systematics of bacteria. *Progress in Botany* **42**, 234–252.

Karlsen, J. and Svendsen, A. B. (1966). Direct GLC of volatile oils in plant material. 1. The monoterpene hydrocarbons of Norwegian spruce needle oil. *Meddel. Norsk Farm. Selskap* **28**, 85–90.

Kartha, A. R. S. and Khan, R. A. (1969). Proportions of $\Delta^{7,8}$-octadecenoic acids in seed fats and ten Umbelliferae species. *Chem. Ind.* 1869–1870.

Katayama, T. (1959). Volatile components of *Laminaria. Nichi-Suisan* **24**, 925–927.

Keen, N. T. (1975). The isolation of phytoalexins from germinating seeds of *Cicer arietinum*, *Vigna sinensis*, *Arachis hypogaea* and other plants. *Phytopathology* **65**, 91–92.

Keller, R. (1925). Parietales and Opuntiales. *In* 'Die Naturlichen Pflanzenfamilien' (Engler, A. and Prantl, K., eds.), 2nd edn., Vol. 21, pp. 154–236. Engelmann, Leipzig.

Kelsey, R. G., Thomas, T. W., Watson, T. J. and Shafizadeh, F. (1975). Population studies in *Artemisia tridenta* ssp. *vaseyana*: chromosome numbers and sesquiterpene lactone races. *Biochem. Syst. Ecol.* **3**, 209–213.

Kern, J. H. (1962). New look at some Cyperaceae mainly from the tropical standpoint. *Advancement Sci.* **19**, 141–148.

Kimura, M. and Ohta, T. (1971). Protein polymorphism as a phase of molecular evolution. *Nature* (*Lond.*) **229**, 467–469.

Kimura, M. and Ohta, T. (1973). Mutation and evolution at the molecular level. *Genetics* **73** (*Suppl.*), 19–35.

King, B. L. (1977a). The flavonoids of the deciduous *Rhododendron* of North America (Ericaceae). *Amer. J. Bot.* **64**, 350–360.

King, B. L. (1977b). Flavonoid analysis of hybridization in *Rhododendron* section Pentanthera (Ericaceae). *Syst. Bot.* **2**, 14–27.

King, B. L., Jones, S. B. and Galle, F. C. (1975). The flavonoid chemistry of our native Azaleas: Uses in classification. *Quart. Bull. Amer. Rhododendron Soc.* **29**, 179–183.

King, K., Jr. and Hare, P. E. (1972). Amino acid composition of planktonic foraminifera: a paleobiochemical approach to evolution. *Science* **175**, 1461–1463.

Kirby, L. T. and Styles, E. D. (1970). Flavonoids associated with specific gene action in maize aleurones, and the role of light in substituting for the action of a gene. *Canad. J. Genet. Cytol.* **12**, 939–940.

Kjaer, A. (1976). Glucosinolates in the Cruciferae. *In* 'The Biology and Chemistry of the Cruciferae' (Vaughan, J. G., Macleod, A. J. and Jones, B. M. G. eds.), pp. 207–220. Academic Press, London and New York.

Kjaer, A. and Larsen, P. O. (1973). Non-protein amino acids, cyanogenic glycosides and glucosinolates. *In* 'Biosynthesis' (Geissman, T. A., ed.), Vol. 2, pp. 71–105. The Chemical Society, London.

Kjaer, A., Madsen, O. and Maeda, Y. (1978). Seed volatiles within the family Tropaeolaceae. *Phytochemistry* **17**, 1285–1287.

Kjaer, A. and Malver, O. (1979). Glucosinolates in *Tersonia brevipes*, Gyrostemonaceae. *Phytochemistry* **18**, 1565.

Kloz, J. (1971). Serology of the Leguminosae. *In* 'Chemotaxonomy of the Leguminosae' (Harborne, J. B., Boulter, D. and Turner, B. L., eds.), pp. 309–366. Academic Press, London.

Kluge, M. and Ting, I. P. (1978). 'Crassulacean Acid Metabolism.' 209 pp. Springer-Verlag, Berlin.

Knobloch, I. W. (1972). Intergeneric hybridization in flowering plants. *Taxon* **21**, 97–103.

Knocke, H. and Ourisson, G. (1967). *Angew. Chem. Inter. Edn.* **6**, 1085.

Koehn, R. K. and Rasmussen, D. I. (1967). Polymorphic and monomorphic serum esterase heterogeneity in catostomid fish populations. *Biochemical Genet.* **1**, 131–144.

Kohne, D. E. (1968). Taxonomic applications of DNA hybridisation techniques. *In* 'Chemotaxonomy and Serotaxonomy' (Hawkes, J. G., ed.), pp. 117–130. Academic Press, London.

Kooiman, P. (1960). The occurrence of amyloids in plant seeds. *Acta Bot. Neerl.* **9**, 208–219.

Kooiman, P. (1969). The occurrence of asperulosidic glycosides in the Rubiaceae. *Acta Bot. Neerl.* **18**, 124–137.

Kooiman, P. (1970). The occurrence of iridoid glycosides in the Scrophulariaceae. *Acta Bot. Neerl.* **19**, 329–340.

Kooiman, P. (1972). The occurrence of iridoid glycosides in the Labiatae. *Acta Bot. Neerl.* **21**, 417–427.

Kowatani, T., Ohno, T. and Kinoshita, K. (1954). On colchicine-induced polyploid insect flower (*Chrysanthemum cinerariaefolium* Bocc.) with special reference to the content of pyrethrin. *J. Pharm. Soc. Japan* **74**, 786–788.

Krauss, G.-J. and Reinbothe, H. (1973). Die freien aminosauren in samen von Mimosaceae. *Phytochemistry* **12**, 125–142.

Krattinger, K., Rast, D. and Koresch, H. (1979). Analysis of pollen proteins of *Typha* species in relation to identification of hybrids. *Biochem. Syst. Ecol.* **7**, 125–128.

Kremer, B. P. (1978). Volemitol in the genus *Primula*: distribution and significance. *Z. Pflanzenphysiol.* **86**, 453–461.

Kubitzki, K. (1968). Flavonoide und Systematik der Dilleniaceen. *Ber. Dtsch. Bot. Ges.* **8**, 238–251.

Kubitzki, K. (1969). Chemosystematische Betrachtungen zur Grossgliederung der Dicotylen. *Taxon* **18**, 360–368.

Kuntzel, H. and Kochel, H. G. (1981). Evolution of rRNA and origin of mitochondria. *Nature* (*Lond.*) **293**, 751–755.

Kupicha, F. K. (1977). The delimitation of the tribe Vicieae and the relationships of *Cicer*. *Bot. J. Linn. Soc.* **74**, 131–162.

Kvenvolden, K. A., Peterson, E. and Brown, F. S. (1970). Racemization of amino acids in sediments from Saanich Inlet, British Columbia. *Science* **169**, 1079–1082.

Lackey, J. A. (1977). A revised classification of the tribe Phaseoleae (Leguminosae: Papilionoideae), and its relation to canavanine distribution. *J. Linn. Soc.* **74**, 163–178.

Ladizinsky, G. (1975). Seed protein electrophoresis of section *Faba* of *Vicia*. *Euphytica* **24**, 785–788.

Ladizinsky, G. (1979). Seed protein electrophoresis in section *Foenum-graecum* of *Trigonella*. *Pl. Syst. Evol.* **133**, 87–94.

Ladizinsky, G. and Hymowitz, T. (1979). Seed protein electrophoresis in taxonomic and evolutionary studies. *Theor. appl. genet.* **54**, 145–151.

Ladizinsky, G. and Johnson, B. L. (1972). Seed protein homologies and the evolution of polyploidy in *Avena*. *Can. J. Genet. Cytol.* **14**, 875–888.

Ladesic, B., Pokorny, M. and Keglevic, D. (1971). Metabolic patterns of L- and D-serine in higher and lower plants. *Phytochemistry* **10**, 3085–3091.

Lam, H. J. (1936). Phylogenetic symbols, past and present. *Acta Biotheretica* **2**, 153–193.

Lam, H. J. (1959). Taxonomy general principles and angiosperms. *In* 'Vistas in Botany' (Turrill, W. B., ed.), pp. 3–75. Pergamon Press, London.

Lance, G. N. and Williams, W. T. (1967). A general theory of classificatory sorting strategies I. Hierarchical systems. *Comp. J.* **9**, 373–380.

Lanjouw, J. (1958). On the nomenclature of chemical strains. *Taxon* **7**, 43–44.

La Roi, G. H. and Dugle, J. R. (1968). A systematic and genecological study of *Picea glauca* and *P. engelmannii* using paper chromatograms of needle extracts. *Can. J. Bot.* **46**, 649–687.

Lattimer, G. L. and Ormsbee, R. A. (1981). 'Legionnaire's Disease.' Marcel Dekker, New York.

Lava-Sanchez, P. A., Amaldi, F. and Posta, A. La (1972). Base composition of ribosomal RNA and evolution. *J. Molec. Evol.* **2**, 44–55.

Lavie, D. and Glotter, E. (1971). The cucurbitanes, a group of tetracyclic triterpenes. *Fortschr. Chem. Org. Naturst.* **29**, 307–456.

Lawler, L. J. and Slaytor, M. (1969). The distribution of alkaloids in New South Wales and Queensland Orchidaceae. *Phytochemistry* **8**, 1959–1962.

Lawrence, G. H. M. (1951). 'Taxonomy of Vascular Plants.' Macmillan Co., New York.

Lawrence, L., Bartschot, R., Zavarin, E. and Griffin, J. R. (1975). Natural hybridisation of *Cupressus sargentii* and *C. macnabiana* and the composition of the derived essential oils. *Biochem. Syst. Ecol.* **2**, 113–119.

Lawrence, W. J. C., Price, J. R., Robinson, G. M. and Robinson, R. (1939). The distribution of anthocyanins in flowers, fruits and leaves. *Phil. Trans. Roy. Soc. Lond.* **230B**, 149–178.

Leach, G. J. and Whiffin, T. (1978). Analysis of a hybrid swarm between *Acacia brachybotrya* and *A. calamifolia*. *Bot. J. Linn. Soc.* **76**, 53–69.

Lebreton, P. (1965). Elements de chimiotaxonomie botanique. 2. Cas de flavonoides chez des Urticales: conclusions generales. *Bull Soc. Bot. Fr.* **111**, 80–93.

Lee, D. W. (1975). Population variation and introgression in North American *Typha*. *Taxon* **24**, 633–641.

Lee, D. W. and Fairbrother, D. E. (1969). A serological and disc electrophoretic study of North American *Typha*. *Brittonia* **21**, 227–243.

Lee, D. W. and Fairbrother, D. E. (1973). Enzyme differences between adjacent hybrid and parent populations of *Typha*. *Bull. Torr. Bot. Club* **100**, 3–11.

Lee, Y. S. and Fairbrother, D. E. (1978). Serological approaches to the systematics of the Rubiaceae and related families. *Taxon* **27**, 159–185.

Legg, P. D. and Collins, G. B. (1971). Inheritance of per cent total alkaloids in *Nicotiana tabacum* L. II. Genetic effects of two loci in Burley 21 × LA Burley 21 populations. *Can. J. Genet. Cytol.* **13**, 287–291.

Le John, H. B. (1971). Enzyme regulation, lysine pathways and cell wall structures as indicators of major lines of evolution in fungi. *Nature* (*Lond.*) **231**, 164–168.

Leone, C. A. (ed.) (1964). Taxonomic Biochemistry and Serology. Ronald Press, New York.

Lester, R. N. (1965). Immunological studies on the tuber bearing *Solanums*. I. Techniques and South American species. *Ann. Bot.* (*London*) **29**, 609–624.

Leventhal, J., Suess, S. E. and Cloud, P. (1975). Non-prevalance of biochemical fossils in kerogen from pre-Phanerozoic sediments. *Proc. Natl. Acad. Sci. U.S.A.* **72**, 4706–4710.

Levin, D. A. (1966). Chromatographic evidence of hybridisation and evolution in *Phlox maculata*. *Amer. J. Bot.* **53**, 238–245.

Levin, D. A. (1967a). Hybridisation between annual species of *Phlox*: population structure. *Amer. J. Bot.* **54**, 1122–1130.

Levin, D. A. (1967b). An analysis of hybridization in *Liatris*. *Brittonia* **19**, 248–260.

Levin, D. A. (1968a). The structure of a polyspecies hybrid swarm in *Liatris*. *Evolution* **22**, 352–372.

Levin, D. A. (1968b). The genome constitutions of eastern North American *Phlox* amphiploids. *Evolution* **22**, 612–632.

Levin, D. A. (1976). Alkaloid-bearing plants: an ecogeographic perspective. *Am. Nat.* **110**, 261–284.

Levin, D. A. (1979). The nature of plant species. *Science* **204**, 381–384.

Levin, D. A. and Schaal, B. A. (1970). Reticulate evolution in *Phlox* as seen through protein electrophoresis. *Amer. J. Bot.* **57**, 977–987.

Levin, D. A. and Schaal, B. A. (1972). Seed protein polymorphism in *Phlox pilosa* (Polemoniaceae). *Brittonia* **24**, 46–56.

Levin, D. A. and York, B. M. (1978). The toxicity of plant alkaloids: an ecogeographic perspective. *Biochem. Syst. Ecol.* **6**, 61–76.

Levy, M. (1976). Altered glycoflavone expression in induced autotetraploids of *Phlox drummondii*. *Biochem. Syst. Ecol.* **4**, 249–254.

Levy, M. and Fujii, K. (1978). Geographic variation of flavonoids in *Phlox carolina*. *Biochem. Syst. Ecol.* **6**, 117–125.

Levy, M. and Levin, D. A. (1971). The origin of novel flavonoids in *Phlox* allotetraploids. *Proc. Natl. Acad. Sci. U.S.A.* **68**, 1627–1630.

Levy, M. and Levin, D. A. (1974). Novel flavonoids and reticulate evolution in the *Phlox pilosa-P. drummondii* complex. *Amer. J. Bot.* **61**, 156–167.

Levy, M. and Levin, D. A. (1975). The novel flavonoid chemistry and phylogenetic origin of *Phlox floridana*. *Evolution* **29**, 487–499.

Lewis, D. H. and Smith, D. C. (1967a). Sugar alcohols in fungi and green plants. I. Distribution, physiology and metabolism. *New Phytol.* **66**, 143–184.

Lewis, D. H. and Smith, D. C. (1967b). Sugar alcohols in fungi and green plants II. Methods of detection and quantitative estimation. *New Phytol.* **66**, 185–204.

Lewis, H. (1957). Genetics and cytology in relation to taxonomy. *Taxon* **6**, 42–46.

Lewis, H. and Lewis, M. E. (1955). The genus *Clarkia*. Univ. Calif. Publ. Bot. **20**, 241–392.

Lewis, H. and Raven, P. H. (1958). Rapid evolution in *Clarkia*. *Evolution* **12**, 319–336.

Lewis, W. H. (1980). Polyploidy in species populations. *Basic Life Sciences* **13**, 103–144.

Lewontin, R. C. (1974). The Genetic Basis of Evolutionary Change. Columbia University Press, New York.

Li, H. L. and Willaman, J. J. (1968). Distribution of alkaloids in Angiosperm phylogeny. *Econ. Bot.* **22**, 239–252.

Li, H. L. and Willaman, J. J. (1972). Recent trends in alkaloid hunting. *Econ. Bot.* **26**, 61–67.

Lincoln, D. E. and Langenheim, J. H. (1976). Geographic patterns of monoterpenoid composition in *Satureja douglasii*. *Biochem. Syst. Ecol.* **4**, 237–248.

Lincoln, D. E. and Langenheim, J. H. (1978). Effect of light and temperature on monoterpenoid yield and composition in *Satureja douglasii*. *Biochem. Syst. Ecol.* **6**, 21–32.

Lincoln, D. E. and Langenheim, J. H. (1979). Variation of *Satureja douglasii* monoterpenoids in relation to light intensity and herbivory. *Biochem. Syst. Ecol.* **7**, 289–298.

Lincoln, D. E., Marble, P. M., Cramer, F. J. and Murray, M. J. (1971). Genetic basis for high limonene-cineole content of exceptional *Mentha citrata* hybrids. *Theor. Appl. Genet.* **14**, 365–370.

Linczevski, J. (1968). 'Novitates Systematicae Plantarum Vascularium.' USSR Academy of Sciences, 298 pp.

Lindley, J. (1830). 'An Introduction to the Natural System of Botany.' London.

Linneaus, K. (1756). Odores medicamentorum. *In* 'Amoenitates Academicae', Vol. 3, p. 183 ff. Lars Salvin's Stockholm, Sweden.

Loening, U. E. (1973). Evolution of nucleic acids. *Pure Appl. Chem.* **34**, 579–590.

Lokki, J., Sorsa, M., Forsen, K. and Schantz, M. V. (1973). Genetics of monoterpenes in *Chrysanthemum vulgare*. I. *Hereditas* **74**, 225–232.

Loomis, W. D. and Croteau, R. (1973). Biochemistry and physiology of lower terpenoids. *In* 'Terpenoids: Structure Biogenesis and Distribution' (Runeckles, V. C. and Mabry, T. J., eds.), pp. 147–186. Academic Press, New York.

Löve, A. and Löve, D. (1957). Drug content and polyploidy in *Acorus*. *Proc. Genet. Soc. Canad.* **2**, 14–17.

Lowry, B., Lee, D. and Hebant, C. (1980). The origin of land plants: a new look at an old problem. *Taxon.* **20**, 183–197.

Lowry, J B. (1972). Anthocyanins of the Podocarpaceae. *Phytochemistry* **11**, 725–731.

Lowry, J. B. (1976). Anthocyanins of the Melastomaceae, Myrtaceae and some allied families. *Phytochemistry* **15**, 513–516.

Lüning, B. (1967). Screening of orchid species for alkaloids. *Phytochemistry* **6**, 857–861.

Luteyn, J. L., Harborne, J. B. and Williams, C. A. (1980). A survey of the flavonoids and simple phenols in leaves of *Cavendishia* (Ericaceae). *Brittonia* **32**, 1–16.

Mabry, T. J. (1970). Infraspecific variation of sesquiterpene lactones in *Ambrosia* (Compositae). *In* 'Phytochemical Phylogeny' (Harborne, J. B., ed.), pp. 269–300. Academic Press, London.

Mabry, T. J. (1973a). The chemistry of geographical races. *Pure Appl. Chem.* **34**, 377–400.

Mabry, T. J. (1973b). The chemistry of disjunct taxa. *Nobel Symposium* **25**, 63–66.

Mabry, T. J. (1976). Pigment dichotomy and DNA–RNA hybridisation data for Centrospermous families. *Plant Syst. Evol.* **126**, 79–94.

Mabry, T. J. (1977). The order Centrospermae. *Ann. Missouri Bot. Gard.* **64**, 210–220.

Mabry, T. J. and Dreiding, A. S. (1968). The Betalains. *In* 'Recent Advances in Phytochemistry' (Mabry, T. J., Alston, R. E. and Runeckles, V. C., eds.), pp. 145–160. Appleton-Century-Crofts, New York.

Mabry, T. J. and Mears, J. A. (1970). *In* 'Chemistry of the Alkaloids' (Pelletier, S. W., ed.),' p. 719. Van Nostrand Reinhold, New York.

Mabry, T. J., Alston, R. E. and Runeckles, V. C. (eds.) (1968). 'Recent Advances in Phytochemistry,' Vol. 1. Appleton-Century-Crofts, New York.

Mabry, T. J., Markham, K. R. and Thomas, M. B. (1970). 'The Systematic Identification of Flavonoids.' Springer Verlag, Berlin.

Mabry, T. J., Neuman, P. and Philipson, W. R. (1978). *Hectorella:* a member of the betalain suborder Chenopodiineae of the order Centrospermae. *Plant Syst. Evol.* **130**, 163–165.

Macleod, A. M. and McCorquodale, H. (1958). Water-soluble carbohydrates of seeds of the Gramineae. *New Phytol.* **37**, 168–182.

Mahato, S. B., Ganguly, A. N. and Sahu, N. P. (1982). Steroid saponins. *Phytochemistry* **21**, 959–978.

Mandel, M., Igambi, L., Bergendahl, J., Dodson, M. L. and Scheltgen, E. (1970). Correlation of melting temperatures and CsCl buoyant density of bacterial DNA. *J. Bacteriol.* **101**, 335–338.

Mann, J. (1978). 'Secondary Metabolism.' Clarendon Press, Oxford.

Manwell, C. and Baker, C. M. A. (1970). 'Molecular Biology and the Origin of Species: Heterosis, Protein Polymorphism and Animal Breeding.' University Washington Press, Seattle.

Marcus, A. (ed.) (1981). 'Proteins and Nucleic Acids.' *In* 'The Biochemistry of Plants, a Comprehensive Treatise,' Vol. 6. Academic Press, New York.

Margoliash, E., Nisonoff, A. and Reichlin, M. (1970). Immunological activity of cytochrome *c*. I. Precipitating antibodies to monomeric vertebrate cytochromes *c*. *J. Biol. Chem.* **245**, 931–939.

Margulis, L. (1981). 'Symbiosis in Cell Evolution.' Freeman, San Francisco.

Markert, C. and Moller, F. (1959). Multiple forms of enzymes: tissue, ontogenetic and species-specific patterns. *Proc. Natl. Acad. Sci. U.S.A.* **45**, 753–763.

Markham, K. R. (1982). 'Techniques of Flavonoid Identification.' Academic Press, London.

Markham, K. R. and Porter, L. J. (1979). Flavonoids of the primitive liverwort *Takakia* and their taxonomic and phylogenetic significance. *Phytochemistry* **18**, 611–615.

Markham, K. R., Mabry, T. J. and Swift, W. J. (1970). Distribution of flavonoids in the genus *Baptisia*. *Phytochemistry* **9**, 2359–2364.

Marks, G. E., McKee, R. K. and Harborne, J. B. (1965). Double chromosome reduction in a tetraploid *Solanum*. *Nature* (*Lond.*) **208**, 359–361.

Marshall, W. R. and Allard, R. W. (1970). Isozyme polymorphisms in natural populations of *Avena fatua* and *A. barbata*. *Heredity* **25**, 373–382.

Martin, J. T. and Juniper, B. E. (1970). 'The Cuticles of Plants.' Edward Arnold. London.

Martinez, M. A. del Pero de and Swain, T. (1977). Variation in flavonoid patterns in relation to chromosome changes in *Gibasis schiedana*. *Biochem. Syst. Ecol.* **5**, 37–43.

Mastenbroek, I., Cohen, C. E. and de Wet, J. M. J. (1981). Seed protein and seedling isozyme patterns of *Zea mays* and its closest relatives. *Biochem. Syst. Ecol.* **9**, 179–183.

Mathis, C. and Ourisson, G. (1963). Etude chimio-taxonomique du genre *Hypericum*. I. Repartition de l'hypericine. *Phytochemistry* **2**, 157–171.

Matter, P. and Miller, H. W. (1972). The amino acid composition of some Cretaceous fossils. *Comp. Biochem. Physiol.* **43B**, 55–66.

Matsubara, H., Hase, T., Wakabayashi, S. and Wade, K. (1980). Structure and evolution of chloroplast and bacterial-type ferredoxins. *In* 'The Evolution of Protein Structure and Function' (Sigman, D. S. and Brazier, M. A. B., eds.), pp. 246–266. Academic Press, New York.

McClintock, D. and Fitter, R. S. R. (1956). 'The Pocket Guide to Wild Flowers.' Collins, London.

McClure, J. W. and Alston, R. E. (1964). Patterns of selected chemical components of *Spirodela oligorhiza* formed under various conditions of axenic culture. *Nature* (*Lond.*) **201**, 311–313.

McClure, J. W. and Alston, R. E. (1966). Chemotaxonomy of the Lemnaceae. *Amer. J. Bot.* **53**, 849–859.

McHale, J. and Alston, R. E. (1964). Utilization of chemical patterns in the analysis of hybridisation between *Baptisia leucantha* and *B. sphaerocarpa*. *Evolution* **18**, 304–311.

McIndoo, N. E. (1945). Plants toxic to insects. *U.S. Dept. Agr. Bur. Entom. Plant Quarantine ET* **661**, 1–286.

McIntosh, L., Paulsen, C. and Bogorad, L. (1980). Chloroplast gene sequence for the large subunit of ribulose bisphosphate carboxylase of maize. *Nature* (*Lond.*) **288**, 556–560.

McMillan, C., Charez, P. I. and Mabry, T. J. (1975). Sesquiterpene lactones of *Xanthium strumarium* in a Texas population and in experimental hybrids. *Biochem. Syst. Ecol.* **3**, 137–141.

McMillan, C., Mabry, T. J. and Charez, P. I. (1976). Experimental hybridization of *Xanthium strumarium* (Compositae) from Asia and America. II. *Amer. J. Bot.* **63**, 317–323.

McNair, J. B. (1935). Angiosperm phylogeny on a chemical basis. *Bull. Torr. Bot. Club* **62**, 515–532.

McNair, J. B. (1941). Energy and evolution. *Phytologia* **2**, 33–49.

McNair, J. B. (1965). 'Studies in Plant Chemistry.' Westernlove Press. Published by the author, 818 South Ardmone Ave., Los Angeles.

Mears, J. A. (1980). Chemistry of polyploids: a summary with comments on *Parthenium*. *In* 'Polyploidy' (Lewis, W. H., ed.), pp. 77–102. Plenum Press, New York.

Mecklenburg, H. C. (1966). Inflorescence hydrocarbons of some species of *Solanum* and their possible taxonomic significance. *Phytochemistry* **5**, 1201–1209.

Meeuse, A. D. J. (1970). The descent of the flowering plants in the light of new evidence from phytochemistry and from other sources. *Acta Bot. Neerl.* **19**, 133–140.

Meier, H. and Reid, J. S. G. (1982). Reserve polysaccharides other than starch in higher plants. *In* 'Intracellular Carbohydrates' (Loewus, F. A. and Tanner, W., eds.), pp. 418–471. Springer-Verlag, Berlin.

Melchert, T. E. (1966). Chemo-demes of diploid and tetraploid *Thelesperma simplicifolius*. *Amer. J. Bot.* **53**, 1015–1020.

Merritt, R. B. (1972). Geographic distribution and enzymatic properties of lactate dehydrogenase allozymes in the flathead minnow, *Pimephales promelas*. *Amer. Natur.* **196**, 173–184.

Merxmuller, H. and Leins, P. (1967). Die Verwandtschaftsbe ziehungen der Krauzblutler und Mohngewachse. *Botan. Jahrb.* **86**, 113–129.

Metcalfe, C. R. and Chalk, L. (1950). 'The Anatomy of the Dicotyledons.' 2 Vols. Oxford University Press, Oxford.

Mez, C. and Ziegenspeck, H. (1926). Der Konigsberger serodiagnostische Stammbaum. *Bot. Arch.* **13**, 483–485.

Miege, J. (ed.) (1975). 'Les Proteines des Graines.' Conservatoire et Jardin botaniques de la Ville de Geneve, Switzerland.

Miller, H. E., Mabry, T. J., Turner, B. L. and Payne, W. (1968). Infraspecific variation of sesquiterpene lactones in *Ambrosia psilostachya* (Compositae). *Amer. J. Bot.* **55**, 316–324.

Mills, J. S. (1973). Diterpenes of *Larix* oleoresins. Phytochemistry **12**, 2407–2412.

Mirov, N. T. (1956). Composition of turpentine of lodgepole × jack pine hybrids. *Can. J. Bot.* **34**, 443–457.

Mirov, N. T. (1963). Chemistry and plant taxonomy. *Lloydia* **26**, 117–124.

Mirov, N. T. (1965). 'The Genus Pinus.' Ronald Press, New York.

Mirov, N. T., Zavarin, E. and Snajberk, K. (1966a). Chemical composition of the turpentines of some Eastern Mediterranean pines in relation to their classification. *Phytochemistry* **5**, 97–102.

Mirov, N. T., Zavarin, E., Snajberk, K. and Costello, A. (1966b). Further studies of turpentine composition of *Pinus muricata* in relation to its taxonomy. *Phytochemistry* **5**, 343–355.

Mitchell, N. D. and Richards, A. J. (1978). Variations in *Brassica oleracea* subsp. *oleracea* detected by the picrate test. *New Phytol.* **81**, 189–200.

Møller, B. L. (1974). Lysine biosynthesis in barley, *Hordeum vulgare*. *Plant Physiol.* **54**, 638–643.

Monod, M., Marie, R. and Feillet, P. (1972). Fractionnement des proteines de quelques varietes de riz par electrophorese en gel de polyacrylamide. *C.R. Hebd. Seances Acad. Sci.* **274D**, 1957–1960.

Moore, D. M., Harborne, J. B. and Williams, C. (1970). Chemotaxonomy variation and geographical distribution of the Empetraceae. *Bot. J. Linn. Soc.* **63**, 277–293.

Moore, D. M., Williams, C. A. and Yates, B. (1972). Studies on bipolar disjunct species II. *Plantago maritima* L. *Bot. Not.* **125**, 261–272.

Morice, I. M. (1957a). Seed fats of some New Zealand Cyperaceae. *Phytochemistry* **14**, 571–574.

Morice, I. M. (1975b). Seed fats of further species of *Astelia*. *Phytochemistry* **14**, 1315–1318.

Mothes, K. (1976). Secondary plant substances as materials for chemical high quality breeding in higher plants. *Recent Adv. Phytochem.* **10**, 385–405.

Müller, K. O. and Börger, H. (1941). Experimentelle Untersuchungen uber die Phytophthora-Resistenz der Kartoffel: Zugleich ein beitrag zum problem der "Erworbenen Resistenz" in Pflanzenreich. *Arb. Biol. Aust.* (*Reichsanst*) *Berlin* **23**, 189–231.

Murata, M., Begg, G. S., Lambrou, F., Leslie, B., Simpson, R. J., Freeman, H. C. and Morgan, F. J. (1982). Amino acid sequence of a basic blue protein from cucumber seedlings. *Proc. Natl. Acad. Sci. U.S.A.* **79**, 6434–6437.

Murray, B. E., Craig, I. L. and Rajhathy, T. (1970). A protein electrophoretic study of three amphiploids and eight species in *Avena. Can. J. Genet. Cytol.* **12**, 651–665.

Murray, B. G. and Williams, C. A. (1973). Polyploidy and flavonoid synthesis in *Briza media* L. *Nature* (*Lond.*) **243**, 87–88.

Murray, B. G. and Williams, C. A. (1976). Chromosome number and flavonoid synthesis in *Briza* L. (Gramineae). *Biochem. Genet.* **14**, 897–904.

Murray, K. E., Shipton, J. and Whitfield, F. B. (1972). Volatile constituents of passionfruit, *Passiflora edulis. Australian J. Chem.* **25**, 1921–1933.

Murray, M. J. and Hefendehl, F. W. (1972). Changes in monoterpene composition of *Mentha aquatica* produced by gene substitution from *M. arvensis. Phytochemistry* **11**, 2469–2474.

Murray, M. J. and Hefendehl, F. W. (1973). Changes in monoterpene composition of *Mentha aquatica* produced by gene substitution from a high limonene strain of *M. citrata. Phytochemistry* **12**, 1875–1880.

Murray, M. J. and Lincoln, D. E. (1970). The genetic basis of acyclic oil constituents in *Mentha citrata* Ehph. *Genetics* **65**, 457–471.

Murray, M. J., Faas, W. and Marble, P. (1972a). Effect of plant maturity on oil composition of several spearmint species grown in Indiana and Michigan. *Crop sci.* **12**, 723–728, 742–745.

Murray, M. J., Lincoln, D. E. and Marble, P. W. (1972b). Oil composition of *Mentha aquatica* × *M. spicata*, F_1 hybrids in relation to the origin of × *M. piperita. Can. J. Genet. Cytol.* **14**, 13–29.

Nachtit, M. and Feucht, W. (1978). Phenolische Sauren bei unterschiedlich stark Wachsenden Arten und Artbastarden der Sektion Eucerasus. *Augen. Botanik* **52**, 321–329.

Nakanishi, K. (1968). Conference on insect-plant interactions. *Bioscience* **18**, 791–799.

Nakanishi, K., Goto, T., Ito, S., Natori, S. and Nozoe, S. (1974). 'Natural Products Chemistry', Vol. I. Academic Press, New York.

Narayan, R. K. J. (1982). Discontinuous DNA variation in the evolution of plant species: the genus *Lathyrus*. *Evolution* **36**, 877–891.

Natarella, N. J. and Sink, K. C. (1974). A chromatographic study of phenolics of species ancestral to *Petunia hybrida*. *J. Hered.* **65**, 85–90.

Naylor, A. W. (1976). Herbicide metabolism in plants. *In* "Herbicides" (Audus, L. J., ed.), 2nd edn., Vol. I, pp. 397–426. Academic Press, London.

Nei, M. (1972). Genetic distance between populations. *Amer. Natur.* **106**, 283–292.

Nevins, D. J., Yamamoto, R. and Huber, D. J. (1978). Cell wall β-D-glucans of five grass species. *Phytochemistry* **17**, 1503–1505.

Nevo, E. (1973). Test of selection and neutrality in natural populations. *Nature* (*Lond.*) **244**, 573–575.

Nicholas, H. J. (1973). Terpenes. *In* 'Phytochemistry. II.' (Miller, L. P., ed.), Van Nostrand-Reinhold, New York.

Nielsen, P. E., Nishimura, H., Liang, Y. and Calvin, M. (1979). Steroids from *Euphorbia* and other latex-bearing plants. *Phytochemistry* **18**, 103–104.

Niklas, K. J. (1976a). Chemical examinations of some non-vascular paleozoic plants. *Brittonia* **28**, 113–137.

Niklas, K. J. (1976b). The chemotaxonomy of *Parkia decipiens* from the lower old red sandstone, Scotland (UK). *Rev. Paleobot. Palynol.* **21**, 205–217.

Niklas, K. J. (1976c). Chemotaxonomy of Prototaxites and evidence for possible terrestrial adaption. *Rev. Paleobot. Palynol.* **22**, 1–17.

Niklas, K. J. and Brown, R. M., Jr. (1981). Ultrastructural and paleobiochemical correlations among fossil leaf tissues from the St. Mavies River (Clarkia) area, Northern Idaho. *Amer. J. Bot.* **68**, 332–341.

Niklas, K. J. and Gensel, P. G. (1977). Chemotaxonomy of some Paleozoic vascular plants. Part II: Chemical characterisation of major plant groups. *Brittonia* **29**, 100–111.

Niklas, K. J. and Giannasi, D. E. (1977a). Flavonoids and other chemical constituents of fossil Miocene *Zelkovia* (Ulmaceae). *Science* **196**, 877–878.

Niklas, K. J. and Giannasi, D. E. (1977b). Geochemistry and thermolysis of flavonoids. *Science* **197**, 767–769.

Niklas, K. J. and Giannasi, D. E. (1978). Angiosperm paleobiochemistry of the Succor Creek Flora (Miocene) Oregon. *Amer. J. Bot.* **65**, 943–952.

Niklas, K. J. and Pratt, L. M. (1980). Evidence for lignan-like constituents in early Silurian (Llandoverian) plant fossils. *Science* **209**, 396–397.

Nordby, H. E. and Nagy, S. (1974). Fatty acid composition of sterol esters from *Citrus* fruit sacs. *Phytochemistry* **13**, 443–452.

Nordby, H. E. and Nagy, S. (1977). Hydrocarbons from epicuticular waxes of *Citrus* peels. *Phytochemistry* **16**, 1393–1397.

Nordenstam, B. (1977). Senecioneae and Liabeae-systematic review. *In* 'The Biology and Chemistry of the Compositae' (Heywood, V. H., Harborne, J. B. and Turner, B. L., eds.), Vol. II, pp. 799–830. Academic Press, London.

Norin, T. (1972). Some aspects of the chemistry of the order Pinales. *Phytochemistry* **11**, 1231–1242.

Northington, D. K. (1974). Chemosystematic studies of genus *Pyrrhopappus*. *Spec. Publ. Mus. Tex. Tech. Univ.* **6**, 1–38.

Novotny, L., Toman, J. and Herout, V. (1968). Terpenoids of the *Petasites paradoxus* and *P. kablikianus* in relation to their phylogeny. *Phytochemistry* **7**, 1349–1353.

Nursten, H. (1970). Volatile compounds: the Aroma of Fruits. *In* 'The Biochemistry of Fruits and their Products' (Hulme, A. C. ed.), Vol. 1, pp. 239–268. Academic Press, London and New York.

Ogawa, H. and Natori, S. (1968). Hydroxybenzoquinones from Myrsinaceae II. Distribution among Myrsinaceae from Japan. *Phytochemistry* **7**, 773–782.

Ogilvie, R. T. and von Rudloff, E. (1968). Biosystematic studies in the genus *Picea* (Pinaceae). IV. The Introgression of white and Engelmann spruce as found along the Bow River. *Canad. J. Bot.* **46**, 901–908.

Ohmoto, T., Ikuse, M. and Natori, S. (1970). Triterpenoids of the Gramineae. *Phytochemistry* **9**, 2137–2148.

Okuda, T. and Mori, K. (1974). Distribution of *manno*-heptulose and sedoheptulose in plants. *Phytochemistry* **13**, 961–964.

Olivieri, A. M. and Jain, S. K. (1977). Variation in the *Helianthus exilisbolanderi* complex: a re-examination. *Madroño* **24**, 177–189.

Ornduff, R. and Bohm, B. A. (1975). Relationships of *Tracyina* and *Rigiopappus* (Compositae). *Madroño* **23**, 53–55.

Ornduff, R., Bohm, B. A. and Saleh, N. A. M. (1973). Flavonoids of artificial interspecific hybrids in *Lasthenia*. *Biochem. Syst.* **1**, 147–151.

Ornstein, L. and Davies, B. J. (1959). 'Disc Electrophoresis'. Distillation Products Industries (Div. of Eastman Kodak Co.).

Ouchterlony, O. (1948). *In vitro* method for testing the toxin-producing capacity of diphtheria bacteria. *Acta path. microbiol. Scand.* **25**, 189–191.

Ourisson, G. (1974). Some aspects of the distribution of diterpenes in plants. *In* 'Chemistry in Botanical Classification' (Bendz, G. and Santesson, J. eds.), pp. 129–134. Academic Press, New York.

Palmer, J. D. and Zamir, D. (1982). Chloroplast DNA evolution and phylogenetic relationships in *Lycopersicon*. *Proc. Natl. Acad. Sci. U.S.A.* **79**, 5006–5010.

Palmer, J. D., Osorio, B., Thompson, W. F. and Nobs, M. A. (1983). Chloroplast DNA evolution in *Atriplex*. Report for 1981–2 of the Carnegie Institute, Washington DC 96–97.

Pant, P. and Rastogi, R. P. (1979). The Triterpenoids. *Phytochemistry* **18**, 1095–1108.

Parfiti, B. D. (1980). Origin of *Opuntia curvospina* (Cactaceae). *System. Bot.* **5**, 408–418.

Paris, R. (1963). The distribution of plant glycosides. *In* 'Chemical Plant Taxonomy' (Swain, T., ed.), pp. 337–358. Academic Press, London.

Parker, W. H. (1976). Comparison of numerical taxonomic methods used to estimate flavonoid similarities in the Limnanthaceae. *Brittonia* **28**, 390–399.

Parks, C. R. and Kondo, K. (1974). Breeding systems in *Camellia* (Theaceae). I. A chemosystematic analysis of synthetic hybrids and backcrosses involving *Camellia japonica* and *C. saluenensis*. *Brittonia* **26**, 321–322.

Parks, C. R., Sandhu, S. S. and Montgomery, K. R. (1972). Floral pigmentation studies in the genus *Gossypium*. IV. Effects of different growing environments on flavonoid pigmentation. *Amer. J. Bot.* **59**, 158–164.

Patterson, C. (1980). Cladistics. *Biologist* **27**, 234–240.

Pauling, L. and Zuckerkandl, E. (1963). Chemical paleogenetics: molecular restoration studies of extinct forms of life. *Acta Chem. Scand. Suppl. No. 1* **17**, 9–16.

Pauly, G. and von Rudloff, E. (1971). Chemosystematic studies in the genus *Pinus contortus* var. *latifolia*. *Canad. J. Bot.* **49**, 1201–1210.

Payne, W. W. (1976). Biochemistry and species problems in *Ambrosia* (Asteraceae-Ambrosieae). *Plant Syst. Evol.* **125**, 169–178.

Payne, W. W., Geissman, T. A., Lucas, A. J. and Saitoh, T. (1973). Chemosystematics and Taxonomy of *Ambrosia chamissonis*. *Biochem. Syst.* **1**, 21–33.

Peacock, D. and Boulter, D. (1975). Use of amino acid sequence data in phylogeny and evaluation of methods using computer simulation. *J. Molecular Biology* **95**, 513–527.

Pedersen, J. A. (1978). Naturally occurring quinols and quinones studied as semiquinones by electron spin resonance. *Phytochemistry* **17**, 775–778.

Pelletier, S. W., ed. (1970). 'Chemistry of the Alkaloids.' Van Nostrand-Reinhold, New York.

Penfold, A. R. and Morrison, F. R. (1927). The occurrence of a number of varieties of *Eucalyptus dives* as determined by chemical analyses of the essential oils. Part I. *Journ. Proc. Roy. Soc. N.S. Wales* **61**, 54–67.

Penny, D., Foulds, L. R. and Hendy, M. D. (1982). Testing the theory of evolution by comparing phylogenetic trees constructed from five different protein sequences. *Nature* (*Lond.*) **297**, 197–200.

Percival, E. (1967). The natural distribution of plant polysaccharides. *In* 'Comparative Phytochemistry' (Swain, T., ed.), pp. 139–158. Academic Press, London.

Percival, E. and McDowell, R. H. (1967). 'Chemistry and Enzymology of Marine Algal Polysaccharides.' Academic Press, London.

Percival, E. and McDowell, R. H. (1981). Algal walls—Composition and Biosynthesis. *In* 'Extracellular Carbohydrates' (Tanner, W. and Loewus, F. A., eds.), pp. 277–316. Springer-Verlag, Berlin.

Percival, M. S. (1961). Types of nectar in angiosperms. *New Phytol.* **60**, 235–281.

Petiver, J. (1699). Some attempts made to prove that herbs of the same make or class for the generality, have the like Vertue and tendency to work the same effects. *Phil. Trans.* **21B**, 289–291.

Phelan, J. R. and Vaughan, J. G. (1976). A chemotaxonomic study of *Brassica oleracea* with particular reference to its relationship to *Brassica alboglabra. Biochem. Syst. Ecol.* **4**, 173–178.

Phillipson, J. D. (1982). Chemical investigations of herbarium material for alkaloids. *Phytochemistry* **21**, 2441–2456.

Phillipson, J. D. and Hemingway, S. R. (1975). Alkaloids of *Uncaria attenuata, U. orientalis* and *U. canescens. Phytochemistry* **14**, 1855–1863.

Piattelli, M. (1976). Betalains. *In* 'Chemistry and Biochemistry of Plant Pigments' (Goodwin, T. W., ed.), pp. 560–596. Academic Press, London.

Piattelli, M. and Minale, L. (1964). Pigments of Centrospermae. II. Distribution of betacyanins. *Phytochemistry* **3**, 547–557.

Pickering, J. L. and Fairbrother, D. E. (1971). The use of serological data in a comparison of tribes in the Apioideae. *In* 'The Biology and Chemistry of the Umbelliferae' (Heywood, V. H., ed.), pp. 315–324. Academic Press, London.

Pimenov, M. G., Sklyar, Y. E., Savina, A. A. and Baranova, Y. V. (1982). Chemosystematics of Section Palaeonarthex of the genus *Ferula. Biochem. Syst. Ecol.* **10**, 133–138.

Plouvier, V. (1963). The distribution of aliphatic polyols and cyclitols. *In* 'Chemical Plant Taxonomy' (Swain, T. ed.), pp. 313–336. Academic Press, London.

Platnick, N. and Funk, V. A. (eds.) (1983) 'Advances in Cladistics II.' Columbia University Press, New York.

Plouvier, V. and Favre-Bonvin, J. (1971). Les iridoides et seco-iridoides: repartition, structure, proprieties, biosynthese. *Phytochemistry* **10**, 1697–1722.

Pokorny, M. (1974). D-Methionine metabolic pathways in Bryophyta: a chemotaxonomic evaluation. *Phytochemistry* **13**, 965–972.

Pollard, C. J. (1982). Fructose oligosaccharides in the Monocotyledons: a possible delimitation of the order Liliales. *Biochem. Syst. Ecol.* **10**, 245–250.

Pollard, C. J. and Amuti, K. S. (1981). Fructose oligosaccharides: possible markers of phylogenetic relationships among dicotyledonous families. *Biochem. Syst. Ecol.* **9**, 69–78.

Poltavtchenko, Y. A., Tkatch, T. N., Tkatch, V. S. and Rudakov, G. A. (1968). Dynamics of the distribution of monoterpenes in Scots pine (*Pinus sylvestris*). *Biol. Nauki* **11**, 71–76.

Ponsinet, G., Ourisson, G. and Oehlschager, A. C. (1968). Systematic aspects of the distribution of di- and triterpenes. *Recent Adv. Phytochem.* **1**, 271–302.

Potter, J. and Mabry, T. J. (1972). Origin of the Texas Gulf Coast Island populations of *Ambrosia psilostachya*: a numerical study using terpenoid data. *Phytochemistry* **11**, 715–723.

Powell, R. A. and Adams, R. P. (1973). Seasonal variation in the volatile terpenoids of *Juniperus scopulorum* (Cupressaceae). *Amer. J. Bot.* **60**, 1041–1050.

Preiss, J. (ed.) (1980). Carbohydrates: structure and function. *In* 'The Biochemistry of Plants: A Comprehensive Treatise,' Vol. 3. Academic Press, New York.

Price, J. R. (1963). The distribution of alkaloids in the Rutaceae. *In* 'Chemical Plant Taxonomy' (Swain, T., ed.), pp. 429–452. Academic Press, London.

Pridham, J. B. (1964). The phenol glucosylation reaction in the plant kingdom. *Phytochemistry* **3**, 493–497.

Pryce, R. J. (1972). The occurrence of lunularic and abscisic acids in plants. *Phytochemistry* **11**, 1759–1761.

Pryor, L. D. and Bryant, L. H. (1958). Inheritance of oil characters in *Eucalyptus*. *Proc. Linn. Soc. N.S. Wales* **83**, 55–64.

Purdie, A. W. and Connor, H. E. (1973). Triterpene methyl ethers of *Cortaderia splendens*. *Phytochemistry* **12**, 1196.

Raffauf, R. F. (1970). 'A Handbook of Alkaloids and Alkaloid-containing Plants.' Wiley, New York.

Raghavendra, A. S. and Das, V. S. R. (1978). The occurrence of C_4-photosynthesis: a supplementary list of C_4 plants reported during late 1974–mid 1977. *Photosynthetica* **12**, 200–208.

Ramshaw, J. A. M. (1982). Structures of Plant Proteins. *Encyclopedia Pl. Physiol. new series*, **14A**, 229–290.

Raynal, J. (1973). Notes Cyperologiques: contribution a la classification de la sous-famille des Cyperoideae. *Adansonia ser.* 2., **13**, 145–171.

Reddi, V. B. and Phipps, J. B. (1972). Free amino acids as taxonomic characters in the tribe Arundinelleae. *Brittonia* **24**, 403–414.

Reddy, M. M. and Garber, E. D. (1971). Genetic studies of variant enzymes. III. Comparative electrophoretic studies of esterases and peroxidases for species, hybrids and amphiploids in the genus *Nicotiana*. *Bot. Gaz.* **132**, 158–166.

Reddy, M. M. and Threlkeld, S. F. H. (1971). Genetic studies of isozymes in *Neurospora*. I. A study of eight species. *Can. J. Genet. Cytol.* **13**, 298–305.

Reichstein, T. (1965). Chemische Rassen in *Acokanthera*. *Planta Medica* **13**, 382–399.

Renold, W. (1970). The chemistry and infraspecific variation of sesquiterpene lactones in *Ambrosia confertiflora*. Ph.D. thesis, University of Texas, Austin.

Rezende, C. M. and Gottlieb, O. R. (1973). Xanthones as systematic markers. *Biochem. System.* **1**, 111–118.

Reznik, H. (1980). Betalains. *In* 'Pigments in Plants' (Czygan, F. C., ed.), 2nd edn., pp. 370–392. Gustav Fischer, Stuttgart.

Rhoades, D. G., Lincoln, D. E. and Langenheim, J. H. (1976). Preliminary studies of monoterpenoid variability in *Satureja douglasii*. *Biochem. Syst. Ecol.* **4**, 5–12.

Rice, R. L., Lincoln, D. E. and Langenheim, J. H. (1978). Palatability of monoterpenoid compositional types of *Satureja douglasii* to a generalist Molluscan herbivore, *Ariolimax dolichophallus*. *Biochem. Syst. Ecol.* **6**, 45–53.

Richardson, P. M. (1978). Lower terpenes in introgressing *Mentha* species. Ph.D. thesis, University of London.

Richardson, P. M. and Young, D. A. (1982). The phylogenetic content of flavonoid point scores. *Biochem. Syst. Ecol.* **10**, 251–255.

Rider, C. C. and Taylor, C. B. (1980). Isoenzymes, outline studies in Biology. Chapman and Hall, London.

Rix, E. M. and Rast, D. (1975). Nectar sugars and subgeneric classification in *Fritillaria*. *Biochem. Syst. Ecol.* **2**, 207–209.

Robeson, D. J. and Harborne, J. B. (1977). Pisatin as a major phytoalexin in *Lathyrus*. *Z. Naturforsch.* **32C**, 289.

Robeson, D. J. and Harborne, J. B. (1980). A chemical dichotomy in phytoalexin induction in the tribe Vicieae of the Leguminosae. *Phytochemistry* **19**, 2359–2366.

Robeson, D. J. and Ingham, J. L. (1979). New pterocarpan phytoalexins from *Lathyrus nissolia*. *Phytochemistry* **18**, 1715–1717.

Roberts, R. J. (1980). Restriction and modification enzymes and their recognition sequences. *Gene* **8**, 329–343.

Robinson, H. (1975). Considerations on the evolution of lichens. *Phytologia* **32**, 407–413.

Robinson, T. (1968). 'The Biochemistry of Alkaloids.' Springer-Verlag, New York.

Robinson, T. (1974). Metabolism and function of alkaloids in plants. *Science* **184**, 430–435.

Robinson, T. (1981). 'The Biochemistry of Alkaloids.' 2nd edn. Springer-Verlag, Berlin.

Rock, H. F. L. (1956). Systematics of *Helenium*. Ph.D. thesis, Duke University, N. Carolina, U.S.A.

Rockwood, D. L. (1973). Variation in the monoterpene composition of two oleoresin systems of loblolly pine. *Forest Sci.* **19**, 147–153.

Rodman, J. E. (1980). Population variation and hybridization in sea-rockets (*Cakile*, Cruciferae): seed glucosinolate characters. *Amer. J. Bot.* **67**, 1145–1159.

Rodman, J. E. (1981). Divergence, convergence and parallelism in phytochemical characters: The glucosinolate-myrosinase system. *In* 'Phytochemistry and Angiosperm Phylogeny' (Young, D. A. and Seigler, D. S., eds.), pp. 43–79. Praeger Publishers, New York.

Rodman, J. E., Kruckeberg, A. R. and Al-Shehbaz, J. A. (1981). Chemotaxonomic diversity and complexity in seed glucosinolates of *Caulanthus* and *Streptanthus* (Cruciferae). *System. Bot.* **6**, 197–222.

Rodriguez, E. (1977). Ecogeographic distribution of secondary constituents in *Parthenium* (Compositae). *Biochem. Syst. Ecol.* **5**, 207–218.

Rodriguez, E., Towers, G. H. N. and Mitchell, J. C. (1976). Biological activities of sesquiterpene lactones. *Phytochemistry* **15**, 1573–1580.

Rodriguez, E., Reynolds, G. W. and Thompson, J. A. (1981). Potent contact allergen in the rubber plant guayule (*Parthenium argentatum*). *Science* **211**, 1444–1445.

Rogers, C. M. (1972). The taxonomic significance of the fatty acid content of seeds of *Linum*. *Brittonia* **24**, 415–419.

Romeike, A. (1978). Tropane alkaloids—occurrence and systematic importance in angiosperms. *Bot. Notiser* **131**, 85–96.

Roo, R. de (1963). La teneur en acide cyanhydrique du trefle blanc tetraploide. *Rev. Agr. Brussele* **16**, 477–479.

Roose, M. L. and Gottlieb, L. D. (1976). Genetic and biochemical consequences of polyploidy in *Tragopogon*. *Evolution* **30**, 818–830.

Rosenthal, G. A. (1982). 'Plant Non-protein Amino and Imino Acids: Biological, Biochemical and Toxicological Properties.' Academic Press, New York.

Rothschild, M. (1972). Some observations on the relationship between plants, toxic insects and birds. *In* 'Phytochemical Ecology' (Harborne, J. B., ed.), pp. 1–12. Academic Press, London.

Rothschild, M. (1973). Secondary plant substances and warning coloration in insects. *In* 'Insect-Plant Relationships' (van Emden, H. F., ed.), pp. 59–83. Blackwell, Oxford.

Rubin, J. L. and Jensen, R. A. (1979). Enzymology of L-tyrosine biosynthesis in mung bean, *Vigna radiata*. *Plant Physiol.* **64**, 727–734.

Ruijgrok, H. W. L. (1968). The distribution of ranunculin and cyanogenetic compounds in the Ranunculaceae. *In* 'Comparative Phytochemistry' (Swain, T., ed.), pp. 175–186. Academic Press, London.

Russell, G. B., Connor, H. E. and Purdie, A. W. (1976). Triterpene methyl ethers of *Chionochloa* (Gramineae). *Phytochemistry* **15**, 1933–1935.

Ryden, L. and Lundgren, J. O. (1979). On the evolution of blue proteins. *Biochimie* **61**, 781–790.

Sachs, J. von. (1890). 'History of Botany (1530–1860).' Authorised translation by H. E. F. Garnsey and I. B. Balfour, Oxford.

Saghir, A. R., Mann, L. K., Ownbey, M. and Berg, R. Y. (1966). Composition of volatiles in relation to taxonomy of American Alliums. *Am. J. Bot.* **53**, 477–484.

Sahai, S. and Rana, R. S. (1977). Seed protein homology and elucidation of species relationships in *Phaseolus* and *Vigna* species. *New Phytol.* **79**, 527–534.

Saleh, N. A. M. and El-Lakany, M. H. (1979). A quantitative variation in the flavonoids and phenolics of some *Casuarina* species. *Biochem. Syst. Ecol.* **7**, 13–15.

Salinas, J., Perez de la Vega, M. and Benito, C. (1982). Identification of hexaploid wheat cultivars on isozyme patterns. *J. Sci. Fd. Agr.* **33**, 221–226.

Salthe, S. N. (1972). 'Evolutionary Biology.' Holt, Rinehart and Winston, Inc. New York.

Sanford, K. J. and Heinz, D. E. (1971). Effects of storage on the volatile composition of nutmeg. *Phytochemistry* **10**, 1245–1250.

Santamour, F. S., Jr. (1972). Flavonoid distribution in *Ulmus*. *Bull. Torrey Bot. Club* **99**, 127–131.

Santamour, F. S., Jr. (1973). Anthocyanins of holly fruit. *Phytochemistry* **12**, 611–615.

Santavy, F. (1970). Papaveraceae alkaloids. *In* 'The Alkaloids' (Manske, R. H. F., ed.), Vol. XII, pp. 333–454. Academic Press, New York.

Santesson, J. (1970). Anthraquinones in *Caloplaca*. *Phytochemistry* **9**, 2149–2166.

Scandalios, J. G. (1969). Genetic control of multiple molecular forms of enzymes in plants: a review. *Biochem. Genet.* **3**, 37–79.

Schantz, M. and Juvonen, S. (1966). Chemotaxonomische untersuchungen in der gattung *Picea*. *Acta Bot. Fennica* **73**, 1–50.

Scheffer, J. J. C., Ruys-Catlender, C. M., Koedam, A. and Svendsen, A. B. (1980). A comparative study of ×*Cupressocyparis leylandii* clones (Cupressaceae) by gas chromatographic analysis of their leaf oils. *Bot. J. Linn. Soc.* **81**, 215–224.

Schildkraut, C. L., Marmur, J. and Doty, P. (1962). Determination of the base composition of deoxyribonucleic acid from its buoyant density in CsCl. *J. Mol. Biol.* **4**, 430–443.

Schopf, J. W. (1970). Precambrian micro-organisms and evolutionary events prior to the origin of vascular plants. *Biol. Rev.* **45**, 319–352.

Schramm, L. C. and Schwarting, A. E. (1961). Alkaloid distribution in Colombian cinchonas. *Lloydia* **24**, 27–40.

Schwarze, P. (1959). Untersuchungen uber die gesteigerte flavonoidproduktion in *Phaseolus*-artbastarden (*Phaseolus vulgaris* × *P. coccineus*). *Planta* **54**, 152–161.

Scogin, R. (1969). Isoenzyme polymorphism in natural populations of the genus *Baptisia* (Leguminosae). *Phytochemistry* **8**, 1733–1737.

Scogin, R. (1972). Proteins of the genus *Lithops* (Airoaceae): Developmental and comparative studies. *J.S. Afr. Bot.* **38**, 55–61.
Scogin, R. (1973). Isoenzyme polymorphism in the genus *Thermopsis. Biochem. System.* **1**, 79–81.
Scogin, R. (1979). Nectar constituents in the genus *Fremontia* (Sterculiaceae): sugars, flavonoids and proteins. *Bot. Gaz.* **140**, 29–31.
Scora, R. W. (1967). The essential leaf oils of the genus *Monarda* (Labiatae). *Am. J. Bot.* **54**, 446–452.
Scora, R. W. and Wagner, W. H. Jr. (1964). A preliminary chromatographic study of eastern American *Dryopteris. Amer. Fern J.* **54**, 105–113.
Scora, R. W., England, A. B. and Bitters, W. P. (1966). The essential oil of *Poncirus trifoliata* (L.) Raf. and its selections in relation to classification. *Phytochemistry* **5**, 1139–1146.
Scora, R. W., Bergh, B. O. and Hopfinger, J. A. (1975). Leaf alkanes in *Persea* and related taxa. *Biochem. Syst. Ecol.* **3**, 215–218.
Seaman, F. C. (1982). Sesquiterpene lactones as taxonomic characters in the Asteraceae. *Bot. Rev.* **48**, 121–595.
Seaman, F. C. and Funk, V. A. (1983). Cladistic analysis of complex natural products: developing transformation series from sesquiterpene lactone data. *Taxon* **32**, 1–27.
Secor, J. B., Conn, E. C., Dunn, J. E. and Seigler, D. S. (1976). Detection and identification of cyanogenic glucosides in six species of *Acacia. Phytochemistry* **15**, 1703–1706.
Seigler, D. S. (1977a). Plant systematics and alkaloids. *In* 'The Alkaloids' (Manske, R. H. F., ed.), Vol. XVI, pp. 1–83. Academic Press, London.
Seigler, D. S. (1977b). The naturally occurring cyanogenic glycosides. *Progr. Phytochem.* **4**, 83–120
Seigler, D. S. and Kawahara, W. (1976). New reports of cyanolipids from sapindaceous plants. *Biochem. Syst. Ecol.* **4**, 263–265.
Seigler, D. S., Dunn, J. E., Conn, E. E. and Holstein, G. L. (1978a). Acacipetalin from six species of *Acacia* of Mexico and Texas. *Phytochemistry* **17**, 445–446.
Seigler, D. S., Simpson, B. B., Martin, C. and Neff, J. L. (1978b). Free 3-acetoxy fatty acids. in floral glands of *Krameria* species. *Phytochemistry* **17**, 995–996.
Selander, R. K. (1976). Genetic variation in natural populations. *In* 'Molecular Evolution' (Ayala, F. J., ed.), pp. 21–45. Sinauer Assoc. Inc., Publisher, Sunderland, Mass.
Sellmair, J., Beck, E., Kandler, O. and Kress, A. (1977). Hamamelose and its derivatives and chemotaxonomic markers in the genus *Primula. Phytochemistry* **16**, 1201–1264.
Semple, J. C. and Semple, K. S. (1978). *Borrichia* × *cubana* (*B. frutescens* × *arborescens*): interspecific hybridisation in the Florida keys. *Syst. Bot.* **2**, 292–301.
Seneviratne, A. S. and Fowden, L. (1968). The amino acids of the genus *Acacia. Phytochemistry* **7**, 1039–1045.
Senser, M. and Kandler, O. (1967). Galactinol, ein Galactosyldonor fur die Biosynthese der Zucker der Raffinosefamilien in Blattern. *Z. Pflanzenphysiol.* **57**, 376–388.
Sharitz, R. R., Wineriter, S. A., Smith, M. H. and Liu, E. H. (1980). Comparison of isozymes among *Typha* species in the Eastern United States. *Amer. J. Bot.* **67**, 1297–1303.

Shaw, C. R. (1965). Electrophoretic variation in enzymes. *Science* **149**, 936–943.
Sheard, J. W. (1977). The taxonomy of the *Ramalina siliquosa* species aggregate (lichenized Ascomycetes). *Can. J. Bot.* **56**, 916–937.
Sheen, S. J. (1970). Peroxidases in the genus *Nicotiana*. *Theor. Appl. Genet.* **40**, 18–25.
Shewry, P. R., Pratt, H. M. and Miflin, B. J. (1978). Varietal identification of single seeds of barley by analysis of hordein polypeptides. *J. Sci. Fd. Agr.* **29**, 587–596.
Shorland, F. B. (1963). The distribution of fatty acids in plant lipids. *In* 'Chemical Plant Taxonomy' (Swain, T., ed.), pp. 253–312. Academic Press, London and New York.
Shulgin, A. T. (1966). Possible implications of myristicin as a psychotropic substance. *Nature* (*Lond.*) **210**, 380–383.
Siddiqui, K. A., Ingversen, J. and Koie, B. (1972). Inheritance of protein patterns in a synthetic allopolyploid of *Triticum monoccum* (AA) and *Aegilops ventricosum* (DDM^vM^v). *Hereditas* **72**, 205–214.
Sieffermann-Harms, D., Hertzberg, S., Borch, G. and Liaaen-Jensen, S. (1981). Lactucaxanthin, an ε,ε-carotene-3,3′-diol from *Lactuca sativa*. *Phytochemistry* **20**, 85–88.
Simmons, D. and Parsons, R. F. (1976). Analysis of a hybrid swarm involving *Eucalyptus crenulata* and *E. ovata* using leaf oils and morphology. *Biochem. Syst. Ecol.* **4**, 97–101.
Sing, C. F. and Brewer, G. J. (1969). Isozymes of a polyploid series of wheat. *Genetics* **61**, 391–398.
Slob, A., Jekel, B., Jong B. de and Schlatmann, E. (1975). On the occurrence of tuliposides in the Liliiflorae. *Phytochemistry* **14**, 1997–2006.
Smith, B. N. and Brown, W. V. (1973). The Kranz syndrome in the Gramineae as indicated by carbon isotopic ratios. *Amer. J. Bot.* **60**, 505–513.
Smith, B. N. and Turner, B. L. (1975). Distribution of the Kranz syndrome among Asteraceae. *Amer. J. Bot.* **62**, 541–545.
Smith, D. B. and Flavell, R. B. (1974). The relatedness and evolution of repeated nucleotide sequences in the genomes of some Gramineae species. *Biochemical Genetics* **12**, 243–256.
Smith, D. M. and Levin, D. R. (1963). A chromatographic study of reticulate evolution in the Appalachian *Asplenium* complex. *Amer. J. Bot.* **50**, 952–958.
Smith, E. L. (1967). The evolution of proteins. *Harvey Lect.* **62**, 231–246.
Smith, H. H. and Abashian, D. V. (1963). Chromatographic investigations on the alkaloid content of *Nicotiana* species and interspecific combinations. *Amer. J. Bot.* **50**, 435–447.
Smith, H. H., Hamill, D. E., Weaver, E. A. and Thompson, K. H. (1970). Multiple molecular forms of peroxidases and esterases among *Nicotiana* species and amphiploids. *J. Hered.* **61**, 203–214.
Smith, R. H. (1964). The monoterpenes of lodgepole pine oleoresin. *Phytochemistry* **3**, 259–262.
Smith, R. M. and Martin-Smith, M. (1978). Triterpene methyl ethers in leaf waxes of *Saccharum* and related genera. *Phytochemistry* **17**, 1307–1312.
Smith, P. M. (1976). 'The Chemotaxonomy of Plants.' Edward Arnold, London.
Snajberk, K., Zavarin, E. and Debry, R. (1982). Terpenoid and morphological variability of *Pinus quadrifolia* and its natural hybridisation with *P. monophylla* in the San Jacinto mountains of California. *Biochem. Syst. Ecol.* **10**, 121–132.
Sneath, P. H. A. and Sokal, R. R. (1973). 'Numerical Taxonomy.' W. H. Freeman, San Francisco.

Solomonson, L. P. and Spehar, A. M. (1977). Model for the regulation of nitrate assimilation. *Nature* (*Lond.*) **265**, 373–375.

Sorensen, N. A. (1968). The taxonomic significance of acetylenic compounds. *Recent Adv. Phytochem.* **1**, 187–228.

Sorensen, N. A. (1977). Polyacetylenes and conservatism of chemical characters in the Compositae. *In* 'The Biology and Chemistry of the Compositae' (Heywood, V. H., Harborne, J. B. and Turner, B. L., eds.), Vol. I, pp. 385–410. Academic Press, London.

Sorensen, P. D., Totten, C. E. and Piatak, D. M. (1978). Alkane chemotaxonomy of *Arbutus*. *Biochem. Syst. Ecol.* **6**, 109–112.

Spensel, V. M., Gaskin, P. and MacMillan, J. (1979). Identification of gibberellins in immature seeds of *Vicia faba* and some chemotaxonomic considerations. *Planta* **146**, 101–106.

Sprague, T. A. (1940). Taxonomic botany with special reference to the angiosperms. *In* 'The New Systematics' (Huxley, S. J., ed.). Clarendon Press, Oxford.

Sprecher, E. (1958). Uber den Einfluss genetischer und klumatischen Faktoren auf die Biosynthese sekundarer Pflanzenstoffe der Weinrante (*Ruta graveolens* L.) *Pharmazie* **13**, 151–153.

Squillace, A. E. (1971). Inheritance of monoterpene composition in cortical oleoresin of slash pine. *Forestry Sci.* **17**, 381–387.

Stace, C. A. (ed.) (1975). 'Hybridisation and the Flora of the British Isles.' Academic Press, London.

Stahl, E. (ed.) (1969). 'Thin Layer Chromatography,' 2nd edn., 1041 pp. George Allen and Unwin, London.

Stangl, R. and Greger, H. (1980). Monoterpene und Systematik der Gattung *Artemisia*. *Pl. Syst. Evol.* **136**, 125–136.

Stanier, R. Y. (1968). Biochemical and immunological studies on the evolution of a metabolic pathway in bacteria. *In* 'Chemotaxonomy and Serotaxonomy' (Hawkes, J. G., ed.), pp. 201–225. Academic Press, London.

Star, A. E., Seigler, D. S., Mabry, T. J. and Smith, D. M. (1975). Internal flavonoid patterns of diploids and tetraploids of two exudate chemotypes of *Pityrogramma triangularis* (Kaulf.). Maxon. *Biochem. Syst. Ecol.* **2**, 109–112.

Stebbins, G. L., Jr. (1950). 'Variation and Evolution in Plants'. Columbia University Press, New York.

Stebbins, G. L., Harvey, B. L., Cox, E. L., Rutger, J. N., Jelenkovic, G. and Yagil, E. (1963). Identification of the ancestry of an amphiploid *Viola* with the aid of paper chromatography. *Amer. J. Bot.* **50**, 830–839.

Stegemann, H. (1979). Characterisation of proteins from potatoes and the index of European varieties. *In* 'The Biology and Taxonomy of the Solanaceae' (Hawkes, J. G., Lester, R. N. and Skelding, A. D., eds.), pp. 279–284. Academic Press, London.

Stein, D. B., Thompson, W. F. and Belford, H. S. (1979). Studies of DNA sequences in Osmundaceae. *J. Molec. Evol.* **13**, 3–21.

Stephen, A. M. (1979). Plant carbohydrates. *In* 'Encyclopedia of Plant Physiology,' new series (Bell, E. A. and Charlwood, B. V., eds.), Vol. 8, pp. 555–584. Springer-Verlag, Berlin.

Stermitz, F. R. (1968). Alkaloid chemistry and systematics of *Papaver* and *Argemone*. *In* 'Recent Advances in Phytochemistry' (Mabry, T. J., ed.), Vol. 1. pp. 161–186. Appleton-Century-Crofts, New York.

Stermitz, F. R., Kim, D. K. and Larson, K. A. (1975). Alkaloids of *Argemone albiflora*, *A. brevicornuta* and *A. turnerae*. *Phytochemistry* **12** 1355–1357.

Stern, D. B. and Lonsdale, D. M. (1982). Mitochondrial and chloroplast genomes of maize have a 12-kilobase DNA sequence in common. *Nature* (*Lond.*) **299**, 698–702.

Stern, K. R. (1961). Revision of *Dicentra* (Fumariaceae). *Brittonia* **13**, 1–57.

Sterner, R. W. and Young, D. A. (1980). Flavonoid chemistry and the phylogenetic relationships of the Idiospermaceae. *System. Bot.* **5**, 432–437.

Stevens, F. C., Glazer, A. N. and Smith, E. L. (1967). The amino acid sequence of wheat germ cytochrome *c*. *J. biol. Chem.* **242**, 2764–2779.

Stewart, W. D. P. (ed.) (1974). 'Algal Physiology and Biochemistry.' 989 pp. Blackwell Scientific, Oxford.

Stillmark, H. (1888). Uber Ricin, ein giftiges Ferment. Diss. Dorpat.

Stinard, P. S. and Nevins, D. J. (1980). Distribution of noncellulosic β-D-glucans in grasses and other monocots. *Phytochemistry* **19**, 1467–1468.

Stoessl, A., Stothers, J. B. and Ward, E. W. B. (1976). Sesquiterpenoid stress compounds of the Solanaceae. *Phytochemistry* **15**, 855–872.

Stone, D. E., Adrouny, G. A. and Adrouny, S. (1965). Morphological and chemical evidence on the hybrid nature of bitter pecan, *Carya* × *lecontei*. *Brittonia* **17**, 97–106.

Stone, D. E., Adrouny, G. A. and Flake, R. H. (1969). New Word Juglandaceae. II. Hickory nut oils, phenetic similarities and evolutionary implications in the genus *Carya*. *Am. J. Bot.* **56**, 928–935.

Storck, R. and Alexopoulos, C. J. (1970). Deoxyribonucleic acid of fungi. *Bacteriol. Rev.* **34**, 126–154.

Stromnaes, O. and Garber, E. D. (1963). A paper chromatographic study of root extracts from 15 species and 6 interspecific hybrids. *Bot. Gaz.* **5**, 363–367.

Stuessy, T. (1980). Cladistics and plant systematics: problems and prospects. *Syst. Bot.* 109–111.

Stuessy, T. F., Irving, R. S. and Ellison, W. L. (1973). Hybridisation and evolution in *Picradeniopsis* (Compositae). *Brittonia* **25**, 40–56.

Sutherland, M. D. and Park, R. J. (1967). Sesquiterpenes and their biogenesis in *Myoporum deserti*. *In* 'Terpenoids in Plants' (Pridham, J. B., ed.), pp. 147–158. Academic Press, London.

Swain, T. (ed.) (1963). 'Chemical Plant Taxonomy.' Academic Press, London.

Swain, T. (ed.) (1966). 'Comparative Phytochemistry.' Academic Press, London.

Swain, T. (1975). Evolution of flavonoid compounds. *In* 'The Flavonoids' (Harborne, J. B., Mabry, T. J. and Mabry, H. eds.), pp. 1096–1129. Chapman and Hall, London.

Swain, T. and Williams, C. A. (1977). Heliantheae-chemical review. *In* 'The Biology and Chemistry of the Compositae' (Heywood, V. H., Harborne, J. B. and Turner, B. L., eds.), pp. 673–698. Academic Press, London.

Sweeley, C. C., Bentley, R., Makita, M. and Wells, W. W. (1963). GLC of trimethylsilyl derivatives of sugars and related substances. *J. Am. Chem. Soc.* **85**, 2497.

Szarek, S. R. and Ting, P. I. (1977). The occurrence of Crassulacean acid metabolism among plants. *Photosynthetica* **11**, 330–342.

Takhtajan, A. L. (1969). 'Flowering Plants: Origin and Dispersal'. Oliver and Boyd, Edinburgh.

Takhtajan, A. (1980). Outline of the classification of flowering plants. *Bot. Rev.* **46**, 225–359.
Tang, C. S., Syed, M. M. and Hamilton, R. A. (1972). Benzyl isothiocyanate content as a possible chemotaxonomic criterion in the Caricaceae. *Phytochemistry* **11**, 2531–2533.
Tanner, W. and Loewus, F. A. (eds.) (1981). Extracellular carbohydrates. *In* 'Encyclopedia of Plant Physiology, New Series,' Vol. 13B. Springer-Verlag, Berlin.
Tateoka, T., Hiraoka, A. and Tateoka, T. N. (1977). Natural hybridisation in Japanese *Calamagrostis*. II. *Calamagrostis langsdorffii* × *C. sachalinensis*, an example of an agamic complex. *Bot. Mag. Tokyo* **90**, 193–209.
Taylor, I. E. P. and Elliott, A. M. (1972). Dehydrogenases in a single population of the moss, *Eurhynchium oreganum*. The effects of dehydration and low temperature on disc electrophoretic enzyme patterns. *Can. J. Bot.* **50**, 375–378.
Taylor, R. J. (1971). Intraindividual phenolic variation in the genus *Tiarella* (Saxifragaceae); its genetic regulation and application to systematics. *Taxon* **20**, 467–472.
Taylor, R. J. (1972). The relationship and origin of *Tsuga heterophylla* and *T. mertensiana* based on phytochemical and morphological interpretations. *Amer. J. Bot.* **59**, 149–157.
Taylor, W. I. and Farnsworth, N. R. (eds.) (1975). 'The Catharanthus Alkaloids.' 323 pp. Marcel Dekker, New York.
Tax, S. (ed.) (1960). 'Evolution after Darwin.' Vol. 1 and 2. University of Chicago Press, Chicago.
Tétényi, P. (1958). Proposition a propos de la nomenclature des races chimiques. *Taxon* **7**, 40–41.
Tétényi, P. (1968). The nomenclature of infraspecific chemical taxa. *Taxon* **17**, 261–264.
Tétényi, P. (1970). 'Infraspecific Chemical Taxa of Medicinal Plants.' Akademiae Kiado, Budapest.
Tétényi, P. (1976). A szaponinok Novenyuilagbeli elterjedesenek kemotaxonomiaja. *Herba Hungarica* **15**, 27–47.
Thakur, M. L. and Ibrahim, R. K. (1974). Biogenesis of flavonoids in flax seedlings. *Z. Pflanzenphysiol.* **71**, 391–397.
Thien, L. B., Heimermann, W. and Holman, R. T. (1975). Floral odours and quantitative taxonomy of *Magnolia* and *Liriodendron*. *Taxon* **24**, 557–568.
Thomas, H. and Jones, D. I. H. (1968). Electrophoretic studies of proteins in *Avena* in relationship to genome homology. *Nature* (*Lond.*) **220**, 825–826.
Thompson, W. F. and Murray, M. G. (1981). The nuclear genome: structure and function. *In* 'Proteins and Nucleic Acids' (Marcus, A., ed.), pp. 1–81. Academic Press, New York.
Thompson, W. R., Meinwald, J., Aneshansley, D. and Eisner, J. (1972). Flavonoid pigments responsible for UV absorption in nectar guide of flowers. *Science* **177**, 528.
Thomson, R. H. (1971). 'Naturally Occurring Quinones.' 2nd edn. Academic Press, London.
Thomson, R. H. (1976). Quinones, nature, distribution and biosynthesis. *In* 'Chemistry and Biochemistry of Plant Pigments' (Goodwin, T. W., ed.), 2nd edn, Vol. 1, pp. 527–559. Academic Press, London.

Thornber, C. W. (1970). Alkaloids of the Menispermaceae. *Phytochemistry* **9**, 157–187.

Thorne, R. (1976). A phylogenetic classification of the Angiospermae. *Evolutionary Biology* **9**, 35–106.

Thorpe, J. P. (1979). Enzyme variation and taxonomy: the estimation of sampling errors in measurements of interspecific genetic similarity. *J. Linn. Soc.* **11**, 369–386.

Thurman, D. A., Boulter, D., Derbyshire, E. and Turner, B. L. (1967). Electrophoretic mobilities of formic and glutamic dehydrogenases in the Fabaceae: a systematic survey. *New Phytol.* **66**, 37–45.

Tilney, P. M. and Lubke, R. A. (1974). A chemotaxonomic study of twelve species of the family Loranthaceae. *J. S. African Bot.* **40**, 315–332.

Timberlake, C. F. and Bridle, P. (1975). Anthocyanins. *In* 'The Flavonoids' (Harborne, J. B., Mabry, T. J. and Mabry, H., eds.), pp. 214–266. Chapman and Hall, London.

Tobolski, J. J. and Hanover, J. W. (1971). Genetic variation in the monoterpene of Scotch pine. *Forest Sci.* **17**, 293–299.

Towers, G. H. N. (1964). Metabolism of phenolics in higher plants and microorganisms. *In* 'Biochemistry of Phenolic Compounds' (Harborne, J. B., ed.), pp. 249–294. Academic Press, London.

Torres, A. M. and Levin, D. A. (1964). A chromatographic study of cespitose Zinnias. *Amer. J. Bot.* **51**, 639–643.

Towle, G. A. and Whistler, R. L. (1973). Hemicelluloses and gums. *In* 'Phytochemistry' (Miller, L. P., ed.), Vol. I, pp. 198–248. van Nostrand Reinhold, New York.

Tschesche, R. and Wulff, G. (1973). Chemie und Biologie der Saponine. *Fortschr. Chem. organ. Naturst.* **30**, 461–606.

Tulloch, A. P. and Hoffman, L. L. (1979). Epicuticular waxes of *Andropogon hallii* and *A. scoparius*. *Phytochemistry* **18**, 267–271.

Turner, B. L. (1956). A cytotaxonomic study of the genus *Hymenopappus* (Compositae). *Rhodora* **58**, 163–186; 208–242; 250–269; 295–308.

Turner, B. L. (1967). Plant chemosystematics and phylogeny. *J. Pure Appl. Chem.* **14**, 189–213.

Turner, B. L. (1969). Chemosystematics: recent developments. *Taxon* **18**, 134–151.

Turner, B. L. (1970). Molecular approaches to populational problems at the infraspecific level. *In* 'Phytochemical Phylogeny' (Harborne, J. B., ed.), pp. 187–206. Academic Press, London.

Turner, B. L. (1977). Chemosystematics and its effect upon the traditionalist. *Ann. Missouri Bot. Gard.* **64**, 235–242.

Turner, B. L. (1981). New species and combinations in *Vernonia* sections *Leiboldia* and *Lepidonia* (Asteraceae). *Brittonia* **33**, 401–412.

Turner, B. L. and Alston, R. E. (1959). Segregation and recombination of chemical constituents in a hybrid swarm of *Baptisia laevicaulis* × *B. viridis* and other taxonomic implications. *Amer. J. Bot.* **46**, 678–686.

Turner, B. L. and Powell, A. M. (1977). Helenieae—systematic review. *In* 'The Biology and Chemistry of the Compositae' (Heywood, V. H., Harborne, J. B. and Turner, B. L., eds.), pp. 699–738. Academic Press, London.

Turrill, W. B. (1942). Taxonomy and phylogeny I–III. *Bot. Rev.* **8**, 247–270, 473–532, 655–707.

Tursch, B., Brackman, J. C. and Daloza, D. (1976). Arthropod alkaloids. *Experientia* **32**, 401–407.

Ulrich, R. (1970). Organic Acids. *In* 'The Biochemistry of Fruits and their Products' (Huhne, A. C., ed.), Vol. I, pp. 89–118. Academic Press, London and New York.

Unger, I. A. and Boucard, J. (1974). Comparison of seed proteins in the genus *Suaeda* (Chenopodiaceae) by means of disc gel electrophoresis. *Amer. J. Bot.* **61**, 325–330.

Vagujfalvi, D. (1973). Changes in the alkaloid pattern of latex during the day. *Acta Bot. Acad. Sci. Hungaricae* **18**, 391–403.

Valadon, L. R. G. and Mummery, R. S. (1971). Carotenoids of Compositae flowers. *Phytochemistry* **10**, 2349–2353.

Valant, K. (1978). Charakteristische flavonoidglykoside und verwandtschaftliche Gliederung der Gattung *Achillea*. *Naturwissenschaften* **65**, 437–438.

van der Vijver, L. M. (1972). Distribution of plumbagin in Plumbaginaceae. *Phytochemistry* **11**, 3247–3248.

van Emden, H. F. (1972). Aphids as phytochemists. *In* 'Phytochemical Ecology' (Harborne, J. B., ed.), pp. 25–44. Academic Press, London.

van Haverbeke, D. F. (1968a). A population analysis of *Juniperus* in the Missouri River Basin. *Univ. Nebraska Studies, new series* **38**, 1–82.

van Haverbeke, D. F. (1968b). A taxonomic analysis of *Juniperus* in the central and northern Great Plains. *Proc. Sixth Central States Forest Tree Improv. Conf.* **6**, 48–52.

Van Hoeven, W., Haug, P., Burlingame, A. L. and Calvin, M. (1966). Hydrocarbons from Australian oil two hundred million years old. *Nature* (*Lond.*) **211**, 1361–1365.

van Rheede van Outshoorn, M. C. B. (1964). Chemotaxonomic investigations in Asphodeleae and Aloineae (Liliaceae). *Phytochemistry* **3**, 383–390.

Vassal, J. (1972). Apport des recherches ontogeniques et seminologiques a l'étude morphologique, taxonomique et phylogenique du genre *Acacia*. *Trav. Lab. Forest. Toulouse* I (VIII) **17**, 1–125.

Vaughan, J. G. (1975). Proteins and taxonomy. *In* 'The Chemistry and Biochemistry of Plant Proteins' (Harborne, J. B. and Van Sumere, C. F., eds.), pp. 281–298. Academic Press, London.

Vaughan, J. G. (1983). The use of seed proteins in taxonomy and phylogeny. *In* 'Seed Proteins' (Daussant, J., Mossé, J. and Vaughan, J. G., eds.), pp. 135–154. Academic Press, London.

Vaughan, J. G. and Denford, K. E. (1968). An acrylamide gel electrophoretic study of the seed proteins of *Brassica* and *Sinapis* species, with special reference to their taxonomic value. *J. Exp. Bot.* **19**, 724–732.

Vaughan, J. G. and Waite, A. (1967). Comparative electrophoretic studies of the seed proteins of certain amphiploid species of *Brassica*. *J. Exp. Bot*. **18**, 269–276.

Vaughan, J. G., Waite, A., Boulter, D. and Waiters, S. (1965). Taxonomic investigations of several *Brassica* species using serology and the separation of proteins by electrophoresis on acrylamide gels. *Nature* (*Lond.*) **208**, 704–705.

Vaughan, J. G., Denford, K. E. and Gordon, E. I. (1970). A study of the seed proteins of synthesised *Brassica napus* with respect to its parents. *J. Exp. Bot.* **21**, 892–898.

Venkataraman, K. (1972). Wood phenolics in the chemotaxonomy of the Moraceae. *Phytochemistry* **11**, 1571–1586.

Vennesland, B., Conn, E. E., Knowles, C. J., Westley, J. and Wissig, F. (eds.) (1981). 'Cyanide in Biology'. Academic Press, London.

Verdcourt, B. (1970). Studies in the Leguminosae-Papilionoideae for the flora of Tropical East Africa. *Kew Bull.* **24**, 379–447.

Vernet, P. (1976). Analyse genetique et ecologique de la variabilite de l'essence de *Thymus vulgaris* L. (Labiee). Ph.D. thesis. Montpellier, France.

Vickery, J. R. (1971). The fatty acid composition of the seed oils of Proteaceae: a chemotaxonomic study. *Phytochemistry* **10**, 123–130.

Vinutha, A. R. and von Rudloff, E. (1968). Gas-liquid chromatography of terpenes, Part XVII. The volatile oil of the leaves of *Juniperus virginiana* L. *Can. J. Chem.* **46**, 3743–3750.

Vogel, H. J. (1963). Lysine biosynthesis and evolution. *In* 'Evolving Genes and Proteins' (Bryson, V. and Vogel, H. J., eds.), pp. 25–40. Academic Press, New York.

Vogel, H. J. (1964). Distribution of lysine pathways among fungi: evolutionary implications. *Am. Nat.* **98**, 435–446.

Voirin, B. and Jay, M. (1978). Apport de la biochimie flavonique a la systematique du genre *Lycopodium*. *Biochem. Syst. Ecol.* **6**, 95–97.

Voirin, B. and Lebreton, P. (1972). Influence de la temperature sur le metabolisme des flavonoides chez *Asplenium trichomanes*. *Phytochemistry* **11**, 3435–3439.

von Rudloff, E. (1962). Gas-liquid chromatography of terpenes. Part V. The volatile oils of the leaves of black, white and Colorado spruce. *Tappi, J. Tech. Assoc. Pulp. Pap. Ind.* **45**, 181–184.

von Rudloff, E. (1966). Gas-liquid chromatography of terpenes. XIV. The chemical composition of the volatile oil of the leaves of *Picea rubens* Sarg. and chemotaxonomic correlations with other North American spruce species. *Phytochemistry* **5**, 331–341.

von Rudloff, E. (1967). Chemosystematic studies in the genus *Picea* (Pinaceae). I, II. *Can. J. Bot.* **45**, 891–901; 1703–1714.

von Rudloff, E. (1968). Gas-liquid chromatography of terpenes. Part XVI. The volatile oil of the leaves of *Juniperus ashei* Buckholz. *Can. J. Chem.* **46**, 678–683.

von Rudloff, E. (1969). Scope and limitations of gas chromatography of terpenes in chemosystematic studies. *Recent Adv. Phytochem.* **2**, 127–162.

von Rudloff, E. (1972a). Seasonal variation in the composition of the volatile oil of the leaves, buds, and twigs of white spruce (*Picea glauca*). *Can. J. Bot.* **50**, 1595–1603.

von Rudloff, E. (1972b). Chemosystematic studies in the genus *Pseudotsuga*. I. Leaf oil analysis of the coastal and Rocky Mountain varieties of the Douglas fir. *Can. J. Bot.* **50**, 1025–1040.

von Rudloff, E. (1975a). Chemosystematic studies of the volatile oils of *Juniperus horizontalis*, *J. scopularum* and *J. virginiana*. *Phytochemistry* **14**, 1319–1329.

von Rudloff, E. (1975b). Volatile leaf oil analysis in chemosystematic studies of North American conifers. *Biochem. Syst. Ecol.* **2**, 131–167.

von Rudloff, E. (1977). Variation in leaf oil terpene composition of Sitka spruce. *Phytochemistry* **17**, 127–130.

von Rudloff, E. and Holst, M. J. (1968). Chemosystematic studies in the genus *Picea* (Pinaceae). III. The leaf oil of a *Picea glauca* × *mariana* hybrid (Rosendahl spruce). *Canad. J. Bot.* **46**, 1–4.

von Rudloff, E. and Nyland, E. (1979). Chemosystematic studies in the genus *Pinus*. III. The leaf oil terpene composition of lodgepole pine from the Yukon Territory. *Canad. J. Bot.* **57**, 1367–1370.

Voss-Foucart, M. F. and Gregoire, C. (1971). Biochemical composition and submicroscopic structure of matrices of nacreous conchiolin in fossil cephalopods (Nautiloids and ammonoids). *Bull. Inst. R. Sci. Nat. Belg.* **47**, 1–42.
Wagner, W. H. (1980). Origin and philosophy of the groundplan-divergence method of cladistics. *Syst. Bot.* **5**, 173–193.
Wallaart, R. A. M. (1980). Distribution of sorbitol in Rosaceae. *Phytochemistry* **19**, 2603–2610.
Wallace, J. W. and Markham, K. R. (1978). Apigenin and amentoflavone glycosides in the Psilotaceae and their phylogenetic significance. *Phytochemistry* **17**, 1313–1317.
Waller, G. R. and Nowacki, E. K. (1978). 'Alkaloid Biology and Metabolism in Plants.' Plenum Publishing Corp., New York.
Walsh, K. A., Ericsson, L. H., Parmelee, D. C. and Titani, K. (1981). Amino Acid sequencing of Proteins. *Ann. Rev. Biochem.* **50**, 261–284.
Waterman, P. G. (1975). Alkaloids of the Rutaceae: their distribution and systematic significance. *Biochem. Syst. Ecol.* **3**, 149–180.
Watson, L. (1965). The taxonomic significance of certain anatomical variations among Ericaceae. *J. Linn. Soc. Bot.* **59**, 111–126.
Watson, L. and Creaser, E. H. (1975). Non-random variation of protein amino-acid profiles in grass seeds and dicot leaves. *Phytochemistry* **14**, 1211–1217.
Watson, R. and Fowden, L. (1973). Amino acids of *Caesalpinia tinctoria* and some allied species. *Phytochemistry* **12**, 617–622.
Watts, R. L. (1971). Proteins and Plant Phylogeny. *In* 'Phytochemical Phylogeny' (Harborne, J. B., ed.), pp. 145–178. Academic Press, London.
Webster, G. L., Brown, W. V. and Smith, B. N. (1975). Systematics of photosynthetic carbon fixation pathways in *Euphorbia*. *Taxon* **24**, 27–33.
Weeden, N. F., Higgins, R. C. and Gottlieb, L. D. (1982). Immunological similarity between a genobacterial enzyme and a nuclear DNA-encoded plastid-specific isozyme from spinach. *Proc. Natl. Acad. Sci. U.S.A.* **79**, 5953–5955.
Weete, J. D. (1972). Aliphatic hydrocarbons of the fungi. *Phytochemistry* **11**, 1201–1205.
Weevers, T. (1943). The relation between taxonomy and chemistry of plants. *Blumea* **5**, 412–422.
Weil, J. H. and Parthier, B. (1981). Transfer RNA and Aminoacyl-tRNA synthetases in plants. *Encyclopedia of Plant Physiology New Series* **14A**, 65–112.
Weimarck, G. (1972). On Numerical Chemotaxonomy. *Taxon* **21**, 615–619.
Weimarck, G. (1974). Population structures in higher plants as revealed by thin-layer chromatographic patterns. *Bot. Notiser* **127**, 224–244.
Weimarck, G., Stewart, F. and Grace, J. (1979). Morphometric and chromatographic variation and male meiosis in the hybrid *Heracleum mantegazzianum* × *H. sphondylium* (Apiaceae) and its parents. *Hereditas* **91**, 117–127.
Weir, B. S., Allard, R. W. and Kahler, A. L. (1972). Analysis of complex allozyme polymorphisms on a barley population. *Genetics* **72**, 505–523.
Wendelbo, P. (1961). An account of *Primula* subgenus *Sphondylia* with a review of the subdivisions of the genus. *Acta Univ. Bergensis Ser. Math. Nat.* no. 11, 49 pp.
Wettstein, R. (1935). 'Handbuch der Systematischen Botanik,' Franz Deutsche, Leipzig.
Wheat, T. E., Whitt, G. S. and Childers, W. F. (1973). Linkage relationships of six enzyme loci in interspecific sunfish hybrids (genus *Lepomis*). *Genetics* **74**, 343–350.

Whiffin, T. (1973). Analysis of a hybrid swarm between *Heterocentron elegans* and *H. glandulosum* (Melastomataceae) *Taxon* **22**, 413–423.
Whiffin, T. (1977). Volatile oils and the study of natural hybridization between *Correa aemula* and *C. reflexa* (Rutaceae). *Aust. J. Bot.* **25**, 291–298.
Whiffin, T. (1981). Analysis of hybridisation between *Eucalyptus pauciflora* and *E. radiata* (Myrtaceae). *Bot. J. Linn. Soc.* **83**, 237–250.
White, A., Handler, P., Smith, E. L. and Stetten, D., Jr. (1959). 'Principles of Biochemistry,' 2nd edn. McGraw-Hill, New York.
Whitney, P. J., Vaughan, J. G. and Heale, J. B. (1968). A disc electrophoretic study of proteins of *Verticillium alboatrum*, *Verticillium dahliae* and *Fusarum oxysporum* with references to their taxonomy. *J. Exp. Bot.* **10**, 415–426.
Whitt, G. S., Childers, W. F. and Cho, P. L. (1973). Allelic expression at enzyme loci in an intertribal hybrid sunfish. *J. Hered.* **64**, 55–61.
Widen, C. J. and Britton, D. M. (1971). Chemotaxonomic investigations on the *Dryopteris cristata* complex in North America. *Canad. J. Bot.* **49**, 1141–1154.
Widen, C. J. and Sorsa, V. (1966). A chromatographic and cytological study of the *Dryopteris spinulosa* complex in Finland. *Hereditas* **56**, 377–381.
Widen, C. J. and Sorsa, V. (1969). On the intraspecific variability of *Dryopteris spinulosa. Hereditas* **62**, 1–13.
Widen, C. J., Sorsa, V. and Sarvela, J. (1970). *Dryopteris dilatata* s. lat. in Europe and the Island of Madeira. A chromatographic and cytological study. *Acta Bot. Fenn.* **91**, 1–30.
Widen, C. J., Widen, H. K. and Gibby, M. (1978). Chemotaxonomic studies of synthesised hybrids of the *Dryopteris carthusiana* complex. *Biochem. Syst. Ecol.* **6**, 5–9.
Widen, K. G., Alanko, P. and Votila, M. (1977). *Thymus serpyllum* L. × *vulgaris* L., morphology, chromosome number and chemical composition. *Ann. Bot. Fennici* **14**, 29–34.
Wieffering, J. H. (1966). Aucubinartige Glucoside und verwandte Heteroside als Systematische Merkmale. *Phytochemistry* **5**, 1053–1064.
Wieffering, J. H. and Fikenscher, L. H. (1974). Aucubinartige Glucoside als systematische merkmale bei Labiaten: Lamiastrum. *Biochem. Syst. Ecol.* **2**, 31–37.
Wildman, S. G., Chen, K., Gray, J. C., Kung, S. D., Kwanynen, P. and Sakano, K. (1975). Evolution of ferredoxin and fraction I protein in the genus *Nicotiana. In* 'Genetics and Biogenesis of Chloroplasts and Mitochondria' (Perlman, P. S., Birky, C. W. and Byer, T. J., eds.), pp. 309–329. Ohio State University Press, Columbus.
Wilkie, K. C. B. (1979). The hemicelluloses of grasses and cereals. *Adv. Carbohydr. Chem. Biochem.* **36**, 215–264.
Wilkinson, R. C., Hanover, J. W., Wright, J. W. and Flake, R. H. (1971). Genetic variation in the monoterpene composition of white spruce. *Forest Sci.* **17**, 83–90.
Willeke, U., Heeger, V., Meise, M., Neuhann, H., Schindelmeiser, I., Vordemfelde, K. and Barz, W. (1979). Mutually exclusive occurrence of trigonelline and nicotinic arabinoside in plant cell cultures. *Phytochemistry* **18**, 105–110.
Williams, A. H. (1955). Phenolic substances of pear-apple hybrids. *Nature* (*Lond.*) **175**, 213.
Williams, C. A. (1975). Flavonoid patterns in some monocotyledonous families. Ph.D. thesis, University of Reading.

Williams, C. A. (1978). The Systematic implications of the complexity of leaf flavonoids in the Bromeliaceae. *Phytochemistry* **17**, 729–734.

Williams, C. A. and Harborne, J. B. (1972). Essential oils in the spiny-fruited Umbelliferae. *Phytochemistry* **11**, 1981–1987.

Williams, C. A. and Harborne, J. B. (1974). The taxonomic significance of leaf flavonoids in *Saccharum* and related genera. *Phytochemistry* **13**, 1141–1149.

Williams, C. A. and Harborne, J. B. (1977a). Flavonoid chemistry and plant geography in the Cyperaceae. *Biochem. Syst. Ecol.* **5**, 45–51.

Williams, C. A. and Harborne, J. B. (1977b). The leaf flavonoids of the Zingiberales. *Biochem. Syst. Ecol.* **5**, 221–229.

Williams, C. A., Harborne, J. B. and Clifford, H. T. (1971). Flavonoid patterns in the monocotyledons: flavonols and flavones in some families associated with the Poaceae. *Phytochemistry* **10**, 1059–1063.

Williams, C. A., Harborne, J. B. and Clifford, H. T. (1973). Negatively charged flavones and tricin as chemosystematic markers in the Palmae. *Phytochemistry* **12**, 2417–2430.

Williams, C. A. and Murray, B. G. (1972). Flavonoid variation in the genus *Briza*. *Phytochemistry* **11**, 2507–2512.

Williams, R. J. (1956). 'Biochemical Individuality: The Basis for the Genetotrophic Concept.' John Wiley and Sons, New York.

Willis, J. C. (1960). 'A Dictionary of the Flowering Plants and Ferns.' 6th edn. Cambridge University Press, Cambridge.

Wilson, A. C., Carlson, S. S. and White, T. J. (1977). Biochemical evolution. *Ann. Rev. Biochem.* **46**, 573–639.

Woese, C. R., Gibson, J. and Fox, G. E. (1980). Do genealogical patterns in purple photosynthetic bacteria reflect interspecific gene transfer? *Nature* (*Lond.*) **283**, 212–214.

Woese, C. R., Magrum, L. J. and Fox, G. E. (1978). Archaebacteria. *J. Molec. Evol.* **11**, 245–252.

Woese, C., Sogin, M., Stahl, D., Lewis, B. J. and Bonen, L. (1976). A comparison of the 16S ribosomal RNA from mesophilic and thermophilic bacilli: some modifications in the Sanger method for RNA sequencing. *J. Molec. Evol.* **7**, 197–213.

Wollenweber, E. (1975). Flavonoid muster als systematisches Merkmal in der gattung *Populus*. *Biochem. Syst. Ecol.* **3**, 35–45.

Wollenweber, E. and Dietz, V. H. (1980). Flavonoid patterns in the farina of Goldenback and Silverback ferns. *Biochem. Syst. Ecol.* **8**, 21–23.

Woodhead, S. and Bernays, E. (1977). Changes in release rates of cyanide in relation to palatability of *Sorghum* to insects. *Nature* (*Lond.*) **270**, 235–236.

Woodin, T. S., Nishioka, L. and Hsu, A. (1978). Comparison of chorismate mutase isozyme patterns in selected plants. *Plant Physiol.* **61**, 949–952.

Wright, C. A. (ed.). (1974). 'Biochemical and Immunological Taxonomy of Animals.' Academic Press, London.

Wright, D. A. and Shaw, C. R. (1970). Time of expression of genes controlling specific enzymes in *Drosophila* embryos. *Biochem. Genet.* **4**, 385–394.

Wyckoff, R. W. G. (1972). 'The Biochemistry of Animal Fossils.' Scientechnica, Bristol, England and Williams and Williams, Baltimore.

Wyler, H., Mabry, T. J. and Dreiding, A. S. (1963). Uber die Konstitution des Randenfarbstoffes Betanin: Zur Struktur des Betanidins. *Helv. Chim. Acta* **46**, 1745–1748.

Yamada, Y., Hagiwara, K., Iguchi, K., Takahasi, Y. and Hsu, H. (1978). Cucurbitacins from *Anagallis arvensis*. *Phytochemistry* **17**, 1798.

Yeh, W. K. and Ornston, L. N. (1980). Origins of metabolic diversity: substitution of homologous sequences into genes for enzymes with different catalytic activities. *Proc. Natl. Acad. Sci. U.S.A.* **77**, 5365–5369.

Yeh, W. K., Shih, C. and Ornston, L. N. (1982). Overlapping evolutionary affinities revealed by comparisons of amino acid compositions. *Proc. Natl. Acad. Sci. U.S.A.* **79**, 3794–3797.

Yeoh, H. H. and Watson, L. (1981). Systematic variation in amino acid compositions of grass caryopses. *Phytochemistry* **5**, 1041–1051.

Yeoh, H. H., Stone, N. E. and Watson, L. (1981). Taxonomic variation in amino acid compositions of ribulose 1,5-bisphosphate carboxylases from grasses. *Biochem. Syst. Ecol.* **9**, 307–312.

Yokoyama, A., Natori, S. and Aoshima, K. (1975). Distribution of tetracyclic triterpenoids of lanostane group and sterols in the higher fungi especially of the Polyporaceae and related families. *Phytochemistry* **14**, 487–497.

Yoshioka, H., Mabry, T. J. and Timmermann, B. N. (1973). 'Sesquiterpene lactones: Chemistry, NMR and Plant Distribution.' University of Tokyo Press, Tokyo.

Young, D. A. and Sterner, R. W. (1981). Leaf flavonoids of primitive dicotyledonous angiosperms: *Degeneria vitiensis* and *Idiospermum australiense*. *Biochem. Syst. Ecol.* **9**, 185–188.

Zanoni, T. A. and Adams, R. P. (1976). The genus *Juniperus* in Mexico and Guatemala: Numerical and chemosystematic analysis. *Biochem. Syst. Ecol.* **4**, 147–158.

Zavarin, E. and Snajberk, K. (1972). Geographical variability of monoterpenes from *Abies balsamea* and *A. fraseri*. *Phytochemistry* **11**, 1407–1421.

Zavarin, E. and Snajberk, K. (1973a). Geographic variability of monoterpenes from cortex of *Pseudotsuga menziesii*. *Pure Appl. Chem.* **34**, 411–434.

Zavarin, E. and Snajberk, K. (1973b). Variability of the wood monoterpenoids from *Pinus aristata*. *Biochem. System.* **1**, 39–44.

Zavarin, E., Mirov, N. and Snajberk, K. (1966). Turpentine chemistry and taxonomy of three pines of southeastern Asia. *Phytochemistry* **5**, 91–96.

Zavarin, E., Cobb, F. W. Jr., Bergot, J. and Barber, H. W. (1971a). Variation of the *Pinus ponderosa* needle oil with season and needle age. *Phytochemistry* **10**, 3107–3114.

Zavarin, E., Lawrence, L. and Thomas, M. C. (1971b). Compositional variation of leaf monoterpenes in *Cupressus macrocarpa*, *C. pygmaea*, *C. goveniana*, *C. abramsiana* and *C. sargentii*. *Phytochemistry* **10**, 379–393.

Zavarin, E., Snajberk, K. and Bailey, D. (1976). Variability in the essential oils of wood and foliage of *Pinus aristata* and *P. longaeva*. *Biochem. Syst. Ecol.* **4**, 81–92.

Zavarin, E., Snajberk, K. and Critchfield, W. B. (1977). Terpenoid chemosystematic studies of *Abies grandis*. *Biochem. Syst. Ecol.* **5**, 81–93.

Zavarin, E., Snajberk, K. and Lee, C. J. (1978). Chemical relationships between firs of Japan and Taiwan. *Biochem. Syst. Ecol.* **6**, 177–184.

Zavarin, E., Critchfield, W. B. and Snajberk, K. (1978). Geographic differentiation of monoterpenes from *Abies procera* and *Abies magnifica*. *Biochem. Syst. Ecol.* **6**, 267–278.

Zavarin, E., Snajberk, K. and Derby, R. (1980). Terpenoid and morphological variability of *Pinus quadrifolia* and its natural hybridisation with *Pinus monophylla* in northern Baja California and adjoining states. *Biochem. Syst. Ecol.* **8**, 225–235.

Zenk, M. H., Furbringer, M. and Steglich, W. (1969). Occurrence and distribution of 7-methyljuglone and plumbagin in the Droseraceae. *Phytochemistry* **8**, 2199–2200.

Zenk, M. H. and Scherf, H. (1964). Distribution of D-tryptophan conjugation mechanisms in the plant kingdom. *Planta* **62**, 350–354.

Zimmerman, F. K. (1972). Allelic interactions at the genetic and protein level. *Biol. Zbl.* **91**, 17–29.

Zobel, B. (1951). Oleoresin composition as a determinant of pine hybridity. *Bot. Gaz.* **113**, 221–227.

Index of Plant Genera and Species

A

Abies, 284, 303, 314, 418
  balsamea, 337
  bifolia, 336
  lasiocarpa, 336
  magnifica, 245
  procera, 245
Abrus, 140
  precatorius 416
Abutilon, 210
  theophrasti, 400
Acacia, 84, 91, 100, 288, 291, 314, 405, 463, 469
  discolor, 176
  heterophylla, 102
  linifolia, 176
  senegal, 469
  tortilis, 469
Acer, 480
  negundo, 400
Achillea, 271
Achlya, 446
Acokanthera oubaio, 125
  schimperi, 125
Aconitum, 435
  napellus, 82
Actinidia arguta, 190
Actinomyces, 444
Adenostyles, 90
Adiantum, 224
Adonis vernalis, 208, 433
Aegialitis annulata, 170
Aegilops, 314, 455
  squarrosa, 417, 423
  speltoides, 456
Aegopodium, 403
Aeonium, 194, 289
Aerobacter aerogenes, 450
Aesculus, 314, 360
Agathis, 115
Agave, 122
  veracruz, 472
Agoseris, 144
Agropyron, 418
Aichryson, 194
Ajuga hipponensis, 206
Albizia julibrissin, 100
Allium, 72
  cepa, 441, 447
  fuscum, 441
  globosum, 441
  karataviense, 441
  porrum, 400
Alluaudia, 435
Aloe, 166, 176, 195, 314
Althaea, 473
Amanita, 80, 307
  muscaria, 98, 161
Ambrosia, 243
  chamissonis, 112
  confertiflora, 244, 251
Ampelocera, 153
Anacyclus, 270
Anagallis, 360
  arvensis, 119
Andropogon, 290
Anemone, 435
Anethum graveolens, 65
Angiopteris, 446
Annona, 314
Anthemis, 129
Anthoxanthum odoratum, 66
Anthriscus, 403
Anthyllis, 372
  vulneraria, 258
Antirrhinum, 139, 403
Aphananthe, 153
Aquilegia vulgaris, 435
Arachis, 360, 366
  hypogaea, 446
Aralia, 185
Arbutus arizonica, 196
  menziesii, 196
  texana, 196
Arceuthos, 117
Archangelisia flava, 82
Argemone, 290
Armeria, 314

Arnica, 202
Artemisia, 283, 314
  absinthum, 109
  dracunculus, 203, 212, 214
  tridentata, 112, 258
Arthrobacter, 449
Arum dioscorides, 67
  maculatum, 68, 400
  orientale, 68
Asclepias, 84, 123
Asparagus officinalis, 473
Asperula, 433
  odorata, 105
Asplenium, 314
Astelia, 291
Aster, 223
Astragalus, 77
Atriplex, 453, 459
Atropa belladonna, 76, 82
Aucuba japonica, 105, 190
Austrocedrus, 117
Avena, 366, 418, 423, 455
Azolla carolina, 346

B
Bacillus subtilis, 346
Baeria, 144
Baptisia, 157, 221, 268, 276, 283, 314, 372, 379
  alba, 278, 322
  lanceolata, 275
  leucantha, 320
  nuttalliana, 274
  perfoliata, 322
  sphaerocarpa, 320
  tinctoria, 322
Beilschmiedia meirsii, 195
Bellis perennis, 403
Berberis, 78, 210
Beta, 366, 423
  vulgaris, 160, 273
Betula, 481
  populifolia, 409
Bidens, 144
Bixa, 297
Blighia sapida, 97, 107
Borrichia, 314
Bothriochloa, 59
Brassica, 314, 403, 412, 418, 423, 473
  alboglabrata, 420
Brassica cont.
  napus, 185, 400, 412
  oleracea, 69, 206, 400, 412
Briza media, 340
Bromus, 211, 433
Bryonia, 360
Bryophyllum, 351
Bulnesia, 411, 418
Bupleurum falcatum, 208
Buxus, 192
Byblis, 170

C
Caesalpinia, 288
Cakile, 314
Calamagrostis, 314
Callitris, 117
Calluna, 297
Camellia, 314
Camissona, 143
Campsis, 314
Camptotheca acuminata, 190
Canavalia ensiformis, 100, 103
Cannabis, 360
  sativus, 400
Capsella, 403
Capsicum, 174, 314, 366, 403
Carica, 71
Carthamus, 366, 418
Carum carvi, 473
Carya, 170, 190, 314
Cassia, 170
Cassiope, 297
Castanea, 481
Castilleja, 314
Castilloa, 212
Casuarina, 314
Catalpa, 105
Catasetum roseum, 65
Catharanthus roseus, 82, 85, 137, 235, 360
Caucalis, 61
Caulanthus, 290
Cavendishia, 314
Celtis, 154, 212, 480
Centrosema, 103
Ceratostigma, 136
Cercidium, 314
Cetrelia, 247
Chamaecyparis obtusa, 117

Chamasaracha, 341
Chenopodium, 360, 418
  fremontii, 240, 258
  vulvaria, 67
Chimaphila, 167
Chionochloa, 124, 290
Chlorella vulgaris, 346
Chrysanthellum, 350
Chrysanthemum vulgare, 326
  cinerariaefolium, 339
  coronarium, 176
Ciboteum splendens, 446
Cicer, 139, 360, 370
Cichorium intybus, 470
Cinchona officinalis, 82, 85, 243
Cirsium, 314, 403
Citrullus lanatus, 101
Citrus, 51, 119, 149, 190, 195, 289, 314
Cladiosporum herbarum, 365
Cladrastis, 158
Clarkia, 23, 24
  biloba, 427
  franciscana, 427
  lingulata, 427
  rubicunda, 427
Claviceps, 80
Clematis, 26, 435
Clostridium, 406, 444
Cochlospermum, 297
Codonocarpus, 163
Coffea arabica, 359
Colchicum autumnale, 78, 82
Coleus, 168
Collinsia, 314
Colocasia esculenta, 405
Columnea, 137
Combretum, 468
Conium maculatum, 76, 78
Corchorus, 314
Coreopsis, 144, 314
Coriandrum sativum, 62
Cornus, 190, 291, 360
  amomum, 431
  racemosa, 431
Correa, 314
Cortaderia, 124, 314
  richardii, 326
  toetoe, 326
Corylus avellana, 206
Corynanthe johimbe, 104
Corynebacterium, 444
Cosmos, 144
Cotoneaster, 177
Cotula, 202
Crataegus, 67, 403
Crocus sativa, 174
Crotalaria, 89
Croton, 115
Cucumis, 360
  sativus, 62
Cucurbita maxima, 347, 400
  pepo, 446
Cupressocyparis, 314
Cupressus, 117, 284, 314
  macnabiana, 60
  sargentii, 60, 265
Cyperus, 171, 351
  pedunculatus, 171
Cytisus, 156, 403
  laburnum, 82
  scoparius, 83, 176

D

Dahlia, 144
Dalbergia, 157, 164, 166, 170
Daphne mezereum, 114
Datura, 366
Daucus, 61, 139, 360, 366
  carota, 50, 55, 447
  glochidiatus, 66
  montanus, 66
Davidia involucrata, 190
Daviesia, 403
Degeneria, 282
Delonix regia, 176
Delphinium, 435
Derris, 282
Dianthus, 366
Dicentra, 22, 24, 314, 322
  canadensis, 223
Dichanthelium, 356
Digitalis, 122, 403
  purpurea, 436
Dillenia, 297
Dionysia, 137, 145
Dioscorea, 122
  bulbifera, 124
Dioscoreophyllum, 407
Diospyros, 166
  kaki, 177

Diplache, 297
Dipterocarpus, 61, 125
Dipteryx, 158
Dodecatheon, 314
Doona, 61, 124
Doronicum, 202
Douglasia, 145
Doxantha unguis-cati, 185
Dracunculus vulgaris, 67
Drosophyllum, 360
Dryopteris, 314
Duboisia, 314, 360

E
Emex, 171
Encelia, 314, 322
Equisetum, 80, 84, 141
  arvense, 405, 446
  sylvaticum, 478
  telmateia, 405
Eranthis, 435
Eriobotrya, 366
Eriophyllum, 145
Eryngiophyllum, 350
Eryngium, 211
Escherichia, 444
  coli, 450, 458
Eucalyptus, 50, 192, 314
  dives, 56
Eupatorium, 314
Euglena, 448
  gracilis, 346
Euphorbia, 119, 191, 291, 350, 360
Euphrasia, 106

F
Fagopyrum esculentum, 400
Fagus, 481
  sylvatica, 206
Ferula, 109
Festuca, 211
Flaveria, 350
Foeniculum vulgare, 66
Fortunella, 124
Fraxinus, 206
Fremontia, 418
Fritillaria, 291
  imperialis, 211
Fucus, 448
Fuerstia, 168

G
Galeopsis, 418
Galium, 360, 433
Gaultheria procumbens, 64
Gaura, 143
Gazania rigens, 176
Geigeria, 109
Genista tridentata, 176
Gentiana, 151, 206
Geranium, 378
Gibasis, 314
Ginkgo biloba, 149, 398, 447
Gironniera, 153
Gleichenia japonica, 405
Glossocardia, 350
Glossogyne, 350
Glycine max, 206
Goodyera, 314
Gossypium, 141, 221, 366, 418, 423
  barbadense, 400
  hirsutum, 446
Gratiola, 119
Greenovia, 194
Gynocardia odorata, 91
Guizotia abyssinica, 400

H
Haplopappus, 360
Hardwickia pinnata, 115
Hazunta, 290
Hectorella, 163
Hedeoma, 314
  drummondii, 225
Helenium, 271
  elegans, 112
  quadridentatum, 112
Helianthus, 327
  annuus, 398, 441
  decapitatis, 177
Helleborus, 122, 314
  foetidus, 66
Helminthosporium carbonum, 364
Heracleum, 314
  mantegazzianum, 402
  sphondylium, 66, 402
Heterocentron, 314
Hevea, 119
Hibiscus, 418, 473
Hieracium, 314, 403
  pillosella, 142

Hordeum, 423, 455
murinum, 417
vulgare, 346, 400, 417
Hydrangea, 481
Hydrastis canadensis, 433
Hydrocybe, 307
Hymenache, 350
Hymenopappus, 21, 31, 32, 378
Hypericum, 164, 172

I
Iberis, 119
Idiospermum, 282, 298
Ilex, 138
Ipomoea, 366
Iris, 146, 314
foetidissima, 66
Isostigma, 350
Itea, 208
Iva, 145, 385

J
Jacaratia, 71
Jarilla, 71
Jateorhiza palmata, 115
Juglans, 164, 167, 170
Juniperus, 117, 286, 303, 314, 485
ashei, 61, 260, 328
horizontalis, 230, 334
saltillensis, 262
scopulorum, 230, 328
virginiana, 61, 218, 246, 328, 337

K
Kalanchoe, 351
Kalmia, 114
Kerria, 213
Khaya, 269
Kigelia africana, 212
Krameria, 291
Krigia, 144

L
Laburnum anagyroides, 176
Lamiastrum, 108
Larix, 314
Larrea, 314, 418
Laser, 109
Laserpitium, 109
Lasthenia, 144, 314, 326, 418
Lathyrus, 97, 102, 139, 156, 370, 441
nissolia, 369
tingitanus, 99, 101
Laurus, 109
Legionnella pneumophila, 440
Lemna, 278, 423
perpusilla, 220
Lens, 139, 198, 370
Lepidium, 473
Leucanthemum vulgare, 214
Leucampyx, 31
Leucothoe, 114
Liatris, 314
aspera, 325
spicata, 325
Lindera, 109
Linum, 291, 366, 473
usitatissimum, 91, 137, 441
Liriodendron, 109, 403
Lobelia inflata, 339
Lolium, 211
multiflorum, 466
perenne, 441
Lonchocarpus, 282
Lonicera, 403
Lotus, 314
corniculatus, 91, 96, 141, 176, 239, 372, 339
tenuis, 96
uliginosus, 96
Ludwigia, 143
Lupinus, 235, 373, 403
albus, 441
Lychnis coronaria, 67
dioica, 206
Lycopersicon, 198, 366, 423, 459
esculentum, 400, 403
Lycopodium, 80, 84
Lysichitum americanum, 68
Lysimachia, 433

M
Machaeranthera gracilis, 133
Machaerium, 153, 166, 170
Magnolia, 63, 87, 403
acuminata, 62
grandiflora, 62
Malus, 67, 149, 156, 314, 366
Manilkara zapota, 207

Marrubium vulgare, 115
Mastixia philippensis, 190
Matthiola incana, 222
Medicago, 140, 368
  sativa, 405, 467
Melandrium, 381
Melanoselinum, 109
Melilotus, 368
  alba, 66
Mentha, 49, 55, 224, 227, 314, 360
  aquatica, 225, 326
  citrata, 225
  longifolia, 341
  spicata, 326
Mentzelia, 418
Michelia, 109
Microcitrus, 124
Microseris, 144
Millettia, 157
Mimulus cupreus, 176
Monanthes, 194
Monarda, 59
Morus, 366
Mucuna, 84, 307
  pruriens, 100
Mundulea, 158
Murraya exotica, 177
Mycobacterium, 444
Myoporum deserti, 258
Myosurus, 435
Myrica, 59, 314
Myristica fragrans, 55, 65

N
Narcissus, 366
  pseudonarcissus, 82
Nemesia, 174
Neolitsea, 109
Nepeta cataria, 51, 106
Nerium, 212
  oleander, 123
Nerisyrenia, 379
Neurospora crassa, 353, 397
Neviusia, 213
Nicotiana, 314, 373, 403, 418, 423
  otophora, 421
  sylvestris, 421
  tabacum, 82, 84, 360, 421
  tomentosiformis, 421
Nigella, 435
  damascena, 400
  sativum, 189
Nyssa, 190
  sylvatica, 431

O
Ocimum, 360
Oenanthe crocata, 200
Oenothera, 143, 423
Ononis, 103, 140, 212
Ophioglossum pendulum, 446
  petiolatum, 446
Ophrys, 62
Opuntia ficusindica, 160
Orchis, 366
Orlaya, 61
Oryza, 366, 418, 423
Osmunda, 452
Oxalis, 314

P
Paeonia, 472
Panicum milioides, 349
Papaver bracteatum, 87
  orientale, 87
  somniferum, 76, 82, 86, 360
Parmelia, 314
  pulla, 250
Parochetus, 140
Parthenium, 145, 287, 314
  hysterophorus, 109, 111
Parthenocissus, 360
Passiflora edulis, 283, 303
Pastinaca, 366, 403
  sativa, 50, 66, 400
Pectis, 350
Pelargonium, 314
Persea, 195, 289, 314, 481
  americana, 208
Petasites, 109, 314
Petroselinum, 360
  crispum, 65
  sativum, 185
Petunia, 314
Phaseolus, 103, 314, 360, 403, 418, 433
  aureus, 400
  vulgaris, 101, 416, 447
Phellodendron, 87

Phlox, 314, 418, 427
carolina, 258
pilosa, 339
Phoenix dactylifera, 473
Phyllocactus hybridus, 160
Phytelephas macrocarpa, 473
Phytolacca americana, 405
esculenta, 405
Phytophthora infestans, 362
Physalis, 314
alkekengi, 177
Picea, 230, 290, 314
engelmannii, 336
glauca, 60, 232, 246, 336
mariana, 60
sitchensis, 336
Pichia kluyverii, 444
Picradeniopsis, 314
Pinus, 50, 57, 58, 114, 154, 238, 314
banksiana, 60, 231
cembroides, 286
contorta, 60, 230
elliotii, 231
lambertiana, 212
palustris, 231
ponderosa, 228
sibirica, 447
sylvestris, 230
taeda, 228
Piscidia, 158
Pisum, 139, 366, 370
sativum, 235, 403, 441, 470
Pittosporum buchananii, 201
Pityrogramma, 314
Plagiobothrys, 167
Plantago, 206, 403, 473
Platanus, 481
Plectranthus, 58, 168
Plumbago, 136, 164, 167
capensis, 132
Podocarpus nakaii, 122
Polygala senega, 122
Polygonatum officinalis, 100
Poncirus, 124
trifoliata, 233
Populus, 314, 326, 406
Potentilla, 224
Primula, 137, 145, 213, 291
elatior, 208
hirsuta, 132
Primula cont.
obconica, 137, 164, 169
sinensis, 137
Prosopis, 314
juliflora, 258
Proteus, 444
mirabilis, 450
vulgaris, 458
Prunus, 314, 403, 463
amygdalus, 90, 92
Pseudomonas, 345, 353, 356
acidovorans, 357
aeruginosa, 357
Pseudorlaya, 61, 66
Pseudotsuga, 303, 314
menziesii, 230, 246
Psilotum, 309
nudum, 446
Pteridium aquilinum, 122
Pterocarpus, 158
Pterocarya, 170
Pterodon, 157
Puccinia menthae, 108
Punica granatum, 80
Purshia tridentata, 119
Pyracantha angustifolia, 177
rogersiana, 177
Pyrrhopappus, 144
Pyrus, 155, 314, 375
domestica, 62

Q
Quercus, 212, 481

R
Ranunculus, 435
Raphanus, 423
Ramalina siliquosa, 246
Remirea maritima, 171
Rhamnus frangula, 166
Rheum, 166, 171
Rhinanthus, 106
Rhododendron, 115, 139, 192, 198, 210, 314, 326
Rhodotorula graminis, 444
Rhodotypos, 213
Rhus vernicifera, 406
Ricinus, 366
communis, 398
Rigiopappus, 314

Robinia, 403
pseudacacia, 411
Roridula, 170
Rosa, 174, 177, 360
Rubia tinctorum, 105, 165
Rudbeckia hirta, 141, 403
Rumex acetosa, 171
alpinus, 166
Ruta, 360

S
Saccharum officinarum, 124, 290, 314
Salix, 314
Salmonella, 444
typhimurium, 450
Salvia, 59, 206, 314
Sambucus, 403
canadensis, 431
nigra, 400
Sanicula, 211
Saprolegnia, 446
Satureja, 228
douglasii, 239, 258
Sauromatum guttatum, 67
Saxifraga, 314
Scirpus articulatus, 171
Secale cereale, 441, 455
Sedum, 195, 360
Selaginella, 149, 291, 355
emiliana, 446
kraussiana, 446
Senecio, 403, 418
jacobaea, 82
Sesamum indicum, 398
Shigella flexneri, 450
Shorea, 124
Silene, 314
Sinapis, 360, 412, 418, 423
Skimmia japonica, 82
Smyrnium, 109
Solanum, 314, 366, 418, 423, 433
crispum, 403
hougasii, 195
polyadenium, 195
tuberosum, 78, 82, 360, 403, 411
Solenostemon, 168
Sorbus aucuparia, 208
Sorghum, 90, 93, 137, 423
Spinacia, 423
oleracea, 400, 405, 407
Spiraea, 213
Spirodela, 278, 423
oligorhiza, 219
polyrhiza, 220
Stachys, 206
Staphylea colchica, 67
Stemonoporus, 125
Stephanomeria, 418
Streptanthus, 290
Streptocarpus dunnii, 165
Streptomyces, 444
Strychnos nuxvomica, 78, 82, 104
Styrax obassia, 208
Suaeda, 418
Symphytum, 314
officinale, 128
Synechococcus, 406
Synthlipsis, 380
Syntrichopappus, 144

T
Tagetes, 360
patula, 141
Takakia, 309
Tanacetum, 360
vulgare, 228
Taraxacum koksaghyz, 177
Taxus, 448
Tersonia, 163
Tetracera, 297
Tetraclinis articulata, 117
Tetragonolobus, 372
Tetragonotheca, 385
Thaumatococcus, 407
Thelesperma simplicifolium, 340
Thermopsis, 157, 278
Thuja orientalis, 117
Thujopsis dolabrata, 118
Thymus, 314
vulgaris, 249, 256
Tiarella, 314
Tilia, 366
Tmesipteris, 309
Toddalia, 87
Torenia, 314
baillonii, 325
fournieri, 325
Torilis, 61
Tracyina, 314
Tradescantia ohioensis, 441

Tragopogon, 314, 418
dubius, 427
mirus, 340
porrifolius, 427
pratensis, 427
Trifolium, 140, 156, 369, 403
campestre, 369
dubium, 369
repens, 91, 239, 339
Trigonella, 66, 140, 291, 364, 368, 418
Tripsacum, 418
Triticum, 314, 392, 409, 423, 455, 467
aestivum, 398, 417, 419, 447, 456
dicoccum, 417, 422
monococcum, 422
spelta, 419
urartu, 417, 456
Tropaeolum, 72, 185, 314
majus, 400
Tsuga, 314
heterophylla, 337
Turgenia, 61
Tussilago, 403
Typha, 327, 409

U
Ulex, 156
europaeus, 176
gallii, 176
Ulmus, 154, 212, 314, 378, 418, 480
Uragoga ipecacuanha, 76, 82
Ursinia, 203, 403

V
Valerianella, 145
Verbascum, 174, 206, 403
Vernonia, 109, 113, 186, 314, 322
altissima, 214
Veronica teucrium, 108
Viburnum, 403, 433
tinus, 212
trilobum, 431
Vicia, 102, 139, 198, 206, 366, 370, 403, 418
angustifolia, 441
faba, 235, 363, 441
Vigna, 103, 366, 403, 418, 433
radiata, 344, 354, 398
Vinca, 212
herbacea, 85
libanotica, 85
Viola, 314
Viscum album, 212, 407
Vitis, 366

W
Welwitschia mirabilis, 213, 351
Widdringtonia, 117
Wolffia, 278, 423
Wolffiella, 278, 423

X
Xanthium, 145, 314
Xanthoxylum, 87
Xylonagra, 143

Z
Zea, 418, 423
mays, 354, 400, 407, 447, 457, 459, 470
Zelkovia, 480
Zinnia, 314, 322

Subject Index

A

Abietic acid, 115
N-Acetyldjenkolic acid, 102
N-Acetylglucosamine, 468
Advantages and disadvantages
  of chemical characters, 39, 41
  of morphological characters, 43
Aesculetin metabolism, 356
Agathic acid, 115, 117
Albizzine, 100, 102
Aliphatic volatiles
  in magnolia flowers, 63
  in orchid flowers, 63
  structures, 62
Alkaloids
  benzylisoquinoline, 87, 311
  biosynthesis, 81
  categories, 78
  chemistry, 78
  chemotaxonomy, 84
  detection, 83
  distribution, 80, 302
  indole, 89
  in *Cinchona*, 86, 244
  in Menispermaceae, 300
  in Rutales, 299
  pyrrolizidine, 89
  surveys, 298
  tropane, 89, 301
  variability, 234
Alkannin, 167
Alkanes
  chemistry, 191
  chemotaxonomy, 193, 289
  detection, 192
  in *Arbutus*, 197
  in Crassulaceae, 194
  in *Solanum*, 196
Allopatric introgression, chemical
  studies of, 327
Aloe-emodin, 164, 171
Amaranthin, 159, 162
Amentoflavone, 309
α-Aminoadipic acid, 346
Aminoid odours in plants, 67
Amygdalin, 92, 96
Amyloids, 472
α-Amyrin, 120
β-Amyrin, 120
Anabasine, 235
Anthochlor pigments, 141
Anthocyanins
  biosynthesis, 134
  chemistry, 131
  chemotaxonomy, 136
  detection, 135
  distribution, 133, 243
  glycosidic patterns, 139
  in Gesneriaceae, 138
  in Vicieae and Trifolieae, 140
Apigenin, 148
  distribution, 296
Apigeninidin, 131, 137
Aromatic volatiles
  in orchid flowers, 65
  in Umbelliferae, 65
  structures, 64
Asperuloside, 105
Atropine, 79, 82
Aucubin, 105
Aureusidin, 142
Aurones, 143
Azaleatin, 147
Azetidine 2-carboxylic acid, 98, 100

B

Bakkenolide-A, 111
Benzoic acid, 363, 366
Benzylisothiocyanate, 71, 72
Berberine, 79, 82
Betalamic acid, 307
Betalains
  biosynthesis, 160, 307
  chemistry, 159
  chemotaxonomy, 306
  detection, 161
  distribution, 161
Betanidin, 159

Betanin, 159
Biochemical characters in plant systematics, 5
γ-Bisabolene, 54, 55
Butein, 142

C

C_4 Photosynthesis
  and Kranz syndrome, 348
  chemotaxonomy, 350
  family distribution, 349
  in Cyperaceae, 351
  pathway, 348
α-Cadinene, 54
Campesterol, 120
Camphor, 52
Canavanine, 100, 103, 289
Capensinidin, 131, 136
Capillen, 199
m-Carboxyphenylalanine, 98
Cardiac glycosides, 123, 125
Δ^3-Carene, 52, 54, 59
α-Carotene, 173, 176
β-Carotene, 173, 176
γ-Carotene, 173
ε-Carotene, 173
Carotenoids
  chemistry, 173
  chemotaxonomy, 175
  detection, 174
Carotol, 54, 55, 62
Carvone, 52
Caryophyllene, 54, 55
Catalposide, 105
Cellulose, 462
  in cereals, 466
Cell wall polysaccharides in fungi, 347, 465
  in cereals, 465
Ceryl alcohol, 192
Chalcones, 143
Chemical characters
  in identification of *Khaya* species, 269
  polymorphism, 238
  races in lichens, 248
  studies of hybrids, 314, 418
  versus morphological characters, 39, 43
Chemical variation
  in biosynthetic pathways, 345
  in conjugation, 355
Chemical variation *cont.*
  in degradation, 355
  in secondary constituents, 216
  in sesquiterpene lactones of *Ambrosia*, 251
  in species of lichens, 246
Chimaphilin, 167
Cholesterol, 121
Chorismate mutase isozymes, 354
Chorismic acid, 354
Chromatographic patterns of plant hybrids, 321
Chrysophanol, 164, 166, 171
Classificatory, systems, 19
  and parallelism, 28
  of Bessey, 25
  of Engler, 25
  of Hutchinson, 26, 27
Codeine, 86
Colchicine, 79, 82
Colourless flavonoids—*see also* Flavonoids
  chemistry, 147
  chemotaxonomy, 153
  chromatographic separation, 152
  detection, 151
  distribution, 149
  spot pattern data, 153
Columbin, 115
Coniine, 79, 82
  biosynthesis, 81
Coreopsin, 142, 144
Corynantheine, 105
Coumarin, 64, 66
Crassulacean acid metabolism, 350
  taxonomic distribution, 351
Crocin, 173
Cucurbitacin D, 121
Cyanidin, 131
  distribution, 293
  glycosides, 132
Cyanogens
  and origin of *Lotus corniculatus*, 96
  biosynthesis, 91
  chemistry, 91
  chemotaxonomy, 94
  detection, 93
  distribution, 92
  polymorphism, 92
Cyanolipids, 92, 95

Cyclitols, 205, 212, 213
  in gymnosperm leaves, 214
Cyperaquinone, 164, 171
Cytisine, 79, 82
Cytochrome *c*, 395
  as a model protein, 396
  matrix of amino acid differences, 398
  numbering of sequence, 393
  phylogenetic tree, 399, 400
  purification, 397
  sequence analysis, 397
  sequence in wheat germ, 392
Cytosine, 447

D
Daidzein, 150
Dalbergione, 164, 169
Decanal, 62
Dehydromatricaria ester, 199, 202
Dehydroquinate hydrolase isozymes, 355
Dehydroquinic acid, 355
Dehydroshikimic acid, 355
Delphinidin, 131
  glycosides, 132
Deoxyribonucleic acids
  and evolution, 454
  basic structure, 443
  chloroplast, 458
  classification, 442
  hybridization, 448
  in bacteria, 445
  in ferns, 446
  in fungi, 446
  in higher plants, 447
  melting curves, 453, 455
  nuclear content, 441
  pyrimidine bases, 447
  reassociation, 452
  relatedness in bacteria, 450
  sequence analysis, 456
Dhurrin, 92
Diallyldisulphide, 72
Diallylmonosulphide, 72
Diaminopimelic acid, 346
Diethylamine, 67
Diferulic acid, 466
Dihydroacacipetalin, 92
Dihydrochalcones, 149
Dillapiole, 64, 65
Dimethyldisulphide, 72
Diosgenin, 121
Di-*n*-propyldisulphide, 72
Diterpenoids
  chemistry, 115
  chemotaxonomy, 117
  detection, 116
  distribution, 115
  in Cupressaceae, 117
Divaricatic acid, 250
Dopa, 100
Dulcitol, 205, 208

E
Ecdysones, 122
Ecdysterone, 121
Ecotypes, 15
Embelin, 164, 169
Emodin, 164, 171
Enzymes
  alcohol dehydrogenase, 426
  chorismate mutase, 354
  dehydroquinate hydrolase, 355
  glutamate dehydrogenase, 347, 426
  lactic dehydrogenase, 347
  restriction endonucleases, 458
  ribulose bisphosphate carboxylase, 406
Ergosterol, 120
Erucic acid, 184, 188
Ethylamine, 67, 68
Ethyl decadienoate, 62
Eugenol, 64, 65

F
Falcarinol, 199, 201
Falcarinone, 199
Farnesol, 54, 55
Fatty acids
  chemistry, 183
  chemotaxonomy, 187
  detection, 186
  distribution in angiosperm seeds, 189
  in Cornaceae, 190
  in Nyssacaceae, 190
Fenchone, 52
Ferredoxin, 404
  affinity tree, 405
  sequence analysis, 405
Flavanones, 149

Flavonoids
  and cladistics, 381
  and data handling, 380
  and polyploidy, 339
  biosynthesis, 134
  colourless, 146
  genetics, 222
  geography, 240
  in *Anacyclus*, 270
  in *Baptisia*, 157, 274, 320
  in fossil plants, 481
  in *Helenium*, 272
  in Lemnaceae, 278
  in *Nerisyrenia*, 379
  in pines, 155
  in primitive plants, 309
  in *Pyrus*, 155, 376
  in Ulmaceae, 154
  surveys, 299
  variability, 219
Fossil
  amino acids, 479
  cutins, 479
  flavonoids, 481
  organic acids, 476
  steroids, 481
Fraction I protein
  subunit variation, 420
  in angiosperms, 423
  in cereals, 422
  in tobacco, 422
D-Fructose, 468
Fructosides
  distribution, 471
Fucosterol, 120
Fundamental characters, fallacy of, 30

G

Galactinol, 205, 209
Galactomannans, 472
Gas chromatography of terpenes, 57
Gel electrophoresis of proteins
  of *Brassica* seed, 412
  of *Typha* pollen, 410
  of umbellifer seed, 415
  of wheat, 409
  techniques, 408
Genistein, 150
Geraniol, 51, 52
  in *Thymus*, 257
Geranylpyrophosphate, 53, 54
Gibberellic acid, 116
Ginkgetin, 150
Glaucolide-A, 111, 113
Gliadin proteins of wheat, 409
Globulins of legume seeds, peptide fingerprints, 414
Glucocapparin, 69
Glucoibervin, 69
Glucosinolates
  biosynthesis, 70
  chemotaxonomy, 71, 290
  distribution, 311
  structure, 69
Glucotropaeolin, 69
O-Glucosyloxycinnamic acid, 64
Glutamic dehydrogenases in fungi, 347
Glycosylflavones, 149, 154
Gossypetin, 142, 144
  8-methyl ether, 142
  7-methyl ether, 145
Grayanotoxin, I, 115
Gums, 467
  chemotaxonomy of, 468
  in *Acacia*, 469
  sugar components, 468
Gynocardin, 92, 95
Hamamelitol, 213
Hardwickic acid, 115
Helenalin, 111, 113
Hemicelluloses
  in lucerne, 467
  in wheat, 467
n-Hentriacontane, 191
Heptadecane, 62
Hirsutidin, 131, 137
Historical periods of systematic biology, 35
Humulene, 54, 55
p-Hydroxybenzoic acid pathways of degradation, 357
6-Hydroxyluteolin, 142, 145
4-Hydroxyphenylpyruvic acid, 352
Hypericin, 164, 172
Hypoglycin A, 98, 101
Hypolaetin, 142, 145

I

Immunoelectrophoresis, 430
Indole, 67

Indicaxanthin, 159
myo-Inositol, 205, 209
Inulins, 470
Ipomeamarone, 363, 366
Iridodial, 105
Iridoids
  biosynthesis, 105
  chemistry, 105
  chemotaxonomy, 107
  detection, 106
  distribution, 105, 311
Isobetanidin, 159
Isoetin, 142
Isoflavones, 149
  biogenetic maps, 158
  variation in legumes, 156
Isokestose, 471
Iso-orientin, 148
  4′-glucoside, 340
Isopelletereine biosynthesis, 81
Isorhamnetin, 147
Isosalipurposide, 142
Isothiocyanates
  chemotaxonomy, 70
  structures, 69
Isovitexin, 148
Isozymes
  banding patterns, 425, 428
  chemotaxonomy, 427
  electrophoresis, 424
  enzyme activities, 426
  species-pair comparisons, 427

J

Juglone, 164, 167, 170

K

Kaempferol, 147

L

Lactic dehydrogenases in fungi, 347
Lactucaxanthin, 179
Lactupicrin, 111
Lathyrine, 101
Levans, 472
Limonene, 51, 52, 53, 54
Limonin, 121, 124
Limonoids, 119
Linalol, 52, 54
Linamarin, 92
Lipids, 183
Loganin, 105, 107
Lotaustralin, 92
Lupanine, 235
Lutein, 173, 176
Luteolin, 148
  7-glucoside, 148, 340
  7,3′-disulphate, 148
  distribution, 296
Luteolinidin, 131, 137
Lycopene, 173
Lysine biosynthesis
  in different phyla, 345
  in fungi, 347
  pathways, 346

M

Maackiain, 369
Malvidin, 131
  glycosides, 132
Mangiferin, 150
Mannitol, 205, 208
Mannose, 468
Marrubiin, 115
Medicarpin, 369, 370
Menthol, 51, 52
Menthone, 51, 52
Methanthiol, 72
D-Methionine metabolism in bryophytes, 358
Methylallyldisulphide, 72
Methylamine, 67
Methylchavicol seasonal variation in *Pinus*, 229
Methyl cinnamate, 64, 65
5-Methylcytosine, 447
Methyl dodecanoate, 62, 63
4-*O*-Methylglucuronic acid, 468
1-Methylmucoinositol, 214
  biosynthesis in gymnosperms, 209
Methyl oleate, 62, 63
Methyl salicylate, 64, 65
α-Methylterthienyl, 200
Mezerein, 115
Microphyllinic acid, 250
Morphine, 81, 86
  biosynthesis, 81
Muscaflavin, 307
Myrcene, 52, 53, 54, 62

Myricetin, 147
  distribution, 295
Myristicin, 64, 65

N
Naringin, 150
Nepetalactone, 51, 105
Nerol, 52
Nerolidol, 54, 55
Nerylpyrophosphate, 53, 54
Nicotine, 79, 82, 84
Nicotinic acid, 359
  arabinoside, 359
  conjugation in suspension culture, 359
  distribution of metabolic pathways, 360
Nitrogen-containing volatiles
  structures, 67
  occurrence, 67
n-Nonacosane, 191
Nonadien-1-ol, 62
Non-protein amino acids
  chemistry, 97
  chemotaxonomy, 100, 288
  detection, 98
  distribution, 98

O
Ocimene, 53
Oestrone, 121
Oleandrin, 121
Oleanolic acid, 120
Oligosaccharides, 205, 206, 211
Orchinol, 363, 366
Organic acids from lignite, 476
Organic sulphides, 72
Orientin, 148
  4′-glucoside, 340
Ouchterlony double-diffusion, 431

P
Parsleyapiole, 64, 65
Parthenin, 111
Parthenolide
  distribution in *Ambrosia*, 255
Patuletin, 142, 145
Pelargonidin, 131
  distribution, 293
Pentadecane, 62, 63
Penta-yne, 199
Peonidin, 131
Perlatolic acid, 250
Petroselinic acid, 184, 189
Petunidin, 131
β-Phellandrene, 54, 59
Phenethylamine, 67
Phloridzin, 150
Phorbol, 115
Phyllocactin, 159, 162
Phylogenetic classification
  of *Clarkia*, 23
  of *Dicentra*, 22
  of *Hymenopappus*, 21
Physcion, 164, 171
Phytoalexins
  biosynthetic pathway, 372
  detection, 365
  fungitoxicity, 372
  induction, 364
  in Leguminosae, 371
  in *Trifolium*, 369
  in *Trigonella*, 368
  in Vicieae, 370
  variation at family level, 366
Phytosterols, 120
α-Pinene, 51, 52, 53, 54, 55, 59, 60
β-Pinene, 51, 53, 54, 55, 59, 60
Pipecolic acid, 101, 103
Pisatin, 150, 363, 370
Plant sampling in chemotaxonomy, 266
Plastocyanin, 402
  affinity tree, 403
  sequence analysis, 402
Plumbagin, 164, 167, 170
Polyacetylenes
  biosynthesis, 200
  chemistry, 198
  chemotaxonomy, 201
  detection, 200
  distribution, 311
  in the Compositae, tribe Anthemideae, 203
Prephenic acid, 352
Pretyrosine, 352
Primin, 164, 169
Proteins
  electrophoresis, 408
  fraction I, 420
  phylogeny, 399
  properties, 392

Proteins *cont.*
sequences, 407
storage, 408
Protein amino acids, 70, 289
from fossils, 477
Prunasin, 92, 95
β-(Pyrazol-l-yl) alanine, 101

Q
Quassinoids, 119
Quebrachitol, 205, 212, 214
Quercetagetin, 142, 145
Quercetin, 147
3-rutinoside, 147
7-glucoside, 147
metabolic fate of, 356
Quinine, 79, 82, 85
variation in *Cinchona*, 244
Quinones
biosynthesis, 166
chemistry, 163
chemotaxonomy, 169
detection, 168
distribution, 165

R
Raffinose, 205, 211
Revised classification of the order Rhoeadales, 71
Ribitol, 205, 208
Ribonucleic acids
base composition of ribosomal, 448
classification, 442
sequence analysis, 456
Rishitin, 363, 366
Rotenone, 150
Rubiadin, 166
Rubixanthin, 173
Ribulose bisphosphate (RuBP) carboxylase, 466—*see also* Fraction I protein)

S
Safynol, 363, 366
Saponins, 122
Sativan, 369
Scyllitol, 205, 212
Secologanin, 105
β-Selinene, 54, 55
Serology
at the family level, 433, 436
in the Didiereaceae, 435
in the Ranunculaceae, 434
methods of analysis, 429
taxonomic applications, 433
Sesquiterpene lactones
chemistry, 110
chemotaxonomy, 112, 286
classification, 110
detection, 111
distribution, 109
in *Ambrosia*, 251
in *Parthenium*, 287
Shikonin, 167
Sinapine, 69
Sinigrin, 69
Sitosterol, 120
Solanine, 79, 82
Sorbitol, 205, 208
Species concept, 9
Squalene, 120
Sterculic acid, 184
Stigmasterol, 120
Stizolobic acid, 307
Stizolobinic acid, 307
Storage proteins
interpretation, 411
separation, 408
taxonomic application, 416
use at generic level, 418
use with hybrids, 418
Strychnine, 79, 82
Sugars
chemistry, 204
chemotaxonomy, 210
detection, 209
in nectars, 210
Sugar alcohols, 205, 208
Sulphurein, 142
Sulphuretin, 142
Sympatric hybridization, chemical studies of, 320

T
Taxonomy
and cytogenetics, 14
and phylogeny, 17
biochemical categories, 16
definition, 7

Taxonomy *cont.*
  experimental categories, 13
  formal categories, 9
  of smell, 50
Tenulin, 111, 113
Terpene volatiles, 51
  biosynthesis, 53, 226, 304
  detection, 55
  developmental variation, 228
  distribution, 51
  environmental variation, 227
  genetic variation, 225
  in Caucalideae, 61
  in junipers, 262, 328
  in *Myrica*, 59
  in *Satureja*, 258
  in spruce, 60, 232, 284
  in thyme, 256
  structures, 52, 54
  surveys, 303
  taxonomic utility, 58
α-Terpineol, 52
  in *Thymus*, 257
Terpinolene, 52
Terthienyl, 199
Thebaine, 81, 86
  biosynthesis, 81
Thin-layer chromatography of terpenes, 56
Thiopropanal *S*-oxide, 72
Thujone, 52
Thymine, 447
Toxins, major classes of, 126
Tricetin, 148
Tricin, 148
Trigonelline, 359
Trimethylamine, 67
Triterpenoids
  chemistry, 119
  chemotaxonomy, 125
  detection, 123
  distribution, 122
  in Dipterocarpaceae, 125
Tryptophan synthesizing enzymes in fungi, 347
Tyrosine biosynthesis
  alternate pathways, 352
  enzymology of, 354
  evolution of, 353

U
Umbelliferose, 205, 207, 211
Uracil, 447
Ursolic acid, 120

V
Vanillin, 64
Vermeerin, 111
Vernolepin, 111, 113
Vernolic acid, 184
Vestitol, 369
Violaxanthin, 173, 176
Vitexin, 148

W
Wyerone, 370
  acid, 199
  epoxide, 370

X
Xanthones, 151

Y
Yellow flavones, 143
Yellow flavonoids
  chemistry, 142
  chemotaxonomy, 143
  detection, 143
  distribution, 142
Yellow flavonols, 143

Z
Zeaxanthin, 173